The Paramount Role of Joints
into the Reliable Response of Structures

NATO Science Series

A Series presenting the results of scientific meetings supported under the NATO Science Programme.

The Series is published by IOS Press, Amsterdam, and Kluwer Academic Publishers in conjunction with the NATO Scientific Affairs Division

Sub-Series

I. **Life and Behavioural Sciences**	IOS Press
II. **Mathematics, Physics and Chemistry**	Kluwer Academic Publishers
III. **Computer and Systems Science**	IOS Press
IV. **Earth and Environmental Sciences**	Kluwer Academic Publishers

The NATO Science Series continues the series of books published formerly as the NATO ASI Series.

The NATO Science Programme offers support for collaboration in civil science between scientists of countries of the Euro-Atlantic Partnership Council. The types of scientific meeting generally supported are "Advanced Study Institutes" and "Advanced Research Workshops", and the NATO Science Series collects together the results of these meetings. The meetings are co-organized bij scientists from NATO countries and scientists from NATO's Partner countries – countries of the CIS and Central and Eastern Europe.

Advanced Study Institutes are high-level tutorial courses offering in-depth study of latest advances in a field.
Advanced Research Workshops are expert meetings aimed at critical assessment of a field, and identification of directions for future action.

As a consequence of the restructuring of the NATO Science Programme in 1999, the NATO Science Series was re-organized to the four sub-series noted above. Please consult the following web sites for information on previous volumes published in the Series.

http://www.nato.int/science
http://www.wkap.nl
http://www.iospress.nl
http://www.wtv-books.de/nato-pco.htm

Series II: Mathematical and Physical Chemistry – Vol. 4

The Paramount Role of Joints into the Reliable Response of Structures

From the Classic Pinned and Rigid Joints to the
Notion of Semi-rigidity

edited by

C.C. Baniotopoulos

Institute of Steel Structures,
Department of Civil Engineering,
Aristotle University of Thessaloniki,
Thessaloniki, Greece

and

F. Wald

Department of Steel Structures,
Czech Technical University,
Prague, Czech Republic

Springer-Science+Business Media, B.V.

Proceedings of the NATO Advanced Research Workshop on
The Paramount Role of Joints into the Reliable Response of Structures
Ouranoupolis, Greece
May 21–23, 2000

A C.I.P. Catalogue record for this book is available from the Library of Congress.

ISBN 978-0-7923-6701-7   ISBN 978-94-010-0950-8 (eBook)
DOI 10.1007/978-94-010-0950-8

Printed on acid-free paper

LIST OF CONTENTS

V ANALYTICAL MODELS FOR JOINTS AND RELIABILITY

FOREWORD

During the 3 days of the NATO Advanced Research Workshop entitled "The paramount role of joints into the reliable response of steel, composite and timber structures. From the classic pinned and rigid joints to the notion of semirigidity", scientists from 23 countries performing research on relevant subjects presented their most recent results, exchanged opinions and ideas and discussed prospects of applications of the proposed theories and methodologies having as ultimate scope more reliable and more safe steel, composite and timber structures. In particular, the important role of the correctly designed and fabricated joints in the safe and reliable response of the aforementioned structures in both local and structure level was extensively discussed. The typology/morphology of these connections was discussed in details taking into account both the conventional pinned and rigid joints and the semi-rigid ones.

The notion of semi-rigid connections in steel, composite and timber structures has been recently introduced in the respective Eurocodes where, these connections, exhibiting a structural behaviour between that of classical pinned and rigid, are characterised as "semi-rigid". The lectures of the present Workshop were related to all the aspects of such joints: from the fundamental notions till the applications of the theory i.e. design, the detailing and the study of the overall response of the respective structures. The structural joints have a major role in the seismic behaviour of framed structures. The local ductility of different types of joints is determined through cyclic experiments and such works have been also discussed during the Workshop. Also, works dealing with the typical cyclic joint characteristics (ductility, cyclic deterioration of strength and stiffness, and energy absorption capacity) were presented.

The goals of this Workshop consisted in a comprehensive survey of all the relevant topics: the definition of semi-rigid structural connections, their classification and their influence to the structural response of the steel, composite and timber structures. In addition, the sources of connection compliance, the application of the component method for characterisation of the joint properties were presented, whereas the notion of rotational capacity, the verification procedures for the available and the required rotational capacity of joints were described in details. In addition, the simulation of the structural response of the joints by means of appropriate numerical methods that take into account all critical phenomena, appearance of frictional or frictionless unilateral contact on the joint interfaces and development of plastification zones within the joints, were also presented. Analysis techniques and design procedures for the beam-to-column, beam-to-beam, column-base-plate connections

ix

and other specific types of conventional or hollow section, simple or moment resistant, structural connections were described. This way, within the present book a useful exchange of views between scientists who perform theoretical, numerical and experimental research on the response of structural connections and engineers who are involved in structural design is taking place.

The reader of the present publication, which summarizes the presentations of the aforementioned Workshop, can see that its main purpose was to bring together high-level experts from slightly different fields, aiming thus to an, as fruitful as possible, exchange of remarks, ideas and conclusions.

Concerning the presentation of the papers of this book, they have been grouped in five sections covering the "Behaviour of structures including joint behaviour", the "Experimental studies of joints and frames", the "Behaviour of earthquake resistant structures including joint behaviour", the "Numerical simulation of the structural response of joints and frames" and "Analytical models for joints and reliability", ranging from theoretical concepts to practical applications.

The editors wish to thank all their colleagues who actively participated to the Workshop and enthusiastically contributed to the preparation of the present book. They also acknowledge with thanks the support of the NATO-Scientific Affairs Division to both the Workshop and the present volume.

C. C. Baniotopoulos (Thessaloniki) and F. Wald (Prague)

WORKSHOP PARTICIPANTS

I. BEHAVIOUR OF STRUCTURES INCLUDING JOINT BEHAVIOUR

EFFECTS OF THE ACTUAL JOINT BEHAVIOUR ON THE DESIGN OF STEEL FRAMES

R. MAQUOI
University of Liège
MSM, Institute of Civil Engineering, B 52/3
Chemin des Chevreuils, 1
B-4000, Liège 1, Belgium

1. General

In a normal building structure, one can usually identify:
- *Primary structural elements*, which constitute the main frame with its joints and the foundation, form the routes by which vertical and horizontal forces are transferred to the ground and provide the frame with resistance and in-plane stability;
- *Secondary structural elements*, such as secondary beams, purlins, ... transfer loads to the primary structural elements and contribute the possible bracing;
- *Other non directly structural elements*, such as sheeting, roofing, cladding, partitions, which transmit the loads to the primary or secondary structural elements.

Usually the elements of the two last categories are designed independently and separately, without significant structural interaction with the main frame; their actions at their supports are then as many loads for this frame.

The design is not conducted on the complex actual frame but well on a model of it, the structural components of which are assumed to obey the rules of the elementary beam theory. This *idealisation* consists in defining a simple *static scheme* where:
- Each member is reduced to its longitudinal axis;
- All the details, which are not directly useful, are removed;
- The imperfections, uncertainties and eccentricities are modelled, with some of them being possibly ignored;
- Some assumptions are made regarding the structural response of joints and constitutive laws.

A properly designed frame has to satisfy both the *ultimate limit states* (ULS) and *service limit states* (SLS). That means that the structure and all its components have, on the one hand, to resist safely the actions (factored loads) and, on the other hand, to comply with deemed-to-satisfy criteria in service conditions (non-factored loads). More especially the internal forces induced by any specified load combination shall be admissible with regard to member/section resistances and the displacements in service

C.C. Baniotopoulos and F. Wald (eds.), The Paramount Role of Joints into the Reliable Response of Structures, 3–16.
© 2000 *Kluwer Academic Publishers.*

conditions within specified limits. On how to rule these checks is normally governed by the standards in force; in this respect, Eurocode 3 [1] and Eurocode 4 [2] aim at becoming "the" references for steel and composite structures respectively.

2. Structural design

Structural design consists in a two-step procedure:
- The *global analysis* aims at the determination of internal forces and displacements in a given structure subjected to a given combination of actions; it is conducted on the ideal static scheme. Basically two types of global analysis can be contemplated: elastic analysis and plastic analysis.
- Several *design checks* of the frame and its components are performed once the global analysis is achieved; the number of them depends on the type of analysis adopted and on the type of cross-section verification, elastic or plastic.

2.1. GLOBAL ANALYSIS

2.1.1. Elastic global analysis

The simplest type of global analysis is the so-called *elastic global analysis*: it assumes implicitly a linear elastic response of the material.

As far as the proportionality between loading and displacement may be considered as an acceptable assumption, the principle of superposition is applicable. Then a *first order elastic analysis* is sufficient to provide the expected results (internal forces/displacements) under both factored and service loads. That is especially allowed [1] when $V_{Sd}/V_{cr} \leq 0,10$ for the load combination under consideration; that is the range of so-called *non-sway frames*, where second order effects are sufficiently small to be fully disregarded. V_{Sd} is the design value of the total vertical load and V_{cr} the elastic critical value of this load for failure in a *sway mode*.

Should the geometric non-linearity due to the so-called $P\text{-}\delta$ effects (element) or $P\text{-}\Delta$ effects (structure) be accounted for, proportionality is lost and the principle of superposition is no more practicable. Then a *second order elastic analysis* is required: the loads are incremented altogether by means of a load multiplier λ growing from 0 up to the desired value γ_F of the load factor at the ULS. The results for the service conditions are those obtained at the appropriate level of loading, i.e. usually at $\lambda=1$. As an alternative to this step-by-step procedure, the sway moments and the internal forces required for the design checks of sections and joints may be evaluated, in accordance with [1], by amplifying simply those resulting from a first order analysis either by a lump factor 1,2 – in the limited range $V_{Sd}/V_{cr} < 0,25$ - or, more generally, by a magnification factor $1/(1-V_{Sd}/V_{cr})$. Both above amplification methods are not equivalent because the buckling lengths to be considered are those relative respectively to the sway mode and to the non-sway mode.

Elastic analysis is thus conducted either in one step (first order) for the combination of factored loads or step-by-step (second order) till the value of the load factor γ_F which has to be reached. In the latter case, some trick is necessary because all the loads are not factored by a same value of the load factor.

Once completed, the elastic analysis is followed by a check of the section/member resistances. Indeed its range of validity, which is restricted to the very first exhaustion of the section resistance wherever in the frame, is neither controlled nor even detected within the method of analysis. Above section resistance shall be understood as the *ultimate* section resistance; depending on the class of the cross-section, it may exceed the elastic resistance.

Elastic analysis is thus allowed even if some yielding occurs within sections but does not account for any plastic redistribution between sections.

2.1.2. Plastic global analysis

In some conditions, the very first exhaustion of any section resistance does not correspond to the actual bearing capacity of the structure: an additional strength reserve is available under the reservation that plastic redistribution between sections is permitted. When this redistribution occurs so as a plastic mechanism is formed, a so-called *plastic global analysis* can be contemplated. Member and joint sections must therefore be able to develop their plastic resistance and to exhibit a sufficient rotation capacity where plastic hinges are likely to develop. Due to the material non-linearity, the principle of superposition is no more valid. Plastic global analysis shall thus be conducted based on the concept of load multiplier increments $\Delta\lambda$.

A *first order plastic analysis* is allowed [1] when $V_{Sd}/V_{cr} \leq 0,10$. That postulates that geometric P-δ and P-Δ non-linear effects are negligible. Once achieved, the plastic analysis provides the first order plastic load multiplier λ_p. The ULS are fulfilled when λ_p is at least equal to the load factor γ_F.

A *second order plastic analysis* is especially required in situations where sway is not sufficiently prevented. It is normally conducted step-by-step by incrementing the load multiplier λ till its ultimate value λ_u, which should be at least equal to the load factor γ_F. Alternatively [1], an evaluation of λ_u is possible, in the range $4 \leq \lambda_{cr}/\lambda_p \leq 10$, by reducing the first order plastic load multiplier λ_p by the factor $1/(0,9+\lambda_p/\lambda_{cr})$ according to the so-called Merchant-Rankine approach. Another simplified method for second order plastic analysis is permitted by [1] in a limited range of application; it consists in amplifying the results of a first order plastic analysis but it gives such a penalty that it would be preferable not be recommend it.

The procedure of plastic global analysis handles both the section resistance and the plastic redistribution. A subsequent check of the member resistance (stability) shall be conducted with due allowance made for the influence of possible plastic hinges on the buckling lengths.

For practice purposes, and especially with a view to make plastic design practicable by hand, a simplified plastic global analysis known as *rigid-plastic analysis* is

6

practicable. The latter disregards the elastic strains compared to the plastic ones and refers to the concept of plastic hinge in contrast to the plastic zone.

Rigid-plastic analysis aims especially to first order calculations by hand. The Merchant-Rankine method enables an approximate account for second order effects from the knowledge of two basic frame strength characteristics λ_p. and λ_{cr}. Elastic-plastic analysis and, possibly, elasto-plastic analysis are computer-oriented methods that need appropriate softwares.

Once achieved, plastic analysis must be followed by a check of the member resistance for in-plane and possible out-of-plane stability.

2.1.3. Synopsis

A synopsis of the methods of global analysis for practice purposes is given in Table 1.

TABLE 1. Methods of global analysis for practice purposes

Global analysis ▼ ▶	First order	Second order		
		Amplified 1st order (I)	Amplified 1st order (II)	General
ELASTIC				
Application range	$V_{Sd}/V_{cr} \leq 0,1$	$V_{Sd}/V_{cr} < 0,25$	None	None
Amplification factor for sway moments and internal forces	Disregarded	$1/(1-V_{Sd}/V_{cr})$	1,2	Included
In-plane buckling lengths	Non-sway mode	Non-sway mode	Sway mode	Non-sway mode
PLASTIC				
Application range	$V_{Sd}/V_{cr} \leq 0,10$		$4 \leq \lambda_{cr}/\lambda_p \leq 10$	None
Account for second order moments and internal forces	Disregarded		$\lambda_u/\lambda_p = 1/(0,9 + \lambda_p/\lambda_{cr})$	Included
In-plane buckling lengths	Non-sway mode		System lengths	Non-sway mode
	Due allowance for the presence of plastic hinges			

2.2. DESIGN CHECKS

The design checks that are still to be performed once the results of the global analysis are got are summarised in Table 2; it might happen that some checks indicated in this table are irrelevant or useless because the tool used for the global analysis yet manages them.

It is while stressing that the extent of these design checks depends on the type of global analysis performed and on the type of cross-section verification.

TABLE 2. Design checks to be performed (inad = if not already done)

Type of global analysis Type of structural elements/Type of checks	Elastic analysis	Plastic analysis
Beams		
• Class of cross-sections		X (inad)
• Rotation capacity		X (inad)
• Serviceability	X	X
• Cross-section resistance	X	
• Local instability	X	X
• Member instability (lateral torsional buckling)	X	X (inad)
Columns /Beam-columns		
• Class of cross-sections		X (inad)
• Rotation capacity		X (inad)
• Serviceability	-	-
• Cross-section resistance	X	
• Local instability	X	X
• Member instability (column buckling, lateral torsional buckling, combined compression and bending)	X	X (inad)
Joints		
• Stiffness	X	X
• Strength	X	X (inad)
• Rotation capacity		X

2.3. BALANCE OF EFFORT FOR ANALYSIS AND FURTHER CHECKS

The choice of the method for global analysis is not only depending on some code specifications but also on personal choices, specific situations, availability of appropriate softwares,...

A particular choice means striking a balance between the volume of effort devoted to global analysis and the volume of effort required for the check of ULS and SLS to be carried out once global analysis is completed. This balance obeys the following

8

elementary rule: the larger the effort made at the stage of global analysis, by using a more sophisticated method, the lesser the effort to be devoted to the further checks.

That is schematically represented in Figure 1.

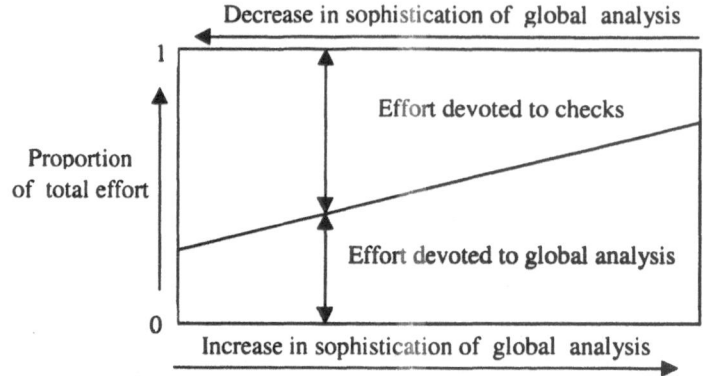

Figure 1. Balance of effort for global analysis and further checks.

3. Structural joints with regard to in-plane stability of resisting frames

In extreme situations, the in-plane stability of resisting frame can be provided:
• Either by the continuity existing between the primary structural elements composing the frame - i.e. between columns and beams and/or between columns and foundation –, resulting in the so-called frame effect (*continuous* construction);
• Or, in the absence of significant continuity, and thus of frame effect, by additional structural elements such as X or K in-plane braces (*pinned* construction).

Joint detailing may be such that none of above cases is realistic. Only a partial continuity exists between the connected elements; that results in a so-called *semi-continuous* construction.

In-plane stability of frames is governed by the in-plane response to horizontal forces. When this response is sufficiently stiff for any additional forces and moments arising from horizontal displacements of the storeys to be disregarded, the frame is said *non-sway*. That means that second order effects may be neglected. When, in contrast, the latter are not negligible, the framework is said *sway*.

It is generally agreed [1] to classify a frame as *non-sway* when it satisfies the following criterion:

$$V_{Sd}/V_{cr} < 0,10$$

where V_{Sd} and V_{cr} are defined above. As an alternative, a regular multi-storey frame may be classified as non-sway when [1] the horizontal displacement δ at the top, relative to the bottom, of each storey, due to both design vertical and horizontal loads (including the effect of frame imperfections), satisfies the criterion:

$$\delta V/hH \le 0,10$$

where H and V are respectively the total horizontal reaction and total vertical reaction at the bottom of the storey under consideration.

It shall be noticed that the classification of a frame is not independent of its loading but is associated to the given combination of actions under consideration

For a given load combination, the general criterion is applied once while the alternative one is applied as many times as there are storeys. Also the use of the alternative criterion is questionable when the beams are not horizontal; that is especially the case when industrial pitch-roof frames.

The joints contribute substantially the global in-plane stiffness of a frame and therefore the in-plane stability. Their detailing is thus likely to effect the frame classification – sway or non-sway - to a great extent. It may happen that, for a given load combination, a continuous frame shall be classified as non-sway but shall be sway if another joint detailing makes the frame significantly less stiff in its plane.

4. Parallel between member sections and joints

From above it results that the joints are basic components of the frame and there is no objective reason for not paying to them a similar attention as the one reserved for long to the member sections.

The member cross-section behaviour may be characterised by an M-ϕ curve, where M is the bending moment at mid-span of a simply supported beam loaded by a point load applied at the same place and ϕ the sum of the rotations at the beam ends. Any joint is characterised by a similar relationship M_j-ϕ, where M_j is the moment transmitted by the joint and ϕ the corresponding relative rotation between the connected member and the rest of the joint. Those two relationships have globally a similar shape with, generally, a more pronounced domain of non-linearity for the joint. Both member and joint sections are thus characterised by their *strength* and their initial *stiffness*.

A member cross-section subjected to a given distribution of direct stresses is said belonging to a certain class (from Class 1 to Class 4). This classification depends on the one hand, of the ability of the section to resist instability in any of its wall elements and, on the other hand, of the consequences such a possible local plate buckling may have on the ability of the cross-section to plastic redistribution. *Ductility* is thus directly related to the amount of rotation that the section needs to develop so as to sustain given internal forces. Therefore the concept of *rotation capacity* is often used and substituted for the one of ductility. Similarly to member cross-sections, joints are classified in terms of ductility or rotation capacity. This classification is a measure of the joint ability to resist premature brittle fracture with due allowance made for the consequences on the global analysis. While the classification of a member section is peculiarly governed by the b/t ratios of the wall elements of this cross-section, the joint classification is basically determined based on geometric and mechanical properties of the joint components.

In Section 2, the interdependence between the global analysis and the subsequent checks was stressed. It appears clearly that not only the members and their cross-sections but also the joints are concerned with it; that strengthens the considerations made above and extends the scope of Figure 1.

5. Joint modelling

For the purposes of global analysis, *continuous construction* implicates a modelling with *continuous joints* while *pinned construction* assumes a modelling with *simple joints*.

A *continuous joint* is normally idealised as a mechanical clamping between the connected members. It is thus moment-resistant and such that any relative rotation between the respective axes of the connected members is prevented. An adequate joint detailing can ensure a continuous joint between members (column splices, beam splices, beam-to-column joints). In contrast, a continuous base joint is more questionable; indeed not only the rotation of the structural member with respect to its foundation but also the rotation of the latter with respect to the neighbouring ground shall be prevented.

A *simple joint* is normally idealised as a mechanical frictionless hinge. It allows for a fully free rotation between the respective axes of the connected members and is therefore unable to resist any bending moment.

Joint detailing makes that, strictly speaking, the joints are never neither continuous, nor simple. The joint response is only such that it approaches one of these extremes. Then, there is no need to consider such joints as semi-continuous for the purposes of global analysis; indeed approximations are allowed so as to consider them as continuous or simple and perform the global analysis accordingly. In contrast, when the difference between the actual and the ideal (simple or continuous) joint behaviour is expected to have a significant impact on the internal forces and/or displacements, a more adequate modelling of the joints is required. The latter results in so-called *semi-continuous construction*; the joints are *semi-continuous joints* and the global analysis shall be consistent with this behaviour.

Recently, much attention was paid to this concept of semi-continuity. The major reason lies in the evolution of the respective material and labour contributions to the global cost of a structure and in the resulting search for labour savings. Very often this prospect is achieved by simplifying the joint detailing, for instance by removing stiffeners which are not necessary for the resistance. There is thus a need for keeping, if not restoring, consistency between the conceptual aspects, the structural modelling and the methods of global analysis.

The practice is still widely to conduct global analysis by assuming either continuous or simple joints. It is a matter of fact that the methods of first order elastic structural analysis (flexibility method, stiffness method, slope-deflection method, and moment distribution method) are usually taught by assuming that the structural joints are modelled as either continuous or simple. The member stiffness/flexibility matrixes and the equilibrium/compatibility equations are usually expressed based on these assumptions. However, in contrast with a widespread belief, only minor arrangements are required to enable the generalisation and the application of these methods to semi-continuous construction. Of course, modern softwares facilitate largely second order elastic analysis and plastic analysis of semi-continuous construction; most of them are henceforth capable of accounting for the semi-continuity of joints.

To which extent, for practice purposes, the global analysis has to reflect or not the semi-continuity of joints is ruled by the so-called *joint classification*.

6. Joint classification with regard to methods for global analysis

For global analysis, a joint shall be classified, similarly to a member section, with regard to its:
- *Initial stiffness,*
- *Strength,*
- *Ductility.*

Which ones of these characteristics are concerned depends on the method of analysis.

The distribution of internal forces resulting from an *elastic global analysis* is basically governed by the relative rigidities of the frame components. Therefore, members and joints shall be characterised by their stiffness only. Indeed, elastic analysis does not at all handle resistance aspects and makes ductility of little importance because its range of application - to be checked afterwards - prevents any plastic redistribution between sections. Strength and ductility rule the section/joint classification and therefore influence the further section/member/joint resistance checks.

A properly conducted *plastic global analysis* makes that strength is nowhere exceeded while it takes due account of a plastic redistribution of the internal forces when necessary. Stiffness, resistance and ductility characteristics altogether rule the process in any generality. However, rigid plastic design is not concerned with member/joint stiffness because neglecting the elastic strains.

As far as joint stiffness is concerned, one has *rigid* joints, *pinned* joints and *semi-rigid* joints (Figure 2.a).

Because a joint is an interface between adjacent members, it may happen that it constitutes a weak section. Regarding the strength classification, it consists simply in comparing the joint *design* moment resistance with two reference resistances that are given by so-called *full-strength boundary* and *pinned boundary*. The resistance of the connected member, fitted with some magnification factor, governs the full strength boundary. The pinned boundary corresponds to 25% of the full-strength boundary. The joint is said *full-strength*, when its resistance exceeds the full-strength boundary, *pinned*, when it is lower than the pinned boundary, and *partial-strength*, when it is not classified as pinned and its design resistance is lower than the full-strength boundary (Figure 2.b).

Joint ductility is involved when yielding contributes the resistance. A *ductile* joint (*Class 1*) allows for a full plastic redistribution of the internal forces within the joint and has a sufficient rotation capacity to allow for a plastic global analysis. A joint with an *intermediate ductility* (*Class 2*) enables a full plastic redistribution within the section but the limited rotation capacity of the joint excludes plastic analysis; its strength is thus limited to its elastic strength whereas either an elastic global analysis shall be performed. A *non-ductile* joint (*Class 3*) allows only for an elastic resistance. For practical applications and when rotation capacity is required, it may be recommended to adopt a joint detailing so as a ductile failure mode is governing the design resistance.

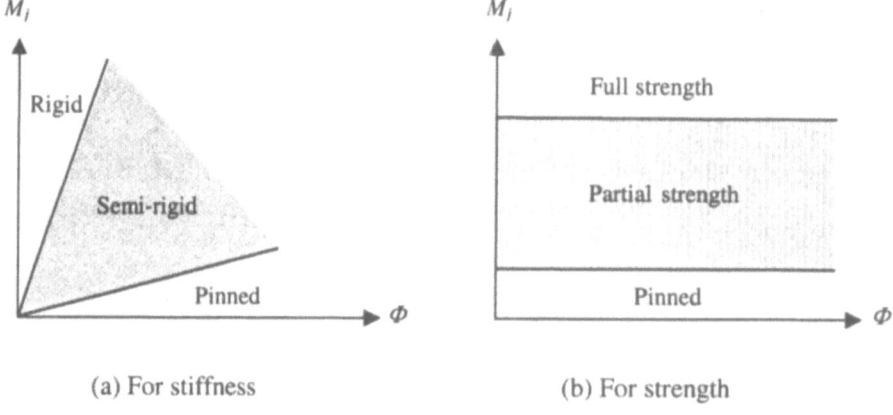

(a) For stiffness   (b) For strength

Figure 2 – Classification for stiffness and strength.

Table 3 gives a synopsis of the joint characteristics that are concerned with each type of global analysis according as the joint modelling.

TABLE 3 - Joint properties, joint modelling and global analysis

Joint modelling	Global frame analysis		
	Elastic	Rigid-plastic	Elasto-plastic
Continuous	Rigid	Full-strength/Ductile	Rigid/Full strength/Ductile
Semi-continuous	Semi-rigid	Partial strength/Ductile	Rigid/Partial strength/Ductile
			Semi-rigid/Full strength/Ductile
			Semi-rigid/Partial strength/Ductile
Pinned	Pinned	Pinned	Pinned

7. Design approaches and methodologies

Once it is agreed that the joints should merit a similar attention as the members, it remains to examine how to manage an appropriate methodology [3].

A preliminary design of the members is required prior to the global analysis of any statically indeterminate structural frame; it is widely practised with a success, which depends mainly on the skill and know-how of the designer. In contrast, a similar preliminary design of the joints is a rather new concept and most designers have still very few experience of the semi-continuity. Substantial progress was achieved recently, which is worthwhile being briefly commented.

The *fabrication phase*, which follows this *design phase*, is likely to influence significantly the global cost of structural frames. In any generality, design concerns not only members but also joints. The people who are responsible for the whole or parts of the relevant design tasks can opt for one of the several approaches, which differ from each other by how much attention is paid to the effect of joint behaviour on the global analysis.

While a single party, who is understandably the fabricator, is in charge of the fabrication tasks, the design activities are carried out:
- Either by a single party - the *engineer* (engineering office) or the *fabricator* - who is responsible for the design of both members and joints;
- Or by two parties: the *engineer* (engineering office), who is responsible for the design of members, and the *fabricator*, who is in charge of the design of joints.

According as engineer and fabricator share or not the responsibilities of the design tasks, three situations – designated as *A* to *C* - are identified (see Table 4).

TABLE 4 – Share of responsibilities for the design and fabrication tasks

Task	Case A	Case B	Case C
Design of members	Engineer	Engineer	Fabricator
Design of joints	Fabricator	Engineer	Fabricator
Fabrication	Fabricator	Fabricator	Fabricator

In *Case A*, the engineer designs basically the members. He specifies also the mechanical requirements to be fulfilled by the joints so as to be consistent with the assumptions made for the global analysis. It is up to the fabricator to design the joints with due attention paid to above requirements, on the one hand, and to specific manufacturing aspects dealing with economy, on the other hand. Due to the corresponding share of responsibilities, it may happen that the joint detailing adopted by the fabricator is sub-optimal; indeed it depends on the member sizing, which was made independently by the engineer at an earlier stage. Should the engineer aim for instance at minimum shape sizes, the fulfilment of safety and serviceability requirements might require some stiffening of the joints. In contrast, rather large shapes for members might result in less complex joints and therefore in a better economy due to labour savings.

In *Case B*, the engineer is responsible for the design of both members and joints. He is then in a position where he can fully handle the interference, for what regards the global analysis, between the mechanical joint properties and the member sizing. Should he decide to do accordingly, the conditions exist for an optimal economy; the latter is likely to be achieved if the designer is fully aware of the basic manufacturing requisites. If not, an increase in the fabrication cost shall be expected.

In principle, *Case C* is ideal with regard to global economy; indeed the design of both members and joints is in the hands of a single party: the fabricator. The latter cannot surely be ignorant of the manufacturing aspects; therefore the economy of the construction depends largely on the fabricator's skill for what regards structural design.

The design methodology refers thus to anyone of the design approaches below:
- *Traditional design approach* - Global analysis is conducted based on: i) a preliminary design of the members, and ii) the assumption that any joint is assumed either simple or continuous. Members and joints are checked in a second step. This approach addresses Case *A, B* or *C*; it was and it is still the most usual in practice.
- *Consistent design approach* - Global analysis is performed based on member and joint properties, which are preliminarily assessed. This approach addresses Case *B* or *C*, and possibly Case *A*.
- *Intermediate design approach* - A single party (Case *B* or *C*) preferably designs the members and the joints.

The application of anyone of these design approaches is a so-called design strategy. Yet it was stressed that consideration of joint properties before starting the global analysis and conducting it accordingly may turn out efficient. The statement of this recommendation is few compared to the handling of the latter; indeed the internal forces in the joints interact with the structural response of these joints. The way to operate is simply the use, for the global analysis, of joint properties, which come out a pre-design of the joints. Several strategies exist in this respect; it is commented below on two of them. The reader especially interested in these matters will usefully refer to the basic reference [3] or to some lectures prepared in the frame of the SSEDTA project [4].

The first strategy proceeds through the so-called good-guess. Based on the revised Annex J of Eurocode 3 [5] and the consideration of the joint configuration only, some first simplified formulae were derived [6], which were somewhat supplemented afterwards [3]. These formulae presuppose some rough choices of the connection detailing parameters for typical joints.

A realistic *approximate* value $S_{j,app}$ of the initial joint stiffness is expressed as:

$$S_{j.app} = E\, z^2\, t_{fc} \big/ C$$

where z is the approximate distance between the compression and tensile force resultants in the joint and $t_{f,c}$ is the thickness of the column flange. Annex J [5] gives guidelines for assessing the lever arm z while values of the C factor are tabled in [3] for different joint configurations and moment loading patterns.

Once the global analysis is completed, the joint is detailed so as to enable the computation of its *actual* initial stiffness $S_{j,ini}$ in accordance with ENV 1993, or by means on an appropriate software. Of course, one must expect some difference between $S_{j,ini}$ and $S_{j,app}$. When this difference does not result in a more than 5% drop in bearing capacity of the frame, it is generally agreed that the global analysis does not need to be restarted. This criterion is satisfied when the actual stiffness $S_{j,ini}$ is comprised in a range determined by so-called lower and upper boundaries [3]:

$$S_{j.ini}^{lower} \leq S_{i.app} \leq S_{j.ini}^{upper}$$

These boundaries are associated to the approximate one, $S_{j,app}$, , and determined based on the beam length L_b and rigidity EI_b; they are slightly different according as the frame is braced or unbraced (see Table 5).

TABLE 5. Boundaries for variance of the initial stiffness

Frame	Lower boundary	Upper boundary
Braced	$S_{j,ini} \geq \dfrac{8 S_{j,app} EI_b}{10 EI_b + S_{j,app} L_b}$	If $S_{j,ini} \leq 8 EI_b / L_b$ then $S_{j,ini} \leq \dfrac{10 S_{j,app} EI_b}{8 EI_b - S_{j,app} L_b}$ else $S_{j,ini} \leq \infty$
Unbraced	$S_{j,ini} \geq \dfrac{24 S_{j,app} EI_b}{30 EI_b + S_{j,app} L_b}$	If $S_{j,ini} \leq 24 EI_b / L_b$ then $S_{j,ini} \leq \dfrac{30 S_{j,app} EI_b}{24 EI_b - S_{j,app} L_b}$ else $S_{j,ini} \leq \infty$

As an alternative to the good guess procedure, reference can be made to design aids, such as tables and softwares. Such tables exist [3, 7]; they provide both the initial stiffness and the strength of several thousands of typical joints described by the shapes used for the connected members, the steel grades and several parameters of the joint detailing. Also information regarding the mode of collapse and the joint classification is given. Using software gives more adaptability; in this respect, [8] is surely the most promising programme. These design aids provide a less rough information than the good guess procedure. They are a quite valuable assistance to the designer, especially when the failure must be ductile. They enable to start, in a very realistic and efficient way, not only elastic analysis but also plastic analysis of semi-continuous frames.

16

8. Conclusions

When performing frame analysis, many joints may be considered as either simple or continuous. There is however a range where account must be taken of their semi-continuity. To distinguish between these three situations, the joints need to be classified, similarly to member sections, with regard to stiffness, strength and ductility. The joint stiffness is likely to influence the classification of the frames and consequently the limits of applicability of first order analysis. When it is required to account for the joint properties in the global analysis, a preliminary assessment of these properties can be conducted according to several methods. Design aids do exist which can assist and help the designer for that purpose.

Acknowledgements

Present paper is largely based on the material developed more thoroughly in the so-called "Designer's manual" [3], which was the main outcome of an ECSC contract. Again the contribution of those who participated this successful issue – A.Souah, I.Ryan and B.Chabrolin (CTICM), J.P.Jaspart and R.Maquoi (University of Liège), K.Weynand (RWTH), M.Steenhuis (TNO) and D.Vandegans and E.Piraprez (CRIF) - is deeply acknowledged.

9. References

1. CEN (1992) *ENV 1993-1-1 - Eurocode 3: Design of steel structures - Part 1-1: General rules and rules for buildings.*
2. CEN (1994) *ENV 1994-1-1 - Eurocode 4: Design of composite structures - Part 1-1: General rules and rules for buildings.*
3. Maquoi, R. and Chabrolin, B. (1998) *Frame design including joint behaviour*, Report N° EUR 18563 EN, ECSC Contract N° 7210-SA/212/320, European Commission, Brussels.
4. European Commission (2000) *Structural Steelwork Eurocodes:Development of a trans-national approach* (see especially modules 2 and 3), CR-Rom, Leonardo da Vinci Programme, University of Sheffield (editor).
5. CEN (1998) *ENV 1993-1-1 - Eurocode 3: Design of steel structures - Part 1-1: General rules and rules for buildings – Annex J – Joints in building frames.*
6. Steenhuis, M. and Gressnigt, N. (1995) Pre-design of semi-rigid joints in steel frames, in F.Wald (editor) *Proceedings of the state of the art, COST C1 Workshop*, Prague, pp. 131-140.
7. CRIF, MSM, ENSAIS and Universita di Trento (1996) *Assemblages flexionnels en acier selon l'Eurocode 3: Outils de calcul pour les assemblages rigides et semi-rigides*, SPRINT Contract RA351, CTICM, Saint-Remy-les-Chevreuse (editor) or CRIF, Liège (editor).
8. MSM, RWTH, ICCS (1998) *CoP The Connection program*, CD Rom, ©1998 ECCS bv.

BEHAVIOR AND DESIGN OF PARTIALLY-RESTRAINED COMPOSITE FRAMING SYSTEMS

A.E. MALECK and D.W. WHITE
Georgia Institute of Technology
Atlanta, GA 30332-0355 USA

1. Introduction

Much progress has been made in recent years towards analytical modeling of the limit states behavior of structural steel frame members, connections and systems. Analysis methods that accurately represent the behavioral effects associated with both system and primary member design limit states have been suggested and referred to by many as "advanced analysis" approaches. By use of advanced analysis, the checking of certain limit-states equations in design standards is superceded. Specifically, in an advanced analysis, the checking of member and system stability is directly included within the analysis. If the global stiffness of the structural model is positive-definite at the design load levels, i.e., if the model is capable of supporting additional load at these levels, then the associated limit states are considered to be satisfied.

Advanced analysis offers significant benefits for the design of "semi-rigid" or "partially-restrained" (PR) frames. This is due to the fact that, for these types of frames, potential reductions in stiffness within the connections must be considered within the design. Proper accounting for these stiffness reductions within design methods based on buckling analysis or column effective length is indirect. Connection stiffnesses are obtained based on an assumed or calculated state of the connections (typically a function of the level of moment in the connections). These stiffnesses are then utilized to determine column buckling loads or effective length factors. The column buckling loads or effective length factors are then utilized for the calculation of nominal strengths of the members under pure axial compression (effects of moment on the member strength being neglected in these calculations). Finally, the member strengths under bending with no axial load are calculated, and the applied axial force and moment from a simplified analysis are compared to a beam-column interaction equation that is anchored by the above axial and bending strengths. The rationality of these beam-column checks is limited when applied to members within redundant structural systems (White & Hajjar, 2000.)

While advanced analysis is the most rational means to assess system behavior, the software capabilities required for this level of analysis are still not readily available in

17

C.C. Baniotopoulos and F. Wald (eds.), The Paramount Role of Joints into the Reliable Response of Structures, 17–30.
© 2000 Kluwer Academic Publishers.

the design office. However, it is possible to achieve a simpler and more rational design methodology with the use of typical second-order elastic analysis software by appropriate accounting for the key behavioral effects that influence the response. This paper presents a "modified elastic" analysis approach that directly accounts for the effects of connection nonlinearity, member distributed plasticity, and geometric imperfections on the distribution of forces within the structural system. It is shown that member and system stability can be adequately checked by usage of the analysis forces from this approach with simplified member design equations that are based on actual member length.

2. Modified Elastic Analysis

2.1. CONNECTION MODELING

When assessing system strength of PR frames, it is important to account accurately for the nonlinear connection response. To this end, rational modeling of the connection nonlinearity in a computationally effective manner is desirable. For simplicity, a portal frame is used here to elucidate the connection behavior within a moment frame subjected to a sequence of design gravity loading, followed by lateral loading. This type of sequential loading has been suggested by many of the current methodologies for PR frame design (Leon, et al. 1998, AISC 1999, Christopher & Bjorhovde 1999). While proportional loading produces less conservative results (Deierlein, 1992), the non-proportional case is generally considered to present a more realistic representation of the load history experienced by the building. A portal frame with PR connections is shown in Fig 1. Fig. 1b shows a hypothetical connection response under gravity load and Fig. 1c shows the connection response under subsequent lateral load.

In this example, the stiffness of both connections decreases with respect to the initial stiffness under gravity load. With the addition of wind load, the windward connection (C1) begins to elastically unload, while the leeward connection (C2) continues to load and decreases in stiffness. Current methods of PR frame design in the US suggest the use of a secant to model the connection stiffness (ASCE, 1998, AISC 1999). As shown in Fig. 2a, this would significantly overestimate the tangent stiffness of connection C2 in the lateral load case, whereas it underestimates the unloading stiffness of connection C1.

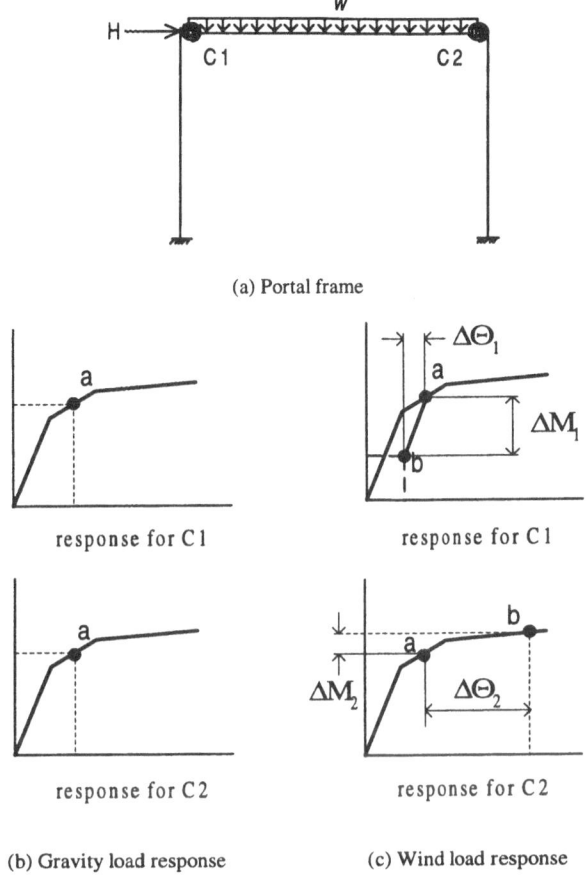

(a) Portal frame

response for C1

response for C1

response for C2

response for C2

(b) Gravity load response

(c) Wind load response

Figure 1. Illustrative portal frame and corresponding connection response.

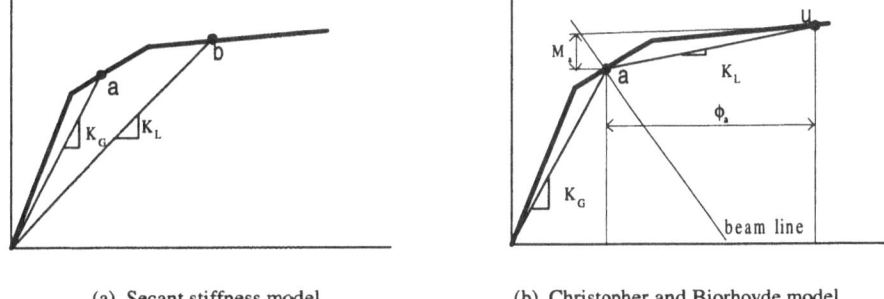

(a) Secant stiffness model

(b) Christopher and Bjorhovde model

Figure 2. Connection models.

Christopher & Bjorhovde (1999) present a method of analysis in which the connection moment and stiffness under each loading is more accurately assessed (Fig 2b.) An initial connection moment, M_a, is estimated based on a beam line analysis at gravity load levels. This is valid to the extent that there is zero or negligible rotation of the column. The effective connection stiffness used for the subsequent lateral loading is then taken as the secant stiffness from point a to the limit state at point u associated with a practical rotation limit based on connection ductility.

While the Christopher and Bjorhovde method more rationally assesses the connection stiffness under non-proportional loading, it requires calculation of unique secant stiffness for each connection based on load level. This paper suggests an alternate approach that utilizes a trilinear connection model. By estimating the portion of the curve on which the connection response will fall, the need for approximate beam line analyses and calculation of a different stiffness for each is avoided. As uniformity in construction tends to lead to reuse of connection details at numerous locations, the simplicity gained by utilizing a single trilinear curve for numerous connections can be substantial.

A response diagram for the connections of the portal frame example is shown in Fig 3. The y intercepts of the lines that define the trilinear connection stiffness model may be easily calculated. Point c is shown as the intercept (or initial moment) for the second portion of the curve. For this example, the connection response under gravity load is assumed to be located at point a. The addition of lateral load changes the connection response on the leeward connection to point b. A new set of reference axes, with an origin at point a, is defined for calculation of the change in the connection moments due to wind load in the connections. In this example, the windward connection, C1, is assumed to elastically unload.

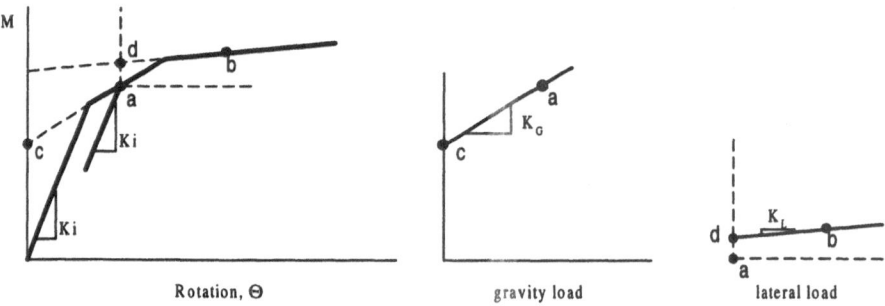

Figure 3. Proposed connection model.

For the gravity load case, the connection is modeled as a linear spring with stiffness K_G, with an initial moment of M_c. This moment is simply the intercept of the line defining the second portion of the moment-rotation curve with the vertical axis (point c in Fig. 3). For the windward connection, C1, the stiffness during the wind load analysis is modeled by a spring of stiffness K_i, where K_i is the appropriate stiffness for elastic unloading, taken as the initial connection stiffness as defined and discussed in (AISC 1999). For the loading connection, C2, a linear stiffness of K_L is used with an initial moment of M_d at point d.

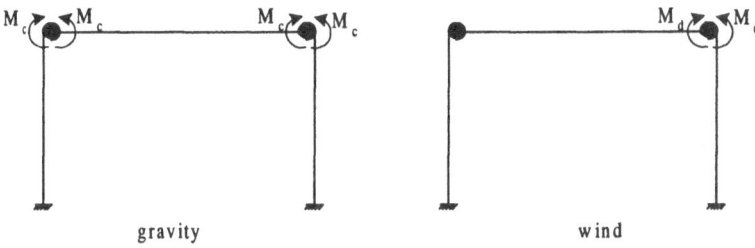

Figure 4. Application of connection y-intercept moments.

While most modern analysis software is capable of modeling linear connection moment-rotation springs, it is less common that these connection models would include initial force or deformation values as input parameters. In this instance, it is possible to model the connection linearly, and account for the "initial" moment (i.e., the y intercepts) by applying these moments to each side of the spring. These may be added as a combination of concentrated external nodal and internal element moments as shown in Fig 4. The corresponding fixed-end forces associated with the internal element moments may be calculated and combined with the external nodal moments shown in Fig. 4 such that only one moment needs to be applied to the nodes in the global frame analysis (along with the reverse of the beam shears corresponding to the fixed-end nodal moments). Fig. 5 shows an elastic, prismatic beam with applied internal moments M_1 and M_2. Based on this configuration, the fixed end moments M_L and M_R are computed as:

$$M_L = (K_b K_{SL} M_2 + K_{SL}(2K_b + K_{SR}) M_1)/A \qquad (1a)$$

$$M_R = (K_b K_{SR} M_1 + K_{SR}(2K_b + K_{SL}) M_2)/A \qquad (1b)$$

where:

$K_b = 2EI_b/L$
K_{SL} = left spring stiffness
K_{SR} = right spring stiffness
$A = 2K_b(K_{SL}+K_{SR}) + K_{SL}K_{SR} - 3K_b^2$

Figure 5. Beam with end connections, subjected to internal concentrated end moments.

2.2 MEMBER STRENGTH ASSESSMENT

Many methods have been suggested by which beam-column member strength may be assessed using actual member length. The most noteworthy of these are the notional load methods used in design codes outside of the US (CEN 1992 ,CSA 1998, SAA 1990). Beam column design in the AISC-LRFD (1999) Specification is based on buckling solutions or the corresponding column effective lengths. All of these methods involve useful but imperfect design approximations. Alternatively, advanced analysis accounts directly and more precisely for system interdependencies in that all phenomena affecting system strength are included in the analysis. With small modifications to common first- or second-order analysis methods, it is possible to more rationally capture the effects of nonlinearity and inelasticity on the distribution of forces within the structural system. If the effects of imperfections due to erection tolerances as well as inelasticity in the columns due to residual stresses, initial out-of-straightness, connection nonlinearity (as previously described) and high axial forces are included in an appropriate approximate fashion within the analysis, the member and system strength can be assessed adequately using the AISC LRFD beam-column interaction equations with actual member lengths.

Figure 6 illustrates a representative sensitive non-redundant benchmark problem, one of a comprehensive parametric studies undertaken by the authors (Maleck 2000). In Fig. 7, advanced analysis beam column interaction curves in terms of the first-order (HL) and second-order (M_2 = HL + PΔ) moment are compared to: (a) a modified second-order elastic analysis (ME) with the AISC LRFD beam-column interaction equations based on actual member length, and (b) a traditional second-order elastic analysis as per AISC LRFD (no geometric imperfections included within the analysis) along with AISC LRFD interaction equations based on effective length. In the ME analysis, an initial out-of-plumb is included in the analysis by use of a notional load of .002P, and member inelasticity is approximated by use of the LFRD column inelastic stiffness reduction factor (τ). Use of the LRFD column curve based on actual member length to assess axial strength inherently accounts for effects of out-of-straightness. Due to the direct inclusion of out-of-plumbness in both the advanced and modified elastic analysis, the corresponding second-order moment curves do not intercept the y-axis. For sensitive, non-redundant benchmark problems with inelastic & out-of-plumbness effects included as described, this approach captures the advanced analysis behavior with less than 5% non-conservative error, and in many cases better reproduces the advanced analysis solution than the current LRFD procedure.

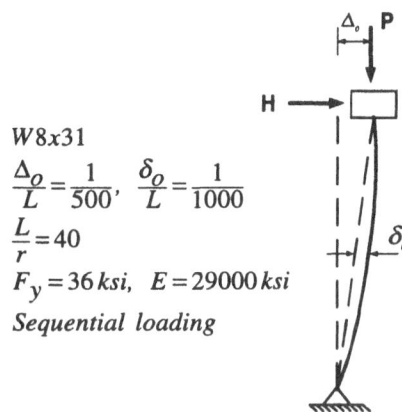

Figure 6. Non-redundant member strength benchmark

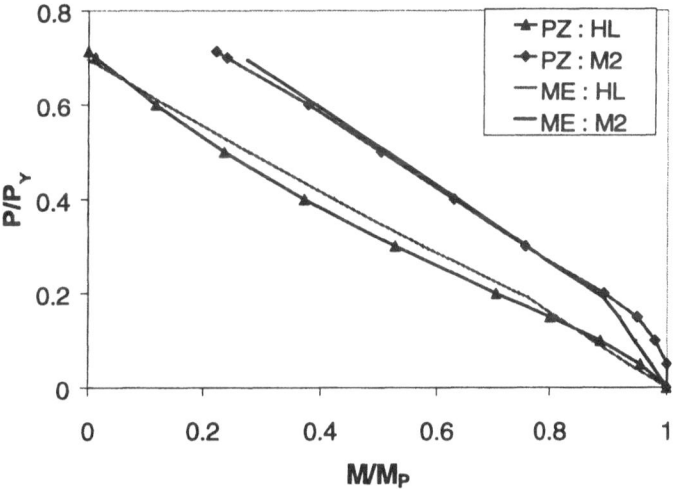

(a) Plastic Zone vs. Modified Elastic

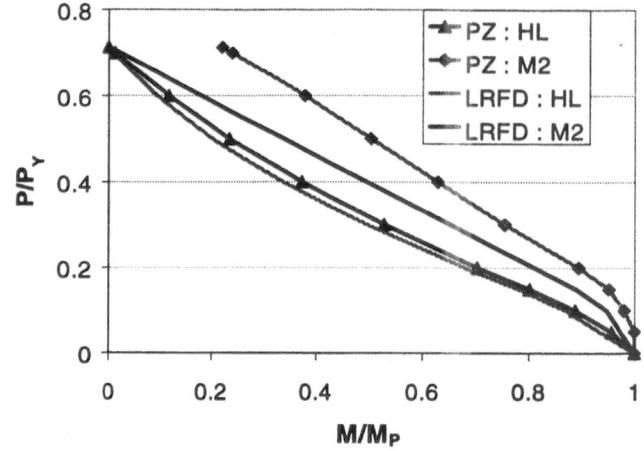

(b) Plastic Zone vs. LRFD w/ K = 2

Figure 7. Member strength benchmark.

2.3 ADDITIONAL MODELING CONSIDERATIONS

While composite beams are indeed non-prismatic, studies by Leon & Forcier (1992) have shown that it is possible to achieve results similar to those achieved in a non-prismatic analysis by using an equivalent prismatic beam stiffness given by:

$$I_{eq} = 0.6I_{LB}^{+} + 0.4I_{LB}^{-} \qquad (2)$$

Where I_{LB}^{+} and I_{LB}^{-} are the positive and negative lower bound moments of inertia, respectively, as defined in the LRFD specification.

When possible, a second-order analysis should be used to directly account for the moment amplification within the system. In a first order analysis, amplification factors may be calculated to determine the amplified moments in the beam columns due to second-order effects. These amplification factors should be determined based on the inelastically reduced column stiffnesses. Subsequently, the amplified beam moments must be calculated based on equilibrium at the beam-column joints.

For many practical framing systems the effect of initial frame imperfections (i.e., out-of-plumbness or lack of verticality) is not significant (Maleck & White, 1999). However, since there are no clear conditions under which these imperfections may be neglected, they should be considered in the analysis. It is preferable to directly model imperfections if possible. In lieu of direct modeling of imperfections, initial out-of-plumbness may be accounted for by the addition of notional horizontal loads of $N = 0.002Q$ where N is the additional horizontal load, and Q is the total story gravity load. Member out-of-straightness is accounted for by use of the AISC-LRFD column strength curves based on an actual member length.

3. Design Example

A four-story, three-bay partially-restrained composite frame originally studied by Forcier (1990) is considered. The design developed by Forcier is based on the LRFD specification. The frame is analyzed and redesigned based on both the suggested design method and by advanced analysis. Design and modeling issues regarding the use of advanced analysis are outlined in Maleck, et. al (1995) with one exception: for composite PR frames, a resistance (ϕ) factor of 0.85 is recommended, corresponding with the AISC-LRFD resistance factors recommended for both composite beams and column strength. The frame is shown in Fig 8. Details of the composite girder and connections details are given below. The connections detail corresponds to the parameters described in AISC Design Guide #8 (Leon, et al 1998)

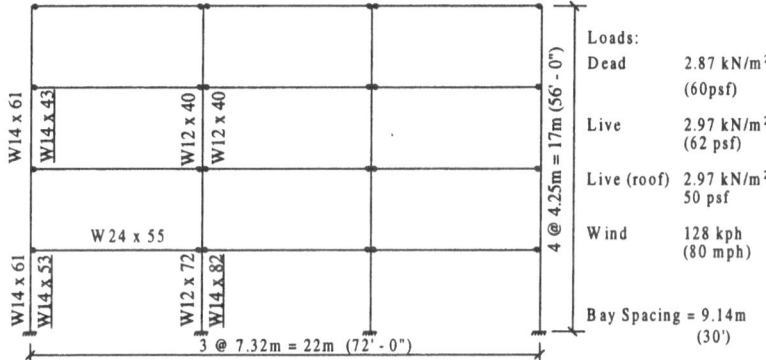

Figure 8. Design example.

Slab:
        lightweight concrete: $f'c = 24{,}100$ kN/m^2 (3.5ksi)
        12.7 cm (5") lightweight concrete slab
          5.1 cm (2") steel deck
Connection:
        $A_{rb} = 10.3$ cm^2 (1.6in^2)
        $F_{yrb} = 414{,}000$ kN/m^2 (60 ksi)
        $A_{sl} = 28.1$ cm^2 (4.36 in^2)
        $A_{wl} = 56.5$ cm^2 (8.75 in^2)
        $Y_3 = 10.16$ cm (4.0 in)

The column designs are constrained such that column sizes extend a minimum of two stories, and upper story column profiles are the same size as or smaller than those in the stories below. Steel girder selection was based on unshored construction loads. The girder selection was made such that no yielding occurred due to wet loading and prior to composite action.

4. Analysis Results

Two load combinations (ASCE 1998) control the design of the above frame:

$$1.2D + 1.6L + 0.5 L_r$$
$$1.2D + 0.5(L + L_r) + 1.3W$$

An advanced analysis of the original design was utilized to perform an initial redesign of the system. The redesign of the frame was based on achieving a more even distribution of yielding within the columns of the system compared to the original design. The lower story internal columns were increased in size, while the exterior columns sizes were reduced. An overall reduction of steel weight was achieved as well as a more even distribution of inelasticity in the system at higher load levels without a substantial decrease in overstrength. The ultimate failure mode of the redesigned frame under lateral load is primarily due to high levels of inelasticity in the leeward connections and at the base of the structure combined with P-Δ effects.

Results of the modified elastic analysis design show the efficiency of the redesign. The column sizes are controlled by the gravity load case. The corresponding AISC-LRFD beam column interaction values using actual member length are shown in Fig. 9. In all instances, the nominal column strength, ϕP_n, is controlled by out-of-plane strength. The modified elastic approach is still conservative with respect to the advanced analysis in which the system achieved an ultimate load factor in the gravity load case of 1.4.

Due to the non-proportional loading, connections on the windward side of the structure elastically unload due to lateral loading, as previously shown in the portal frame example (Fig 1). The higher stiffness in the elastically unloading connections assists in the lateral stability of the system. Fig. 10 shows the connection response under the gravity and lateral load portions of the modified elastic analysis.

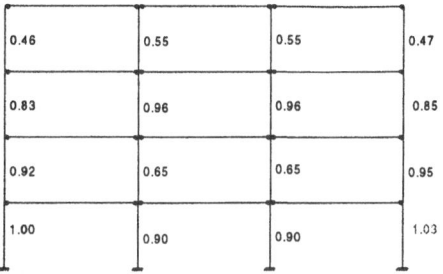

0.46	0.55	0.55	0.47
0.83	0.96	0.96	0.85
0.92	0.65	0.65	0.95
1.00	0.90	0.90	1.03

Figure 9. Beam-column interaction values under gravity load.

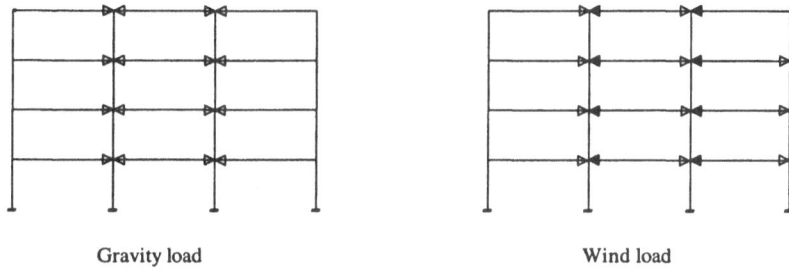

Gravity load   Wind load

◁   Loading, 2nd branch of trilinear curve
•   Elastic
◀   Elastic unloading

Figure 10. Connection response.

Use of the modified elastic analysis method requires that the connections denoted by open triangles be modeled on the second portion of the trilinear curve for the gravity analysis. This typically requires the addition of externally applied moments in those connections as shown in Fig. 4. Due to repetition of the same connection detail within the design, the calculation of only one connection curve and y-intecept are required. Under wind load, only three connections (on the exterior leeward side) require addition of external moments. All other connections either continue loading on the same portion of the nonlinear curve or unload elastically.

To assess the accuracy of the modified elastic approach, a comparison of the second-order load-deflection curves for the advanced (dashed) and modified elastic analysis (solid) is shown in Fig. 11. Top story displacement is normalized with respect to the maximum service deflection limit H/500. In the modified elastic analysis, connection nonlinearity, column inelasticity and initial imperfections are modeled as previously described. Comparison of the analyses shows less than 5% error in the lateral displacements predicted by the modified elastic approach. The initial displacement at

28

zero lateral load is due to the P-Δ effects associated with the initial out-of-plumbness. Fig. 12 shows the load-deflection response predicted by the advanced analysis for loading of the frame to its limit load under wind loading with the design gravity loads held constant. This frame is typical of many partially-restrained composite building frames in that there is substantial reserve strength for lateral loading due to the fact that the columns are controlled by the gravity load combination and the steel beam sizes are controlled by wet loading conditions during construction. Also, the modified elastic analysis design method is often conservative simply because the design loading is limited to that which causes the most critical component within the structural system reach one of its limit states. This ignores the beneficial inelastic redistribution which may occur within a redundant structural system.

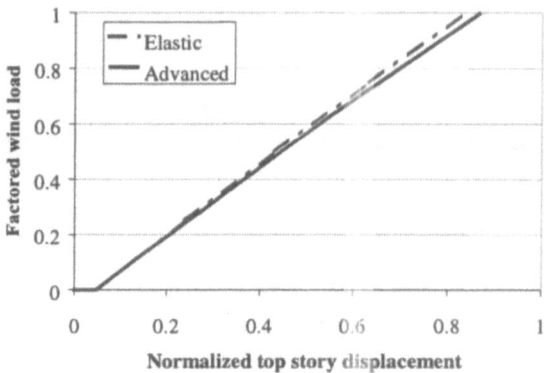

Figure 11. Comparison of load versus lateral deflection at factored wind load level.

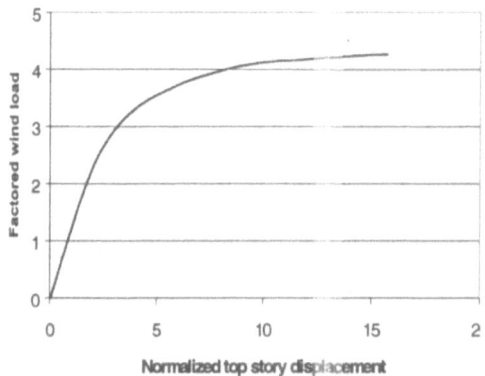

Figure 12. Complete load-deflection up to collapse, from advanced analysis.

5. Conclusions

While advanced analysis offers the most comprehensive means of accounting for system nonlinearity and inelasticity, it is possible within the level of computational capabilities available in a typical design office to rationally account for these effects in a simplified analysis-design methodology. The connection model presented can be used with first or second order analysis. Also, the proposed approach for handling the connection nonlinearity can be utilized directly within the present AISC LRFD procedures which involve calculation of buckling loads or column effective lengths. The method offers results comparable to, yet conservative with respect to, an advanced analysis

6. References

AISC (1999) *Load and Resistance Factor Design Specification for Steel Buildings*, Manual of Steel Construction - Load and Resistance Factor Design, Third Edition, American Institute of Steel Construction, Inc., Chicago, IL.

AISC (1993) *Code of Standard Practice for Steel Buildings and Bridges*, Manual of Steel Construction - Load and Resistance Factor Design, Second Edition, American Institute of Steel Construction, Inc., Chicago, IL.

ASCE Task Committee on Design Criteria for Composite Structures in Steel and Concrete (1998) Design Guide for Partially Restrained Composite Connections, *Journal of Structural Engineering*, ASCE, **124**(10), 1099-1114.

ASCE (1998) *ASCE 7-98: Minimum Design Loads for Buildings and Other Structures*, American Society of Civil Engineers, Reston, VA.

CEN (1992), "*ENV 1993-1-1 Eurocode 3, Design of Steel Structures, Part 1.1 - General Rules and Rules for Buildings*," European Committee for Standardization, Brussels, 1992.

Christopher, J. E. & Bjorhovde, R. (1999) Semi-Rigid Frame Design Methods for Practicing Engineers, *Engineering Journal*, AISC, **36**(1), 12-28.

CSA (1998) *Limit States Design of Steel Structures, CAN/CSA-S16.1*, Canadian Standards Association, Toronto, Ontario, Canada.

Deierlein, G.G. (1992) An Inelastic and Design System for Steel Frames with Partially Restrained Connections, *Connections in Steel Structures II: Behavior Strength and Design*, Bjorhovde, R. Colson, A., Haaijer, G., and Stark, J. W. B., eds., AISC, Chicago, IL, 408-415.

30

Leon, R.T. & Forcier, G.P. (1992) Parametric Study of Composite Frames, *Connections in Steel Structures II: Behavior Strength and Design,* Bjorhovde, R. Colson, A., Haaijer, G., and Stark, J. W. B., eds., AISC, Chicago, IL, 152-159.

Forcier, G. P. (1990), "A Parametric Study of Composite Frame Behavior," Master's Thesis, School of Civil Engineering, University of Minnesota, Minneapolis, MN.

Leon, R. T., Hoffman, J. J. and Staeger, T.(1998) *Partially Restrained Composite Connections,* Steel Design Guide Series 8, American Institute of Steel Construction.

Maleck, A.E. and White, D.W. (1998) Effects of Imperfections on Steel Framing Systems. *Proceedings, Annual Technical Session,* Structural Stability Research Council, Gainsville, FL, 43-52.

Maleck, A.E., White, D.W. and Chen, W.F. (1995), Practical Application of Advanced Analysis in Steel Design, *Structural Steel,* Proceedings 4th Pacific Structural Steel Conf., Vol. 1, Steel Structures, 119-126.

Maleck, A.E., (2000) *Design and Analysis of Composite Partially-Restrained Steel Framing Systems,* Ph.D. Dissertation, Georgia Institute of Technology, *in preparation.*

SAA (1990), *AS4100-1990, Steel Structures, Standards Association of Australia,* Australian Institute of Steel Construction, Sydney, Australia.

White, D.W. & Hajjar, J.F. (2000), Stability of steel frames: the case for simple elastic and rigorous inelastic analysis/design procedures, *Engineering Structures* **22**, 155-167.

PREDICTION OF ULTIMATE LOAD OF STEEL FRAMES WITH SEMI-RIGID CONNECTIONS

M. IVÁNYI
Budapest University of Technology and Economics, Department of
Bridges and Structures
H-1111, Budapest, Mûegyetem rkp. 5-7, Hungary

1. Introduction

1.1. LOAD-DEFORMATION RESPONSE OF FRAMES

The beams, columns, and beam-columns do not occur in isolation, but many of them joined together make up a structural frame. This frame is the skeleton which supports the loads which the structure is called upon to support.

The purpose of frame analysis is to determine the limits of structural usefulness of a given frame and to compare the predicted performance with the required one. Such an analysis is part of the design process, wherein adjustments and new analyses are made until the predicted performance matches as closely as possible the design requirements.

In this article the methods of determining the maximum load capacity and the deformation response of frames will be examined. This topic is a vast one and we shall only be able to cover a small portion of it. Emphasis will be placed on basic behaviour and the discussion will center around very simple examples.

Frame behaviour is characterized by the relationship between the loads, as they vary during the loading history, and the resulting deformations. A typical load-deflection curve is shown in Fig. 1. The relationship is non-linear from the beginning because of second-order geometric effects (that is, the forces produce deformations which in turn influence the forces). After the elastic stage is reached, the slope of the curve is further reduced, and finally the slope becomes zero at the maximum load P_M.

A curve, such as in Fig. 1 gives the value of the maximum load which can be carried by the frame, as well as the magnitude of the deformation corresponding to any load intensity. Furthermore, at least the ascending branch of the curve is in stable equilibrium. In design the curve can be used to check if (1) the ratio P_M/P_W (where P_W is the actual or working load) is sufficiently near a specified load-factor (as determined by judgment or prescribed by a code or a specification), and (2) the deflection at working load v_W is less than or equal to a specified maximum value. In the design

C.C. Baniotopoulos and F. Wald (eds.), The Paramount Role of Joints into the Reliable Response of Structures, 31–46.
© 2000 *Kluwer Academic Publishers.*

operation we try to match these requirements with the structural behaviour, each time adjusting the structure until the requirements are met.

Ideally it would be desirable to construct a load-deflection curve for each structure. We then could obtain the various items of information which we are interested in. Unfortunately we are only able to construct load-deflection curves for very simple structures. For more complex frames we need to introduce assumptions which will permit us eventually to obtain bounds for the value of P_M.

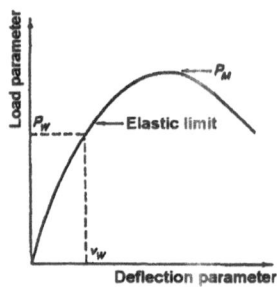

Figure 1. Load–deflection curve of a structure.

Two significant "target" models are used, as follows.

(A) Interactive Plastic Hinge. The traditional concept of plastic design of steel structures is based on the assumption that under gradually increasing static loads plastic zones develop and grow in size and number, and eventually cause unstricted, increasing deflections; thus loading to the onset of ultimate limit state of the structure. The concept was first introduced by *Kazinczy* [20] by establishing concept of the "plastic hinge". Some basic questions are still discussed. Among them are the effects of the difference between ideal-plastic constitutive law and actual behaviour of steel material and the consequense of local instability (plate buckling; lateral buckling). The element of the bar is considered to be built up of plate elements (following the pattern of steel structures) instead of a compact section. Then the behaviour of the "plastic hinge" can be characterized by tests with simple supported beam (Fig. 2). Based on these tests a yield-mechanism for the bar-element can be introduced, giving basis for a mechanism curve: defining thus the descending branch of the moment-rotation diagram. We introduce the concept of "interactive plastic hinge" which can substitute the classical concept of plastic hinge in the traditional methods of limit design, but can reflect the effect of phenomena like strain-hardening, residual stresses, plate buckling and lateral buckling [16], and role of semi-rigid connections.

(B) Semi-rigid Connection. The type of beam-to-column connection used is a primary determinant of the behaviour of the frame. The construction types are defined in terms of connection rotational stiffness and moment resistance as represented by a moment-rotation (M-θ) diagram (Fig. 3). Generally, accurate M-θ diagrams can only be obtained experimentally by tests. Fig. 3 contains M-θ diagrams for typical connections and indicates how they would normally be classified. The non-linearities in behaviour

which are shown are the result of yielding of the connection components or local regions of the connected members, or slip of the fasteners.

Corrections are so diverse and so complex that large amounts of experimental and analytical data on connection deformation must be collected and systematized before reliable semi-rigid frame analysis can become common practice [23].

Local and lateral bucklings are so diverse and so complex that large amounts of experimental and analytical data must be collected and systematized thus reliable frame analysis become practice [2].

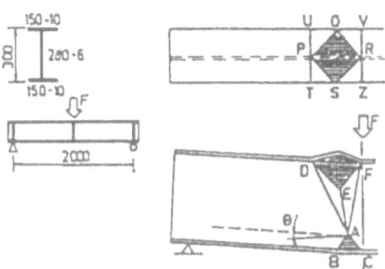

Figure 2. Local buckling of plate elements of I coss-section.

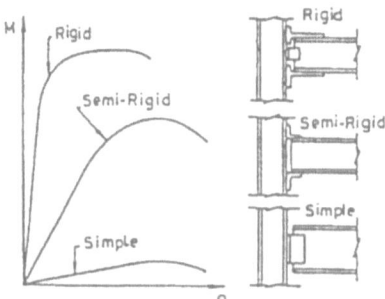

Figure 3. Moment–rotation curves for connections.

1.2. EFFECT OF SOFTENING PHENOMENON

Studying the effect of softening phenomenon one should keep in mind that the load-displacement diagram of the structure may be of an ascending type even if the of the given member section or semi-rigid connections are of a descending type.

In the theory of plasticity, when deriving the condition of plasticity or some other physical relationships, *Drucker's* postulate for stability is applied, by assuming stable materials [4].

It should be noted that *Drucker's* postulate is not a natural law but a criterion of classification [5], the materials very often do not correspond to the assumptions of stable materials, or structural elements may behave in an unstable way, while, at the same time, their material is of a stable state.

Maier [21] was the first to treat the problem of the effect of the unstable state of certain members on the behaviour of a triangulated structure. Again it was *Maier* who in 1966 re-introduced the subject and investigated a structure consisting of compressed members and rigid beams where load-displacement diagram of individual members contained stable and unstable parts.

Maier and *Drucker* [22] re-examined the original *Drucker* postulate applied when determining the condition of plasticity since the original postulate is suitable for the determination of the convexity and normality of the condition of plasticity in case of stable materials only.

When studying the load bearing capacity of steel structures, the problem of unstable material or softening material, according to *Drucker's* postulate does not appear since

34

the strain-hardening of the steel material may increase in a major way the plastic load bearing capacity of steel structure. However, as it has been known for a long time, the final collapse of steel structures is caused – in a high percentage of cases – by instability (plate buckling, flexural-torsional buckling), or semi-rigid phenomena that may occur in the cross section or in a structural unit (Fig. 4).

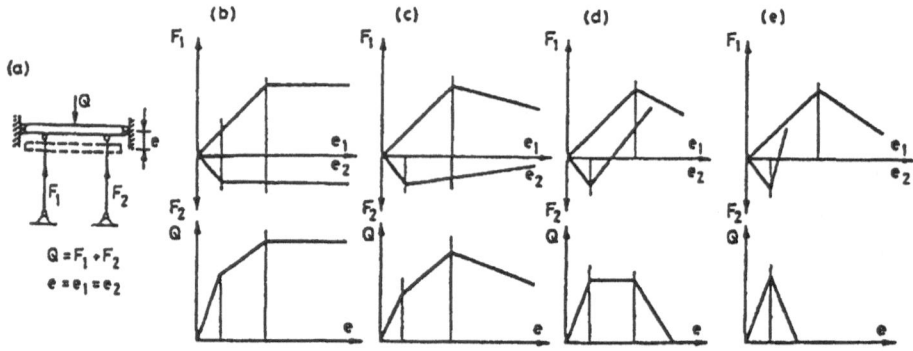

Figure 4. Behaviour of simple structure.

Concerning steel structures the properties of plastic hinges over and above the usual elastic-ideally plastic-hardening behaviour may be complemented with the effect of instability (flexural-torsional buckling) developing in the given structural unit (environment of the plastic hinge), or in the periphery of semi-rigid connections.

This type of inelastic or interactive hinge describes the behaviour of the structural unit and at the same time, also satisfies the criteria of unstable or softening structural unit, according to *Maier-Drucker's* postulate.

When determinig the plastic load bearing capacity of steel structures the interactive hinge of softening has so far not been considered or applied. The effects of the stability phenomena causing the softening character (flexural-torsional buckling, plate buckling, semi-rigid connections) can be taken into account indirectly with the aid of construction rules. In principle, mathematical programming allows the investigation of more complex steel structures, too, however, it is less suitable for designing practice. The author [14] has suggested a procedure that is taking into account the softening character of the inelastic hinge in the form of an interactive zone. The softening character of the interactive zone is caused by the buckling of the component plates, a phenomenon that can be studied with the help of the yield mechanism.

The purpose of this article is to describe (1) how the load-deflection curve of frames may be constructed in as exact a manner as possible and (2) to describe approximate methods whereby the load-deflection curve, and particularly P_M, can be estimated.

2. Investigation of Plate Buckling with the Aid of Yield Mechanism

In the course of plate experiments, if the thickness/width ratio is small the plate does not lose its load-bearing capacity with the development of plastic deformation but is able to take the load causing yield until a deformation characteristic of the plate occurs;

it is even able to take a small increase in load. In the course of the process "crumplings" (buckling) can be observed on the plate surface. These "crumplings" are formed by a yield mechanism, with the plastic moments acting in the linear plastic hinges (peaks of waves) not constant but ever-increasing due to strain-hardening. The yield mechanism performed by "crumpling" extends to the component plates of the bar. The description of its behaviour is obtained, from among the extreme-value theorems of plasticity, with the aid of the theorem of kinematics.

Thus, in the course of our investigations, an upper limit of load bearing has been determined. However, to be able to assess the results, the following have to be considered: on one hand, the yield mechanisms are taken into account through the "crumpling" forms determined experimentally; and on the other hand, the results of theoretical investigations are compared with the experimental ones.

2.1. YIELD MECHANISM FORMS BASED ON EXPERIMENTAL RESULTS [15,17]

The different forms of yield mechanisms can be determined on the basis of experimental results. The yield mechanism forms of an I-section bar can be classified according to the following critieria.

(a) According to the manner of loading.

(b) According to the positions of the intersecting lines of the web and the flanges, the so-called "throat-lines"; thus,

(i) the evolving formation is called a planar yield mechanism if the two "throat-lines" are in the same plane after the development of the yield mechanism.

(ii) the evolving formation is called a spatial yield mechanism if the two "throat-lines" are not in the same plane after the development of the yield mechanism.

2.1.1. *Bending moment constant along the members axis*

(a) Planar yield mechanism: The buckled form of the bent specimen and the chosen yield mechanism formation are shown in Fig. 5a. As an effect of moment M, a rotation θ develops.

As an effect of M, tension and compression regions develop.

The symbol of the yield mechanism is $(MC)_P$, where C stands for the constant bending moment.

(b) Spatial yield mechanism: The form of the spatial yield mechanism in the case of a bent rod is indicated in Fig. 5b. The rod ends are assumed to be hinge-supported in both main inertia directions. The yield mechanism models the buckling of the component plates of the bent member, the lateral buckling of the beams as well as their interaction.

The symbol of this yield mechanism is $(MC)_S$.

36

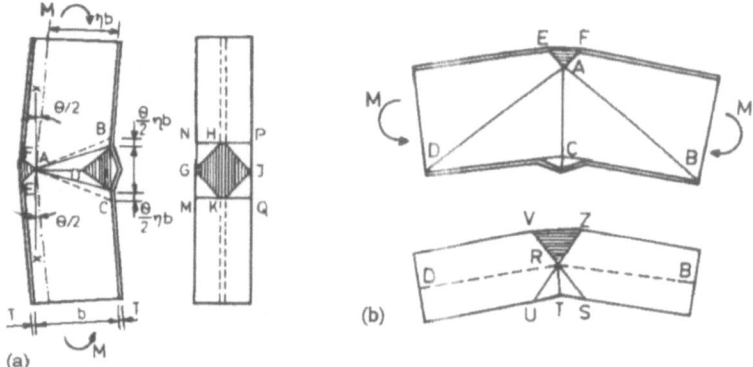

(a)

(b)

Figure 5. Planar *(a)* and spatial *(b)* yield mechanism of bent member
with bending moment constant along the member axis.

2.1.2. *Bending moment varying along the rod axis*

In the case of a varying bending moment along the member axis, it is assumed that the "crack" of the web plate of the I-section in the cross-section of the concentrated force is hindered by the thickness of the web plate or by the ribs.

Climenhaga and *Johnson* [3] assumed yield mechanism forms similar to those introduced in the preceding paragraph for the investigation of buckling occurring in the steel beam part of a composite steel-concrete construction.

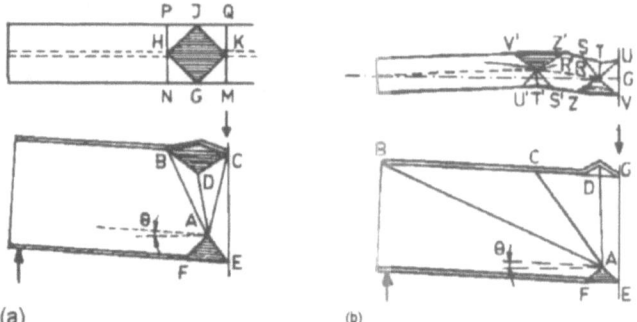

(a)

(b)

Figure 6. Planar *(a)* and spatial *(b)* yield mechanism of bent member
with bending moment varying along the member axis.

(a) Planar yield mechanism: The buckled form of a bent specimen and the selected yield mechanism are shown in Fig. 6a. As an effect of the moment, a rotation θ develops.

Because of the clamping of the cross-section EC, the yield mechanism loses its symmetric character.

The symbol of the yield mechanism is $(MV)_P$ where V stands for the varying moment.

(b) Spatial yield mechanism: The form of the spatial yield mechanism in the case of a bending moment varying along the rod axis is shown in Fig. 6b. As an effect of the moment, a rotaton θ develops.

The symbol of the yield mechanism is $(MV)_S$.

2.1.3. *Yield mechanism of the component plates of an I-section member*

Yield mechanism formations have been determined for different stresses. On the basis of the experimental results it is expedient to decompose these yield mechanism formations into the yield mechanism formations of the component plates of an I-section rod, as certain component plate formations appear in other yield mechanisms too.

To classify the yield mechanisms of component plates, the following division has been used.

(a) Flange plate, if the plate is supported along one line.

(b) Web plate, if the plate is supported at the unloaded ends.

    (bi) axial forces and bending $(W-1) - (W-6)$

    (bii) transverse forces transmitted directly through the web $(W-11-12-13)$

    (bii) transverse forces only on one side of the web panel $(W-21-22-23)$

    (biii) tension fields on the web panel $(W-30; W-40)$

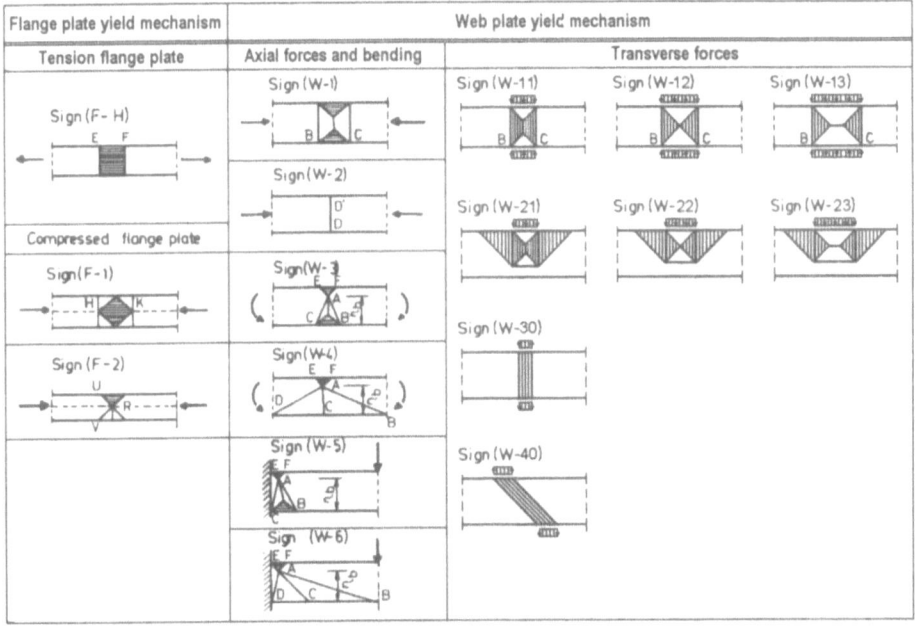

Figure 7. Yield mechanisms of the component plate elements of an I-section member and beam-to-column joints.

Fig. 7 shows the yield mechanisms of the component plates where F is the flange plate, W is the web plate; the odd numbers refer to the planar yield mechanisms and the even ones to the spatial yield mechanisms.

38

2.1.4. *Yield Mechanism of Joint Configurations*

Yield Mechanism of Single-sided Joint Configurations. The main sources of deformability of joint configuration which must be contemplated in a beam-to-column major joint are:

– the connection deformability $M_b - \theta_c$ characteristic;

– the column web panel shear deformability $V_{wp} - \gamma$ characteristic;

– the local buckling of column web panel.

In the case of the yield mechanism formations in Fig. 8a, the effect of the beam local buckling cross-section, column web shear panel and patch loading has also been taken into account.

The symbol of this yield mechanism is $(SSJ)_P$.

Yield Mechanism of Double-sided Joint Configurations. The main sources of deformability of joint configuration which must be contemplated in a beam-to-column major joint are:

– the left hand side connection deformability $M_{b1} - \theta_{c1}$ characteristic;

– the right hand side connection deformability $M_{b2} - \theta_{c2}$ characteristic;

– the column web panel shear deformability $V_{wp} - \gamma$ characteristic;

– the local buckling of column web panel.

The yield mechanism formation is in Fig. 8b. The symbol of this yield mechanism is $(DSJ)_P$.

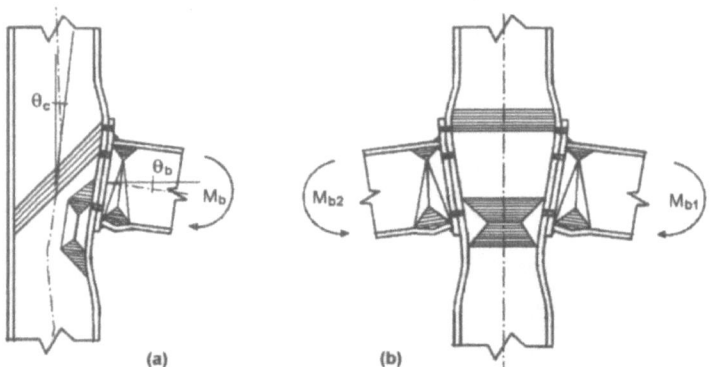

Figure 8. Single-sided *(a)* and double-sided *(b)* joint configuration.

2.2. "JOINING" THE YIELD MECHANISMS OF COMPONENT PLATES

The "joining" of the yield mechanisms of component plates depends on the positions of the so-called "throat-lines" of the yield mechanism chosen on the basis of the experimental results.

In cases pertaining to planar yield mechanisms, this "joining" is to be realised in a linear manner, with a linear plastic hinge: the length of the linear plastic hinge is governed – due to the properties of the chosen yield mechanism – by the length of the yield mechanism of the compression flange plate (F-1). In the case of spatial yield mechanisms, the "joining" should be realised at one or more points.

The relationships between the component plate yield mechanisms and the "joining" of the component plates have been given by *Iványi* [15] who later gave the basic relationships of partial cases [6,12,13].

2.3. MODEL OF THE INTERACTIVE HINGE

The plastic load-bearing investigation assumes the development of rigid-ideally plastic hinges; however, the model describes the inelastic behaviour of steel structures but with major constraints and approximations. There are some effects with the consideration of which the behaviour of the steel material and the I-section member can be taken into account more realistically.

(1) When determining the load-displacement relationship of an I-section member, the symbol of the elastic state is E and if the so-called "rigid" state is assumed instead of the elastic one, the symbol of the rigid state is R.

(2) The effect of residual stress and deformation is characterised by a straight line for ease of handling. The symbol used when taking the residual stress and deformation into consideration is O.

(3) Strain-hardening is one of the important features of the steel material; S indicates that it has been accounted for.

(4) The effect of buckling of the I-section member component plates on the rod element and beam-to-column joint configuration load-displacement relationship has been investigated; this is indicated by L.

The models that take the above effects into consideration in the investigation of load-displacement (relative displacement) relationships of an I-section are called "interactive" ones.

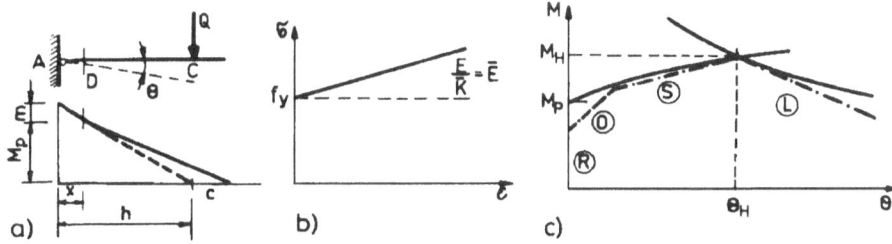

Figure 9. Model of interactive hinge.

The model of the interactive hinge taking into consideration the effect of rigid – residual stress – strain-hardening – plate bucklings can be described with the aid of the "equivalent beam length" suggested by *Horne* [8] (Fig. 9a). The material model employed in the investigations is shown in Fig. 9b. The effect of the residual stresses and deformations is substituted by a straight line. The effect of strain-hardening can be determined with the help of the rigid-hardening (*R-S*) model. The buckling of the I-section member component plates is described by the yield mechanism curve, which is substituted by a straight line.

Fig. 9c indicates the load-displacement relationship of the (*R-O-S-L*) interactive hinge. The substitution by straight lines is justified to simplify the investigations. In the (*R-O-S*) sections the intersections are connected while in section L the moment-rotation relationship is substituted by a tangent that can be drawn at the apex.

3. Analysis of Steel Frames with Global Bar Elements

Matrix methods are available for computer determination of stresses in plane bar systems [24]. These methods are relying either on the force or on the displacement method, this latter has been applied in the program.

Simpler cases involve the bar element in Fig. 10a permitting fast, easy computations mainly on an elastic material model. Bar stiffness matrix $\hat{\mathbf{K}}$ is common knowledge; stiffness values are obtained by solving basic problems of hyperstatic beams.

Our goal seemed to be better achieved by applying complex bar element (Fig. 10b):

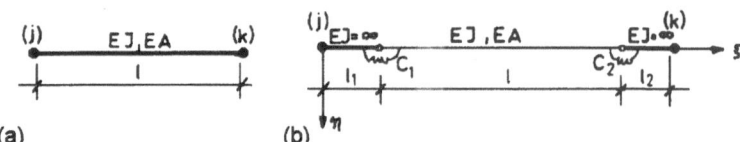

Figure 10. Sinple *(a)* and complex *(b)* bar element.

Two end parts of the bar, of lengths ℓ_1 and ℓ_2, are infinitely rigid (maybe $\ell_1 = \ell_2 = 0$); the middle part is elastic. Rigid and elastic bar parts are connected by a rotation spring each, able to rotation θ in the structure plane alone. Stiffnesses, i.e. spring constants are c_1 and c_2 [1,2].

Details of the method which gives the connection between the unity end deformations and the relevant stress resultant can be found in the literature [17].

Spring characteristics are of the general form in Fig. 9, as interactive hinge. Sections have different spring constants $c = \Delta M / \Delta\theta$ indicating the given section of the elasto-plastic behaviour or of the stability condition of the bar past. The characteristic is strictly monotonous for θ but not for M. Namely there is a peak followed by a descending path of the curve.

4. A Simple Approximate Method

Numerous approximate engineering methods are introduced in the literature [11], from which as one of the possibilites we are going to deal with the extension of the Mechanism Curve Method. The Mechanism Curve Method – above the determination of the plastic load bearing capacity – can be applied to take the effect of finite deformations and strain hardening of steel into consideration.

4.1. MECHANISM CURVE METHOD

Horne [8] proposed the use of the simple rigid-plastic-rigid relationship in order to take into account the effect of strain-hardening on the collapse load of a structure.

Change of geometry due to elastic-plastic deformations tends to decrease the ultimate load bearing capacity of steel frames in comparison with the plastic collapse load. This tendency is counteracted by the strain-hardening properties of steel. The rigid-plastic-rigid theory of structural behaviour is found to be an adequate mean to assess the stiffness of a structure immediately on the formation of the last hinge in a plastic hinge mechanism.

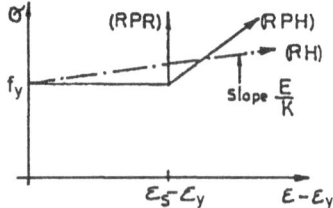

Figure 11. Strain-hardening models.

Different strain-hardening theories can be used during the analysis (Fig. 11):
– rigid-plastic-rigid (*RPR*) model [8],
– rigid-plastic-hardening (*RPH*) model [9],
– rigid-hardening (*RH*) model.
This treatment uses the rigid-hardening (*RH*) hinge model.
The summations could be included in the rigid-plastic work equation, which then becomes

$$\lambda \left(\sum_i Q_i u_i + \sum_k N_k L_k \phi_k^2 \right) = \sum_j (M_{pj} + m_j) \theta_j . \tag{1}$$

4.2. APPROXIMATE ENGINEERING METHOD TO TAKE THE EFFECT OF PLATE BUCKLING INTO CONSIDERATION [15]

In the field of plastic design of steel structures the effect of plate buckling can be taken into consideration by the so called indirect method. During analysis it should be determined that the ratio of plate element dimensions of the section should be less than the ratio given in the specification; in this case buckling of plate elements do not occur until mechanism formation. Such kind of direct method can be applied to eliminate the disturbing effect of plate buckling, but not to analyze – at least only to predict – the effect of plate buckling in regard with a given structure. We extend the category of hardening plastic hinges by taking the effect of plate buckling into consideration. Such hinge model can be the basis of an Approximate Engineering Method, that – without

42

analyzing the full load history –, with simple methods can directly take the effect of plate buckling into consideration.

Fig. 9c shows the linear interaction of moment-rotation of interactive hinge that contains the effects of strain-hardening and plate buckling. The essence of Approximate Engineering Method is that the two effects are separated and the interactive hinge of the structure is put together from two separate components (Fig. 12):

(1) Strain-hardening component: (S)

(2) Plate buckling component: (L)

With the assumed two hinge components the values of the load parameter for the chosen mechanism of the framework can be determined as a function of finite deformations.

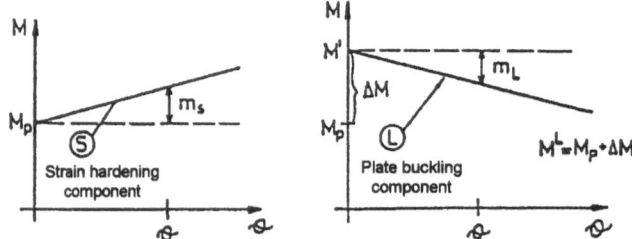

Figure 12. Separated components of interactive hinge.

The expression for (1) strain-hardening component is:

$$\lambda_{(S)}\left(\sum Qu + \sum NL\phi^2\right) = \sum M_p\theta + \sum m_S\theta ; \qquad (2)$$

$$\lambda_{(S)} = \frac{\sum M_p\theta + \sum m_S\theta}{\left(\sum Qu + \sum NL\phi^2\right)}. \qquad (3)$$

To write down the expression for the (2) plate buckling component it should be assumed that the interactive hinge characteristic curve contains rigid, plate buckling effects, so rigid behaviour goes up to the value of $M' = M_p + \Delta M$ first, then a linearly decreasing change is taken into consideration due to the effect of plate buckling. Because of the shape of the characteristic curve belonging to the (2) plate buckling component, external and internal capacities and works are written similarly to the (1) strain-hardening component, except the sign of the increment $md\theta$

$$\lambda_{(L)}\left(\sum Qu + \sum NL\phi^2\right) = \sum M\theta - \sum m_L\theta ; \qquad (4)$$

$$\lambda_{(L)} = \frac{\sum M\theta - \sum m_L\theta}{\sum Qu + \sum NL\phi^2}. \qquad (5)$$

Load parameter $\lambda_{(S)}$ takes the effect of strain-hardening into consideration, while load parameter $\lambda_{(L)}$ that of plate buckling.

From the displacement given by the intersection of the two curves; the reduction-like change of state is due to the effect of plate buckling (Fig. 13). In connection with the results it should be emphasized, that – similarly to plastic load bearing capacity analysis – the expression – taking the two separa te components into consideration – assumes the structure motionless till the moments M_P and M' in the hinges form.

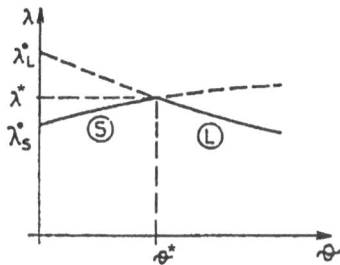

Figure 13. Load parameter and displacement curve.

The axial forces in bars are assumed to be proportional to the external loading. Equivalent cantilever length h for the interactive hinge can be determined by the moment diagram from plastic load bearing capacity analysis.

5. Evaluation of Load Bearing Capacity of Steel Frames

5.1. TEST PROGRAM

The experimental research project was carried out in the Laboratory of the Department of Bridges and Structures, Budapest University of Technology and Economics.

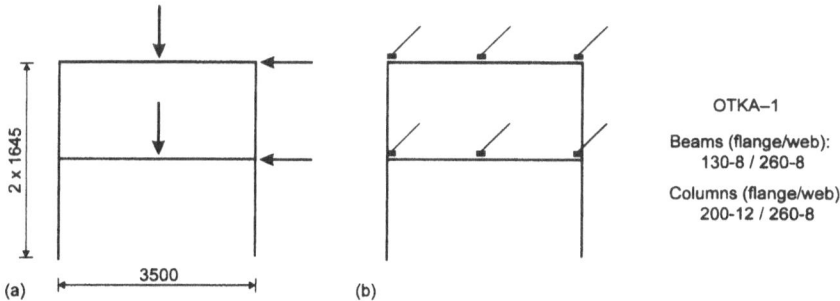

Figure 14. The tested frame: *(a)* main geometry and *(b)* lateral restrains.

An overall view of the testing arrangement is shown in Fig. 14. The frames examined are two-storey single-bay ones. Both the columns and the beams are welded I sections. Pairs of test frames are identical. Columns are connected to a rigid steel base

44

element by two bolts through an end plate (layout generally regarded as pinned joint in the practice).

Beams and columns are connected with flush end plate joints (Fig. 15). In frame OTKA-1, the connections are strengthened with single-sided additional web plates. These were found to be necessary on the basis of an analysis of joint behaviour according to Revised Annex J of Eurocode 3.

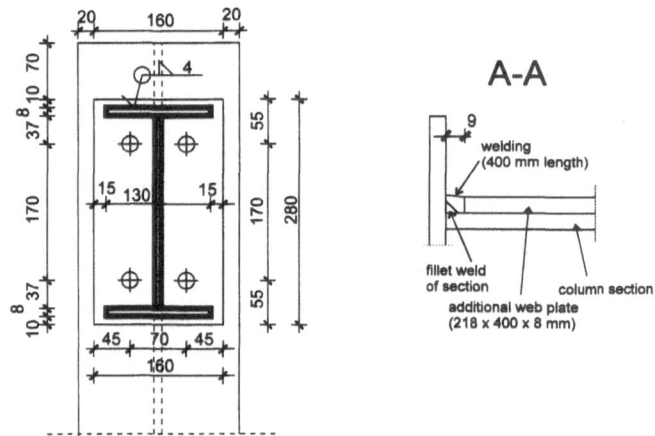

Figure 15. Beam-to-column connection and the detail of the additional web plate.

In order to avoid lateral-torsional buckling, lateral restraints are applied to the frame at the beam-to-columnjoint locations and at the mid-spans of the beams, see Fig. 14.

The frame is loaded by two vertical concentrated loads at the mid-spans of the beams, and two horizontal loads applied at one side of the frame in the levels of the beams (Fig. 14). The two vertical loads are increased and decreased proportionally using three hydraulic jacks (one larger to the lower beam and two smaller and identical to the upper) connected into one oil circuit. Because of the slight difference between the pressure surfaces of the larger jack on one hand and the two smaller jacks on the other, the lower beam was loaded by a concentrated load 89% in magnitude of the load on the upper beam. The vertical loads are applied through so-called gravity load simulators [7], devices which ensure the verticality of the loads within certain limits of lateral displacements of the points of application of the loads. The horizontal loads are applied using one hydraulic jack through a simply supported vertical beam, which ensures the applied load to be equally distributed between the two beam levels. The direction of these horizontal loads is reversible.

5.2. RESULTS OF THEORETICAL AND EXPERIMENTAL INVESTIGATION

Concerning the experimental frame OTKA-1, the relation of load-deflection curve developes according to Fig. 16. The Approximate Engineering Method is presented on test frame OTKA-1 (Fig. 16). The comparison shows that the Approximate Engineering Method gives a satisfactory results for the maximum loads and the descending state path of whole structure as well; and at the same time the analysis can be done at the "desk of the designer".

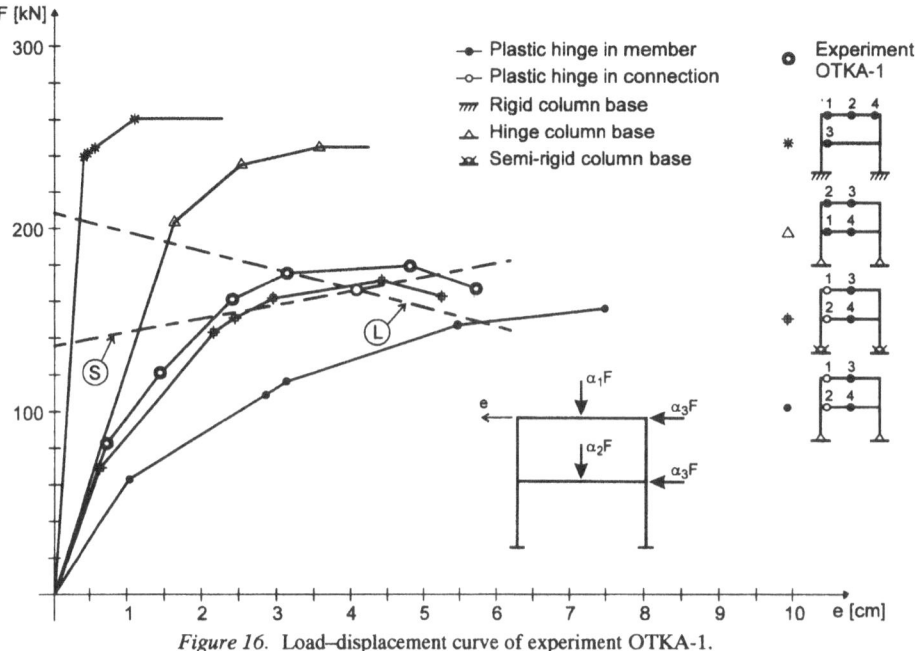

Figure 16. Load–displacement curve of experiment OTKA-1.

6. References

1. Baksai, R. (1983) Plastic analysis by theoretical methods of the state change of steel frameworks, Diploma work, Technical University of Budapest, (In Hungarian).
2. Baksai, R., Iványi, M. and Papp, F. (1985) Computer program for steel frames taking initial imperfections and local buckling into consideration, *Periodica Politechnica Civil Engineering* **29:3-4**, pp. 171-185.
3. Climenhaga, J.J. and Johnson, P. (1972) Moment-rotation curves for locally buckling beams, *Journal of Structural Division ASCE* **98**, ST6.
4. Drucker, D.C. (1951) A more fundamental approach to plastic stress-strain relations, In *Proceedings of lst U.S. Natl. Congress of Applied Mechanics*, ASME, pp. 487.
5. Drucker, D.C. (1964) On the postulate of stability of material in the mechanics of continua, *Journal de Mechanique* **3** (Paris), pp. 235.
6. Gioncu, V. and Petcu, D. (1997) Available Rotation Capacity of Wide Flange Beams and Beam-Columns, *Journal of Constructional Steel Research* **43:1-3**, pp. 161-217.
7. Halász. O. and Iványi, M. (1979) Tests with Simple Elastic-Plastic Frames, *Periodica Polytechnica Civil Engineering* **23:3-4**, pp. 151-182.

8. Horne, M.R. (1960) Instability and the plastíc theory of structures, Transactions of the EIC **4**; 31.

9. Horne, M.R. and Medland, J.C. (1966) Collapse loads of steel frameworks allowing for the effect of strain-hardening, in *Proceedings of Inst. of Civil Engineers* **33**, pp. 381-402.

10. Horne, M.R. and Merchant, W. (1965) *The stability of frames*, Pergamon Press.

11. Horne, M.R. and Morris, L.J. (1981) *Plastic design of low-rise frames*, Constrado Monographs, Granada.

12. Iványi, M. (1979) Yield mechanism curves for local buckling of axially compressed members, *Periodica Polytechnica Civil Engineering* **23:3-4**, pp. 203-16.

13. Iványi, M. (1979) Moment-rotation characteristics of locally buckling beams, *Periodica Polytechnica Civil Engineering* **23:3-4**, pp. 217-230.

14. Iványi, M. (1980) Effect of plate buckling on the plastic load carrying capacity of frames, in *Proccedings of IABSE 11th Congress*, Vienna.

15. Iványi, M. (1983) Interaction of stability and strength phenomena in the load carrying capacity of steel structures. Role of plate buckling (In Hungarian), DSc. Thesis, Hung. Ac. Sci., Budapest.

16. Iványi, M. (1985) The model of "interactive plactic hinge", *Periodica Politechnica Civil Engineering* **29:3-4**, pp. 121-146.

17. Iványi, M. (1992) Ultimate Load Behaviour of Steel-Framed Structures, *Journal of Constructional Steel Research* **21:1-3**, pp. 5-42.

18. Iványi, M. (2001) *Stability of Steel Frames. Imperfection, Semi-rigidity, Softening*, Akadémiai Kiadó, Budapest (Forthcoming).

19. Iványi, M. and Varga G. (1999) A New Series of Full Scale Tests with Semi-Rigid Connections, in M. Iványi and D. Dubina (eds.), *Proceedings of 6th Int. Conference on Stability and Ductility of Steel Structures*, Timisoara, Romania, 9-11 Sept. 1999, Elsevier.

20. Kazinczy, G. (1914) Experiments with fixed-end beams (In Hungarian), *Betonszemle* **2**, pp. 68.

21. Maier, G. (1961) Sull'equilibrio elastoplastico delle strutture reticolari in presenza di diagrammi forze-elongazioni a trotti desrescenti, *Rendiconti, Istituto Lombasdo di Scienze e Letture, Casse di Scienze A*, Milano **95**, pp. 177.

22. Maier, G. and Drucker, D.C. (1966) Elastic-plastic continua containing unstable elements obeying normality and convexity relations, *Schweizerische Bauzeitung* **84:23**, pp. 1.

23. Nethercot, D. (1985) Utilisation of experimentally obtained connection data in assessing the performance of steel frames, in Wai-Fah Chen (ed.), *Connection Flexibility and Steel Frames*, pp. 13-37.

24. Szabó, J. and Roller, B. (1971) *Theory and analysis of bar systems* (In Hungarian), Mûszaki Könyvkiadó, Budapest.

EFFECTIVE LENGTH FACTOR CONSIDERING SEMI-RIGID CONNECTIONS

Z. B. PETKOV & D. T. BLAGOV
University of Architecture, Civil Engineering and Geodesy,
Division of Structural Mechanics,
№ 1 Hristo Smirnenski blvd., 1421 Sofia, Bulgaria

1. Introduction

Most of the connections in steel structures are designed as semi-rigid mainly because of technological reasons. On the other hand such connections allow the structural response to be controlled [1]. Structures with semi-rigid connections deserve special attention since the global flexibility may considerably be increased. The behaviour limits should be determined using advanced methods for analysis and design. The problem of making the performance more predictable becomes a subject of intensive research work during the last decade [2], [3].

The determination of effective column length factor in frames is only one of the problems that can be outlined. It is expected that buckling would considerably be influenced by the partial rigidity of the connections and it is aimed this measure to be numerically evaluated. Referring to frames with rigid connections this procedure can significantly be simplified by making use of alignment chart method [4], [5]. According to this method the determination of effective column length factor (K-factor) is performed considering a subassemblage subjected to axial loading and prescribed boundary conditions. The influence of imposed conditions on the K-factor is studied in [6] and [7]. Although the method is approximate it is shown [4] that the accuracy of the results is satisfactory and alignment chart procedure is implemented in some modern design codes [1] mainly for steel columns. Some other applications to reinforced concrete columns are shown in [8] and [9]. The principal advantage of this method is found in simplicity and small amount of data that should be prepared for the K-factor evaluation.

The determination of K-factor in semi-rigid frames can be carried out using traditional buckling analysis and finite element method. The most attractive idea from purely practical point of view is to make an attempt with application of alignment chart method including semi-rigid connections. Kishi, Chen and Goto [10] developed a modified alignment chart procedure, where beams are flexibly connected to the columns using

47

rotational springs. Such connection implies rotational flexibility (R-flexibility). The basic constitutive relationship that can be used is *moment-rotation*. It is shown that the error associated with alignment chart method remains in acceptable limits when connections with R-flexibility are used. Details towards development of a beam member with rotational springs at both ends are given in [11]. The failure mechanism with plastic hinges in the R-links is likely to dominate the response. Numerical modelling of *moment-rotation* relationship by means of connection semi-rigidity can be found in [12] and [13].

Most of the connections are designed to dissipate energy when the elastic limits are exceeded. Seismically resistant structures are designed to develop plastic deformations and some of them dissipate energy through beam-to-column connections unless the rotational capacity is still not exceeded [14]. In some provisions of [1] the connections that yield can be treated as semi-rigid if secant stiffness is used. Gomes *et al.* [3] show that a low strength joint may be classified as nominally pinned even when its initial stiffness is large, see Figure 2. Kishi, Chen and Goto [10] show how semi-rigidity can be incorporated into alignment chart procedure. Goel and Leelataviwat [15] propose another dissipative mechanism, based on vertical flexibility (V-flexibility) in the plastic range. This connection is located in the mid-span of the beams. Structural behaviour is characterised by truly strong columns – ductile beams. This is achieved by web opening near the mid-span. The flexibility of this connection is also increased in plastic range and may result in significant changes in the effective length of the columns. The influence of V-flexibility on the K-factor obviously requires additional research. It is the purpose of this paper to clarify the problem how the K-factor is influenced by V-flexibility. Another objective is to evaluate the K-factor assuming interaction between R- and V- flexibility. It is also aimed to extend the application of alignment chart method including various types of semi-rigid connections. To achieve these purposes, the chart method is modified similarly to [10], but including V-flexibility in stability equations. Accuracy of the method is improved by changing the traditional boundary conditions for the upper and lower columns included in the subassemblage. The corresponding modification factors for braced and unbraced frames are derived and used in stability equations. The application of the theoretical results is numerically illustrated by 3D plot of the K-factor in terms of connection flexibility. Accuracy analysis and conclusions follow calculations.

2. Theoretical Background of the Study

The study presented herein is theoretically based on the alignment chart method, described in [4], [5], [6] and [7]. According to this approach the effective length factor of a selected column member can be determined approximately considering only a small subassemblage extracted from the entire frame. Subassemblage is consisting of a central column and its adjacent members, see Figures 3 and 6. The following assumptions are valid throughout the analysis:

1. The frame whose effective column length factor is to be determined is regular in horizontal direction and in elevation. It is shown [4] that this method is applicable even

if this assumption is not strictly satisfied. Rectangular multibay multistorey frame can be suggested in general.

2. The axial force effects in the beams are negligibly small.

3. The behaviour of all members remains within the elastic limits.

4. Each beam has rotational springs at both ends (R-connections) and a single vertical spring (V-connection) in the mid-span, see Figures 4 and 5. The spring stiffness remains constant throughout the analysis and connection flexibility is parametrically evaluated. The moments of flexibly connected beams with columns are modified to follow the prescribed column nodal rotations.

5. The columns are loaded as shown in Figure 6 and have identical stability functions.

6. Second order theory of elastic buckling is applied.

7. Beams may have different material and section properties; the connection stiffness may also be different for each connection. For the sake of simplicity, however, in the section, where numerical results are obtained and discussed, all beams have identical properties; stiffness of the connections follows the same assumption.

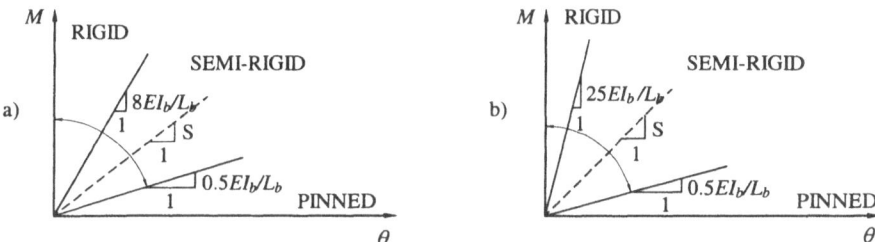

Figure 1. Classification of beam-to-column connections by stiffness according to the Eurocode 3 – Annex J (revised): a) unbraced frames and b) braced frames. The region studied in the paper is denoted by arrow.

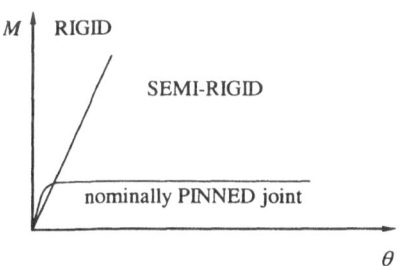
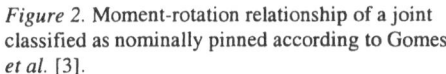
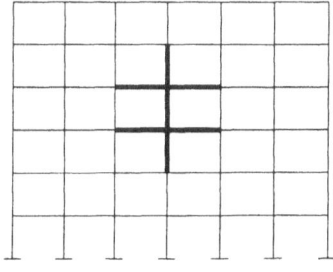

Figure 2. Moment-rotation relationship of a joint classified as nominally pinned according to Gomes *et al.* [3].

Figure 3. Frame, regular in horizontal direction and in elevation. Subassemblage needed for application of alignment chart method.

In Eurocode 3 [1] connections are classified regarding their strength and their stiffness. The stiffness classification is clear from Figure 1. It is shown (see Gomes *et al.* [3]) however that if plastic rotations are adopted the use of initial stiffness as a unique description parameter is not correct. In other words the initial stiffness of the connection is not enough to classify the connection properties. If the connection is specified as low strength and ductile then basing on its secant stiffness the connection should be considered as nominally pinned, see Figure 2.

Let us denote by ξ and η dimensionless parameters being defined as a ratio between beam stiffness and corresponding spring stiffness, so that

$$\xi = \frac{6}{c}\left(\frac{EI}{L}\right)_b , \qquad \eta = \frac{12}{\bar{c}L_b^2}\left(\frac{EI}{L}\right)_b \qquad (1)$$

The subscript 'b' is used to denote beam quantities such as Young's modulus E, moment of inertia I and length L. The constant c represents the rotational spring stiffness, whereas $\bar{c}$ is the vertical spring stiffness, see Figures 4 and 5. Note that if $\xi=0$ and $\eta=0$ the connection is rigid. When ξ and η tend to infinity the connection does not resist at all and can be specified as fully flexible. Figure 5 represents a frame whose beams have both types of connections.

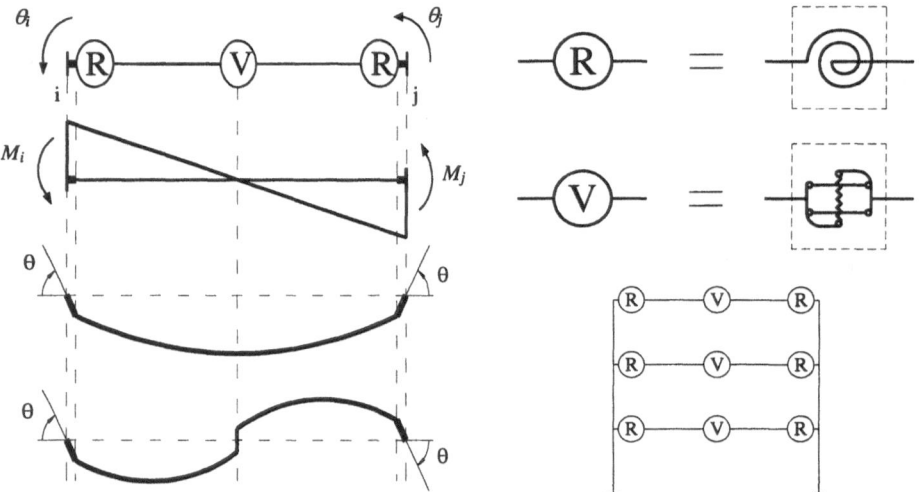

Figure 4. Symmetric and anti-symmetric modes of beam deformations assuming R- and V-semi-rigid connections.

Figure 5. Two types of semi-rigid connections: connection R with rotational flexibility and connection V with vertical flexibility, applied to simple one bay three storey frame [15].

The relationship between beam end moments, M_i and M_j, and beam end rotations, θ_i and θ_j, can be expressed as follows:

$$\begin{Bmatrix} M_i \\ M_j \end{Bmatrix} = \left(\frac{EI}{L}\right)_b \begin{bmatrix} \mu & \chi \\ \chi & \mu \end{bmatrix} \begin{Bmatrix} \theta_i \\ \theta_j \end{Bmatrix} \qquad (2)$$

The deflected beam and some of the notations mentioned above are shown in Figure 4. Also, location of the R- and V-connections could be seen from the same figure. The following expressions are used to simplify the above relationship:

$$\mu = \frac{6(2 + \xi + \eta)}{(3 + \xi)(1 + \xi + 2\eta)} \; ; \quad \chi = \frac{6(1 - \eta)}{(3 + \xi)(1 + \xi + 2\eta)} \tag{3}$$

Following [4] and [5], the typical deflected line considering braced frame is characterised by $\theta_i = \theta$, $\theta_j = -\theta$ and describes symmetric mode of beam deflection. It follows from Equation (2) that

$$M_i = \left(\frac{EI}{L}\right)_b (\mu - \chi).\theta$$

where

$$\mu - \chi = \frac{6}{3 + \xi}.$$

Since unbraced frame [4], [5] is related to anti-symmetric mode of beam deflection, it is suggested that $\theta_i = \theta_j = \theta$. The corresponding result for the end moment following from Equation (2) is

$$M_i = \left(\frac{EI}{L}\right)_b (\mu + \chi).\theta$$

where

$$\mu + \chi = \frac{6}{(1 + \xi + 2\eta)}.$$

Similarly to [10], the effects of semi-rigid connections can be accounted for by multiplying the beam stiffness with modification factors. Consequently, the beam stiffness is reduced and the most convenient form regarding further steps is

$$\left(\frac{EI}{L}\right)_b^* = \left(\frac{EI}{L}\right)_b . \frac{1}{\alpha} \tag{4}$$

where the asterisk is a modification symbol. The coefficient α is specified as follows:

$$\alpha = 1 + \frac{\xi}{3} \qquad \text{for braced frames}$$

(5)

$$\alpha = 1 + \xi + 2\eta \qquad \text{for unbraced frames}$$

It is seen from Equation (4) that if a single beam is subjected to either symmetric or anti-symmetric boundary conditions the beam stiffness is generally reduced as a result of semi-rigidity of the connections. Furthermore, the alignment chart method requires considering non-sway and sway modes of deformations of a single subassemblage, see Figure 6. Note that deflected lines in the both states are associated with imposed boundary conditions for the rotations and horizontal displacements used in frames with rigid connections [4], [5], [6], [10]. In this case the usual boundary conditions for upper and lower column considering unbraced frames are taken as $\theta_C = \theta_B$ and $\theta_D = \theta_A$. In general, considering semi-rigid frames, the angles θ_C and θ_D are suggested to differ from θ_B and θ_A respectively.

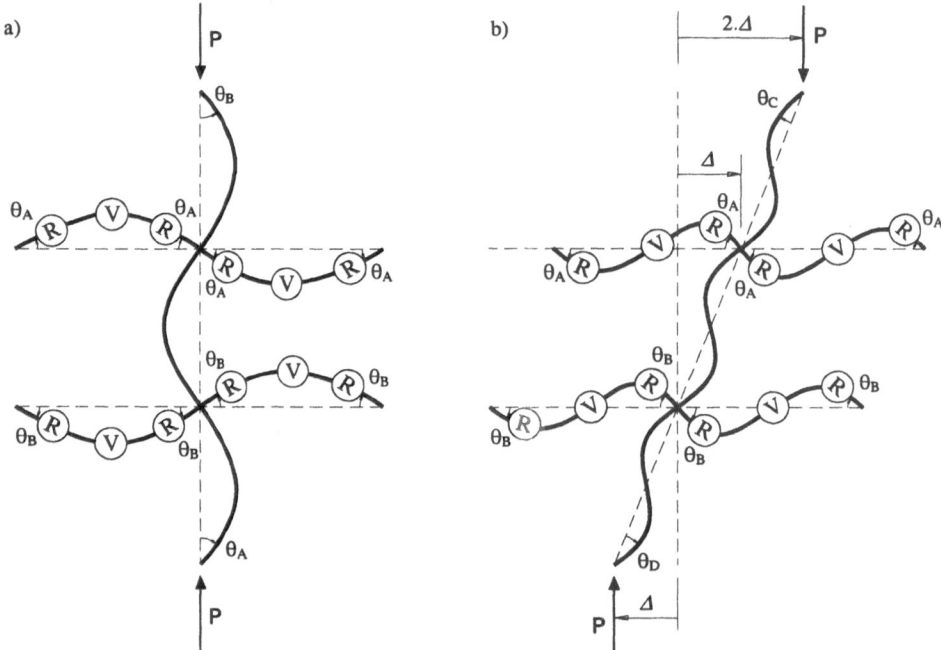

Figure 6. Structural subassemblages with the imposed deformations used for evaluation of *K*-factor in a) braced and b) unbraced frames.

Following the alignment chart method [4], [5] the modified joint stiffness ratios G_A^* and G_B^* are then defined in terms of modification coefficient α as

$$G_A^* = \frac{\sum\limits_{A}\left(\dfrac{EI}{L}\right)_c}{\sum\limits_{A}\left(\dfrac{EI}{L}\right)_b^*} = \alpha_A G_A \quad \text{and} \quad G_B^* = \frac{\sum\limits_{B}\left(\dfrac{EI}{L}\right)_c}{\sum\limits_{B}\left(\dfrac{EI}{L}\right)_b^*} = \alpha_B G_B \tag{6}$$

where

$$G_A = \frac{\sum\limits_{A}\left(\dfrac{EI}{L}\right)_c}{\sum\limits_{A}\left(\dfrac{EI}{L}\right)_b} \quad , \quad G_B = \frac{\sum\limits_{B}\left(\dfrac{EI}{L}\right)_c}{\sum\limits_{B}\left(\dfrac{EI}{L}\right)_b} \tag{7}$$

where G_A and G_B are the joint stiffness ratios introduced for frames with rigid connections. The subscript 'c' indicates quantities assigned to the columns. Note that Equation (6) allows different beams and connections to be used in joints A and B, see Figure 6. Obviously beams with different section properties and connections can easily be implemented.

The stability equations obtained basing on the boundary conditions shown in Figure 6 are written in the form [4], so that

for braced frame

$$\frac{G_A^* G_B^*}{4}\left(\frac{\pi}{K}\right)^2 + \frac{G_A^* + G_B^*}{2}\left[1 - \frac{\left(\dfrac{\pi}{K}\right)}{tg\left(\dfrac{\pi}{K}\right)}\right] + \frac{2tg\left(\dfrac{\pi}{2K}\right)}{\left(\dfrac{\pi}{K}\right)} - 1 = 0$$

and for unbraced frame, assuming that $\theta_C = \theta_B$ and $\theta_D = \theta_A$,

$$\frac{G_A^* G_B^*\left(\dfrac{\pi}{K}\right)^2 - 36}{6\left(G_A^* + G_B^*\right)} - \frac{\left(\dfrac{\pi}{K}\right)}{tg\left(\dfrac{\pi}{K}\right)} = 0$$

For more general case when θ_C and θ_D are not equal to θ_B and θ_D respectively, the 1stability equation for unbraced frames is presented in the Appendix.

3. Numerical Results

In the preceding section it is shown theoretically that the column K-factor of semi-rigid frames with a given connection stiffness can be determined using the current alignment chart, known from [4] and [5]. In this section the application of the above formulae is numerically illustrated. Numerical analysis is performed assuming beam length L_b=6.0 m and column length L_c=3.60 m. All beams have identical section properties, the same assumption is accepted for the columns. However, the algorithm is more general and allows beams and columns with different properties to be used. The connection stiffness is included by ξ and η parameters. Both parameters are varied and their influence on the K-factor is then evaluated.

The limits of ξ and η are determined referring to Figures 1 and 2 respectively. Recognising that for rotational connection $s = c$, and taking into account Equation (1), it follows that the parameter ξ ranges between 0 and 12, see Figure 1. For the vertical connection Equation (1) and Figure 1 can be used to define the limits of parameter η. It is recognised that in this case

$$s = \bar{c} L_b^2 / 4.$$

This implies that $0 \le \eta \le 6$. Note that according to Figure 1 in calculations rigid connections are treated as semi-rigid and that is the reason why the lower limit of ξ and η is taken to be zero. Note also that the upper limits of ξ and η for braced and unbraced frames are related to the lowest value of bounding stiffness $0.5(EI/L)_b$. Two cases are further considered as typical for the design practice: *first case* – the ratio I_c/I_b is taken to be 0.6 (strong beams, weak columns), and *second case* – the above ratio is taken equal to 3.141 (strong columns, weak beams). Following the results given by Equations (5) and (6), the modified joint stiffness ratios become

$$G_A^* = G_B^* = G\left(1 + \frac{\xi}{3}\right) \quad \text{for braced frames and}$$

$$G_A^* = G_B^* = G(1 + \xi + 2\eta) \quad \text{for unbraced frames,}$$

where, following the Equation (7), G=1.0 in the first case and G=5.236 in the second case.

Here ξ and η are characterising the semi-rigidity of the corresponding type of connection. The case $\xi\cong0$ and $\eta\cong0$ represents rigid connection. Both parameters are varied and the K-factor is determined for each ξ and η pairs. The results are plotted in Figure 7 for braced frame and in Figure 8 for unbraced frame.

It is seen that, considering braced frames, the V-type of semi-rigid connection does not influence the K-factor, see Figure 7. This is clearly indicated by Equation (5) and by the above equation, where modification factor does not depend on η parameter. The K-factor results are slightly affected by R-type of semi-rigid connections considering frames with strong columns-weak beams, see Figure 7. It can be observed that in this case the K-factor is close to unity and ranges in a quite narrow band of values. The opposite design philosophy for frames with strong beams-weak columns obviously provides more sensitive results for the K-factor with respect to R-semi-rigidity and it is illustrated in the same figure.

Let us consider now Figure 8 showing the results for unbraced frames. The R- and V-semi-rigidity exhibit considerable effect on the K-factor. The tendency of growing up the K-factor when flexibility is increased is clearly indicated in both cases of different design philosophies. When ξ and η reach their upper bounds the K-factor is increased approximately three times comparing to the case of rigid connections. Note that this result is obtained assuming that R- and V-flexibility are simultaneously included. If V-type of connections are not used, the increase of the K-factor due to R-flexibility is expected to be approximately two times greater than the corresponding result for rigid connections, see K-ξ plane of Figure 8. Similar result is observed if only V-type of connections is used, see Figure 8, K-η plane. Interaction effects between R- and V-semi-rigidity may lead to unacceptable increase of the K-factor. In general, it can be concluded that unbraced frames are more sensible to connection flexibility regarding the K-factor.

4. Accuracy analysis

It is established [10] that the error associated with the alignment chart method considering semi-rigid frames with R-flexible connections is increased. It is expected from engineering insight that the use of more than one type of flexible connections will lead to greater errors in the K-factor. The validity of the results derived by alignment chart method should be compared to the results obtained by finite element method considering the whole frame. Herein, six storey six bays steel frame is studied for this purpose. All columns have identical cross sections; all beams have also identical section properties. All R-connections have identical stiffness obtained in accordance with Equation (1); the same is valid for V-connections. For the sake of simplicity external load is represented by nodal forces, assigned to the roof nodes of the frame. The computer software ANSYS [16] is used to carry out the finite element solution. It is reasonable this solution to be used as exact.

If $\theta_C = \theta_B$ and $\theta_D = \theta_A$ are used as boundary conditions recommended for unbraced frames with rigid connections in [4] and [5], the error between approximate and exact result for the K-factor becomes very large. In this case Kishi, Chen and Goto [10] recommend reducing the error by varying the values of θ_C and θ_D or, in other words, to use boundary conditions, different from these, mentioned above, see Figure 6. The stability equation corresponding to modified column rotations is given in [10] and briefly discussed in the Appendix of the present paper. The basic parameter that accounts for the modified boundary conditions is denoted by β and defined in the Appendix.

It is seen from Tables 1 and 2, that for braced frames $\beta=1$ ($\theta_C = \theta_B$ and $\theta_D = \theta_A$) provides results with satisfactory accuracy and there is no need to change boundary conditions for the columns. In contrast with this observation for unbraced frames, it seems reasonable to find out such a value of β, that makes the error smaller. Considering now both design concepts mentioned in the Tables. It is seen that the K-factor is strongly affected by β if flexible connections are included. It is recommended $\beta=1.04$ to be used regarding both concepts, see Table 1 and Table 2. This value of β is taken in order to minimise the average error in both tables.

TABLE 1. Accuracy analysis results for strong beams-weak columns

$G_{Ac2}=$	Braced						Unbraced					
$G_{Bc2}=$	$\xi=0, \eta=0, \alpha=1$			$\xi=12, \eta=6, \alpha=25$			$\xi=0, \eta=0, \alpha=1$			$\xi=12, \eta=6, \alpha=25$		
β	Al. Ch.	ANSYS	Error,%	Al. Ch.	ANSYS	Error,%	Al. Ch.	ANSYS	Error,%	Al. Ch.	ANSYS	Error,%
1	0.7743	0.7469	3.67	0.9302	0.9219	0.90	1.317	1.261	4.44	4.625	3.617	27.87
1.1	-	-	-	-	-	-	1.307	1.261	3.65	3.461	3.617	4.51
1.08	-	-	-	-	-	-	1.309	1.261	3.81	3.621	3.617	0.11
1.05	-	-	-	-	-	-	1.312	1.261	4.04	3.911	3.617	8.13
1.04	-	-	-	-	-	-	1.313	1.261	4.12	4.026	3.617	11.31
1.03	-	-	-	-	-	-	1.314	1.261	4.20	4.153	3.617	14.82

TABLE 2. Accuracy analysis results for weak beams-strong columns

$G_{Ac2}=$	Braced						Unbraced					
$G_{Bc2}=$	$\xi=0, \eta=0, \alpha=1$			$\xi=12, \eta=6, \alpha=25$			$\xi=0, \eta=0, \alpha=1$			$\xi=12, \eta=6, \alpha=25$		
β	Al. Ch.	ANSYS	Error,%	Al. Ch.	ANSYS	Error,%	Al. Ch.	ANSYS	Error,%	Al. Ch.	ANSYS	Error,%
1	0.9329	0.9052	3.06	0.9850	0.9702	1.53	2.271	2.140	6.12	10.42	6.728	54.88
1.1	-	-	-	-	-	-	2.121	2.140	0.90	4.558	6.728	47.61
1.08	-	-	-	-	-	-	2.148	2.140	0.37	4.968	6.728	35.43
1.05	-	-	-	-	-	-	2.191	2.140	2.38	5.878	6.728	14.46
1.04	-	-	-	-	-	-	2.206	2.140	3.08	6.320	6.728	6.46
1.03	-	-	-	-	-	-	2.222	2.140	3.83	6.882	6.728	2.29

5. Conclusions

The usual practice in making general conclusions requires large number of examples to be carried out and then studied and analysed. However some tendencies can be clarified analytically or will appear in all numerical examples as a rule. The following conclusions can be summarised:

1. Alignment chart method is a reliable tool for K-factor evaluation in semi-rigid frames. Its application can successfully be extended in this area taking into account that the stiffness of the connections is finite and flexibility can be accounted for by some modifications. There exist possibilities to improve the accuracy of the method by varying the boundary conditions.

2. For semi-rigid braced frames with R- and V-connections the influence of V-flexibility is negligibly small. The influence of R-flexibility on the K-factor is greater in frames with strong beams-weak columns.

3. For semi-rigid unbraced frames R-flexibility and V-flexibility affect the K-factor. This influence is found to be large enough to be taken into account in the design with big care although the dissipative properties of the structure are improved.

4. Interaction effect between R- and V-type of connections may lead to considerable increase of the K-factor. The use of combined semi-rigid connections should be considered with care from stability point of view.

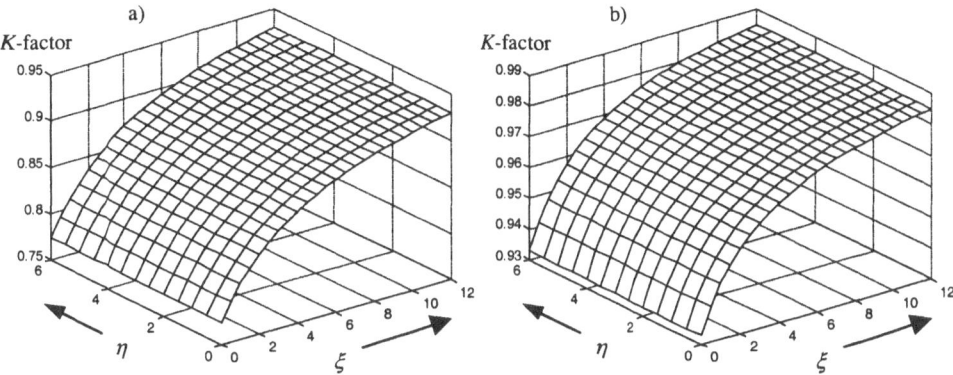

Figure 7. K-factor versus ξ and η parameters considering braced frames:
a) $G_A=G_B=1.0$ (strong beams-weak columns); b) $G_A=G_B=5.236$ (strong columns-weak beams).

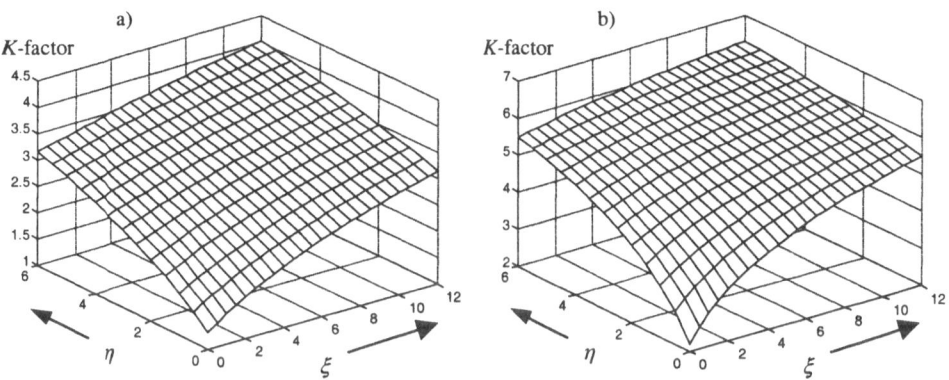

Figure 8. K-factor versus ξ and η parameters considering unbraced frames:
a) $G_A=G_B=1.0$ (strong beams-weak columns); b) $G_A=G_B=5.236$ (strong columns-weak beams).

6. Appendix

The formulae given bellow are taken from [10] and are used to improve the accuracy of the K-factor in semi-rigid unbraced frames. The stability equation considering alignment chart method has the following more general form of presentation:

$$\text{Stability equation}: \quad \det \begin{vmatrix} a_{11} & a_{12} & a_{13} \\ a_{21} & a_{22} & a_{23} \\ a_{31} & a_{32} & a_{33} \end{vmatrix} = 0$$

where

$$a_{11} = s_{ii} + \frac{6}{G_A^*} - G_{Ac1} \frac{s_{ij}^2}{s_{ii}} \qquad a_{22} = s_{ii} + \frac{6}{G_B^*} - G_{Bc3} \frac{s_{ij}^2}{s_{ii}}$$

$$a_{12} = G_{Ac2}s_{ij}$$

$$a_{23} = -\left(s_{ii} + s_{ij}\right)\left(1 - G_{Bc3}\frac{s_{ij}}{s_{ii}}\right)$$

$$a_{13} = -\left(s_{ii} + s_{ij}\right)\left(1 - G_{Ac1}\frac{s_{ij}}{s_{ii}}\right)$$

$$a_{31} = a_{32} = s_{ii} + s_{ij}$$

$$a_{21} = G_{Bc2}s_{ij}$$

$$a_{33} = v^2 - 2\left(s_{ii} + s_{ij}\right)$$

Stability functions are denoted according to [4] as s_{ii} and s_{ij} and can be found in the same reference. It is also noted that

$$G_{Aci} = \frac{\left(\dfrac{EI}{L}\right)_{ci}}{\displaystyle\sum_A\left(\dfrac{EI}{L}\right)_c} \quad ; \qquad G_{Bci} = \frac{\left(\dfrac{EI}{L}\right)_{ci}}{\displaystyle\sum_B\left(\dfrac{EI}{L}\right)_c} \quad ; \qquad i = 1,2,3$$

For the paper purposes it is assumed $\beta = G_{Ac2} = G_{Bc2}$.

7. References

1. Eurocode 3: Design of Steel Structures. Part 1.1. Revised Annex J: Joints in Building Frames (1996) *ECCS Committee TC10 - Structural Connections*.
2. COST C1 (1994) Semi-rigid Behaviour of Civil Engineering Connections, *Proceedings of the Second State of the Art Workshop*, Prague, 26-28 October.
3. COST C1 (1998) Control of the Semi-rigid Behaviour of Civil Engineering Structural Connections, *Proceedings of the International Conference*, Liege, 17-19 September.
4. Chen, W.F., and Lui, E.M. (1987) *Structural Stability*, Elsevier.
5. Galambos, T.V. (1968) *Structural Members and Frames*, Prentice-Hall.
6. Duan, L., and Chen, W.F. (1989) Effective Length Factor for Columns in Unbraced Frames, *J. of Struct. Engng.*, Vol. **115**, № 1, 149-165.
7. Duan, L., and Chen, W.F. (1988) Effective Length Factor for Columns in Braced Frames, *J. of Struct. Engng.*, Vol. **114**, № 10, 2357-2370.
8. Duan, L., King, W.S., and Chen, W.F. (1993) K-Factor Equation to Alignment Charts for Column Design, *ACI Struct. J.* May-June, 242-248.
9. Hu, Y.X., Zhou, R.G., King, W.S., Duan, L., and Chen, W.F. (1993) On Effective Length Factor of Framed Columns in the ACI Buildings, *ACI Struct. J.*, March-April, 135-143.
10. Kishi, N., Chen, W.F., and Goto, Y. (1997) Effective Length Factor of Columns in Semi-rigid and Unbraced Frames, *J. of Struct. Engng.*, Vol. **123**, № 3, 313-320.
11. Chen, W.F. (ed.) (1987) *Joint Flexibility in Steel Connections*, Elsevier.
12. Chen, W.F., and Kishi, N. (1989) Semi-rigid Steel Beam-to-Column Connections: Database and Modelling, *J. Struct. Engng.*, Vol. **11**, № 1, 105-119.
13. Kishi, N., and Chen, W.F. (1990) Moment-Rotation Relations of Semi-rigid Connections with Angles, *J. of Struct. Engng.* **7**, 1813-1835.
14. Aribert, J.M., Dubina, D., and Grecea, D. (1995) Parametrical Study on a New Method for Q-Factor Evaluation, in F.M. Mazzolani and V. Gioncu (eds.), *STESSA, Conference Proceedings*, E&FN Spon, Romania, pp. 394-401.
15. Goel, S.C., and Leelataviwat, S. (1998) Seismic Design by Plastic Method, *Engineering Structures*, Vol. **20**, № 4-6, 465-471.
16. ANSYS User's Manual for Revision 5.0 (1994) Swanson Analysis Systems, Inc., Houston.

PRACTICAL DEMONSTRATION OF THE USE OF JOINT FLEXIBILITY IN STEEL FRAME DESIGN

R. AROCH

Slovak University of Technology, Department of Steel and Timber Structures, Radlinskeho 11, 813 68 Bratislava, Slovakia

1. Introduction

It is a common approach to design steel structures with either rigid and full-strength connections or nominally pinned connections. However, connections have a significant influence on the fabrication costs and thus at the overall costs of a steel structure. Further, the connections influence the distribution of internal forces and moments in the structure, so they influence the profile dimensions of beams and columns. The semi-rigid and partial-strength connections approach therefore gives the opportunity of balancing the material cost of the beams against the labour cost of the connections. It is possible to represent real frame behaviour only with consideration of the particular connection details, from which follows, that an economical design is possible only with concurrent design of members and their connections. The connections influence the design of members and, on the other way round, the members influence the design of their connections. A significant saving of material and labour is possible, thus lowering the total cost of the structure.

Economy studies in various countries have shown possible benefits from the use of the concept of semi-rigid joints ([1], [2], [3], [7], [8] and [11]). From these studies it can be concluded that the possible savings due to semi-rigid design can be 20-25% in case of unbraced frames and 5-9% in case of braced frames ([12] and [13]).

This paper illustrates the importance of taking account of the real behaviour of joints in steel frame design. It demonstrates on practical design example of a braced and unbraced frame the influence and advantage of the active use of joint flexibility in steel frame design. The examples are solved according to STN P ENV 1993-1-1 with partial safety factors according to the Slovak National Application Document ([4], [5], [10]).

2. Software Package

For the analysis and design we have used the IDA NEXIS 32 software package (marketed in Western Europe under the name ESA-Prima Win), which is a fully

C.C. Baniotopoulos and F. Wald (eds.), The Paramount Role of Joints into the Reliable Response of Structures, 59–68.

integrated application. Besides the structural analysis and member design modules it has also a connection design module. The technology and calculation methods are based on Annex J of ENV 1993-1-1. The programme covers the design of pinned, rigid and semi-rigid frame connections [6].

3. Design of a braced frame

The geometry of a four-bay, four-storey braced frame is shown in Figure 1. The spacing of the frames is 6 m and the used steel grade is S235. Characteristic values of actions are: a) dead weight of the steel structure (γ_f = 1.1), b) dead load 3 kN/m^2 (γ_f=1.15), c) imposed load 2 kN/m^2 (γ_f = 1.3), d) wind load (wind zone IV, 0.55 kN/m^2 ,γ_f = 1.2), e) snow load (snow zone II, 0.7 kN/m^2 ,γ_f = 1.4) according to STN 73 0035 [9].

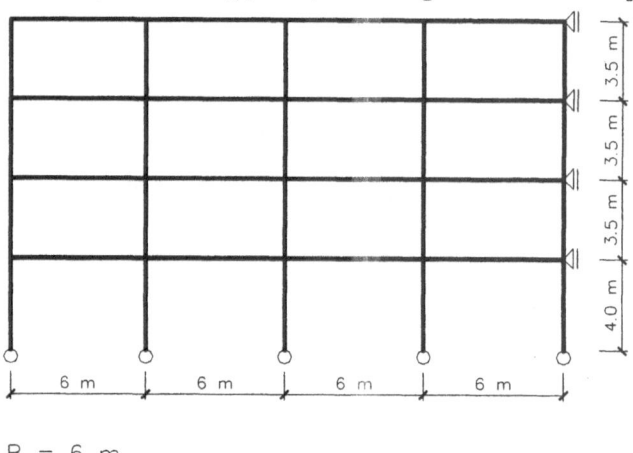

B = 6 m

Figure 1

The frame was designed with welded rigid joints, with bolted semi-rigid joints (extended end-plates) and with bolted pinned joints (short end-plates). The designed joints are depicted in Figure 2. Their design properties (moment resistance $M_{j,Rd}$ and initial rotational stiffness $S_{j,ini}$) are given in Table 1.

Table 1

Joint	Welded Rigid Joints		Bolted Semi-Rigid Joints		Bolted Pinned Joints	
	$M_{j,Rd}$ [kNm]	$S_{j,ini}$ [MNm/rad]	$M_{j,Rd}$ [kNm]	$S_{j,ini}$ [MNm/rad]	$V_{j,Rd}$ [kN]	$S_{j,ini}$ [MNm/rad]
SB - EC	279.4	∞	67.1	26.9	162.2	0
SB - IC	279.4	∞	75.9	29.0	162.2	0
RB - EC	218.3	∞	60.7	22.4	159.1	0
RB - IC	218.3	∞	68.6	24.0	159.1	0

Key: SB - Storey Beam,   RB - Roof Beam,   EC - External Column,   IC - Internal Column

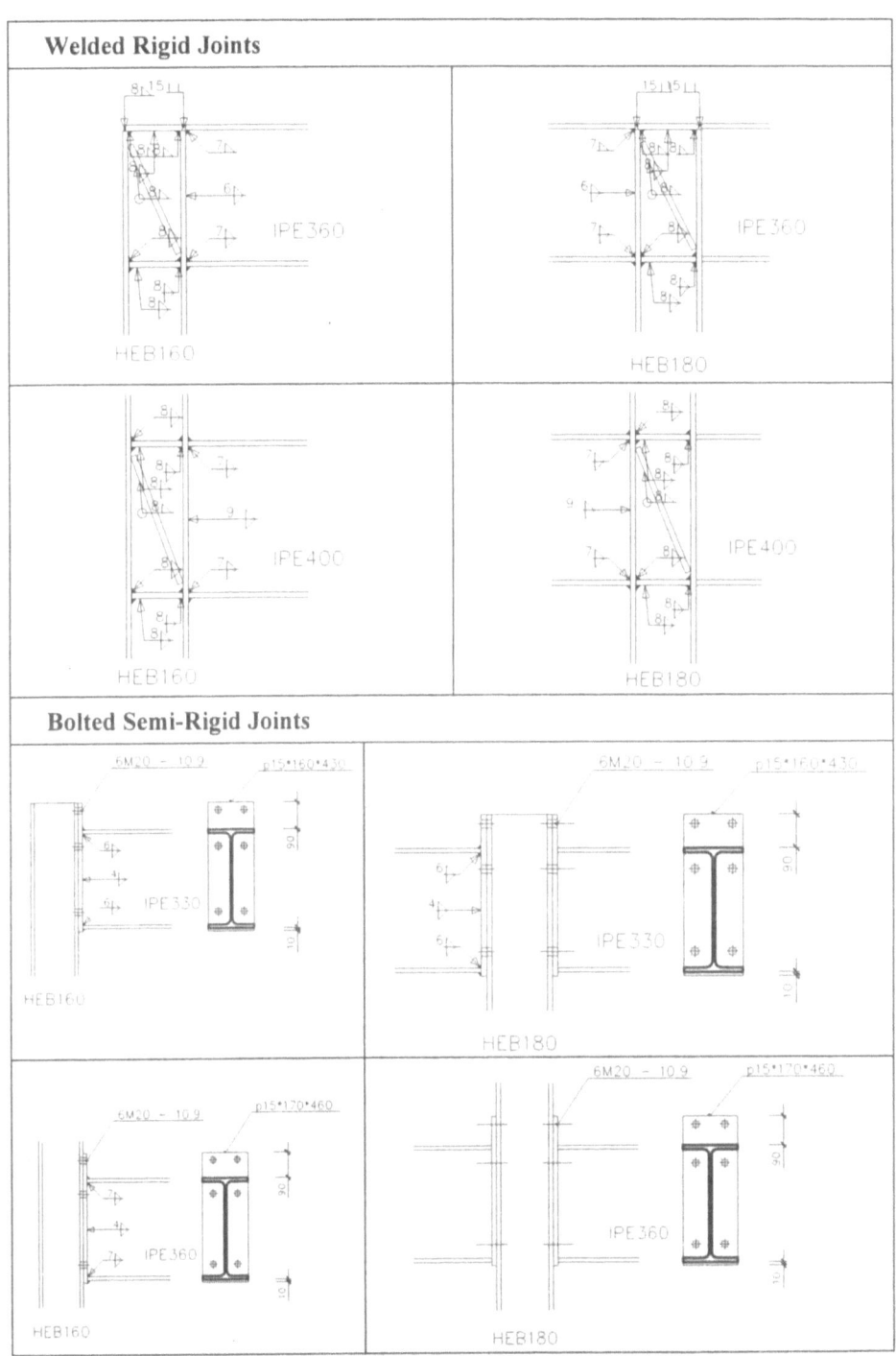

Figure 2

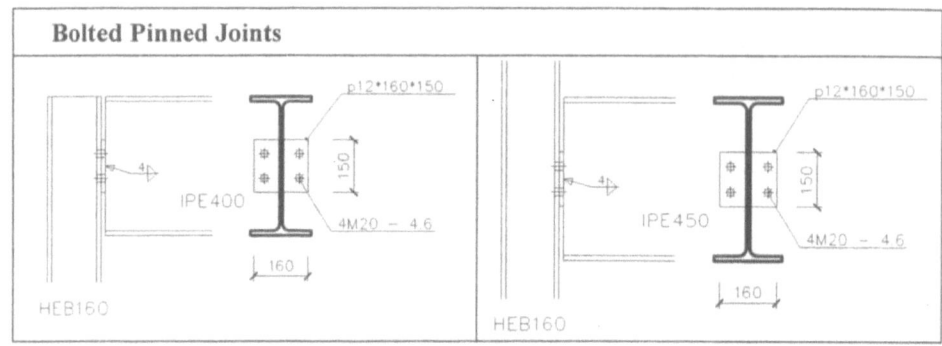

Figure 2 (cont.)

Elastic global analysis was used for the global analysis. The rotational stiffness of joints was taken as $S_{j,ini}$ / 2 in the global analysis. Plastic design of members and joints was used. Table 2 shows the maximal hogging and sagging beam moments. We can see how the rotational joint stiffness influences the distribution of bending moments in the beams. By balancing the hogging and sagging moments we can find an optimal solution.

Table 2

Maximal Beam Moments [kNm]	Welded Rigid Joints	Bolted Semi-Rigid Joints	Bolted Pinned Joints
SB - EC hogging	-47.2	-43.2	0
SB - IC hogging	-131.2	-65.4	0
SB - sagging	86.4	115.2	167.2
RB - EC hogging	-28.6	-28.1	0
RB - IC hogging	-95.6	-51.2	0
RB - sagging	65.6	84.1	122.9

Key: SB - Storey Beam,   RB - Roof Beam,   EC - External Column,   IC - Internal Column

External and internal columns were designed with a constant cross-section over their length. Summary of the results of the designs is shown in Table 3.

Table 3

Sections Review	Welded Rigid Joints	Bolted Semi-Rigid Joints	Bolted Pinned Joints
EC - External Columns	HEB 160	HEB 160	HEB 160
IC - Internal Columns	HEB 180	HEB 180	HEB 180
SB - Storey Beams	IPE 400	IPE 360	IPE 450
RB - Roof Beams	IPE 360	IPE 330	IPE 400
Total Frame Weight	9612 kg	8756 kg	10642 kg
Difference	90 %	82 %	100 %

Key: SB - Storey Beam,   RB - Roof Beam,   EC - External Column,   IC - Internal Column

4. Design of an unbraced frame

The geometry of the two-bay, five-storey unbraced frame is shown in Figure 3. The spacing of the frames is 6 m and the used steel grade is S235. Characteristic values of actions are: a) dead weight of the steel structure ($\gamma_f = 1.1$), b) dead load of floors 3 kN/m^2 and roof 1.5 kN/m^2 ($\gamma_f = 1.15$), c) imposed load 2 kN/m^2 ($\gamma_f = 1.3$), d) wind load (wind zone IV, 0.55 kN/m^2 ,$\gamma_f = 1.2$), e) snow load (snow zone II, 0.7 kN/m^2 ,$\gamma_f = 1.4$) according to STN 73 0035.

An equivalent geometric imperfection in the form of an initial sway imperfection

$$\phi = \frac{1}{200}\left(0.5 + \frac{1}{3}\right)^{0.5}\left(0.2 + \frac{1}{5}\right)^{0.5} = 0.00289 \text{ was used in the frame analysis. The limit}$$

for the horizontal deflection under characteristic values of actions has been taken as $\frac{1}{500}H = 36$ mm and greatly influenced the design of the unbraced frame.

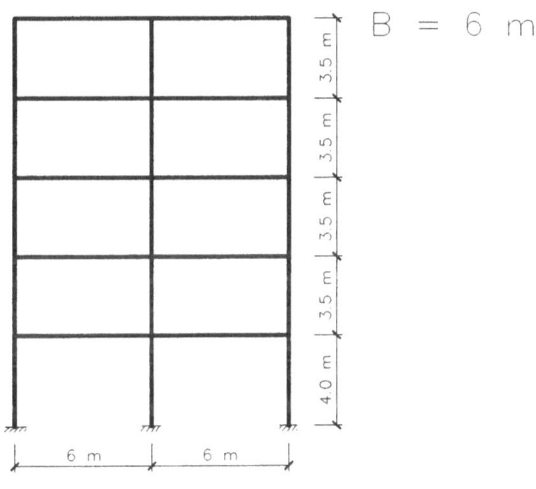

Figure 3

The frame was designed with welded fully rigid joints, with welded semi-rigid joints and with bolted semi-rigid joints (extended end-plates). The designed joints are depicted in Figure 4. Their design properties (moment resistance $M_{j,Rd}$ and initial rotational stiffness $S_{j,ini}$) are given in Table 4.

64

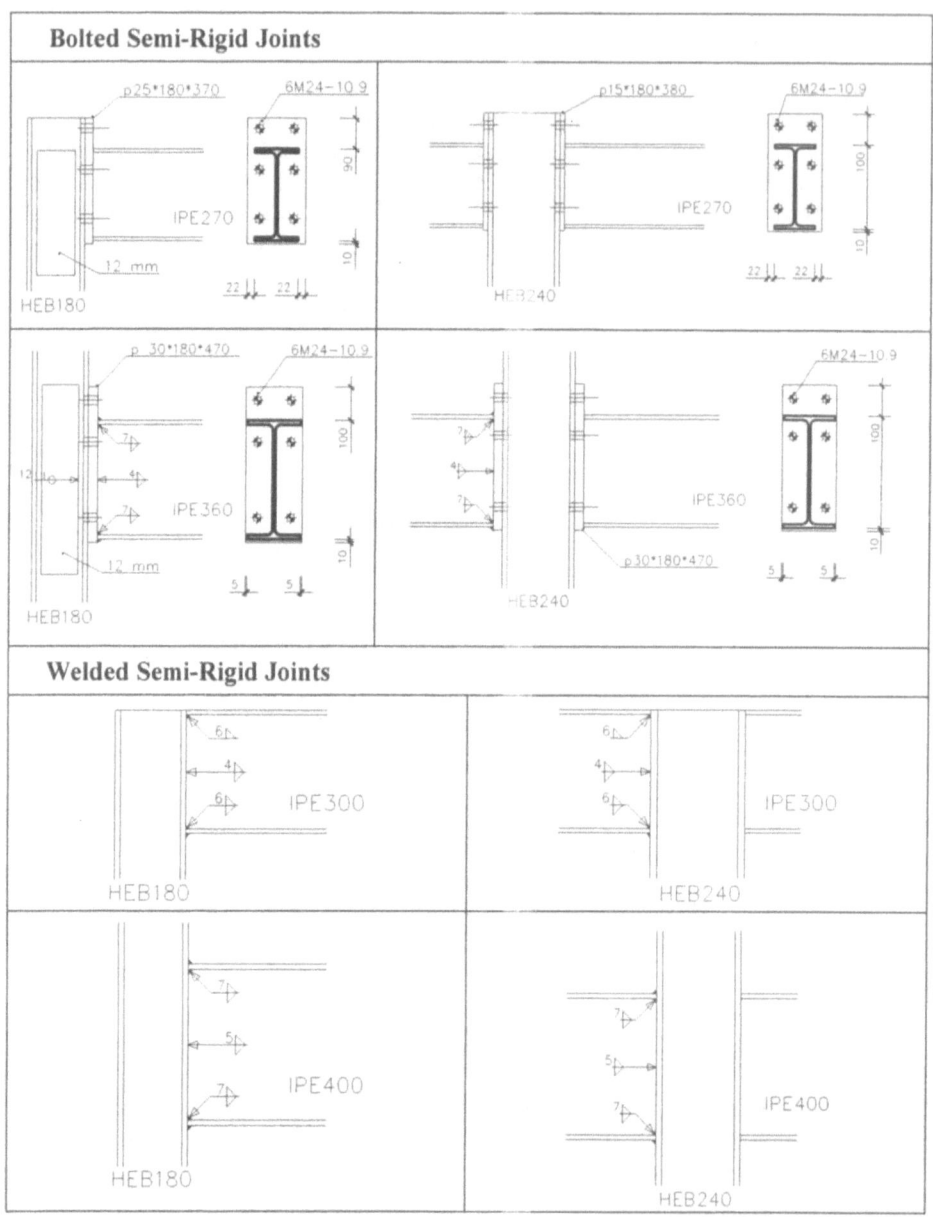

Figure 4

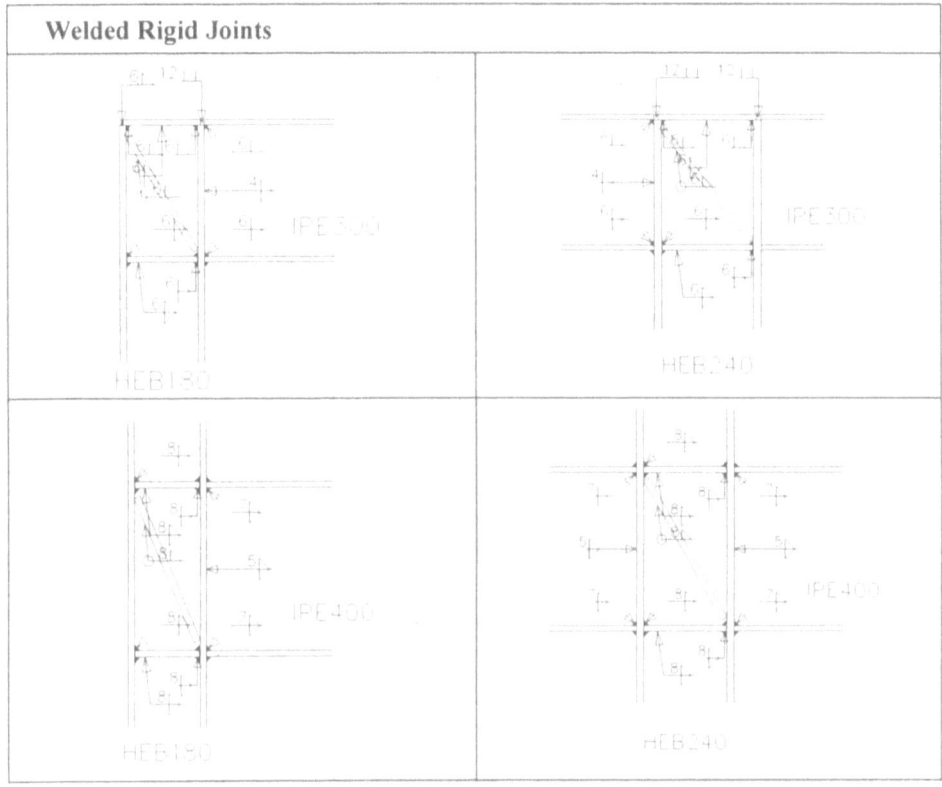

Figure 4 (cont.)

Table 4

Joint	Welded Rigid Joints		Welded Semi-Rigid Joints		Bolted Semi-Rigid Joints	
	$M_{j,Rd}$ [kNm]	$S_{j,ini}$ [MNm/rad]	$M_{j,Rd}$ [kNm]	$S_{j,ini}$ [MNm/rad]	$M_{j,Rd}$ [kNm]	$S_{j,ini}$ [MNm/rad]
SB - EC	279.4	∞	86.8	42.8	125.4	45.8
SB - IC	279.4	∞	142.6	60.8	143.5	45.1
RB - EC	134.1	∞	41.9	24.0	77.3	25.3
RB - IC	134.1	∞	63.0	31.8	79.0	20.1

Key: SB - Storey Beam,   RB - Roof Beam,   EC - External Column,   IC - Internal Column

Table 5 shows the maximal hogging and sagging beam moments. Table 6 shows the maximal horizontal deflection under characteristic values of actions.

Results of the designs are shown in Table 7.

Table 5

Maximal Beam Moments [kNm]	Welded Rigid Joints	Welded Semi-Rigid Joints	Bolted Semi-Rigid Joints
SB - EC hogging	-76.4	-72.2	-83.4
SB - IC hogging	-184.0	-120.0	-115.7
SB - sagging	83.7	107.6	101.9
RB - EC hogging	-29.5	-25.5	-27.7
RB - IC hogging	-60.1	-42.1	-39.2
RB - sagging	33.1	44.0	42.3

Key: SB - Storey Beam,   RB - Roof Beam,   EC - External Column,   IC - Internal Column

Table 6

Maximal Horizontal Deflection [mm]	Welded Rigid Joints	Welded Semi-Rigid Joints	Bolted Semi-Rigid Joints
Top of the building	22.0	30.9	36.0

Table 7

Sections Review	Welded Rigid Joints	Welded Semi-Rigid Joints	Bolted Semi-Rigid Joints
EC - External Columns	HEB 180	HEB 180	HEB 180
IC - Internal Columns	HEB 240	HEB 240	HEB 240
SB - Storey Beams	IPE 450	IPE 400	IPE 360
RB - Roof Beams	IPE 300	IPE 300	IPE 270
Total Frame Weight	7572 kg	7031 kg	6515 kg
Difference	100 %	93 %	86 %

Key: SB - Storey Beam,   RB - Roof Beam,   EC - External Column,   IC - Internal Column

5. Conclusion

In the case of the braced frame, the material saving for the alternative with rigid joints was 9.7% in comparison with the one with pin connections. It is, however, necessary to make complicated and costly joint stiffening, that will decrease the total cost saving of the rigid jointed frame. In the case of a semi-rigid jointed frame we obtain a material saving of up to 17.7% compared to the frame with pinned joints. This difference is already considerable, even though the total cost will also decrease by something, because the fabrication of a semi-rigid joint is usually more expensive than the fabrication of a pinned one.

In the case of the unbraced frame, the material saving for the alternative with welded semi-rigid joints was 7.1 % in comparison to a rigid jointed frame. In the case of a bolted semi-rigid jointed frame we obtain a material saving of up to 14% compared to the frame with rigid joints. The total cost saving of the frame with semi-rigid joints will even increase, because there is no need to make complicated and costly joint stiffening. So, when minimum cost of steel structures is of interest, we can use two different strategies:

- in the case of braced frames - we can reduce the profile dimensions, i.e. reduce the material costs, but increase the fabrication cost of joints in comparison with a frame with simple joints,
- in the case of unbraced frames - we can simplify the joint detailing, i.e. reduce the fabrication costs. The connections can be designed without expensive stiffeners.

By an appropriate choice of joint types and their properties we can influence the magnitude of critical loads and buckling lengths, what, in this case, did not influence the design of column sections. We can affect the distribution and possible redistribution of internal forces and moments from the beams to the columns and vice versa, as well as the frame deformation. We can not only reduce the beam profiles but also decrease the cost of fabrication and thus find an optimal design with the lowest overall cost.

This paper was published under the Grant Research Project VEGA No 1/7139/20, which is financially supported by the Research Grant Agency of the Ministry of Education of the Slovak Republic and by the Slovak Academy of Sciences.

References

1. Anderson, D., Colson, A., Jaspart, J.P. and Wald, F. (1993) Ekonomické styčníky pro ocelové rámy, *Stavební Obzor* 10, 287-293.
2. Anderson, D., Colson, A., Jaspart, J.P. (1996) *Connections and Frame Design for Economy*, ECCS No77, Brussels, Belgium.
3.Ároch, R. (1998) Využitie poddajnosti styčníkov pri výpočte rámových konštrukcií (The Use of Joint Stiffness in Calculating Frame Structures), *Proceedings of the Workshop of Experts on Steel Structures*, Banská Štiavnica.
4. ENV 1993-1-1 (1992) *Eurocode 3, Design of Steel Structures, Part 1.1: General Rules and Rules for Buildings*, Commission of the European Communities, European Prenorm, Brussels, Belgium.
5. ENV 1993-1-1/A2 (1998) *Eurocode 3, Design of Steel Structures, Part 1.1: General Rules and Rules for Buildings, Amendment A2*, Commission of the European Communities, European Prenorm, Brussels, Belgium.
6. Rammant, J.P., Van Isacker, F., Van Loock, C. and El Masri, N. (1999) Expert System for the Design of Steel Connections, *Proceedings of the Conference Eurosteel '99*, ČVUT, Praha.
7. Steenhuis, C.M. (1992) Frame Design and Economy, *Proceedings of the State of the Art Workshop on Semi-Rigid Behaviour of Civil Engineering Structural Connections*, ed. by A. Colson, COST C1, Strasbourg, France, 549-559.

68

8. Steenhuis, C.M., Stark, J.W.B. and Gresnight, A.M. (1997) Steel Structures: Cost Effective Connections Beat Minimum Weight Design, *Progress in Structural Engineering and Materials*, Vol. **1**, 8-24.

9. STN 73 0035 (1986) *Zatížení stavebních konstrukcí* (Actions on Structures), Vydavatelství Úřadu pro normalizaci a měření, Praha.

10. STN P ENV 1993-1-1 (1994) *Eurocode 3 with National Application Document*, Úrad pre normalizáciu, metrológiu a skúšobníctvo SR, Bratislava, Slovakia.

11. Wald, F. and Sokol, Z. (1999) *Navrhování styčníků* (Design of Joints), Vydavatelství ČVUT, Praha.

12. Weynand, K., Jaspart, J.-P. and Steenhuis, M. (1998) Economy Studies of Steel Building Frames with Semi-Rigid Joints, *Journal of Constructional Steel Research*, **46**:1-3, Paper No.63, Elsevier Science Ltd.

13. Weynand, K., Feldmann, M. and Sedlacek, G. (1997) Nachgiebige Anschlüsse im Stahlbau: Modell nach Eurocode 3, Sicherheits- und Wirtschaftlichkeits-untersuchungen, *Stahlbau* **66**, Heft 11, 770-781.

DEVELOPMENT, DESIGN AND TESTING OF JOINTS IN COMPOSITE STRUCTURES

A.K. KVEDARAS
Vilnius Gediminas Technical University, Saulėtekio alėja 11,
LT-2040 Vilnius, Lithuania, akve@st.vtu.lt

1. Introduction

There are examined concrete-filled steel tubular structures assigned to use in structural framework, in which though joints and connections acting forces should be given over all main components of composite member uniformly. Therefore the structural solutions of joints and connections should have related with that specific character. Some possible connection solutions are given in a work [1].

Connections of built-up double-chord concrete-filled steel tubular columns are discussed in a work [2]. Both parallel chords together are connected by the chats of box-shape sections made from double channels and welded by their ends to the external steel shells of those chords. Tests confirmed a great strength of welded joints. The stepped cross arm is done quite high because it has to take action of pressure from the chord of internal side of upper past of column. Besides, on another shorter column's chord through knife pins the crane girder is supported. The stepped cross arm was tested under such load and good results were received: the connection is deforming only a little and is stiff.

Structural solutions of spatial frame of building of universal destination are discussed in works [3] and [4]. In this building the centrifuged concrete-filled steel tubular columns, composite steel-concrete cast-in-situ rafters and hollow pre cast concrete slabs are used. All that structures are connected together in a rigid connection into spatial load bearing framework. Because of that a great economical effect may be gained.

2. Laced concrete-filled steel tubular columns

For long and comparably under small action of loads columns the compact concrete filled steel tubular cross-sections are insufficiently useful because such circumstances require an increased cross-section and utilization more concrete and steel. The same situation arises with beam-columns especially if their eccentricities are large. In such cases more suitable are the laced columns consisting of two or more concrete filled steel tubular chords and lacing.

69

C.C. Baniotopoulos and F. Wald (eds.), The Paramount Role of Joints into the Reliable Response of Structures, 69–76.
© *2000 Kluwer Academic Publishers.*

70

One example of structural solution of such columns is presented on Fig. 1. They may be analysed as laced steel uniaxially compressed elements or beam-columns due to main principles of codes but taking into account peculiarities connected with behaviour of concrete filled steel tubular chords. The design examples of such columns are presented in [1]. On the other hand in work [5] it is assigned a place for analysis of stress behaviour of laced concrete filled steel tubular columns too.

From concrete filled steel tubular members may be collected not only the columns but also more complicated plane or space structural systems: trusses, arches, frames, grids, towers, masts, etc. Usually the composite concrete filled members are used as the compression elements in such systems, but their tension elements are either solely steel or composite concrete filled, but prestressed ones. Although the research data exist [1] showing quite great efficiency of the tension concrete filled steel tubular members also without of prestressing.

The laced composite columns of double chords made of external thin-walled steel tubes and internal hollow concrete core cast by centrifugal force were used in projects of hot-water supply boiler-houses. The full length of those columns was 13.95, 15.75 and 19.05 meters. A longitudinal steel gantry consisting from the slats made of I-sections or RHS connected the two chords of those laced columns. For the first and second buildings of the Petrasiunai dolomite quarry the effective laced composite columns of above mentioned types were designed instead of typical precast reinforced concrete laced columns. Such type of the

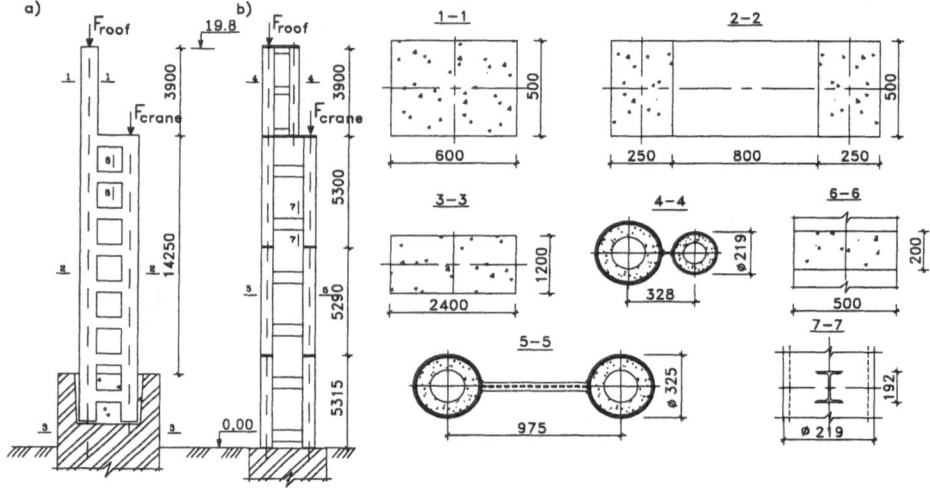

Fig.1. The geometry of the compartment pre cast stepped reinforced concrete column KDX-61 and cast-in-situ concrete post footing with anchor cell (a) and hybride stepped compositesteel tubular column (b) and their cross-sections used for building of dolomite quarry

composite structures were designed at first time for the mill buildings with the height of premises more than 20 meters and with the 20-30 tons of carrying power overhead cranes. The laced composite columns for the above mentioned buildings were designed as stepped shafts with the intermediate-lacing members as slats made from the rolled members. As the step, the lateral bracket beams of technological balconies for the buildings of boiler-houses might be seated from the both sides of chords of the deeper portion of the laced composite column. On the steps of laced columns of the buildings of dolomite quarry the steel crane girders were seated. Thinner portion of the shaft extends above the step to support roof loads. The step is at different distances below the top. That depends on the height of the columns applied in a certain building. The longer external chord of the stepped column supports also the wall's panels.

The stepped laced composite columns for buildings of the dolomite quarry differ from the above mentioned composite columns, used for boiler-houses, with the external dimensions of cross-sections of hollow composite chords (ø325x4.5mm against ø219x3.5mm), with the type and dimensions of cross-section of gantry slats, with the dimensions of gantry slats' spacing and with the kind of structural solution of connections, joints and splices.

Stepped composite columns were checked for direct stress and bending about major and minor axes to be hinged at a top and fixed at a base. The deep section of the lower shaft resists bending below the step. For transmitting the bending moments from the upper shaft the junction of the two-laced composite shafts was strongly done with the field splices.

3. Space framework of multistory office building

There is given a characteristic of space system of composite load bearing structures designed and realized in large office building in Vilnius [12]. A great economical effect is gained because of development here of spun concrete filled steel tubular columns, composite girders and pre cast reinforced concrete slabs with rigid connections between them.

The preliminary structural solution provided for pre cast cast-in-situ framework with concrete filled circular steel tubes of external diameter $\varnothing$ 530 mm as columns and concrete filled twin-H-girders, 30S2 and 35S2, as top frame members.

Redesign of above-mentioned structures provided for use a principle of structural interaction connection between the pre cast cast-in-situ composite structures and pre cast reinforced concrete floor slabs. Columns of framework are designed as hollow concrete filled circular steel tubes of external diameter of $\varnothing$ 325 mm and $\varnothing$ 426 mm due to lengths and loads of members. Lengths of columns were usually through one story and were varying from 3400 up to 4500 mm. Thickness of steel tube walls were 4.0 mm, only in some cases 6.0 mm. Thickness of concrete core walls were 30-50 mm, due to the column length and to value of actions on columns.

The girders of framework are designed as composite members consisting of built-up twin-channel section (No. 22 and No. 27) reinforced with inserted cages of steel reinforcement and concrete filled after. These built-up twin-channel beams are continuous and are provided for bearing of erection loads including the weight of pre cast reinforced concrete hollow core slabs.

All connections between the heads of hollow concrete filled steel tubular columns and continuous built-up twin-channel girders are welded. All welds are flat. The necessary reliability of connections between columns and continuous composite girders is provided for joining the external steel shells of composite columns with welded reinforcing bars and the girder of steel reinforcement through the column-girder connection with additional reinforcing bars too.

The connection of girder and upper composite column is realized through the erection steel ring superimposed on girder channels and fixed by welding. After putting on the same channels the concrete slabs the column-girder connection and space between the channels and edges of concrete slabs have to be concreted. Then the lower end of upper column has to be put into the erection ring and welded. The one of column-girder connections is shown on Fig. 2.

4.   Composite members of framework for farm and mill buildings

For construction of farm-buildings in Lithuania during the long time the reinforced concrete structures are usually used. The portal frame consists of pre cast reinforced concrete columns and three-hinged arch with main tie and joists or pre cast reinforced concrete beams. Spans usually are 9 – 18 meters. Height of premises is 5.0 - 7.0 m. Because such buildings have an

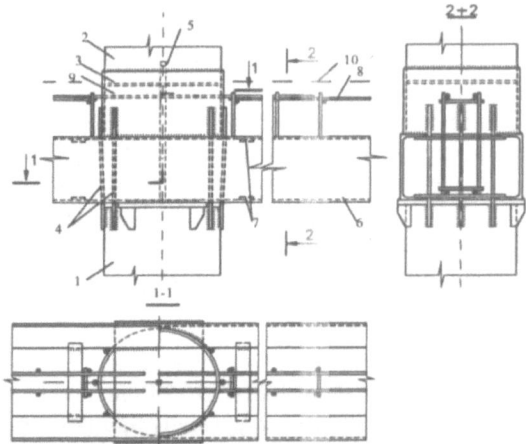

Fig. 2. The scheme of rigid connection of hollow concrete filled circular steel tubular column and continuous composite girder for multistory office building,
1 - lower composite column; 2 - upper composite column; 3 - erection ring for upper column; 4 - joining the external steel shells of lower and upper columns by welded reinforcing bars; 5 - pipe for aeration of the column core hollows; 6 - steel twin-channels of composite girder; 7 - batten-plates of girder as partial shear connectors of concrete core; 8 - cages of steel reinforcement of girders; 9 - joining cages additional longitudinal steel reinforcing bars going through the connection; 10 - upper level of filled concrete

extremely big weight, the designing was carried out to propose project of more effective building including more lightweight framework and roof and wall claddings. One of designed buildings [8] of such kind is presented by fragment of main load bearing structural framework shown on Fig. 3.

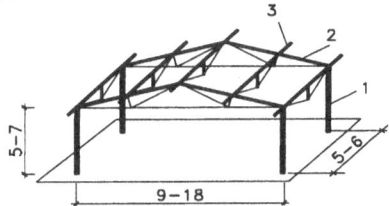

Fig.3 Fragment of the structural scheme of farm building. 1 - one-stem hollow concrete filled steel tubular columns; 2 - hybrid composite-steel roof trusses with hollow concrete filled steel tubular chords; 3 - hybrid composite-steel joists

For one of farm buildings with dimensions of it lay-out 12x33 m, the columns were made of supper-thin walled circular steel tubes of external diameter ∅325x1.5 mm with 40 mm thick hollow concrete core. The upper chords of roof trusses and joists were made from external supper-thin walled circular steel tubes with external diameter ∅219x1.5 mm and 30 mm thick hollow concrete core. The other tension members of trusses and joists are done of A-II grade reinforcing steel bars. The compression members are made of rolled steel channels or angles. For determination of statistical estimates of mechanical properties of materials used in members of investigated trusses and joists, they were divided into a number of small specimens, which were distributed into separate groups and tested. Statistical estimates of investigated structural systems included also the stresses of members obtained by structural analysis of trusses or joists taking into account the stochastic nature of loads. For structural systems of above-mentioned building-hybrid composite-steel trusses and joists - the simplified version of safety estimations have been used. There the resistance is accepted as fixed accidental function and the statistical estimates of mechanical properties are assumed being invariable. Probabilistic methods in stability analysis of hybrid composite-steel roof trusses and joists used proved the great reliability of such systems.

5.   Nomenclature of specimens, test procedures and results

Tests were indispensable for choosing such structure for field splices of column chords that would be most technological and reliable and with the strength not less than one of composite chords. The splices of chords were used because of the peculiarities of technology of concrete mixes centrifuging and needs columns' erection on building site. Tests on the stepped laced composite columns for the buildings of both types – boiler-houses and dolomite quarry – were carried out in a horizontal position by the equipment which was specially developed for those two cases. At first case the two stepped laced composite columns of 13.95 m in real length and in second case the one such column of 19.8 m in real length have been tested. These full-scale testing allowed checking also behaviour of joints between the hollow concrete-filled chords and different shapes used for lacing slats and other members. The laced tested columns were simulating joints – rigid or semi-rigid – for multi-story single span plane frame.

The large scale of tests of hollow composite specimens was carried out [9]. Parameters of test specimens varied from Ø 140 mm to Ø 325 mm, thickness of steel shell – from 1.5 mm to 8.0 mm, cylindrical mean strength of centrifuged concrete of internal hollow cores – from 12 MPa till 80 MPa. Tests with the short hollow composite specimens under uniaxial compression are witnessing a high level of their critical deformations. Longitudinal strains were reaching the level of 500.10^{-5} and transversal – 250.10^{-5}. These values depend on thickness of steel shell, on ratios between the thickness of steel shell and concrete core and between the strength of concrete core and steel shell. The fixed process of interaction on the contact surface between the steel shell and concrete core and observed high level of ultimate longitudinal and lateral strains very significantly influences on behaviour of joints and connections.

The large-scale tests of hollow composite beam-columns were carried out too. Initial eccentricities of longitudinal compression forces acting on these members were varying from 0.25 to 0.85 of external diameter of composite members. The slenderness ratios of beam-columns were not very high (about 12), but deflections at the critical load were found being 1/125 – 1/75 of elements' length. The critical longitudinal strains of more compressed part of cross-section of composite member exceeded $(450-900)\ 10^{-5}$. At the tension part of cross-section these strains were equal to $(180-600)\ 10^{-5}$. That means the strains of composite beam-columns being close twice higher than those of uniaxially compressed members.

Tests of roof trusses shown on Fig.3 according to [10] enabled fixing of complex Stress State in their hollow composite members – upper chords and especially in nodes. Some values of strains measured on the external surfaces of steel shells of composite members exceeded the values of strains corresponding to the beginning of yielding of steel.

According to the obtained results of tests of laced composite column [11] conclusion may be gained about containing of large reserves of strength and stiffness under action of designed factored load on such column. An experimental investigation of the fragment of bottom crane part of laced composite spun column with traverse gives largest support force on traverse and composite chord equal to 1,500 kN. The yield of web of traverse has started at 900 kN value of acting compression force. At the largest value of acting force any sign of yield in fillet welds of joints and in the steel of chord's shell have not been fixed. No signs of rupture of concrete at that moment have been fixed also. In this case a resistance of traverse connection is twice-trice higher of largest design factored pressure of over5head crane girders through the end bearing stiffeners and the reliability of connections is more than sufficient.

An experimental investigation of fragment of laced composite column with two UC cross-bars gives the largest lateral force supplied at the middle of spun of UC cross-bar (slat) equal to F = 800 kN, which corresponds to the bending moment M = 68.5 kNm and shear force equal to Q = 400 kN. Failure of the fragment takes place because of local buckling of the web of UC slat. The fillet welds of joints between the UC slats and the steel shell of composite spun chords of laced column restrained the ultimate load, in the welds just did not appear the signs of yielding. In the steel shell and concrete core of composite chords did not appear the ultimate values of forces and strains. The yield of steel shell corresponds to the value of bending moment equal M = 47,1 kNm and shear strength Q = 275 kN. During yielding of the web the bending moment is 1.2 times and the shear force is 4.4 times higher than the maximum factored values of those forces of the slats. The test failure value of the point loads acting on the upper chords of hybrid composite-steel roof truss was 49.9 kN. The characteristic of these loads was equal top to 20.8 kN. The test failure values of bending moments of two hop-perches were 32.5 and 23.1 kNm, the ultimate values of them being 1.3 and 1.1 respectively and limited by ultimate value of deflection.

6. Conclusions

Research carried out with the hollow concrete filled circular steel tubular structures have showed a specific behaviour under load action significantly different from concrete filled steel tubes in which concrete core is solid. Because of that design procedures also are different from those used for composite members with solid concrete core and usually approved by some national codes of practice or Eurocode 4. Structures, which are discussed here in, are designed on the base of discussed principles and enabled to save a lot of building materials without loosing of safety and reliability. So is mainly because of possibility to use interaction occurring between the external steel shell and the hollow internal concrete core and developing in loaded structure, especially if it is tubular.

Recommended design method is corroborated by abundant new data of experimental and full-scale tests of solid and hollow concrete filled steel tubular members and their systems. These data obviously show results depending on the model of stress-strain Limit State of stub members and their connections.

It is shown that the critical longitudinal strains of concrete filled steel tubular members tested by uniaxial compression are great of value, which amounts to 500.10^{-5}. These strains depend on the wall thickness of steel shell, on ratios of strengths and wall thickness of steel shell and concrete core. That is confirmed by the test results of physical, mechanical, rheological and technological properties and interaction factors of members, their cross-sectional components and materials.

An increased stiffness of concrete-filled members allows construct joints of greater strength and safety. Interaction occurring between the components of concrete-filled members allows use higher levels of loading on connections and splices too. In some cases the joints in composite structural systems may be assessed not only as hinged or rigid, but also as semi-rigid. That may more precisely reflect the real behaviour in such structural systems, in which the concrete filled steel tubular members, especially hollow, are used.

7. References

1. Kvedaras A. (1985) *Detailing and Manufacture of Concrete Filled Metal Tubular Structures,* Publish. Ed. Council of the Lithuanian Ministry for Higher Education, Vilnius (in Lithuanian).
2. Kvedaras A. and Sapalas A. (1991) Investigations of Laced Composite Spun Strut and it Connections, in *Proceedings "Structures and Foundations" of 1st Intl Conf. "Modern Building Materials, Structures and Techniques in Construction",* Technika, Vilnius, pp. 25-35.
3. Kvedaras A., Pranevicius J., Sapalas A., Sapalas K. and Valiūnas B. (1995) The Composite Steel-Concrete Space Structural System, in *Proceedings of 4th Intl Conf. "Modern Building Materials, Structures and Techniques* 2, Technika, Vilnius, pp. 102-107 (in Lithuanian).
4. Kvedaras A. and Sapalas A. (1998) Efficient mixed space structural systems in Lithuania, in *Proceedings of Nordic Steel Construction Conference 98* 1, Nork Stalforbund, Oslo, pp. 477-488.
5. Kikin A.J., Sanzharovsky R.S. and Trull V.A. (1974) *Concrete Filled Steel Tubular Structures,* Strojizdat, Moscow. (In Russian).
6. Kvedaras A. and Sapalas K. (1994) Development of Lattice Hollow Composite Columns and Connections, in T.Jávor (ed.), *Steel-Concrete Composite Structures,* Expertcentrum, Bratislava, pp. 214-217.
7. Kvedaras A. and Mykolaitis D. (1995) On the Stability of Steel and Composite Concrete Filled Steel Tubular Beam-Columns, in M.Iványi (ed.), *Stability of Steel Structures 1995, Budapest: Further Direction in Stability Research and Design* 2, Akadémiai Kiadó, Budapest, pp.223-230.
8. Kvedaras A., Mykolaitis D. and Sapalas A. (1996) Projects of Hollow Concrete Filled Steel Tubular Structures, in J.Farkas and K.Jarmai (eds.), *Tubular Structures VII,* A.A.BALCEMA Publishers, Rotterdam / Brookfield, pp. 341-348.
9. Kvedaras A., Sapalas A. and Valiunas B. (1995) The Outlook and Survey of Research and Applications of Spun Composite Structures in G. Marciukaitis (ed.), *Transactions of Vilnius Tech. Univ. "Constructional Structures"* 20, Technika, Vilnius, pp. 19-30.
10. Juozaitis J., Kvedaras A., Mykolaitis D. and Valiūnas B. (1995) Light Composite Roof Trusses for Storehouses, in G. Marciukaitis (ed.), *Transactions of Vilnius Tech. Univ. "Constructional Structures"* 20, Technika, Vilnius, pp.62-71 (in Lithuanian).
11. Kvedaras A. and Sapalas A. (1997) Full-Scale Testing of Composite Structures for Estimation of Their Safety, in K.S. Virdi, F.K. Garas, J.L. Clarke, G.S.T. Armer (eds.), *Structural Assessment: The role of large and full-scale testing,* E & FN SPON, London, pp. 482-490.
12. Kvedaras A. and Sapalas A. (1998) Research and Practice of Concrete-Filled Steel Tubes in Lithuania, in Journal of Steel Constructional Research 49 (1999), Elsevier, Dordrecht, pp. 197-212.

APPLICATION OF THE THEORY OF SEMI-RIGID JOINTS INTO STEEL AND COMPOSITE STRUCTURES

A. PAVLOV
Research and Design Institute Promstalkonstruktsiya
13, Sadovaya Samotechnaya, 103473, Moscow, Russia

1. Introduction

Semi-rigid joints are playing a leading role in the behaviour of building structures. Today it is possible to note that the basis researches for semi-rigid joints are completed. But until now the practical applications for all new knowledge are rare. The reason is that practical engineers are not trained to use non-linear calculations of structures. Therefore the support like adequate tables, handbooks, simplified formulae and software's is missing.

INTAS Project 96/2154 "Slim floor construction" has been realised in 1997 — 1999 and sponsored by the International Association for the promotion of co-operation with scientists from the New Independent States of the former Soviet Union (NIS). It was developed through the co-operation of the following organisations:

— University of Innsbruck, Austria. Professor F. Tschemmernegg is the team leader and the Co-ordinator of the Project,

— University of Trento, Italy, Professor R. Zandonini is the team leader,

— Research and Design Institute Promstalkonstruktsiya, Moscow, Russia. Professor V. Kalenov is the team leader,

— Moscow State University of Civil Engineering, Russia. Professor M. Belyi is the team leader.

The proposals for practical design have been developed in the frames of INTAS Project 96/2154. There are:

— Simplified formulae to calculate the stiffness and strength of each component of a slim floor connection for rectangular hollow column sections, circular hollow column sections and H-shaped column sections,

— Design tables for joints of rectangular and circular hollow column sections include the stiffness and strength values in dependence of joint parameters,

— Global analysis tables for beams considering a joint stiffness (semi-rigidly supported beams) include the values of bending support moments, maximal bending span moments and maximal deflections.

This paper gives the survey of some marked results.

77

C.C. Baniotopoulos and F. Wald (eds.), The Paramount Role of Joints into the Reliable Response of Structures, 77–86.
© 2000 *Kluwer Academic Publishers.*

78

2. Stiffness and strength simplified formulae (circular hollow column sections)

Institute of Steel, Timber and Mixed Building Technology of University of Innsbruck, Austria [1] has developed the sophisticated mechanical models for stiffness and plastic design resistance calculation for the compression and shear region of rectangular column hollow sections (steel and composite).

Basing on the sophisticated mechanical models the simplified formulae have been developed to be brought into ENV1994-1-1/Annex J. The proposed solutions are guided by the resistance formulae given for H-shaped sections in EC3- Revised Annex J and the Fourth Draft of EC4-Annex J.

The relatively simple formulae have been gained with the help of comprehensive parameter studies out of complex formulae of Müller [1]: for stiffness by Herzog [2] and for resistance by INTAS 96/2154 teams. The given values are related to the realistic points L and S without any transformations.

Figure 1 lists the relevant components and groups of Innsbruck model considering all conventional and advanced joints. With this model the "moment — rotation" curve can be simulated taking into consideration all component influences $F_{i,R}$ — $W_{i,R}$.

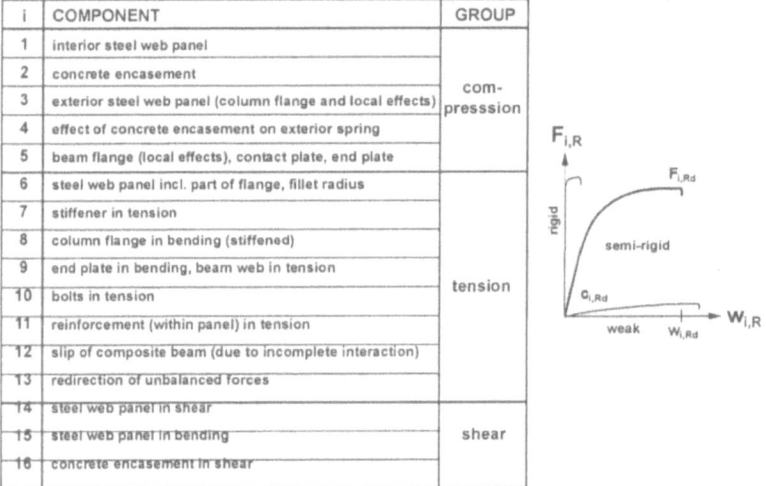

i	COMPONENT	GROUP
1	interior steel web panel	com-presssion
2	concrete encasement	
3	exterior steel web panel (column flange and local effects)	
4	effect of concrete encasement on exterior spring	
5	beam flange (local effects), contact plate, end plate	
6	steel web panel incl. part of flange, fillet radius	tension
7	stiffener in tension	
8	column flange in bending (stiffened)	
9	end plate in bending, beam web in tension	
10	bolts in tension	
11	reinforcement (within panel) in tension	
12	slip of composite beam (due to incomplete interaction)	
13	redirection of unbalanced forces	
14	steel web panel in shear	shear
15	steel web panel in bending	
16	concrete encasement in shear	

Figure 1. Innsbruck component model

The formulae for stiffness and design resistance calculation for some components and for circular hollow column sections are presented in Table 1. Joint location on the structure is shown in Figure 2.

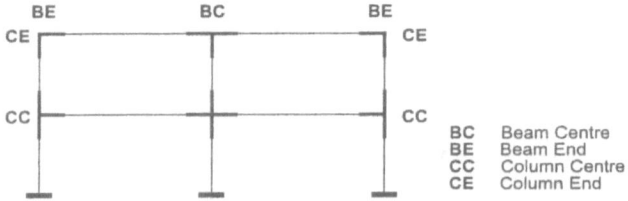

Figure 2. Joint location within the structure

TABLE 1. Stiffness and design resistance calculation (circular hollow column sections)

No. comp.	Notations	Stiffness, $C_{0,Rd}$	Design Resistance, $F_{pl.Rd}$ or $V_{pl.Rd}$	Notice
1+3	$\beta=\dfrac{b}{d}$; E_a — Young modulus of steel; $f_{y,a,k}$ — yield strength of steel.	$\dfrac{35E_a t^3}{d^2(1-\beta)^{0,75}}$	$\left(7,39+\dfrac{0,60}{1-\beta}\right)t^2\dfrac{f_{y,a,k}}{\gamma_a}$	
2+4	$\beta=\dfrac{b}{d}$; $b_{eff}=b_0+5t$; $b_0=t_b+2\sqrt{2}a$; $\psi=0,76(6,4292-0,1078f_{c,k}[N/mm^2])$ $\Psi\geq2.44$ or $\Psi\cdot f_{c,k}\geq72$ E_{cm} — Young modulus of concrete; $f_{c,k}$ — characteristic compressive strength of concrete encasement.	$\dfrac{E_{cm}d}{4}\left(\dfrac{b_0}{d}\right)^{0,4}$	$\psi\cdot b\cdot b_{eff}\,\dfrac{0,85f_{c,k}}{\gamma_c}$	
14+15	A_a — area of the circular section; $A_v=Rt$; G_a — shear modulus of steel; $f_{y,a,k}$ — yield strength of steel.	$\dfrac{1}{1,9}\dfrac{A_a G_a}{z}$	$\dfrac{3,21f_{y,a,k}A_v}{\sqrt{3}\gamma_a}$	NO Column web bending
		$\dfrac{1}{2,6(3,8)}\left[\dfrac{A_a G_a}{z}\left(\dfrac{d}{z}\right)^{0,5}\right]$	$Min\left(\begin{array}{c}\dfrac{3,21f_{y,a,k}A_v}{\sqrt{3}\gamma_a}; \\[2mm] \dfrac{f_{y,a,k}}{\gamma_a}\cdot\dfrac{d^3-(d-2t)^3}{3z}\end{array}\right)$	Incl. Column web bending, CC (CE)-Joints
16.	$v=0.55\left(1+2\dfrac{N_{Sd}}{N_{pl,zo,Rd}}\right)\leq11$; E_{cm} — Young modulus of concrete encasement; $f_{c,k}$ — characteristic compressive strength of concrete encasement.	$\dfrac{1}{40(80)}\left[E_{cm}d\left(\dfrac{d}{z}\right)^{0,8}\right]$ $\dfrac{1}{45(95)}\left[E_{cm}d\left(\dfrac{d}{z}\right)^{0,8}\right]$	$v\,\dfrac{0,85f_{c,k}}{\gamma_c}A_c\sin\theta$, where $A_c=0,8\pi\dfrac{(d-2t)^2}{4}\cos\theta$ $\theta=\arctan\dfrac{d-2t}{z}$	NO Column web bending, CC (CE)-Joints Incl. Column web bending, CC (CE)-Joints

3. Design tables for joints and their components (circular hollow column sections)

A joint scheme type is shown in Figure 3.

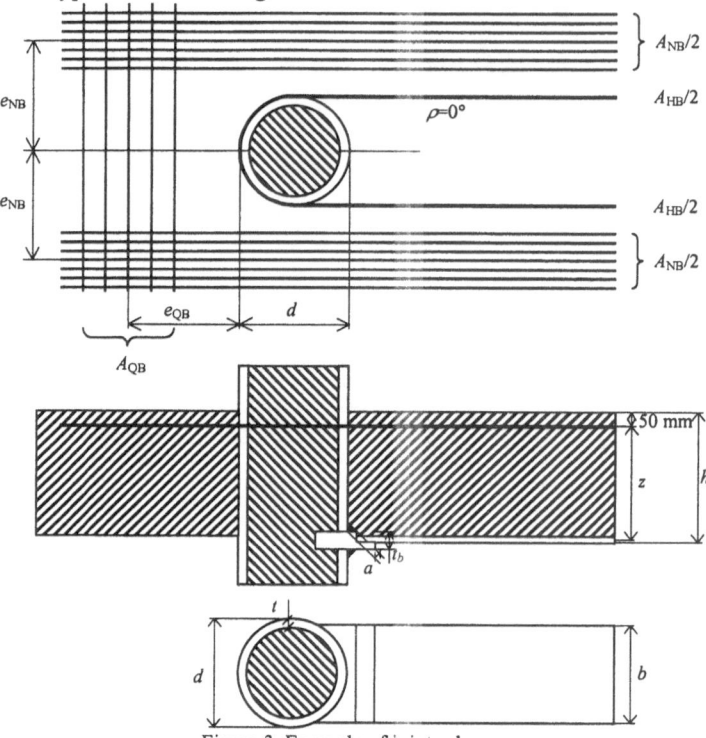

Figure 3. Example of joint scheme

The design tables have been developed for the following joint parameters:
- circular hollow column sections:

d, mm	t, mm			
139,7	4,0	6,3	10,0	12,5
159,0	5,0	8,0	12,5	16,0
168,3	5,0	8,0	`12,5	16,0
193,7	6,2	8,0	12,5	16,0
219,1	6,3	10,0	16,0	20,0
244,5	6,3	10,0	16,0	25,0
273,0	6,3	10,0	16,0	25,0
323,9	8,0	12,5	20,0	30,0
355,6	8,0	12,5	20,0	30,0
406,4	10,0	16,0	25,0	30,0
457,0	8,0	12,5	20,0	30,0
508,0	12,5	20,0	25,0	30,0
559,0	12,5	20,0	25,0	30,0
610,0	12,5	20,0	25,0	30,0

- column steel grade:

- Fe 360 ($f_{y,a,k}$=235 N/mm², γ_a=1,1, E_a=210000 N/mm², G_a=80000 N/mm²).
- Fe 510 ($f_{y,a,k}$=355 N/mm², γ_a=1,1, E_a=210000 N/mm², G_a=80000 N/mm²).

- column and slab concrete grade:
  - C 20/25 ($f_{c,k}$=20,0 N/mm², γ_c=1,5, E_{cm}=29000 N/mm², α=0,85).
  - C 30/37 ($f_{c,k}$=30,0 N/mm², γ_c=1,5, E_{cm}=32000 N/mm², α=0,85).
  - B 40 ($f_{c,k}$=35,3 N/mm², γ_c=1,5, E_{cm}=32500 N/mm², α=0,85).

- reinforcement steel grade :
  - B St 550 ($f_{y,s,k}$=550 N/mm², γ_s=1,15, E_s=210000 N/mm²).

- μ=1 — edge joint and μ=0 — internal joint under balanced moment.
- ρ=0°.
- $d_s = d_{HB}$
- e_{NB}=(0,75 — 1,25)d.
- e_{QB}=(0,75 — 1,25)e_{NB}.
- $h \approx z$+50 mm.
- t_b=20 mm, a=4 mm and t_b=50 mm, a=10 mm.
- $\beta = \dfrac{b}{d} = 0,5$ and $\beta = \dfrac{b}{d} = 0,9$.

- z=z_{min} and z=z_{max}.

Five steps have been realized to calculate stiffness and strength of each joint:

1) Stiffness $C_{0,co}$ and strength $F_{pl,co,Rd}$ calculation for compression zone on the basis of formulae from [1].

2) Choice of reinforcement diameter d_{NB}=d_{QB} and d_{HB} for tension zone.

Assumptions:

- tension strength:
  - $F_{L,t,Rd(1)}$=$F_{pl,co,Rd}$ to make the optimal use of the component's strength.
  - $F_{L,t,Rd(1)}$=0,5 $F_{pl,co,Rd}$; in that case the reinforcement is decisive for the joint failure with the effect of ductile failure mode.
  - if the joint has to be stiffer it is necessary to increase the reinforcement diameter in the tension zone.

- there are seven rebars of secondary longitudinal reinforcement with A_{NB} at each side (n_{NB}=7).

- there is one loop of main longitudinal reinforcement with A_{HB} at each side for μ=0 and one loop of main longitudinal reinforcement with A_{HB} for μ=1 (n_{HB}=1).

- there are five rebars of transversal reinforcement with A_{QB} (n_{QB}=5).

- $A_{(1)} = A_{NB(1)} + A_{HB(1)}$

- $A_{HB(1)}$=0,4 $A_{NB(1)}$.

- possible reinforcement diameter in Russia (mm): 3, 4, 5, 6, 7, 8, 9, 10, 12, 14, 16, 18, 20, 22, 25, 28, 32, 36, 40. If any reinforcement diameter is not used in some European country, it is necessary to use the corresponding values A_{NB}, A_{HB} and A_{QB} with another reinforcement diameter and to change number of rebars.

- if $A_{NB} > 2 \cdot 1,1 \cdot \dfrac{\gamma_s}{\gamma_c} \cdot \dfrac{\alpha \cdot f_{c,k}}{f_{y,s,k}} \cdot A_c \cdot \sin(\delta)$ then reinforcement diameter is chosen

  from the condition $A_{NB} = 2 \cdot 1,1 \cdot \dfrac{\gamma_s}{\gamma_c} \cdot \dfrac{\alpha \cdot f_{c,k}}{f_{y,s,k}} \cdot A_c \cdot \sin(\delta)$.

3) Stiffness $C_{L,t}$ and strength $F_{L,t,Rd}$ calculation for tension zone on the basis of formulae from [1].

4) Stiffness $S_{L,ini}$ and strength $M_{L,Rd}$ calculation for L point of the joint.

5) Stiffnesses $S_{S,ini,CC}$, $S_{S,ini,CE}$ and strengths $M_{S,CC,Rd}$, $M_{S,CE,Rd}$ calculation for the shear panel in the S point of the joint on the basis of formulae from [1].

Design table example is shown in Table 2.

TABLE 2. Design table example

d	t	$d_{NB}=d_{QB}$	d_{HB}	A_{NB}	A_{HB}	A_{QB}	β	t_b	a	z	L point		S point			
											$S_{L,ini}$	$M_{L,Rd}$	$S_{S,ini}$		$M_{S,Rd}$	
											$\mu=1$	$\mu=0$	CC joints	CE joints	CC joints	CE joints
mm	mm	mm	mm	mm²	mm²	mm²	-	mm	mm	mm	MNm	kNm	MNm		kNm	
273,0	16,0	10	22	1100	760	785	0,5	20	4	110	6,8	10,0	97,8	62,4	58,5	115,3	115,3

4. Design tables for beams considering the real joint stiffness

4.1. CONSTANT BEAM STIFFNESS $EI_b=EI_{b.hog}=EI_{b.sag}$ AND DIFFERENT JOINT STIFFNESSES S_1 AND S_2

Figure 4 shows the principal moment and deflection distribution in the beam.

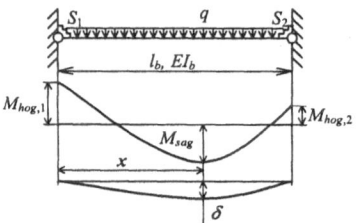

Figure 4. Principal moment and deflection distribution in the beam
with constant beam stiffness and different joint stiffnesses

Denote $\gamma_{j,1} = \dfrac{EI_b}{l_b \cdot S_1}$ and $\gamma_{j,2} = \dfrac{EI_b}{l_b \cdot S_2}$.

The formulae to define the elastic internal bending moment distribution are the following:

$$M_{hog,1} = -\gamma_{hog,1} \cdot \frac{q \cdot l_b^2}{12}, \tag{1}$$

where

$$\gamma_{hog,1} = \frac{6 \cdot \gamma_{j,2} + 1}{12 \cdot \gamma_{j,1} \cdot \gamma_{j,2} + 4 \cdot (\gamma_{j,1} + \gamma_{j,2}) + 1}. \tag{2}$$

$$M_{hog,2} = -\gamma_{hog,2} \cdot \frac{q \cdot l_b^2}{12}, \tag{3}$$

where

$$\gamma_{hog,2} = \frac{6 \cdot \gamma_{j,1} + 1}{12 \cdot \gamma_{j,1} \cdot \gamma_{j,2} + 4 \cdot \left(\gamma_{j,1} + \gamma_{j,2}\right) + 1}. \tag{4}$$

M_{sag} is the maximal value of the span moment.

$$M_{sag} = \frac{q \cdot l_b^2}{8} - \left(\frac{\gamma_{hog,1} + \gamma_{hog,2}}{2}\right) \cdot \frac{q \cdot l_b^2}{12} = \gamma_{sag} \cdot \frac{q \cdot l_b^2}{8}, \tag{5}$$

where

$$\gamma_{sag} = 1 - \frac{\gamma_{hog,1} + \gamma_{hog,2}}{3}. \tag{6}$$

x is the distance corresponding to the maximal value of the span moment M_{sag}.

$$x = \gamma_x \cdot l_b, \tag{7}$$

where

$$\gamma_x = \frac{\gamma_{hog,1} - \gamma_{hog,2} + 6}{12}. \tag{8}$$

δ is the deflection in the point x of the maximal value of the span moment M_{sag}.
$\delta \approx \delta_{max}$. Maximal error is equal to 1,41%.

$$\delta = \gamma_\delta \cdot \frac{5 \cdot q \cdot l^4}{384 \cdot EI_b}, \tag{9}$$

where

$$\gamma_\delta = 3,2 \cdot \gamma_x \cdot \left(\gamma_x^3 - \frac{\gamma_{hog,1} + \gamma_{hog,2} + 6}{3} \cdot \gamma_x^2 + \gamma_{hog,1} \cdot \gamma_x + \frac{3 - 2 \cdot \gamma_{hog,1} - \gamma_{hog,2}}{3}\right). \tag{10}$$

Design tables include the values of $\gamma_{hog,1}$, $\gamma_{hog,2}$, γ_{sag}, γ_x and γ_δ depending on the factors $\gamma_{j,1}$ and $\gamma_{j,2}$. Each table has the dimension 61×61 and the factors $\gamma_{j,1}$ and $\gamma_{j,2}$ are varying from 0,0001 to 1000.

3D diagrams of $\gamma_{hog,1}$ and γ_δ values depending on the factors $\gamma_{j,1}$ and $\gamma_{j,2}$ and basing on the design tables are presented, as examples, in Figures 5 and 6.

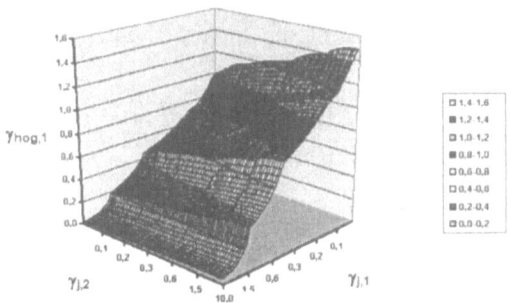

Figure 5. 3D diagram of $\gamma_{hog,1}$ values depending on the factors $\gamma_{j,1}$ and $\gamma_{j,2}$.

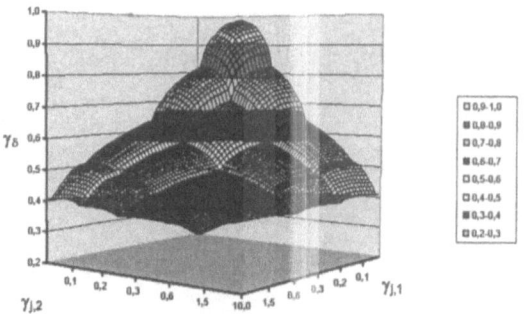

Figure 6. 3D diagram of γ_δ values depending on the factors $\gamma_{j,1}$ and $\gamma_{j,2}$.

4.2. DIFFERENT BEAM STIFFNESSES $EI_{b.hog}$ AND $EI_{b.sag}$ AND EQUAL JOINT STIFFNESS $S=S_1=S_2$

Figure 7 shows the principal moment and deflection distribution in the beam.

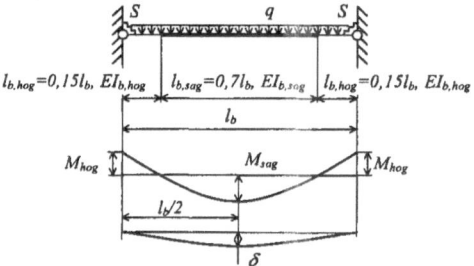

Figure 7. Principal moment and deflection distribution in the beam
with different beam stiffnesses and equal joint stiffness

Definitions:

$$l_{b,hog} = 0{,}15 \cdot l_b, \; l_{b,sag} = 0{,}7 \cdot l_b, \; \gamma_j = \frac{EI_{b,hog}}{l_b \cdot S}.$$

Denote $\gamma_b = \dfrac{EI_{b,sag}}{EI_{b,hog}} \cdot \dfrac{l_{b,hog}}{l_{b,sag}} = \dfrac{3}{14} \cdot \dfrac{EI_{b,sag}}{EI_{b,hog}}$.

The formulae to define the elastic internal bending moment distribution are the following:

$$M_{hog} = -\gamma_{hog} \cdot \frac{q \cdot l_b^2}{12}, \tag{11}$$

where

$$\gamma_{hog} = \frac{0{,}81 \cdot \gamma_b + 1{,}255}{13{,}333 \cdot \gamma_j \cdot \gamma_b + 2 \cdot \gamma_b + 1}. \tag{12}$$

M_{sag} is the maximal value in the middle of the beam span.

$$M_{sag} = \frac{q \cdot l_b^2}{8} - \gamma_{hog} \cdot \frac{q \cdot l_b^2}{12} = \gamma_{sag} \cdot \frac{q \cdot l_b^2}{8}, \quad (13)$$

where

$$\gamma_{sag} = 1 - \frac{2}{3} \cdot \gamma_{hog}. \quad (14)$$

The best elastic internal bending moment distribution, when

$$-M_{hog} = M_{sag} = \frac{q \cdot l_b^2}{16}. \quad (15)$$

$$\text{Or } \gamma_{hog} \cdot \frac{q \cdot l_b^2}{12} = \frac{q \cdot l_b^2}{16}, \text{ or } \gamma_{hog} = 0,75. \quad (16)$$

From formula (17): if $\gamma_b = \frac{3}{14}\left(or \frac{EI_{b,sag}}{EI_{b,hog}} = 1\right)$ then $\gamma_{j,best} = 0,167$, if

$\gamma_b = \frac{3}{7}\left(or \frac{EI_{b,sag}}{EI_{b,hog}} = 2\right)$ then $\gamma_{j,best} = 0,048$, if $\gamma_b = \frac{9}{14}\left(or \frac{EI_{b,sag}}{EI_{b,hog}} = 3\right)$ then

$\gamma_{j,best} = 0,010$, for $\gamma_b = \frac{6}{7}\left(or \frac{EI_{b,sag}}{EI_{b,hog}} = 4\right)$ the best elastic internal bending moment

distribution is not possible.

δ is the deflection in the middle of the beam span. $\delta = \delta_{max}$.

$$\delta = \gamma_\delta \cdot \frac{5 \cdot q \cdot l_b^4}{384 \cdot EI_{b,sag}}, \quad (17)$$

where

$$\gamma_\delta = \frac{7}{373} \cdot \left[51,5175 - 39 \cdot \gamma_{hog} + \left(9,585 - 18 \cdot \gamma_{hog}\right) \cdot \gamma_b\right]. \quad (18)$$

Some values of γ_{hog}, γ_{sag} and γ_δ depending on the factor γ_j for the factor

$\gamma_b = \frac{3}{7}\left(or \frac{EI_{b,sag}}{EI_{b,hog}} = 2\right)$ are presented in Table 3.

TABLE 3. Values of γ_{hog}, γ_{sag} and γ_δ depending on the factor γ_j for the factor $\gamma_b = \frac{3}{7}\left(or \frac{EI_{b,sag}}{EI_{b,hog}} = 2\right)$

γ_j	10	2	1	0,75	0,5	0,4	0,3	0,25	0,2	0,15	0,1	0,08	0,048 Best!	0,03	0,01	0,001
γ_{hog}	0,027	0,121	0,212	0,261	0,340	0,387	0,449	0,488	0,534	0,590	0,660	0,692	0,752	0,790	0,837	0,860
γ_{sag}	0,982	0,920	0,859	0,826	0,773	0,742	0,701	0,675	0,644	0,606	0,560	0,538	0,499	0,473	0,442	0,427
γ_δ	1,015	0,933	0,854	0,811	0,742	0,701	0,647	0,613	0,573	0,524	0463	0,435	0,383	0,350	0,309	0,288

86

The diagrams of γ_{hog}, γ_{sag} and γ_δ values depending on the factor γ_j for the

factor $\gamma_b = \dfrac{3}{7}\left(or\ \dfrac{EI_{b,sag}}{EI_{b,hog}} = 2\right)$ are presented in Figure 8.

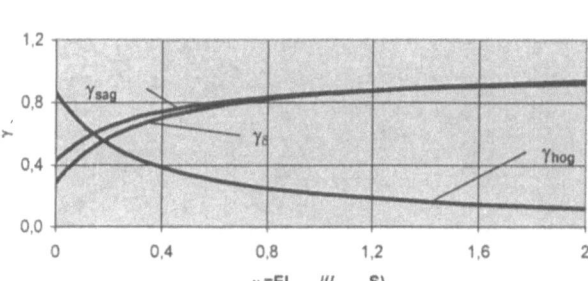

Figure 8. Diagrams of γ_{hog}, γ_{sag} and γ_δ depending on the factor γ_j

for the factor $\gamma_b = \dfrac{3}{7}\left(or\ \dfrac{EI_{b,sag}}{EI_{b,hog}} = 2\right)$

5. Conclusion

Taking the stiffness and strength values of the joints it is easy to get the moment distribution within the beams supported by these joints. It makes the use of semi-rigid construction easier and safer thus reducing the possibility of design errors.

6. Acknowledgement

The author thanks the INTAS organization for the financial support of the whole project, Co-ordinator of the project Prof. Tschemmernegg, his colleagues Dipl.-Ing. Dr.techn. G. Huber and Dipl.-Ing. D. Rubin and all team leaders for their important contribution to this research.

7. References

1. Müller, G. (1998) *Das Momentenrotationsverhalten von Verbundknoten mit Verbundstützen aus Rohrprofilen*, Dissertation am Institut für Stahlbau und Holzbau der Universität Innsbruck.
2. Herzog, M. (1998) *Steifigkeitskalibrierung von Komponenten für Knoten in Mischbauweisen mit Hohlprofilstützen für ENV-1994-1-1, Annex J*, Diplomarbeit am Institut für Stahlbau und Holzbau der Universität Innsbruck.

WIND EFFECTS ON LARGE ANTENNAS AND TELECOMMUNICATION TOWERS: ANALYSIS AND DESIGN OF MAIN COMPONENTS AND JOINTS

C. BORRI, M. BETTI & P. BIAGINI
Dept. of Civil Engineering, University of Florence
3, S.Marta, 50139 Firenze, Italy

1. General

The paper introduces the structural concept for the design and verification of the 175m tall, self-supporting steel truss tower for MW-broadcasting of RAI (Radiotelevisione Italiana). The very slender tower, built in flat country region in the south of Rome, is one of the first realised examples of a new broadcasting technology, which makes it necessary to set up a double bundle of emitting cables (non-structural ones), both inside and outside the tower shaft. Although not having any structural function, the actual dynamic behaviour of the tower may be sensitively influenced by the cables, giving a determinant contribution to the structural damping. Static, stability and dynamic analysis have been performed for the correct prediction of the structural response, for its dimensioning and design; for T.D. analyses, the actual wind loads have been considered by means of generating artificial fields of cross-correlated wind velocity processes over the whole structure.

2. Introduction

The new RAI MW tower at S. Palomba (Rome) is the tallest trussed self-supporting steel tower in Italy with a total height of 175 m, with a multi-pyramidal-trunk shape, and a square ground dimension of 25.8 m (Fig. 2.1). The tower is strengthened by means of stiffening platforms at different heights; it rises at an elevation of 131m a.s.l., on a flat country ground without any important obstacle in the surroundings.

The carrying structure is made of a trussed framework, whose elements, mainly based on both, single L and multi-L-profiles, are joined by bolted connections. The elements are realised by means of following cross-section arrangements (s. Fig. 2.2) XL-arrangement (for the main corner legs up to the height of 148 m), DL-arrangement (main diagonal elements, single L, for the upper corner legs up to 82.5 m).

All structural elements are in galvanised steel (type Fe510C) and painted after the mounting.

C.C. Baniotopoulos and F. Wald (eds.), The Paramount Role of Joints into the Reliable Response of Structures, 87–102.
© 2000 *Kluwer Academic Publishers*.

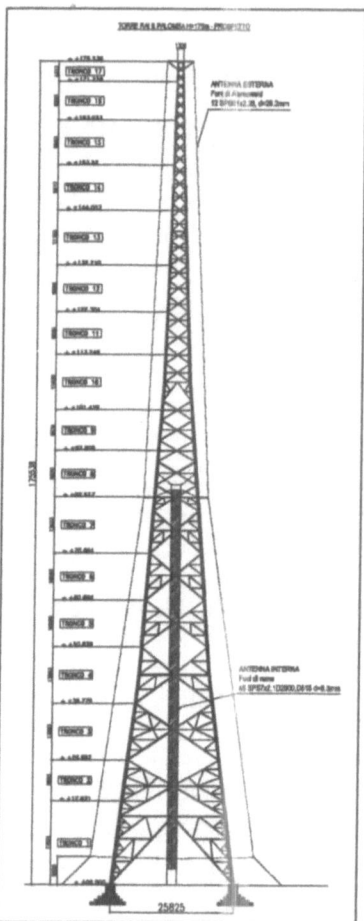

Figure 2.1: Side view of the tower

Beyond the dimensional characteristics described before, the main peculiarity of the structure concerns the system of transmission. This has been carried out by means a double bundle of cables: the first one is external to the tower and the second one inside (internal). They are connected at the base to a tuning cabin. The external bundle has been made by 12 copper wires with an *alumoweld* cover, for a global diameter of 20.2 mm, which is connected to the structure in correspondence to the summit crown and to the horizontal platform at the height of 82.5 m. It's anchored at the foundation with an appropriate concrete foundation.

The inside bundle is composed by 48 copper wires with a diameter of 6.3 mm: they originate from the base of the tower and reach the height of 82.5 m.

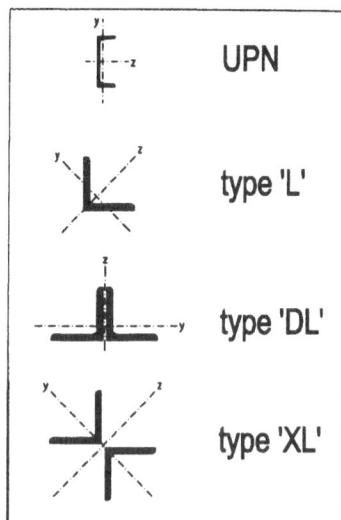

Figure 2.2: Different arrangements used for the element's cross-section

The effects of these elements have been examined in the following static analyses (Sect. 2.1), in terms of static loads (both wind & ice coating are added to the dead weight). In addition, the actual dynamic effect has been fully considered in the T.D. analyses (s. Sect. 2.2), showing a beneficial effect the structural behaviour of the tower.

3.   Performed Analyses

3.1. STATIC AND STABILITY ANALYSIS

The RAI tower at S. Palomba is a very innovative example of tall trussed tower, since it is one of most slender tower ever been erected in Italy, with a slenderness ratio of 6.8. The structural analysis has been carried out by modelling the structure through a FEM model: the complete scheme required as much as 1,110 structural joints, for a total of 6,660 degree of freedom, using 2,045 beam elements (for the corner legs and other passing main connections) and 912 truss elements.

The static analysis has been developed in two phases: first, stresses, deformations and overall displacements have been assessed by performing a series of linear elastic analysis. In the second phase, a large displacement nonlinear static analysis has been carried out, from which we obtained a load-displacement path. The aim of this nonlinear analysis is to verify the outcomes from the previous linear analysis.

The static analysis, linear and nonlinear, of the structure has been carried out using the loads prescribed by the Italian Rules [1] and RAI specifications. Two incident incoming wind directions have been taken into consideration as much as peculiar attention has been devoted to the ice formation.

The first direction (DIR.1 +45°) is wind acting in the diagonal direction of the section of the tower. The second (DIR.2 +90°) is wind acting in the normal direction of the section of the tower.

The Italian Rules foresee the determination of the pressure of the wind, according to the following relation:

$$q = q_{ref} \cdot c_e \cdot c_p \cdot c_d \qquad (1)$$

where c_e, c_p e c_d are, respectively, the coefficient of exposure (variable with the altitude), the coefficient of pressure and the dynamic coefficient, while q_{ref} is the reference kinetic pressure given by the following expression:

$$q_{ref} = \frac{v_{ref}^2}{1,6} \qquad (2)$$

in which v_{ref} is the maximum speed reference of the wind. It is measured at 10 meters of altitude on a ground of II° category: it is a value averaged on 10 minutes, valued with reference to a recurrence period of 50 years. The coefficient of exposure c_e has been given from the following expression:

$$c_e = k_r^2 \cdot c_t \cdot \ln\left(\frac{z}{z_0}\right) \cdot \left[7 + c_t \cdot \ln\left(\frac{z}{z_0}\right)\right] \qquad z \geq z_{min} \qquad (3)$$

$$c_e = c_e(z_{min}) \qquad z < z_{min} \qquad (4)$$

where k_r, the parameter of exposition, c_t, the coefficient of topography and z_0 the roughness have been called respectively 0.2, 1 and 0.1. For the coefficient of pressure c_p the Italian Rule has a specific reference to the truss towers with square or rectangular section , for which the value of this coefficient is :

$$c_p = \begin{cases} 2,4 \text{ for towers with tubular elements with round section} \\ 2,8 \text{ for towers with elements without round section} \end{cases}$$

The last coefficient, the dynamic coefficient c_d, has been considered as 1,2.
Finally the loads condition that we have investigated are the following:

1) Italian Rules (D.M.16-1-1996), normal direction
2) Italian Rules (D.M.16-1-1996), diagonal direction
3) 70 km/h without ice, normal direction (D.M.14-2-1992)
4) 70 km/h without ice, diagonal direction (D.M.14-2-1992)
5) 160 km/h with ice, normal direction (D.M.14-2-1992)

6) 160 km/h with ice, diagonal direction (D.M.14-2-1992)
7) 200 km/h without ice, normal direction (D.M.14-2-1992)
8) 200 km/h without ice, diagonal direction (D.M.14-2-1992)

Fig. 3.1 shows the deformed for the condition of load with wind acting in normal direction. This deformed is, obviously, similar for all loads condition (diagonal wind or normal wind) due to the structural behaviour of the tower as a cantilever beam.

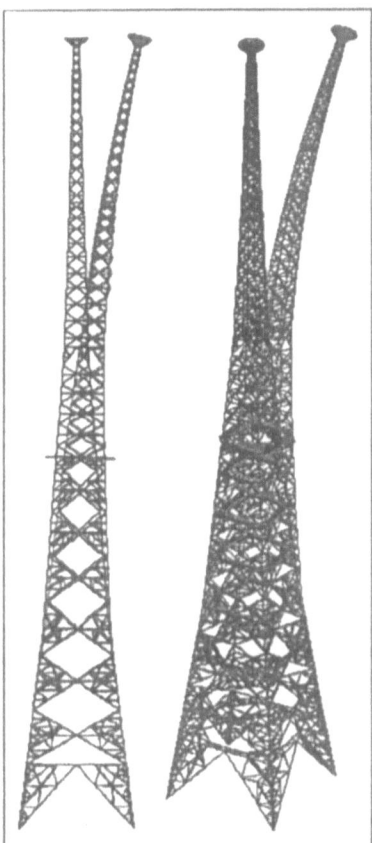

Figure 3.1: Linear static analysis: deformed shape

Fig. 3.2 shows the horizontal displacement, function of the height, of the geometric axis of the tower under the normal wind load. This displacement has been calculated making the averages of the displacements measured in correspondence of four corner legs of each height for various wind speeds.

The greatest deformations originate from the load configuration due to wind speed of 160 km/h plus ice coating. The heavy condition appears to be wind at 200 km/h but this does not produce the maximum displacements since the presence of the external cables of the antenna modifies the wind load. The presence of 12 mm ice cover on the external cables duplicates the surface of exposure to the wind of the cables so that the maximum displacements originate from the load condition at 160 km/h with ice.

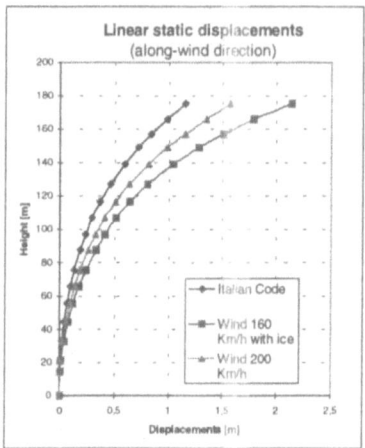

Figure 3.2: Linear static analysis: displacements

The RAI's prescribed load conditions result definitely more onerous than those due to the Italian Rules. While the maximum top tower displacement, relative to the load condition of Italian Rules, amounts to 1.15 m, the maximum top tower displacements due to the load conditions with wind at 200 km/h (without ice) and 160 km/h (with ice) reach respectively 1.569 and 2.142 m.

By means of the geometrically nonlinear analysis (large displacements), departing from the undeformed configuration and proceeding with steps of equal load of 10% of the total load, we have reconstructed the load-displacement path. The algorithm used in the solution is the classical one of Newton-Raphson with update of the stiffness tangent matrix.

Fig.s 3.3 and 3.4 show load-displacement path for wind speed at 160 km/h and wind according to the D.M. 1996. Both diagrams have been built with reference to joint 1 to altitude 175 meters (Fig. 3.3 shows the horizontal displacements, Fig. 3.4 shows the vertical displacements).

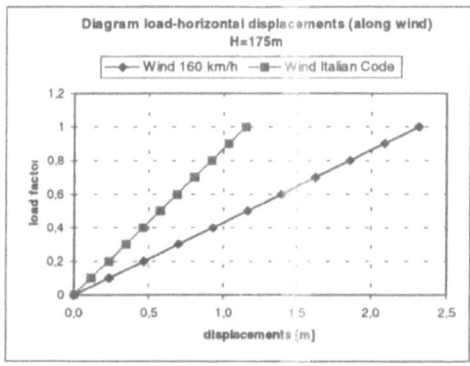

Figure 3.3: Non linear static analysis: horizontal displacements

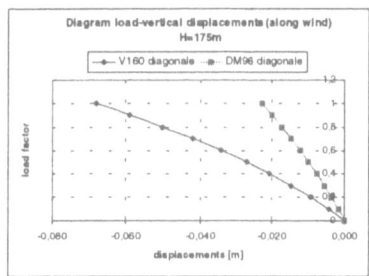

Figure 3.4: Non linear static analysis: vertical displacements

The result of nonlinear analyses showed that the structural behaviour of the tower is linear until the design loads with non relevant variations both for the displacement and for the stress.

a global stability analysis has been performed to assess the safety degree of the structure towards the instability phenomena, looking for a certain number of critical shapes up to the whole instability of one or more corner legs. The classical linear stability analysis is based on the imposition of the following condition:

$$\det (K_e + \lambda K_g)=0 \qquad (5)$$

where K_e is the stiffness tangent matrix of the structure, while K_g is the geometric stiffness matrix as a function of the stress state. From the above linear eigenvalue problem (5), the critical load external factors and the corresponding critical shapes are achieved.

The stability analysis has pointed out some initial critical shapes related to the beam of the platform at 44.5 m of altitude (critical local shapes of ovalisation of the stiffening platform) and subsequently the critical shape relative to the instability of the main corner legs. Tab. 3.1 resume the critical parameters found for the configurations of load with wind from D.M. 16-1-1996 and wind at 160 km/h with ice. These critical parameters correspond to the instability of the compressed corner leg. Both load configurations are referred to diagonal wind. In this situation the structure presents one of the corner legs strongly compressed.

TABLE 3.1: Critical shapes and critical load factors in stability analysis

N° critical shape	Load factor	
	Wind 160 km/h + ice	Wind D.M. 1996
1	4,53	6,38
2	4,73	6,65
3	5,43	7,73
4	5,53	=
5	5,72	=

94

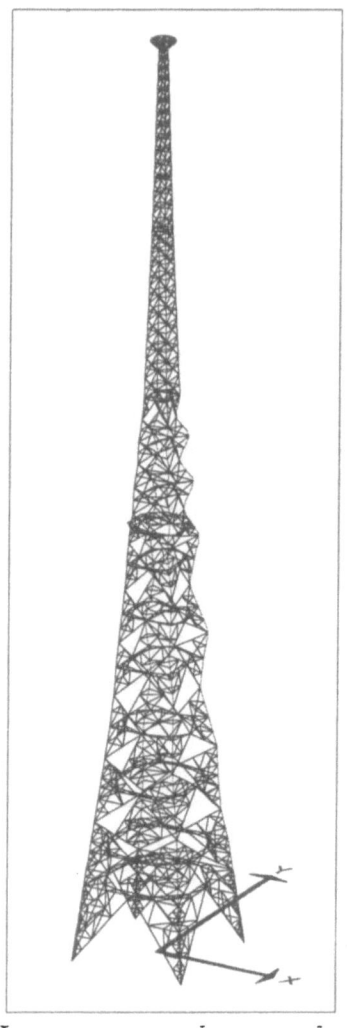

Fig. 3.5 represents the critical shape associated to the instability of the compressed legs (with load multiplier 5,72). It represents the most important critical shape since it is concerning the instability of the main structural elements.

3.2 DYNAMIC ANALYSIS

Considering the importance and the peculiarity of examined structure we also studied the dynamical behaviour carrying out a dynamic analysis in to the time domain (TD)

In order to perform these analyses, it's necessary to know the dynamic properties of the structure. The assessment of the main frequencies has been done using the same model and the same package adopte in the static analysis. Fig. 3.6 shows the first three dynamic fundamental shapes, while Tab. 3.2 reports the related frequencies.

TABLE 3.2: Eigenfrequencies and eigenmodes of oscillation

Nr.	Whitout ice		With ice coating		Type
	Period [s]	Freq. [Hz]	Period [s]	Freq. [Hz]	
1	1.7	0.57	2	0.48	1°
2	1.7	0.57	2	0.48	1°
3	0.6	1.50	0	1.33	2°
4	0.6	1.50	0	1.33	2°
5	0.4	2.37	0	2.13	

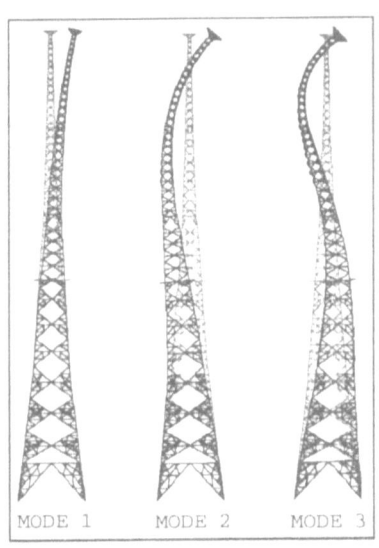

MODE 1   MODE 2   MODE 3

The T.D. analysis results quite cumbersome from the computational point of view, it is also necessary the definition of some parameters to characterise the structure and to make reliable the resolution algorithm. It has been used the Newmark algorithm which requires the definition of two parameters; δ end α. It has been used for these parameters, in accordance with Newmark, the value that guarantees the unconditional stability of the algorithm (δ=1/2 and α=1/4). The structural damping has been estimated according to the Rayleigh method. The damping matrix [D] of the whole structure is calculated in the following way:

$$D = \alpha \cdot M + \beta \cdot K \tag{6}$$

96

where α and β are two constants estimated according to the value of the structural damping of the first two modal shape. In our case they are assumed in reference to the first and third frequencies with the following values: $\xi_1 = 1\%$ and $\xi_3 = 3\%$.

The global time length of the analysis has been fixed in 300 seconds, with a temporal scanning of 0.04 seconds.

The load has applied the structure by nodal forces, to various altitudes (17 levels), in correspondence of the four corners of the tower's section. In this way, thanks to the presence of the platforms and of the strong horizontal beams, the forces apply efficiently on the whole structure. The applied load has assessed in the following way:

$$F(t) = \frac{1}{2} \cdot \rho \cdot A \cdot \left[\overline{U} + u(t)\right] \cdot C_p \qquad (7)$$

where A is the frontal surface invested from the wind and C_P is the pressure coefficient of the Italian Rules. In this way the outcomes of the static analysis and dynamic analysis could be compared. The load time histories have been generated by means of a numerical model developed at the Department of Civil Engineering of Florence. This uses the method of the auto-regressive filters. This technique allows us to arbitrarily select the target spectrum, which characterises the generated process. A very important characteristic of the model is the cross-correlation structure by means of a correlation function specifically chosen. On the whole 68 time histories of wind speed (4/each of the 17 levels) have been generated with a length of 300 sec and time interval of 0.04 sec. For example, in Fig.s 3.7 and 3.8 we report, respectively, the time-speed and the correspondent spectrum for the corner node at 106.900 m.

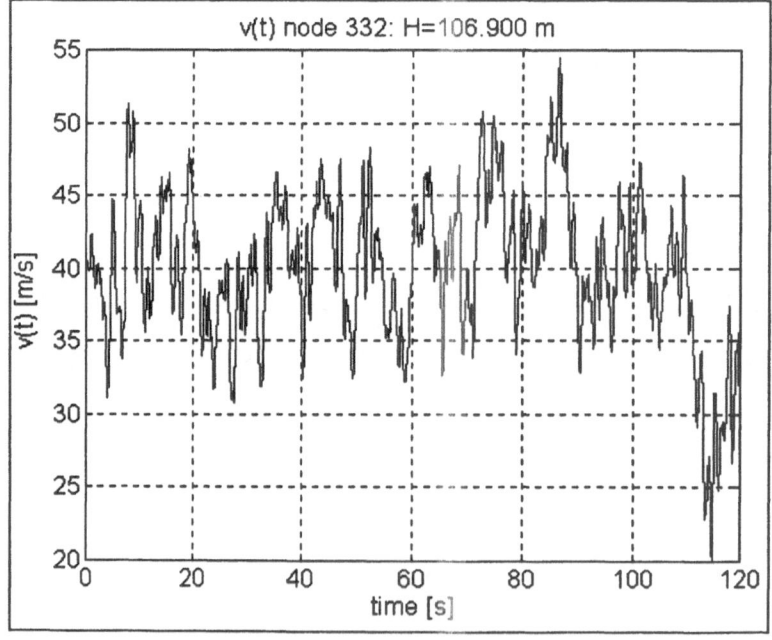

Figure 3.7: Diagram time-velocity

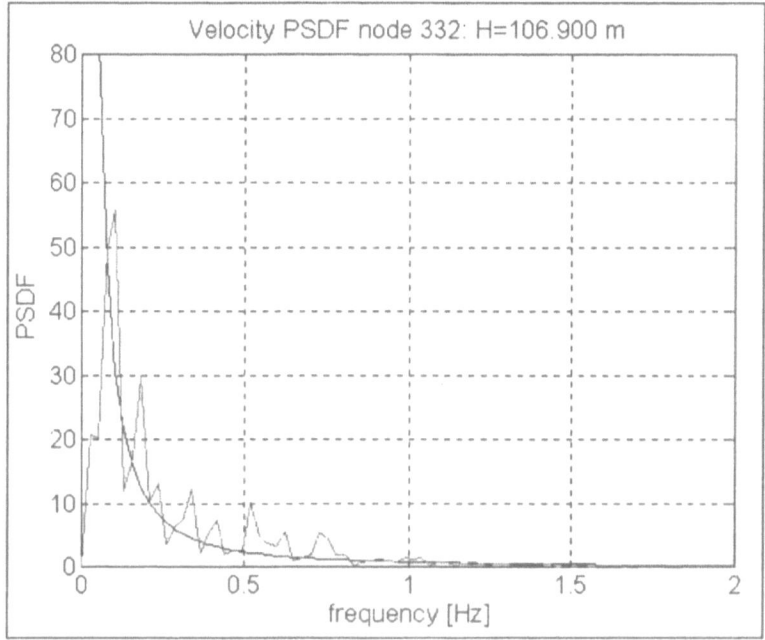

Figure 3.8: Velocity PSDF

The dynamic analysis in the time domain has been also executed on two different models of the tower: The first has been realised without the external cables. The second one with the external cables modelled by truss element with initial pre-tension. The target is to investigate the influence of this cable to the structural behaviour of the tower.

Fig. 3.9 shows of the time response of the structure of a node of the tower crown.

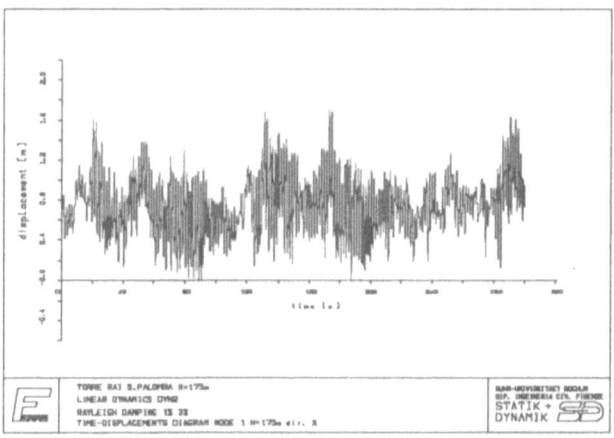

Figure 3.9: Time-history of along-wind displacements at the top

Fig. 3.10 shows the response spectrum calculated for the time-history to Fig. 3.9.

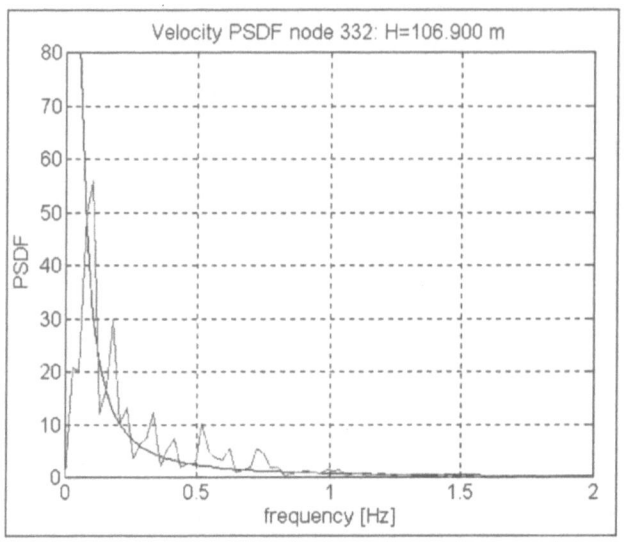

Figure 3.10: Time-Domain analysis: response spectrum

The evident two peaks of Fig. 3.10 are related to the first two frequencies of the structure. With R1 and R2 it has been indicated the areas under the curves (these represent the contributions of the resonance responses to the general response of the system). The B area is a background response i.e. the contribution of the response almost static to the total response. Since the area under the spectrum line represents the variance of the process, we have the following relationship:

$$\sigma_x{}^2 = B + R_1 + R_2$$

The T.D. analysis of the response has furnished the following values for the present time-process:

$$\sigma_x{}^2 = 0.0381;$$
$$B = 0.01418;$$
$$R_1 = 0.0239;$$
$$R_2 = 1.55E\text{-}05;$$

Therefore the contribution of the various frequencies to the answer of the system is the following:

$$\frac{B}{\sigma_x^2} = 37.2\% \,, \frac{R_1}{\sigma_x^2} = 62.7\% \,, \frac{R_2}{\sigma_x^2} = 0.1\%$$

Therefore, these results confirmed that the structural behaviour of the tower is essentially due to the flexural behaviour as a cantilever where the first frequency dominates the structural response.

4. Additional Verification and Structural Connection

Particular attention was paid to the realisation and to the criterions of verification of the connections between the main structural elements (main corner legs and main diagonal elements).

Figure 4.1: Central joint of the bracing trusses

During the design process it was very important to realise out-of-plane bracing trusses (s. Fig. 4.1), in order to assure global stability of the tower and to reduce the slenderness ratio of the most stressed diagonals. The out-of-plane trusses reach an height of 48.5 m from ground level.

Figure 4.2: Bolted joint of corner beam

Many structural connections were analysed, in order to guarantee safety, reliability and durability.

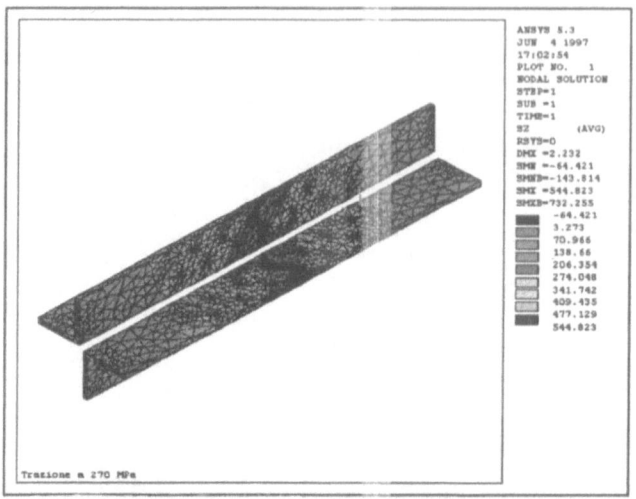

Figure 4.3: Numerical model of the corner beam bolted joint: distribution of tensions

The cross-section-dimensioning criterion follows the Italian Rules (D.M.9/1/1996 and the code of practice for the application of the D.M.14/2/1992 that used Stress Available Method).

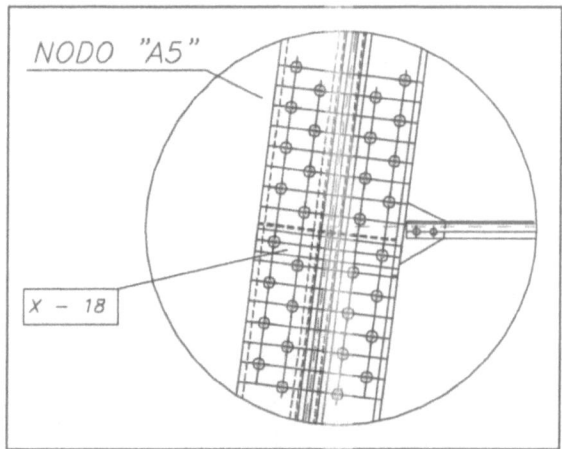

Figure 4.4: Node A5: joint between two profiles XL 250x28

Connections between main corner legs (s. Fig. 4.2) are investigated with the Italian Rules. These connections are designed under the condition that they are able to transfer the maximum action (i.e. function of the cross-section area of the corner legs, with the yielding tension in the profiled XL250x28 is 270 Mpa).

A numerical model (s. Fig. 4.3) for this connections it has also been performed. By means of this numerical model the stress and strain behaviour for the corner beam and the plate between the XL-arrangement has been investigated.

The stress values produced by the external loads are arising in a range of values for which the Fe510 C steel has a linear elastic behaviour then the elastic linear model adopted in the numerical analysis is suitable for the description of the stress state. Moreover the stress between profile and bolts reaches high value but always within the given limits.

Particular attention was also paid to truss checking of horizontal platform which don't have shown any problems for the deflection in the horizontal plane under the live load. It has also verified the hypothesis of rigid platform assumed in the calculation model. It has been found that the stiffening of each single horizontal platform is suitable for a correct redistribution of the actions of the wind. The relative displacements to the median point of the side compared to those of the corner node are less than 3 mm.

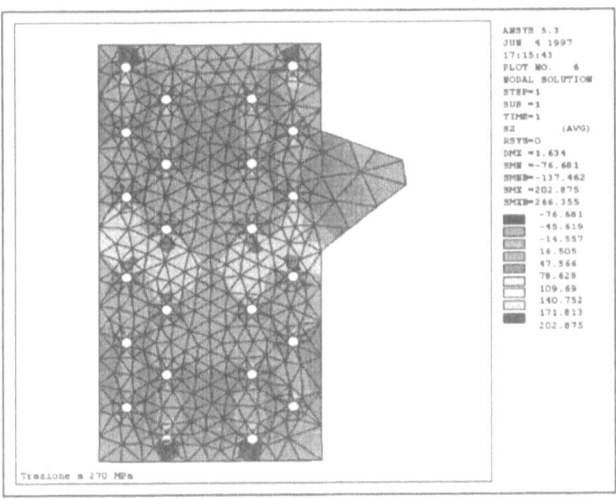

Figure 4.5: Numerical model of the internal joint plate: stress distribution

5. Conclusions

Line-like slender systems (trussed towers) require particular attention to the study of the structural answer. Specifically attention must be pointed on an accurate comparison between the available analysis methods. In this specific case the dynamic analysis has confirmed the outcome of the static analysis. Moreover the global stability analysis showed how the structural reliability is suitably assured with regard to the live loads.

The dynamic interaction effect between trussed tower and cable due to the oscillation of the external cables has been fully considered and it showed a beneficial effect on the structural damping.

Finally, specific attention has been focused in joint design of the main elements. The connection of the corner legs has been designed to completely guarantee the continuity of the cross-section and a linear elastic analysis has been carried out in order to have a local check of the corner legs stress behaviour. The analysis showed a substantial agreement with the normative checks and the numerical results.

References

1. Augusti G., Borri C. & Gusella V. (1990) Simulation of wind loading and response of geometrically nonlinear structures, *Int. Journal. on Structural Safety.* **8**, 161-179.
2. Augusti G. & Borri C. (1991) On the response of flexible structures to wind loading: models and simulations, in W.B. Krätzig et al., *Structural Dynamics*, A. Balkema Publishers, Rotterdam, **2**, 1107-1114.
3. Borri C. & Zahlten W. (1998) Time-domain simulation of the nonlinear response of cooling tower shells subjected to stochastic wind loading, *Int. Journal on Engineering Structures*, **20**, 881-889.
4. Borri C., Crocchini F. & Mirto F. (1998) La torre RAI O.M.-175m di S.ta Palomba, Atti del 5. Convegno Naz. di Ingegneria del Vento, IN-VENTO-98, Perugia, 81-93.
5. Borri C. & Facchini L. (1999) Some recent developments in modeling turbulent wind loads and dynamic response of large structures, in *Structural Dynamics*, Fryba & Naprstek Eds, Balkema Publishers, **1**, 3-12.

II. EXPERIMENTAL STUDIES OF JOINTS AND FRAMES

APPLICATION OF DEFORMATION CRITERIA IN BOLTED CONNECTION STRENGTH DESIGN

V. KALENOV
Research and Design Institute Promstalkonstruktsiya
13, Sadovaya Samotechnaya, 103473, Moscow, Russia

1. Introduction

The Limit States Design of Steel Structures has been introduced in Russia since 1955. The main ideas of Eurocode 3 or Load and Resistance Factor Design in USA are similar to those ones of Limit State Design. This method is being developed in Russia constantly. At present it involves three basic concepts, namely: ultimate capacity, total serviceability and serviceability for normal operation.

Ultimate capacity (Limit State of group 1A) implies an attainment of maximum by the load-deformation curve, i.e. in practice the loss of ultimate capacity means a failure of the structure or of its part. The total serviceability loss (Limit State of group IB) means impossibility of structure further service because considerable deformations or displacements has developed though the structure hasn't destroyed. The loss of the serviceability for normal operation is referred to the Limit State of group II. The normal operation presents a stationary process governed by mean statistical loads according to the operation conditions. When group II reaches its limit state the structure conserves its load carrying capacity and its operation may continue at greater deformations and loads. However, normal operation may become difficult. The statistical ensuring level of the group IA limit state must be maximum one, group IB – smaller one, group II – minimum one.

The limit states criteria are force ones for group IA, deformation or force ones for group IB and usually deformation and rarely force ones for group II. The force criteria reflect well-defined qualitative phenomena: failure, initiation of yielding, emergence of cracks, etc. The deformation criteria are set as limit of only quantitative variation of deformations, displacements and crack displacements. The deformation criteria are less clear and less defined than the force ones. However, the well-founded substitution of the force criterion by higher deformation one may be quite justified and profitable. In recent years the deformation criteria has found a wide use for designing bolted connections, and in particular, for bearing and friction-bearing type connections. It's more correct, because the deformations are limit of serviceability and its loss occurs before the loss carrying capacity. The practical realization of this approach became possible after carrying out the wide complex of the experimental and theoretical researches.

The condition of Limit State by the deformation criterion for bearing and friction-bearing connections could be written as

C.C. Baniotopoulos and F. Wald (eds.), The Paramount Role of Joints into the Reliable Response of Structures, 105–114.
© 2000 *Kluwer Academic Publishers.*

$$\Delta(N) \le \Delta_{ult}, \qquad (1)$$

where

Δ_{ult} — ultimate permissible shear displacements for structure or its elements in dependence on service conditions;

Δ — shear displacements of bolted connections due to external load — N.

Analysis of the well-known researches shows that the bolted connection behavior depends on such factors:

$$C(\Delta, N) = f(C_1, C_2, C_3, C_4), \qquad (2)$$

where

C_1 — characteristics of friction effects (slip coefficient; initial bolt tension, type of steel and so on);

C_2 — technological characteristics of manufacture and assemblage of connections (the methods of the edges, holes formation of the connected elements, surface preparation and so on);

C_3 — geometrical and mechanical characteristics of bolts and connected elements;

C_4 — characteristics affecting on the uneven distribution of the external load between the connection fasteners.

The first two groups of factors (C_1, C_2) have been studied well and existing design methods adequately reflect their influence on the actual connection behavior.

The well-known design methods of bearing and friction-bearing types of connections in most cases don't take into account criterion of deformation, which determines correspondence of the chosen type of connection to real behavior of the structure and values of the connection load capacity. The results of the experimental and theoretical researches of single and many-bolted bearing and friction-bearing types of connections are presented below.

2. Deformation of bolts subjected to shear

The object of researches: the bolts M16, M20 and M24, grades 5.8, 8.8, 10.9 and 11.9. The investigations of bolts have been carried out in special devices that imitated single-bolted connections with two-shear planes (Figure 1). In order to expect bearing deformation, all the elements of the device have been made of very high strength steel with tensile strength about 2000 MPa and hardness about 60 HRC. Ten bolts of each diameter and grade have been tested.

Figure2, for example, shows experimental relationship between shear forces to one plane and shear displacements of tested high strength bolts.

By regressive analysis methods of experimental data the empirical relationships (corresponding to low limit of confidence interval with probability 95%) between the shear forces and shear displacement have been obtained. The knowledge of bolts shear deformation allows to take account of it in analysis of tested connection total displacements.

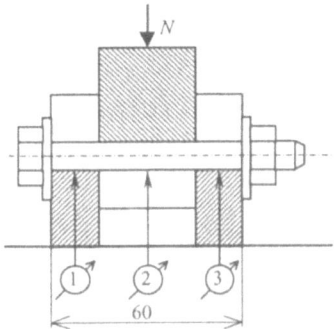

Figure 1. Shear-test scheme of bolts

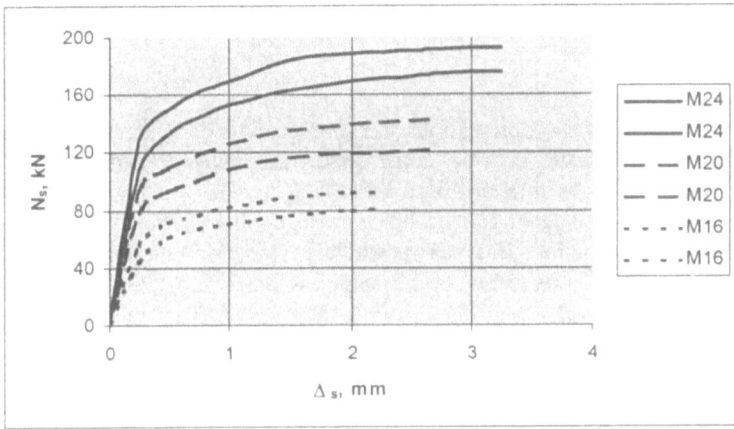

Figure 2. Shear force per one plane and bearing displacements for bolts 8.8.

3. Experimental researches of single-bolted bearing connections

The experimental researches have been carried out on single-bolted connections with 5.8, 8.8, 10.9 and 11.9 bolts. Schemes of tested specimens are given in Figure3. <u>Two-shear connections.</u> The cover plates 25 mm thickness have been made of very high strength steel with tensile strength 2000 MPa and hardness 60 HRC to exclude bearing deformation. Filler plate and tested plate had the same thickness. Six high-strength bolts M24 have fastened device elements. To prevent slip between device elements the bolts have been prestressed (bolt preload is about 270 kN). <u>One-shear connections.</u> The tested elements of these connections have been manufactured of steel of the same thickness and strength. Bolt holes have been drilled for all connections, their diameters exceeded the bolt diameters from 0,5 to 4,0 mm. Every part of tests included series of specimens with specific combinations of thickness (t=4÷20 mm), tensile strength of tested plate steels (σ_u=370÷610 MPa), diameters of bolts (d=16, 20, 24 mm), end and edge distances (e, e'). 592 specimens have been tested.

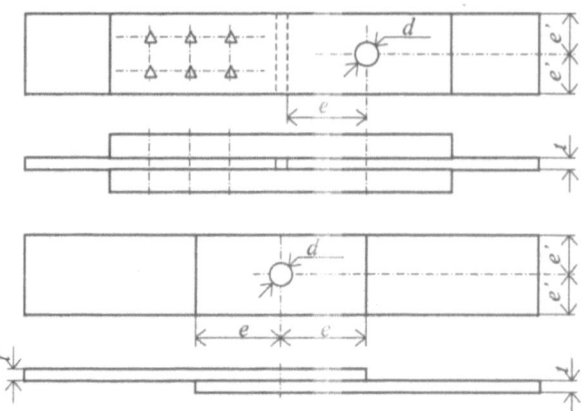

Figure 3. A specimen for bearing connections tests.

Analysis of the test results has shown:
- Typical feature of the connections is rapid growth of bearing deformations (hole elongation), but the elastic bearing deformations range is insignificant. Bearing displacements reach 12 — 14 mm at failure loads (Figure 4).
- Bearing force — N, that can be transferred by the connection at reaching some definite bearing deformations, increase practically linearly with the increase of end distance from $1,5d$ to $(3 \div 3,5)d$. When end distance is more than $(3 \div 3,5)d$ the applied load reaches the maximum value and doesn't change later on (Figure 5). It's well seen that the increase of end distance from $2d$ to $3d$ allows increasing transferred bearing forces $1,25$ — $1,35$ times when other conditions are equal.
- Bearing forces that can be transferred by the connection increase practically directly proportionally with increase bolt diameters, thickness of connected elements and tensile strength of their steel.

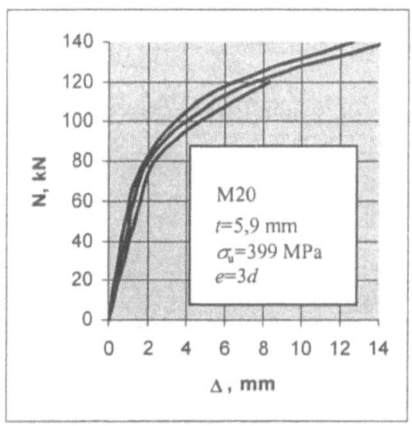
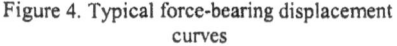
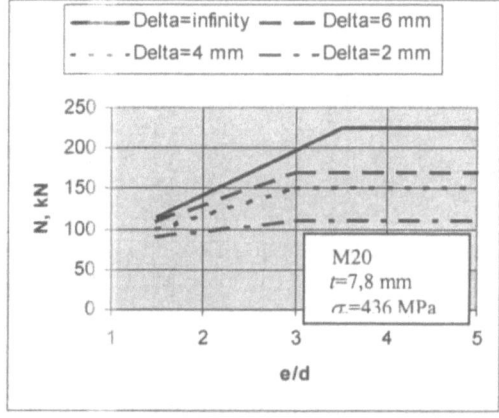

Figure 4. Typical force-bearing displacement curves

Figure 5. Relationships of shear forces and end distances.

By regressive analysis methods of experimental data the empirical formula was obtained:

$$N = k_e dt \sigma_u f(\Delta),\qquad(3)$$

where

N — bearing forces (kN);

k_e — coefficient which takes into account end distance and is equal to: $k_e=1,0$ at $e>3,0d$; $k_e=0,4+0,2e/d$ at $1,5d \le e \le 3,0d$;

$f(\Delta)$ — coefficient which takes into account bearing deformation of each connected elements (mm) and is equal to:

— For bolts grades 5.8 and 8.8

$$f(\Delta) = \begin{cases} 0,147\Delta,\ \text{at } 0 < \Delta \le 0,8\,\text{mm} \\ 0,085 + 0,044\Delta - 0,004\Delta^2\,,\ \text{at } 0,8 < \Delta \le 5,0\,\text{mm} \end{cases},\qquad(4)$$

— For bolts grades 10.9 and 11.9

$$f(\Delta) = \begin{cases} 0,127\Delta,\ \text{at } 0 < \Delta \le 0,8\,\text{mm} \\ 0,067 + 0,045\Delta - 0,0033\Delta^2\,,\ \text{at } 0,8 < \Delta \le 5,0\,\text{mm} \end{cases},\qquad(5)$$

Δ — bearing displacements of each connected elements (mm).

The expressions (4) and (5) are intended for calculation values of coefficient $f(\Delta)$ corresponding to low limit of confidence interval with probability 95,0%.

4. Experimental researches of single-bolted friction-bearing connections

Two types of connections with high-strength bolts M24 were tested: single-shear and two-shear connections. Each of the tested specimens consisted of plates of the same thickness and strength. The thickness of connected plates was varied from 4 to 16 mm. The connected elements were manufactured of carbon and low-alloy steel with tensile strength from 370 to 630 MPa. The difference between hole and bolt diameters has been $(2,0 \div 2,5)$ mm. The width of tested connections was specified 8d in order to decrease the net section longitudinal deformations. The end distance was set equal to $3,5d$. Control of bolt pretension — B_0 and change of bolts tension during tests was carried out by four resistance strain gages attached to the bolt body. 28 specimens have been tested in all.

The analysis of bolt behavior has indicated that at the early stage of displacements of connected elements from 0,15 to 0,3 mm (the initial point of slip) the loss of bolt pretension is about $5 \div 10\%$. During major slip between the connected elements when the hole clearance is taken up and the bolt is in bearing, bolt pretension doesn't change practically. As the applied load is increased the bolt pretension diminishes with the increase of bearing displacements.

By regressive analysis methods of experimental data the empirical formula for calculated of axial bolt tension was obtained:

$$B_i = B_0\, f_t(\Delta) = B_0(0,85 - 0,024\Delta).\qquad(6)$$

The values of bolt tension calculated by this formula correspond to low limit of confidence interval with probability 95%. Assuming that friction coefficient is a constant value, it is possible to separate the total forces transferred by the connection to the forces transferred by friction N_F and bearing N_Δ (of course, if the bearing strength of forces transferred by friction N_F and bearing N_Δ (of course, if the bearing strength of connected elements is the critical parameters with respect to the shear strength bolts of connection):

$$N = N_\Delta + N_F = k_e dt\sigma_u\, \mathrm{f}(\Delta) + \mu n_s B_0 \mathrm{f}_t(\Delta),\tag{7}$$

where

μ-friction coefficient;
n_s-number of friction planes.
Figure 6 illustrates this fact.

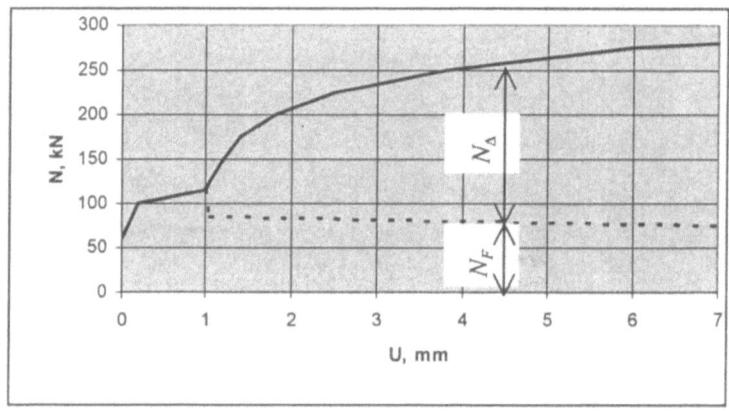

Figure 6. Diagram of single bolted friction-bearing connection behavior.

5. Researches of many-bolted connections

In real many-bolted connections there are the deviations of distances between the centers of the bolt holes from their nominal values because of manufacture imperfections. As a result bolts come into bearing not simultaneously. The question is what must be the necessary values of connection displacements so that all bolts should come into bearing and whether the first bearing bolts have sufficient shear deformation capacity to provide coming into bearing of all remaining bolts. The necessary many-bolted connection displacements were determined by methods of probability theory.

Deviations of distances between hole centers — δ were considered as random values distributed by the normal law. It was assumed that after the major slip equal to the difference between diameters of hole and bolt (d_0-d) one of the bolts of connection came into bearing with the hole surface. It was necessary to determine the probability of that at least l bolts of the connection with confidence $P^*=0,95$ would come into bearing — P_l. This probability can be found from equation:

$$1-(1-P_l)^n - P_l(1-P_l)^{n-1} - \ldots - C_n^l P^{n-l}(1-P_l)^{n-l-1} = 0,95,\tag{8}$$

where

n — number of bolts in connection;
C — combinations of n elements taken l at time.

The formula to determinate displacements necessary the bolts to come into bearing has the following form:

$$\Delta_{bl} = \frac{2}{3} z \delta, \tag{9}$$

where z — is to be found by the value of normal function of distribution from the following expressions:

$$\Phi^*(z) = \frac{1}{2n} \int_{-\infty}^{z} e^{-\frac{t^2}{2}} dt, \ \Phi^*(z) = \frac{P_l + 1}{2}. \tag{10}$$

As a result we have obtained numerical values of shear displacements of bolt bodies and bearing displacements of connected elements which are necessary for every bolt of connection come into bearing. Figure 7 gives an example of three-bolted connection behavior. The displacements necessary for all the bolts to come into bearing are calculated by formulae (8) — (10).

Figure 7 shows that ultimate force which can be transferred by the connection with non-simultaneous bolts bearing — N_{ns} is smaller than the ultimate force with simultaneous bolts bearing — N_s.

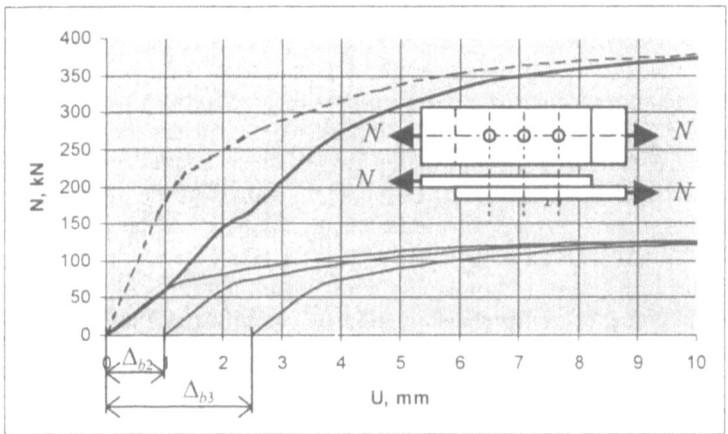

Figure 7. Diagram of three-bolted connection behavior.

It is necessary to introduce additional coefficient in the design formula:

$$\gamma_b = \frac{N_{ns}}{N_s} \text{ or } \gamma_b = \frac{f(\Delta) + f(\Delta - \Delta_{b2}) + f(\Delta - \Delta_{b3}) + \dots + f(\Delta - \Delta_{bn})}{n f(\Delta)}. \tag{11}$$

Numerical analysis has shown that coefficient $\gamma_b = 0,9$ satisfies only in case of connections with ultimate deviations of distance between the hole centers in group $\delta = \pm 0,7$ mm. When the values δ are bigger, coefficient γ_b sharply reduces (at $\delta = \pm 1,5$ mm, $\gamma_b = 0,65$). The results of this problem solution have been confirmed by the tests of two- and six-bolted full size connections.

6. Application of mathematical simulation methods to researches of friction-bearing connections behavior

Experimental researches of many-bolted friction-bearing connections real behavior require to test many full-size specimens to obtain correct evaluation of great number random factors influence. These circumstances determine the necessity to formulate a problem of mathematical simulation of many-bolted connections real behavior. The problem intend for study of many-bolted connections, considering one or two shear planes, variable quantity of bolts in row and line, variable rigidity of connected elements, variable distance between bolts, elastic-plastic stage of connected elements deformation. The connection transfer's longitudinal axial load - N.

The problem is solved by statistical simulation of Monte-Carlo method for many times' statistically indefinite and physically nonlinear system with variable links. Links have preset probabilistic characteristics.

The connection behavior simulation has solved the following theoretical tasks:

1. Design procedure of bolt displacements in friction-bearing connection during major slip.

2. Design procedure of clearances between bolt and wall of hole in the connection, considering real diameters of bolts and holes and deviations of distances between the holes centers in every connected element with account to their orientation and signs.

3. Special algorithm of program passes for boundary task solution of many-bolted connection strength considering random piece-linear diagrams of every bolt to shear and connected elements to bearing. This algorithm determines:

- forces increments in bolts and increments of shear displacements and bending of bolts bodies and bearing of connected elements for arbitrary step of load — ΔN;

- forces increments in parts of connected plates between bolts and increments of length of these intervals for arbitrary step of load — ΔN.

4. Piece-linear diagrams of bolt strength to shear, diagrams of connected elements to bearing and diagrams of bolt pretension and bearing deformations have been suggested after numerous tests.

5. Computer universal program "Bolt" for mathematical simulation of many-bolted connections behavior, transferring axial forces has been worked out.

Initial data for connections are the following: number of solving tasks for statistics collection, number of bolt rows and bolts in every row, distances between the bolt rows, type of connection (one shear or two shear plane), thickness of connected elements, tensile strength and steel yield stress of connected elements, factors of hardening, axial rigidities of plate, external load, nominal bolt and hole diameters, deviations of distances between hole centers in every plate, friction coefficients between surfaces of connected elements and between elements and washers, bolt preload.

The result of program are the evaluation of average value and standard deviation by the first, end and middle bolt rows of connection for the following factors: displacements of connected elements, forces transferring by bolts and bearing of connected elements, forces transferring by means of contact surfaces of elements, tensile forces in bolts. These factors were considered as load functions and were given out for printing with step $N_{max}/12$, where N_{max} is a preset load to the connection.

The example of design of the friction-bearing type connection is presented in Figure 8.

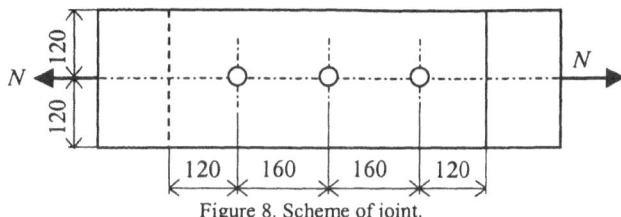

Figure 8. Scheme of joint.

Initial data of connection:
- number of solving problems for statistic collection — 30;
- number of bolt rows — 3;
- tensile strength — 0,48 kN/mm²;
- yield stress — 0,33 kN/mm²;
- thickness of each elements — 10 mm;
- axial rigidity of plates — 500000 kN;
- hardening factor — 0,04;
- maximum load to the connection — 1050 kN;
- number of bolts in a row — 1;
- distance between bolt rows — 160 mm
and the next parameters:

	Nominal	Mean	Standard
1. Bolt diameter (mm)	24,000	23,740	0,090
2. Hole diameter (mm)	26,000	26,650	0,085
3. Deviations of distance between the hole centers (mm)	0,0	0,0	0,350
4. Friction coefficient between plates	0,350	0,420	0,023
5. Friction coefficient between plate and washer	0,100	0,120	0,007
6. Bolt tensile force (kN)	271,04	298,14	9,03

Diagrams of the following relationships for bolt of the row 1 (evaluation of mathematical expectation and symmetrical region two standards wide) are given in Figure 9:
- load (N) — displacements of connected elements (Δ);
- load (N) — forces, transferred by friction of contact surfaces (T).

114

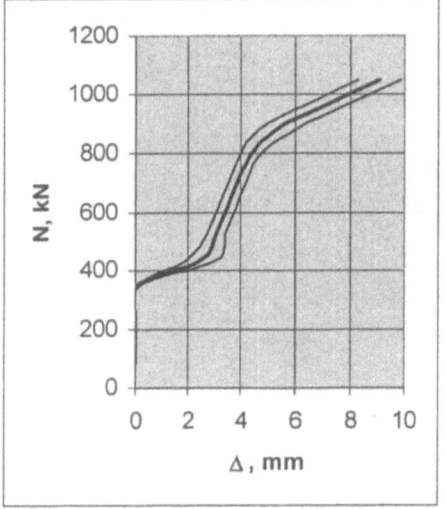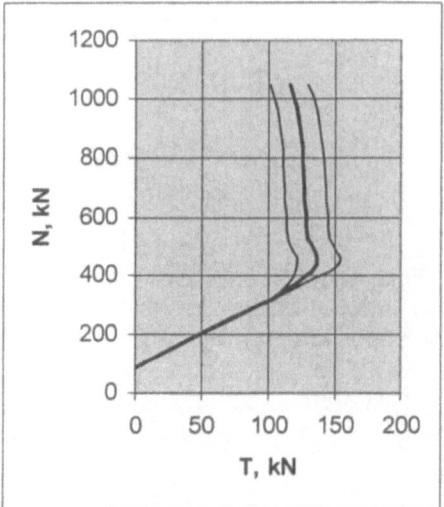

Figure 9. Diagrams of experimental relationships.

7. Conclusion

Researches of bearing and friction-bearing types of connections allowed working out "Design Recommendations of bolted connections subjected to shear based on deformation criterion of limited states". Design of the connections according to the proposed Recommendations provides more complete use of their load capacity and reduction of bolt number 1,5 — 1,8 times in comparison with traditional methods of design.

LARGE - SCALE EXPERIMENTAL TESTS
ON STEEL AND COMPOSITE FRAMES

A. KOZŁOWSKI
Rzeszów University of Technology, Civil Engineering Department
35-959 Rzeszów, Poznańska 2. Poland

1. Introduction

Global analysis of each structure is conducted for assumed model of this structure. Models of structures are nowadays much improved – many influences, previously omitted, are taken into account, as: spatial structure behaviour, influence of floors and walls, imperfections, and lastly real behaviour of frame joints (semi-rigid joints). Global analysis with account of semi-rigid joints is troublesome because of non-linear joint characteristics. For plane building frames, main joints characteristic is moment-rotation relationship (M-ϕ curve). Analytical modelling of these curves is based on experimental tests conducted mainly on isolated joints models, consisted of short column and beam elements connected by joint [1], [2], [3]. For many years, a lot of joint experimental tests were conducted, data bases of joints test results were created (e.g. SERICON, SCDB) and analytical models of joints were developed (linear, multilinear, polynomial, power, exponential etc.). These models, as well as these obtained using "component method" included in Annex J of EC 3 [4], are used in global frame analysis. Only one objective way for verification joints models seems to be experimental test of frames in natural scale [5], [6], [7]. Because of high cost of such tests, they are made very seldom. The paper deals with steel and composite, sway frame test conducted in natural scale on site.

2. Description of the Tested Structure

Tested frame is a part of the new built Laboratory for Building Department of Rzeszów University of Technology. It is two-storey one bay building of dimensions: 8,1 x 18,0 m and height of 7,2 m. Main structure of building consists of steel frames spacing 6,0 m with composite steel-concrete floors (fig. 1). Columns were designed as HEB 200, beams IPE 300 (fig. 2), beam-to-column connections were bolted flush end plate with 4 M 20 10.9 bolts and 15 mm end-plate. Column bases were designed with 2 M 20 anchor bolts. During steel frame tests only steel structure was assembled with bracing bars prevented from frame stability. Tests were repeated after concreting the first floor.

C.C. Baniotopoulos and F. Wald (eds.), The Paramount Role of Joints into the Reliable Response of Structures, 115–128.
© 2000 *Kluwer Academic Publishers.*

Figure 1. View of tested frame

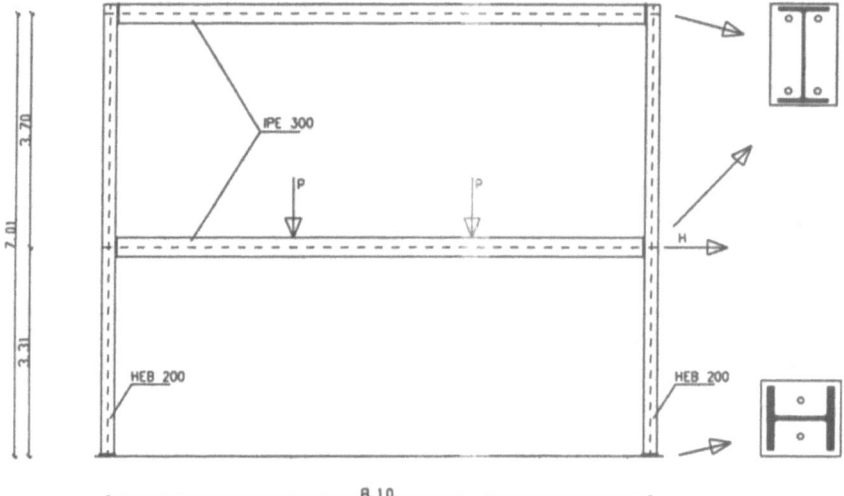

Figure 2. Scheme of steel frame

3. Test Program of the Steel Frame

Tests had to be non-destructive with understanding regard, limited to elastic range. Test loading was calculated on the assumption to not exceed stresses in beam of 2/3 of yield limit and permissible deflection. Vertical loads P were two concentrated forces of maximum values 50 kN spacing at third point along the beam span. These loading were applied by double-nutted bolts connected to φ 30 steel bars anchored to foundation and beam. Horizontal load H of maximum value of 10 kN was applied by hand power winch connected to frame joint and by steel rope to the wall of nearby building (fig. 3).

Figure 3. Equipment for application of horizontal load

Forces were measured by tensometer gauge. Before tests, beams and column dimensions were measured to get real sections properties. Behaviour of end-plate joints may be influenced by range of bolts pretensioning. To investigate this effect, tests were conducted three times, in the following stages:
- I stage; full bolt pretensioning by calibrated wrench to moment 640 Nm,
- II stage; half bolt pretensioning to moment of 320 Nm,
- III stage; hand tightening.

For stage I, the following loading cycles were applied (in brackets codes of loading are presented):
- vertical loading P = 10 kN in order to remove the slackening in the assembled frame and to ascertain proper functioning of strain gauges and transducers,
- release to "0" (P0/H0),
- vertical loading increased step by step to values: P = 10 kN, 20 kN, 30 kN, 40 kN and 50 kN (P10/H0, P20/H0, P30/H0, P40/H0, P50/H0),
- horizontal loading H = 10 kN with vertical loading P = 50 kN (P50/H10),
- release of horizontal loading to "0" (P50/H0),
- release all loading to "0" (P0/H0),
- horizontal loading H = 10 kN (P0/H10),
- release to "0" (P0/H0).

For stages II and III, the following loading cycles were applied:
- vertical loading P = 25 kN and P = 50 kN (P25/H0, P50/H0),
- horizontal loading H = 10 kN for P = 50 kN (P50/H10),
- release of horizontal loading to "0" (P50/H0),
- release of all loading to "0" (P0/H0),
- horizontal loading H = 10 kN (P0/H10),
- release to "0" (P0/H0).

During the tests the following measurement were made:
- strains in beam and column sections by means of tensometer strain gauges,

118

- relative rotations of joints by means of dial gauges (fig.3),
- rotations of beam ends by geodetic equipment,
- lateral drifts of frame by geodetic device,
- deflections of beams.

Arrangement of measuring points is shown in fig. 4.

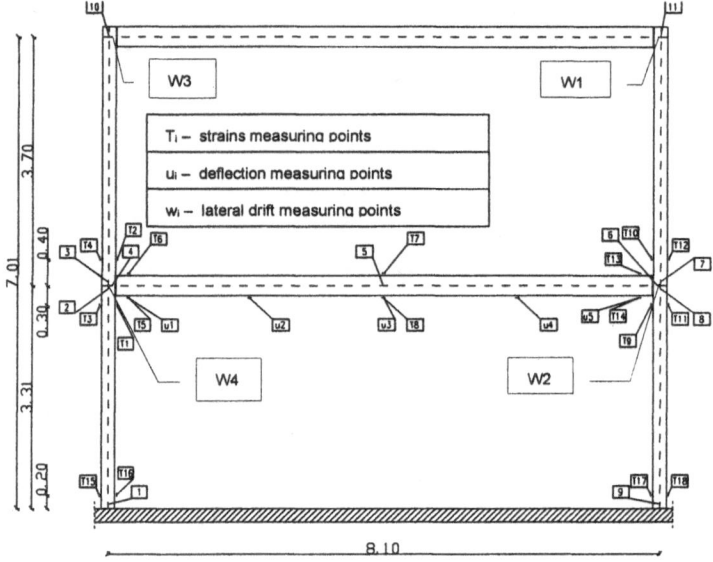

Figure 4. Arrangement of measuring points

4. Steel Frame Test Results

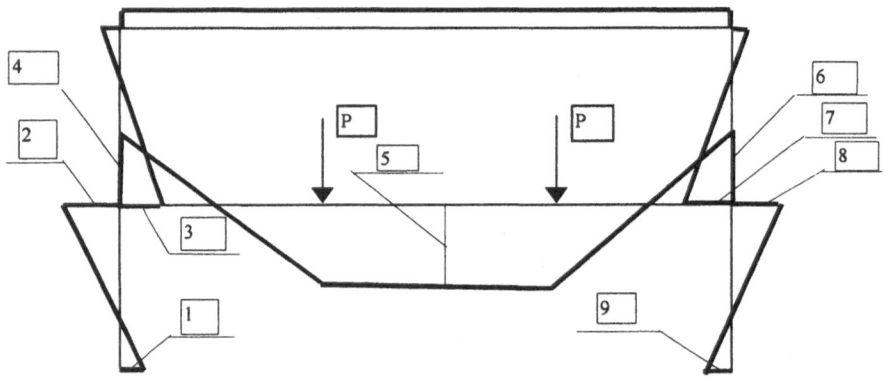

Figure 5. Bending moment diagram for vertical loading

The main test results were the distribution of moments in frame, frame deflections, and joints rotation. Bending moment diagram for vertical loading is shown in fig. 5.
Moments were calculated on the bases of measured strains in beam and column sections and actual element properties. Frame elements section characteristics calculated on the basis of measured section dimensions are collected in table 1.

TABLE 1. Actual properties of frame section

Point	A [cm²]	J [cm⁴]	W [cm³]
1	76,86	5661,1	563,4
2	74,97	5538,8	551,3
3	76,16	5601,5	557,5
4	55,46	8568,8	567,2
5	55,93	8609,8	570,4
6	56,16	8636,6	572,2
7	74,55	5504,6	548,8
8	75,13	5543,1	552,6
9	74,50	5498,3	548,2

Bending moments and vertical deflections of beam exposed to vertical loading P = 50 kN are presented in table 2.

TABLE 2. Bending moments M [kNm] and deflections f [mm] for vertical loading P = 50 kN

P	Stage I test M	f	Stage I Robot M	f	Stage II test M	f	Stage II Robot M	f	Stage III test M	f	Stage III Robot M	f
1	10,28		12,85		9,78		12,79		8,64		11,21	
2	-34,67		-33,17		-34,58		-31,96		-29,14		-28,24	
3	29,34		24,63		28,38		23,74		25,88		20,86	
4	-64,21		-57,81		-63,16		-55,70		-55,21		-49,09	
5	71,79	27,4	74,68	25,08	73,14	27,7	76,82	26,01	79,80	30,0	82,23	28,6
6	-60,08		-53,93		-58,12		-52,86		-51,89		-45,98	
7	27,94		24,1		26,87		23,61		24,53		20,61	
8	-31,52		-29,82		-31,12		-29,25		-27,24		-25,37	
9	9,99		11,44		10,98		10,82		10,54		9,46	

Table 3 contains values of bending moments for horizontal loading H = 10 kN.

TABLE 3. Bending moments in frame for horizontal loading H = 10 kN, in kNm.

Point	Stage I test	Stage I Robot	Stage II test	Stage II Robot	Stage III test	Stage III Robot
1	-9,92	-10,0	-10,12	-10,09	-10,41	-10,44
2	9,52	6,58	9,45	6,47	8,72	6,13
3	2,37	1,35	2,50	1,61	2,69	1,82
4	7,70	5,06	6,96	4,86	6,13	4,41
5	0,58	0,11	0,23	0,07	0,99	0,09
6	-7,13	-4,83	-6,89	-4,71	-5,55	-4,13
7	-2,65	-1,7	-2,59	-1,75	-2,84	-1,98
8	-10,21	-6,53	-9,63	-6,46	-8,51	-6,11
9	10,64	9,97	10,01	10,08	8,99	10,41

Lateral drifts of frame for horizontal loading H = 10 kN are collected in table 4. Relative joints rotation ϕ_j and the rotation of beam ends ϕ_b are presented in table 5. Initial stiffnesses obtained in steel frame tests are collected in table 6.

TABLE 4. Lateral drift of steel frame [mm] for horizontal loading H = 10 kN

P	Stage I		Stage II		Stage III	
	test	Robot	test	Robot	test	Robot
W1	4,6	4,92	5,0	5,05	5,3	5,5
W2	3,0	3,01	3,1	3,1	3,2	3,29
W3	4,3	4,93	4,6	5,06	4,9	5,51
W4	2,8	3,05	2,9	3,07	3,0	3,26

TABLE 5. Rotations of joints ϕ_j and beam ends ϕ_b [mrad]

Loading	$\phi_{j,\,left}$	$\phi_{j,\,right}$	$\phi_{b\,left}$	$\phi_{b\,right}$
Stage I				
P10/H0	0.838	1.093	1.551	0.96
P20/H0	2.011	2.404	3.306	3.083
P30/H0	3.128	3.279	4.23	4.235
P40/H0	3.911	4.098	6.841	6.715
P50/H0	4.813	4.909	7.539	8.358
P50/H10	4.131	5.355	8.157	8.281
Stage II				
P25/H0	2.961	3.224	3.556	3.699
P50/H0	5.15	5.191	6.943	7.519
P50/H10	5.084	5.847	7.517	7.175
Stage III				
P25/H0	4.246	4.536	5.624	5.755
P50/H0	6.76	7.268	9.953	10.067
P50/H10	6.536	7.869	9.65	8.55

TABLE 6. Initial stiffness of steel joints obtained in frame tests [kNm/rad]

Stage	left joint	right joint
I	13 765	11 267
II	11 578	10 190
III	7 665	6 728

Stiffness of column-base obtained from test was 35348 kNm/rad.

5. Test of Frame with Composite Floor

One month after concreting first floor (fig. 6), frame test was repeated. Section of composite beam close to joint is presented in fig. 7and view of reinforcement of floor before concreting is shown in fig. 8. The same equipment was used for vertical and horizontal loading as in steel frame tests but greater vertical loading, which reaches 100 kN for composite frame.

Figure 6. View of building during test of frame with composite floor

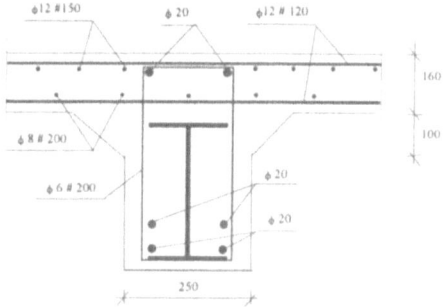

Figure 7. Section of composite beam

The following loading cycles were applied:
- vertical loading P = 10 kN in order to remove the slackening in the assembled frame and to ascertain proper functioning of strain gauges and transducers,
- release to "0" (P0/H0),
- vertical loading increased step by step to values: P = 20 kN, 40 kN, 60 kN, 80 kN and 100 kN (P20/H0, P40/H0, P60/H0, P80/H0, P100/H0),
- horizontal loading H = 10 kN with vertical loading P = 100 kN (P100/H10),
- release all loading to "0" (P0/H0),
- horizontal loading H = 10 kN (P0/H10),
- release to "0" (P0/H0).

During test, the same measuring as for steel frame were done.

Moments in frame with composite first floor were collected in table 7, beam deflections and joint rotations in table 8.

TABLE 7. Bending moments [kNm] for frame with composite floor

P	P20/H0		P40/H0		P60/H0		P80/H0		P100/H0		P100/H10	
	test	Robot	test	Robot	test	Robot	test	Robot	test	Robot	test	Robot
1	1,54	2,15	2,92	3,98	4,89	6,25	6,47	8,39	8,33	10,65	16,04	19,08
2	-5,46	-5,86	-9,91	-10,88	-14,77	-17,04	-19,91	-22,8	-25,06	-28,9	-34,02	-37,45
3	3,69	4,51	6,81	8,37	10,9	13,11	14,67	17,57	19,22	22,31	18,91	21,9
4	-9,15	-10,37	-16,72	-19,25	-25,67	-30,18	-34,54	-40,47	-44,28	-51,3	-52,31	-59,35
5	45,44	44,2	84,56	81,9	132,9	128,5	178,8	172,1	225,29	218,5	225,37	218,4
6	-8,77	-10,15	-16,03	-18,8	-25,13	-29,5	-32,16	-39,5	-43,64	-50,17	-38,42	-44,48
7	3,45	4,42	6,23	8,18	10,52	12,84	13,81	17,22	18,94	21,85	19,52	22,15
8	-5,32	-5,73	-9,8	-10,6	-14,61	-16,63	-18,35	-22,28	-24,7	-28,25	-18,08	-20,02
9	1,74	2,29	3,27	4,26	5,37	6,67	7,08	8,94	8,92	11,36	2,12	3,04

TABLE 8. Deflection u_3 [mm] and joint rotation ϕ_j [mrad] for frame with composite floor

P	P20/H0		P40/H0		P60/H0		P80/H0		P100/H0		P100/H10	
	test	Robot	test	Robot	test	Robot	test	Robot	test	Robot	test	Robot
u_3	3,27	1,84	4,42	3,41	5,67	5,35	6,91	7,17	8,39	9,10	8,38	9,09
$\phi_{j.left}$	0,175	0,198	0,319	0,365	0,491	0,569	0,675	0,777	0,859	0,982	0,969	1,279
$\phi_{j.right}$	0,211	0,248	0,382	0,449	0,613	0,704	0,828	0,954	1,163	1,206	1,073	1,021

Initial joint stiffness for left joint was established as 52 281 kNm/rad, for right joint as 41 521 kNm/rad.

6. Test of Steel Joint on Isolated Specimen

Test was conducted on single joint model FG, in natural scale. This joint model was prepared from the same steel and profiles as frame elements, in the same factory. Drawing of joint specimen is shown in fig. 9, and scheme of test arrangement on fig. 10. View of specimen FG on test rig is shown in fig. 11. Bolts were fully tightening to moment of 640 Nm.

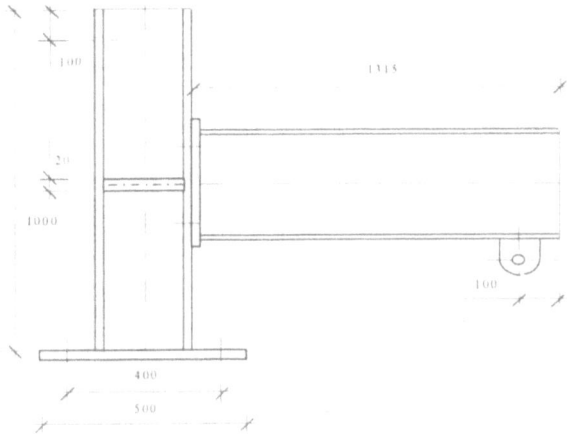

Figure 9. Drawing of FG specimen

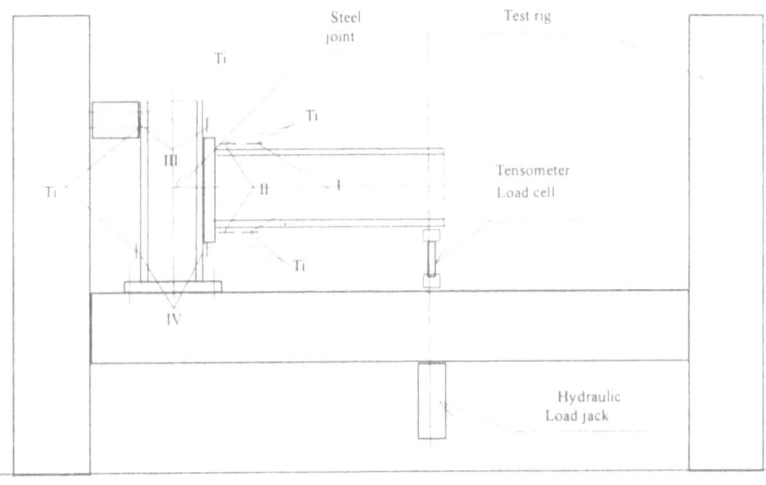

Figure 10. Test rig and specimen set-up

During test the following measurement were done:
- strains in beam and column sections (fig. 10 - points T_i),
- rotation of web panel ϕ_{pan},
- rotation of connection ϕ_c,
- rotation of beam end ϕ_b.

Global joint rotation ϕ_j was calculated as: $\phi_j = \phi_{pan} + \phi_c$.
Values of joint rotations under loading are collected in table 9.
Initial stiffness of joint obtained from test was 16 100 kNm/rad.

124

Figure 11. View of test set-up

TABLE 9. Results of FG specimen test

Load P	$P_{real.}$ *	M_j	ϕ_c	$\phi_{pan.}$	ϕ_j
KN	KN	KNm	mrad	mrad	mrad
0	0,0	0,0	0,0	0,0	0,0
10	10,07	13,24	0,522	0,38	0,902
20	20,71	27,23	0,966	0,73	1,696
30	31,45	41,36	1,566	1,054	2,62
40	42,08	55,34	2,22	1,453	3,673
50	52,82	69,46	3,60	1,847	5,45
60	63,56	83,58	6,618	2,115	8,733
0	0,0	0,0	0,527	0,413	0,94
60	63,56	83,58	6,73	2,115	8,845
70	74,19	97,56	8,869	3,181	12,05

* real values of load obtained from tensometer lad cell

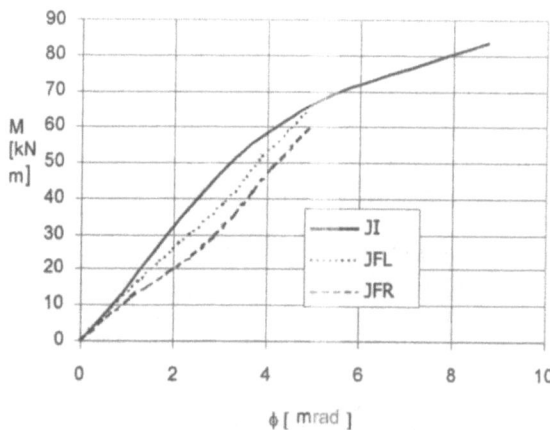

Figure 12. Comparison of M - ϕ curves for steel joint tested on isolated specimen (JI) with joints tested in frame (JFL - left joint) and (JFR - right joint)

Values of initial stiffness and design moment capacity calculated according to Annex J of EC 3 [4] are:

$S_{j.ini} = 17\ 300$ kNm/rad,

$M_{Rd} = 67,8$ kNm.

Comparison of M - ϕ curves obtained for steel joints in frame test and obtained in isolated test is shown in fig. 12.

7. Test of Isolated Composite Joint

Test was conducted on single specimen FZ, which was the real copy of joints used in frame. Drawing of specimen FG is shown in fig.13.

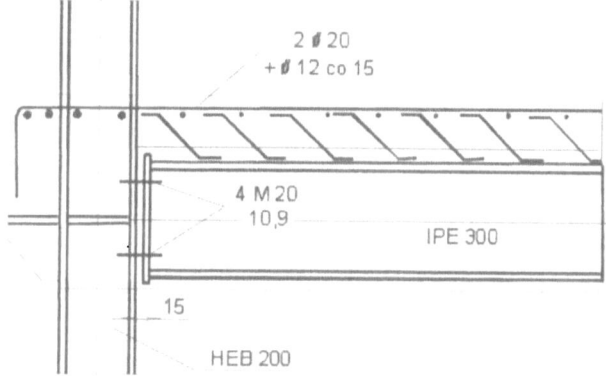

Figure 13. Drawing of FZ specimen

Figure 14. View of specimen FZ on test rig

126

Width of concrete plate was 1250 mm. Test was conducted on the same test rig as steel joint. View of specimen on test rig is shown in fig. 14. The main aim of test was to measure joint rotations, to find initial stiffness and create M - φ curve. Results of test are presented in table 10. P_{real} means real joint loading obtained from tensometer load gauge.

TABLE 10. Results of FZ specimen test

Load P	P_{real}	M_j	$φ_j$
KN	KN	KNm	mrad
0	0,0	0,0	0,0
20	20,25	25,11	0,541
40	40,6	50,34	0,915
60	59,83	74,19	1,288
80	78,79	97,7	1,702
100	99,4	123,25	2,286
0	0,0	0,0	0,637
100	100,42	124,52	2,543
120	118,31	146,71	3,722
0	0,0	0,0	1,788

Initial stiffness of joint from test had value of 48 400 kNm/rad, which is close to these obtained in frame test. M - φ curves for joints tested in frame JFR (rigth joint) and JFL (left joint) and obtained in isolated test WW are compared in fig. 15.

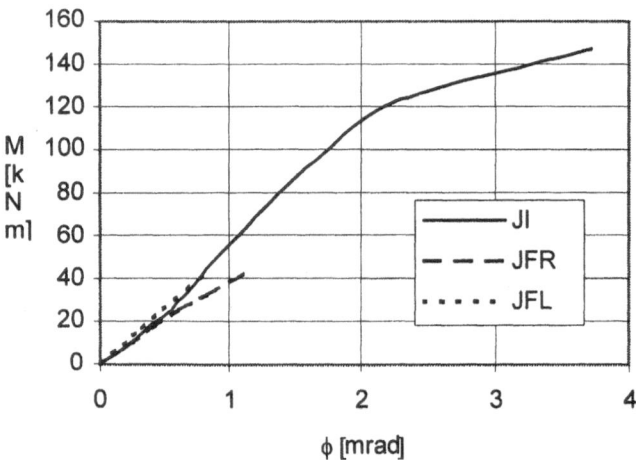

Figure 15. Comparison of behaviour of joint tested on isolated specimen (JI) and joints tested in frame (JFL - left joint) and (JFR - right joint)

8. Comparison of Behaviour of Steel and Composite Joints

Comparison of joint characteristics for steel (JS) and composite (JC) joints tested in frame is shown in fig. 16. It is seen that composite joints posses much higher stiffness

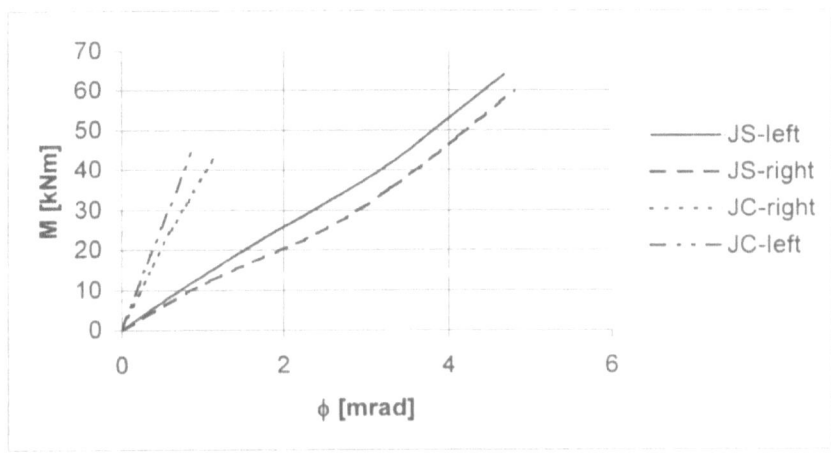

Figure 16. M - φ curves obtained from tests for steel (JS) and composite (JC) joints

(3,8 times) than steel joints. Bending moments in composite joints have smaller values than in steel joints, in spite of bigger loading applied to composite frame. This is because of much higher stiffness of composite beam, which takes more loading.

9. Comparison with Computer Calculation Results

As can be seen in fig. 16, steel and composite joints show linear behaviour for applied loading. For global analysis of tested frames, computer program with elastic springs in joints was used (program ROBOT V6). Spring stiffnesses for beam-to-column and column bases connections were taken the same as obtained in tests (table 6). Actual beam and column section properties (table 1) and real values of loading were introduced into program. Results of computer program calculations are presented in tables 2, 3, 4 for steel frames, and in tables 7, 8 for frame with the composite floor.

128

10. Conclusions

Results of the full-scale tests on steel frame with flush end-plate joints and with different bolts preloading as well as results from test on the same frame with composite first floor have been reported. The results show the effect of bolt preloading on the joint stiffness and frame behaviour.

Stiffness of column bases, made with only two bolts and traditionally considered as pinned, is rather large and strongly influence the frame behaviour.

Stiffness of composite joint, obtained by adding 0,7 % of reinforcement in the upper layer of concrete slab, is 3,8 times larger than that for bare flush end-plate joint.

Larger stiffness of composite beam makes the deflections of beam lower and moments in joints smaller than in steel frame, even for bigger loading. It confirms the conclusion that frames with composite floors are very effective and reasonable type of structure.

Presented test results may be used for verification of computer programs for global frame analysis.

Preliminary comparison between test results and results from computer analysis, conducted with account of linear joints behaviour, shows quite good agreement.

11. References

1. Nethercot, D.A.. (1985) Steel Beam-to Column Connections – A Review of Test Data and Their Applicability to the Evaluation of the Joint Behaviour of the Performance of Steel Frames. *CIRIA Project 338*. London.
2. Chen, W.F. (1988) *Steel Beam-to-Column Building Connections*. Elsevier Science Publishers. London & New York.
3. Bródka, J., Kozłowski, A.. (1996) *Stiffness and Strength of Flexible Joints*. Rzeszów University Of Technology Publishers. Rzeszów-Białystok.
4. ENV 1993-1-1. (1992) Eurocode 3: Design of Steel Structures. Part 1.1. General Rules and Rules for Buildings. *CEN*.
5. Benussi, F., Nethercot, D.A., Zandonini, R. (1996) Experimental Behaviour of Semi-Rigid Connections in Frames, in *Connections in Steel Structures III: Behaviour, Strength and Design*, Elsevier Science Publishers.
6. Shanmugam, N.E., Yu, C.H., Liew, R. (1995) Large-Scale Testing of Steel Sway Frames, in S. Kitipornchai, G.J. Hancock, M.A. Bradford (eds), *Structural Stability and Design*.
7. Ivanyi, M., Hegedus, L., Ivanyi, M. Jr., Varga, D.(1998) Failure Tests of Two-Storey Steel Frames. *Journal of Constructional Steel Research*, , 46: 1-3.

CYCLIC TESTS ON BOLTED STEEL DOUBLE-SIDED BEAM-TO-COLUMN JOINTS

D. DUBINA, A. STRATAN, A. CIUTINA
University "Politehnica" of Timisoara, Faculty of Civil Engineering and
Architecture, Department of Steel Structures and Structural Mechanics
Str. Stadion nr.1, 1900-RO, Timisoara, Romania

1. Introduction

The key points in the behaviour of Moment Resisting Steel Frames (MRSF) are the beam-to-column joints, located near the dissipative zones. These should posses adequate rotation capacity and resistance in order to resist the earthquake action. Consequently, the recent American design codes [1] include special provisions for classification of MRSF in terms of beam-to-column joint plastic rotation capacity and resistance.

Laboratory tests performed in Timisoara on double-sided beam-to-column joints [2], [3], have shown important differences in the behaviour of joints subjected to gravitational loads (balanced moments), and horizontal seismic loads (unbalanced moments). Anti-symmetrical loading has led to a smaller moment capacity and stiffness, and a substantial increase in the rotational ductility of the joint, as compared to symmetrically loaded joints. The difference of joint behaviour was given by the column web panel, which is not subjected to shear in case of symmetrically loaded joints.

An alternative to the "standard" European column cross-section (built up of hot-rolled I or H profiles) is the use of X-shaped cross-sections, made out from two hot-rolled profiles welded along the median axis or built-up sections made out from welded plates, as is shown in Figure 1.

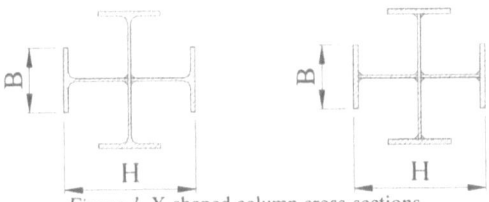

Figure 1. X-shaped column cross-sections.

Generally, this type of cross-section is used for space moment resisting frames, due to similar stiffness on both directions and convenient three- and four-way moment connections. The use of X-shaped columns brings important changes in the behaviour of the beam-to-column joints. If transversal stiffeners are used, the effective shear area of

C.C. Baniotopoulos and F. Wald (eds.), The Paramount Role of Joints into the Reliable Response of Structures, 129–138.
© 2000 *Kluwer Academic Publishers.*

130

the column web is increased due to column flanges parallel to the web. The increase of shear area introduces two effects in the joint behaviour subjected to seismic forces: an increase of connection stiffness, and an increase of the moment capacity.

Joints to X-shaped columns have not been tested particularly in the past. This paper summarises the program of experimental tests carried out at the Laboratory of Steel Structures from the Faculty of Civil Engineering and Architecture, The "Politehnica" University of Timisoara.

2. Testing Program

2.1. TESTING SPECIMENS

The testing program comprised six joint specimens, as follows:

- three joints under symmetrical loading (series BX-SS) – see Figure 2a, from which one joint tested under monotonic loads (specimen BX-SS-M), and the other two under cyclic loads (specimens BX-SS-C1 and BX-SS-C2).

- three joints tested under anti-symmetrical loading (series BX-SU), - see Figure 2b: one under monotonic loads (specimen BX-SU-M) and the other two under cyclic loads (specimens BX-SU-C1 and BX-SU-C2).

Eight bolts M20 gr 10.9 and an extended end plate of 20mm thickness compose the bolted connection. Transversal stiffeners of 14mm on the column panel web have been used. The bolts have been prestressed according to the Romanian Standard for steel structures (torque moment of 64 daNm). However, in order to account the influence of the prestressing grade, the last joint of each series has been prestressed at half the prestressing value required by the code.

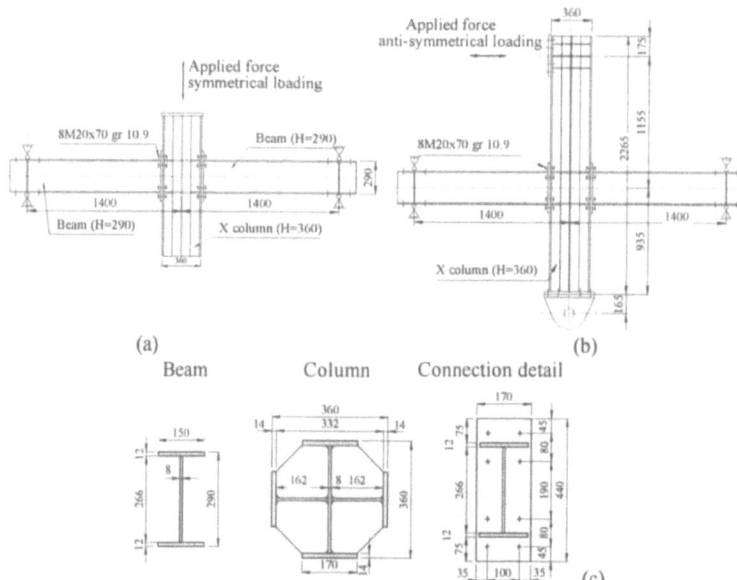

Figure 2. Joint configuration for symmetrical loading (a), anti-symmetrical loading (b) and joint detailing (c).

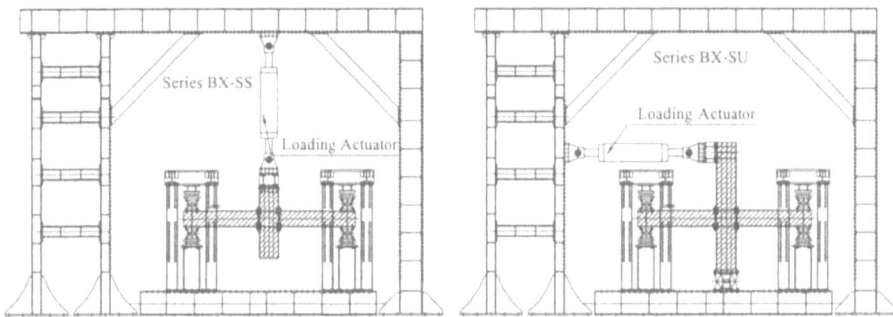

Figure 3 Joint set-up for series BX-SS and BX-SU

Figure 3 shows the joint set-up for series BX-SS (symmetrical loading) and series BX-SU (anti-symmetrical loading)

2.2. LOADING PROCEDURE

Tests were performed in accordance with the Recommendations of the European Convention for Structural Steelwork (ECCS) – [5]. The first specimen from each series was tested monotonically.

In the case of cyclically loaded joints, the ECCS procedure was applied, where the yielding displacement was determined in the monotonic tests. The loading was applied in displacement control.

2.3. MATERIALS

S235 Steel was used for designed specimens. Tensile tests were performed on samples extracted from component elements of specimens for obtaining the actual mechanical characteristics of the steel. In TABLE 1 are given the mean values for the yield strengths.

Results of the coupon tests match fairly well to the mill certificates for beam flanges, column flanges and stiffeners, while the yield strength for the end plates, beam and column webs display important differences. This is true, especially in the case of end plates, the steel being rather S355 grade.

TABLE 1. Yield strength for plates used in joint fabrication [N/mm²]

Plate	t=8 mm	t=12 mm	t=14 mm	t=20 mm
f_y mill certificates	258.0	303.0	258.0	235.0
f_y coupon tests	316.2	310.1	295.20	372.5

3. Experimental Results

The experimental results of the tests are mainly linked to the moment-rotation relationship for each joint; in fact, this curve describes best the joint behaviour. TABLE 2 gives the main results for all three series of tested joints. Failure of a specimen was considered when the induced force falls by 50% of the maximum force, attained for that specimen. The moment was computed at the column face. The experimental results

132

were monitored in terms of maximum bending moments, ultimate rotations, the dissipated energy and the type of failure.

3.1. MONOTONICALLY LOADED SPECIMENS

In Figure 4 are presented the moment-rotation curves for the monotonic specimens.

BX-SS-M specimen shows a normal behaviour, with the resistant bending moment increasing monotonically, and gradual stiffness degradation after attaining the yielding. The plastification of column flanges and end plates was initiated in the tensioned zone of the joint. At the attainment of the maximum moment, a bolt situated in the extended part of the end plate (tension zone) failed. The maximum rotation obtained by in this case was of 43.20 mrad.

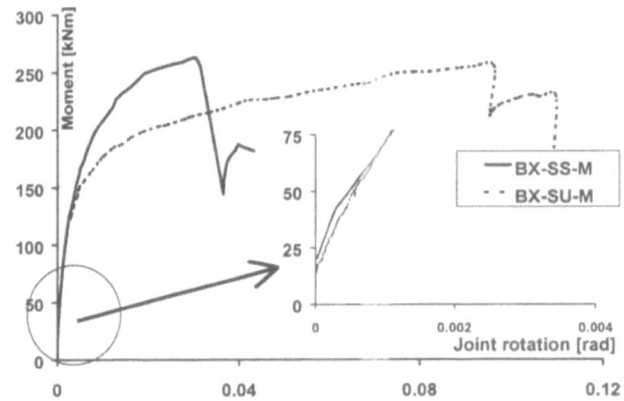

Figure 4. Moment-Rotation relationships obtained for specimens BX-SS-M and BX-SU-M

(a)   (b)
Figure 5. Failure of the specimens BX-SS-M (a) and BX-SU-M (b)

In case of BX-SU-M specimen, the maximum moment was of comparable magnitude to that obtained for specimen BX-SS-M. However, the mean joint rotation increased more than double, as compared to the specimen BX-SS-M (105.5 mrad for BX-SU-M specimen). In this case, two bolts, from the extended part of the end plate failed. For each bolt failure, a substantial drop in the moment resistance occurred, as it can be seen in Figure 4. However, compared to symmetrical monotonic joint, a difference existed in the failure of the anti-symmetrical one, and it was introduced by the column panel zone that was visibly sheared. This fact can be seen in Figure 5, where are shown the failures of the two specimens.

The initial stiffness was similar for both specimens, as it can be seen in the elastic chart extracted from the main one. However, according to the EC3-Annex J calculations, the theoretical initial stiffness is significantly smaller in the case of anti-symmetrically loaded joints (see TABLE 2).

3.2. BX-SS CYCLIC SPECIMENS

The cyclically tested joints, BX-SS-C1 and BX-SS-C2, have been tested by ECCS procedure, (four cycles in elastic range, then three cycles for each even multiple of yielding displacement). They have shown close values of the maximum bending moments to the ones resulted in case of monotonic specimen, but the failure was different. After attainment of the maximum moment, a sudden degradation of the behaviour was noticed due to the failure of beam to end plate fillet welds. However, the cracks initiated in the most stressed zones, i.e. the outer fillet beam-to-end plate weld and propagated through the Heat Affected Zone (HAZ) into the end plate or the beam flange. These cracks generated a rapid decrease in the connection moment, as it is shown in Figure 6a, for specimen BX-SS-C1. Generally, the maximum rotations obtained for the cyclically loaded joints were dramatically reduced as compared to the monotonic test, due to premature weld failure. Moreover, in case of these specimens, no softening branch could be observed in the moment-rotation curve.

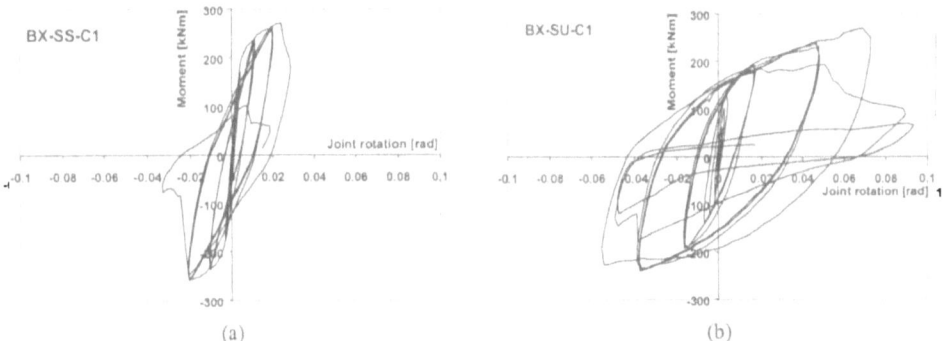

(a)   (b)

Figure 6. Moment-Rotation relationship for cyclic loaded joints: (a) symmetrical loaded joints; (b) anti-symmetrical loaded joints

Figure 8a shows the joint failure of the specimen BX-SS-C2. Although the load was applied symmetrically, only one connection failed, due to steel or weld imperfections, or load eccentricity. The amount of total energy dissipated by the two specimens is rather small, especially in the case of specimen BX-SS-C2 – see Figure 7.

3.3. BX-SU SERIES

In [1] and [2] is relived the fact that for usual column cross-sections (hot-rolled or built-up I or H profiles), the maximum moment in the joint obtained in anti-

134

symmetrically loaded joints is substantially smaller as compared to the maximum moment obtained in the symmetrically loaded joints, while the joint rotation is much higher. This fact is only partially confirmed by the tests made on X-shaped column cross-section joints. Actually, for the anti-symmetrical joints the drop in moment capacity is very small (about 5% in average), while the increase in maximum joint rotation is quite important (more that 150%). On the other hand it can be noticed the fact that cycles in anti-symmetrical loading are more stable, presenting also a softening branch, as can be seen in Figure 6b for specimen BX-SU-C1.

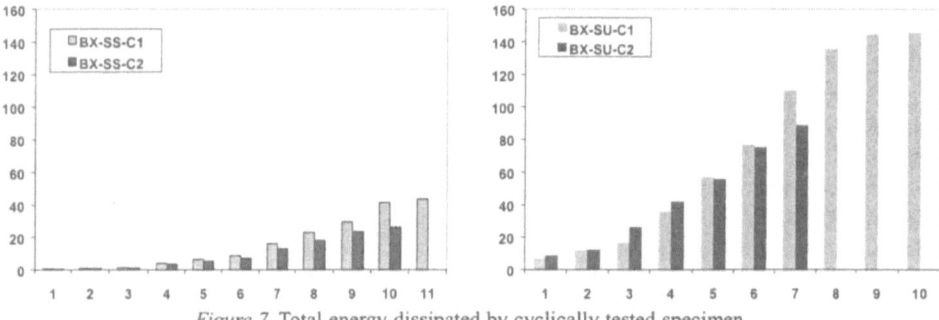

Figure 7. Total energy dissipated by cyclically tested specimen

The initiation of failure for the specimens BX-SU-C occurred in a similar manner as for the monotonic test (by column flange and end plate plastification and also by shear deformation of the column web panel), but as for BX-SS-C series, the failure occurred by cracks initiated in the beam-to-end plate welds or in the HAZ. In Figure 8b is shown the failure of the right connection for the specimen BX-SU-C1.

From the point of view of the total energy dissipated, it can be said (Figure 7) that, generally, the anti-symmetrically loaded joints dissipated an amount of energy at least double as compared to the symmetrically loaded joints (including the monotonic tests). The difference in dissipated energy is attributed to the plastic panel zone deformation working in shear. For the specimen BX-SS-C2, the dissipated energy is significantly smaller compared to BX-SS-C1 specimen, as it resisted fewer cycles.

(a)   (b)

Figure 8. BX-SS-C2: Weld failure (a); BX-SU-C1: End plate failure (b)

A very important factor that can be noticed in case of BX-SU series, is the way of deformation of the column panel zone. For usual column cross sections (H or I profile), the shear deformation of the column web panel is shown in Figure 9a. On X-shaped column cross-sections, the column flanges parallel to the web act as a web stiffener, as it is schemed in Figure 9b. This fact is revealed by the measurements recorded both for the general web shear deformation, compared to the second-axis column flange deformation.

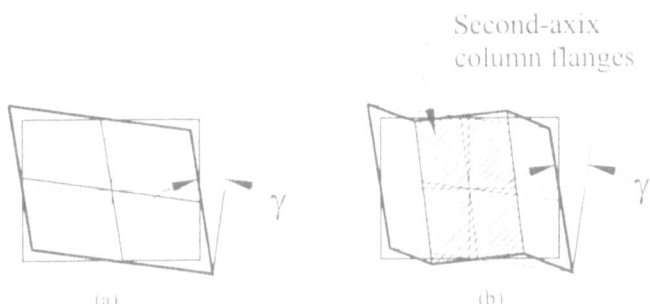

Figure 9. Panel zone deformation for usual column cross sections (a), and X-shaped cross section (b).

3.4. WELD FAILURE INSPECTIONS

Due to the fact that generally the beam-to-end plate fillet welds had an unsatisfactory behaviour under cyclic loading, they have been investigated in detail.

Laboratory tests have shown that generally, there was a lack of penetration between the fillet welds into the base material, as it is shown in Figure 10a, this being an important factor in weld failure. A second factor that accelerated the cracking process was the plate lamellar tearing (see Figure 10b). This was due to non-metallic inclusions during the hot-rolling of the plate. All these factors, combined to other welding defects and weld non-metallic inclusions gave rise to premature non-ductile weld failure.

In order to improve the behaviour of beam-to-column connections under cyclic loading, it is crucial to realise good quality welds. Therefore, some recommendations have to be taken into account in design and manufacturing process:

- full penetration welds should be used if load reversals are expected for the connected structural members
- notch-tough weld rods should be used for welding material
- use of base material with guaranteed through-thickness quality, to reduce the tendency to lamellar tearing

Figure 10. Lack of weld penetration (a) and plate exfoliation for beam-to-end plate weld (b)

4. Comparison of the Experimental Tests to the EC 3 Annex J Calculations

TABLE 2 comprises the results of monotonic and cyclic tests, expressed in terms of dissipated energy, maximum and yielding values of connection moments and of joint rotation, as well as the initial stiffness $S_{j,ini}$. For the cyclically tested joints the values are given for both branches of the moment-rotation diagram.

Analytical evaluation of joint properties (initial stiffness and moment capacity) has been performed according to EC3 – Annex J [6]. Measured geometrical and mechanical characteristics have been used in computations, the partial safety factors being set to 1.0.

EC3 – Annex J provides models for strong axis joints connecting I and H section members only. X-shaped column cross-section could be considered only by some sort of approximation. The most important difference between the behaviour of I-shaped and X-shaped column cross-sections is the increase of the panel zone shear area in case of X-shaped columns, due to presence of flanges parallel to the considered web, if transversal stiffeners are used. In this way, the effective shear area of the column panel zone is increased by the area of the two column flanges parallel to the considered web (see Figure 11a).

EC3-Annex J allows an increase in the shear area of the column web by means of supplementary web plates, on one or both sides of the column web. The shear area is increased by $b_s \cdot t_{wc}$ only (where b_s is the width of the web plate and t_{wc} is the thickness of the column web), regardless if one or two web plates are used (see Figure 11b). Due to presence of transversal stiffeners, it is more appropriate to use the "full shear area" idealisation of the joint for Annex J computation model. Anyway, the joint characteristics (initial stiffness and moment capacity) have been computed according to both approaches, and are presented in TABLE 2.

TABLE 2. Experimental and EC3 Annex J results for tested joints.

Specimen	Tot. energy kNm rad	θ_{max}^+ mrad	θ_{max}^- mrad	M_{max} kNm	M_{min} kNm	$S_{j,ini}^+$ KNm/rad*10³	$S_{j,ini}^-$	θ_y^+ mrad	θ_y^-	M_y^+ KNm	M_y^-
					Symmetrically Loaded Joints						
EC3-full A_s	---	---		---		*55.64*		*2.97*		*165.40*	
EC3–red. A_s	---	---		---		*55.64*		*2.97*		*165.40*	
BX-SS-M	9.01	43.20		263.34		48.03		3.26		180.79	
BX-SS-C1	43.75	28.0	21.0	271.6	259.1	55.91	59.60	3.26	2.60	197.2	188.0
BX-SS-C2	26.28	17.4	18.1	261.8	259.8	71.24	63.5	2.66	2.39	194.8	206.8
					Anti-symmetrically Loaded Joints						
EC3-full A_s	---	---		---		*32.99*		*4.76*		*156.91*	
EC3–red. A_s	---	---		---		*25.19*		*4.24*		*106.77*	
BX-SU-M	21.02	105.5		258.36		51.50		2.28		137.66	
BX-SU-C1	145.54	72.5	55.3	269.4	240.6	35.07	29.08	3.77	4.44	153.1	161.2
BX-SU-C2	88.37	39.2	46.8	240.1	236.6	27.82	40.53	5.54	3.37	179.8	161.2

* full A_s – see Figure 11a; red. A_s – see Figure 11b

The joints characteristics computed by Annex J are the same in the case of symmetrically loaded joints, the column panel being not sheared in this case. It should be noted that the computations according to Annex J are rigorously valid for monotonically loaded joints only, cyclic loading being not covered by the EC3 procedure.

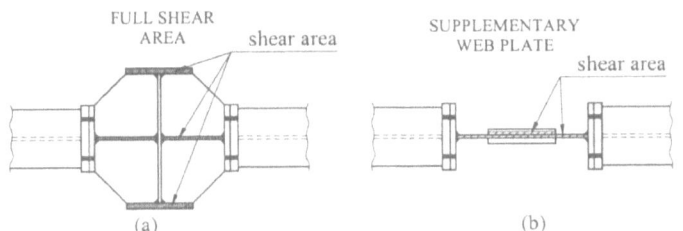

FULL SHEAR AREA   shear area   SUPPLEMENTARY WEB PLATE   shear area

(a)   (b)

Figure 11. Full shear area approach (a), and supplementary web plate approach (b)

The conventional values of yielding moment and rotation are computed according to the ECCS procedure [5]. It should be mentioned that the value of the experimental initial stiffness is quite sensible to the procedure used to determine it. Here, the initial stiffness was computed by means of a regression analysis between the points of $0.15 \cdot M_{max}$ and $0.30 \cdot M_{max}$, points that are within the elastic branch. The monotonic moment-rotation curves and the envelopes of the cyclic curves, as well as the EC3 – Annex J predictions are presented in Figure 12 for both joint series. Comparing experimental values obtained for symmetrical loaded joints to the EC3 results, it can be observed important differences in the yielding moment, the experimental values being greater by about 10-20% than the analytical predictions. Generally, by these values, the initial stiffness for symmetrical loaded joints was close to the one given by Annex J, with the exception of specimen BX-SS-C2.

For the anti-symmetrically loaded joints, the analytical predictions by Annex J in which the "full shear area" is considered for the column web panel, show closer values to the test results. Also, it can be observed from TABLE 2 that the initial stiffness for the monotonic test is considerably higher compared to that of cyclic tests and results of Annex J. On the other hand, the yielding moment for the same specimen is smaller that the one given by Annex J. The experimental results of cyclically loaded joints have shown closer values to the analytical predictions.

It should be noted that Annex J offers few provisions for the rotation capacity of the designed joints. Provided the design of extended end plate connections is governed by column flange or end plate resistance, the joint is considered to be enough ductile for plastic analysis. Tested specimens comply with these provisions of Annex J, therefore they fulfil the code requirement for ductility. Anyway, this requirement is very ambiguous and should be supported by relevant values.

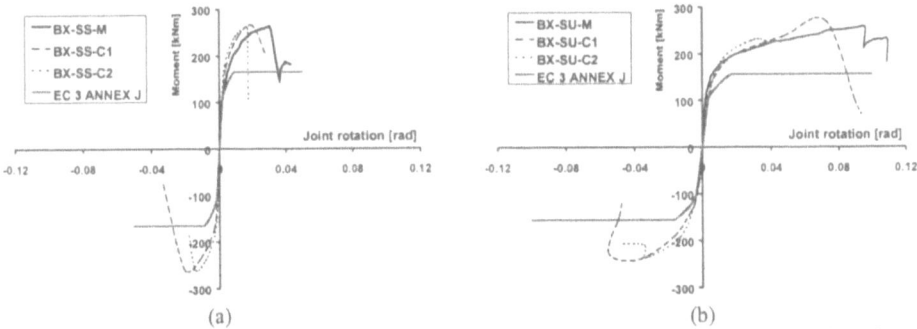

(a)   (b)

Figure 12. Envelope and monotonic moment-rotation relationships for BX-SS (a) and BX-SU (b) series.

138

As a main conclusion, it can be said that the joint characteristics computed by Annex J have shown results confident to the experimental ones, with the remark that "full shear area" should be considered for the column web panel in case of X-shaped columns with transverse stiffeners.

5. Conclusions

The use of X-shaped columns makes possible a convenient design of three- and four-way connections for space moment resisting frames. Also, it brings important advantages in the joint behaviour under anti-symmetrical loading over usual I or H shaped columns. Column flanges parallel to the considered web lead to a natural stiffening of the column panel zone. The increase in the panel zone shear area reduces significantly the drop in moment capacity for anti-symmetrically loaded joints with respect to symmetrical ones, but reduces in some extent the initial stiffness. Anyway, the stiffened panel zone participates to the plastic mechanism, assuring a significantly increased ductility of anti-symmetrically loaded joints.

Cyclic loading introduces differences between the type of failure. While for monotonic tests the failure was mainly by bolt failure and column flange/ end plate deformations, in the case of cyclic tests, it was by brittle failure of fillet welds. Therefore, fillet welds are not recommended in zones with load reversals. Similar tests performed in the laboratory having full-penetration beam-to-end plate welds have shown improved behaviour under cyclic loading, especially in terms of joint ductility (about 50% higher values of ultimate rotation).

The pretensioning rate does not affect significantly the initial stiffness and moment capacity of the joint. However, higher values of ultimate rotations have been observed in the case of fully pretensioned specimens. Anyway, further researches are necessary to confirm this observation.

Analytical model of EC 3 Annex J provides a reliable prediction for behaviour of I beam to X shaped columns extended end plate connections, but an appropriate modelling for the increased shear area of the panel zone should be used.

6. References

1.   AISC 97 (1997) *Seismic Provisions for Structural Steel Buildings*. American Institute of Steel Construction, Inc. Chicago, Illinois, USA
2.   Dubina D., Grecea D., Ciutina A., Stratan A. (2000). *Influence of Connection Typology and Loading Asymmetry, Chapter 3.2 in: Moment Resistant Connections of Steel Building Frames in Seismic Areas -*, (Mazzolani F.M. ed.) E&FN SPON (London).
3.   Dubina, D. Ciutina, A.L., Stratan, A. (in press). Influence of Loading Asymmetry on the Cyclic behaviour of Beam-to-Column Joints, *Proceedings of the Third International Conference Behaviour of Steel Structures in Seismic Areas*, STESSA 2000, 21-24 August 2000, Montreal, Canada.
4.   Elnashai, A.S. et al. (1996). *Experimental and Analytical Investigation into the Seismic Behaviour of Semi-rigid Steel Frames*. ESEE Research Report No. 96 - 7 December 1996, Imperial College, London.
5.   ECCS (1985). *Recommended Testing Procedures for Assessing the Behaviour of Structural Elements under Cyclic Loads*, European Convention for Constructional Steelwork, Technical Committee 1, TWG 1.3 – Seismic Design, No.45
6.   ENV 1993-1-1. (1997) *EUROCODE 3: Part 1.1. Revised Annex J: Joints in Building Frames*. Approved Draft: January 1997; Brussels: CEN, European Committee for Standardisation.

EXPERIMENTAL INVESTIGATION OF LOAD CARRYING CAPACITY OF THE JOINTS INTO AN ALUMINIUM DOME

D. DAKOV, I. TOTEV & O. GANCHEV
Department of Steel and Timber Structures
University of Architecture, Civil Engineering and Geodesy,
Sofia, Bulgaria

1. Introduction

The present experimental investigation aims at determining the real load carrying capacity and stiffness of the asymmetric joints for assembling the basic rafters of an aluminium dome. For simulating real loading a special test platform was designed. The analyses of the test results proved that the mechanical connections have sufficient reliability and offer a possibility of saving expensive welding works.

There is a great variety of commercially available patented joint connections for assembling aluminium domes. Besides the structural efficiency, a fundamental principle in the structural arrangement of the joints is the ease of manufacturing with a sufficient degree of reliability. The mechanical connections with hooking or bolt connections are preferable, whereas the welding works are not desirable.

In the present paper the joint connection for assembling the basic rafters of aluminium dome with a span of 19,2m (patent of "Schüco") is investigated. The structure of the dome consists of two hemispheric parts with diametere of 15,6m connected by a short cylindrical vault of 3,6m (Fig. 1).

The structural composition was made of meridian ribs and oval hoops with a box section. The profiles were completed with aluminium-magnesium alloy $ALMgSiO_5F_{22}$ to DIN 1748. In relation to DIN 4113 Teil 1, for this alloy yield sterngth $\sigma_y = 160MPa$ and tensile strength $\sigma_u = 215MPa$ were expected.

The specific conditions for erecting the dome without interrupting operations into the building demanded manual transferring, keeping and fixing of all the structural members, wheras combining in advance was impossible.

The dome was investigated as a 3D-strucutre with continuous ribs, which were hinged, connected with the supporting ring. The ribs were with a polygonal outline and for obtaining the rigid connection in the angle of folds, spatial mechanical connection has been provided.

Unfortunately, load carrying capacity and stiffness of the connection are not given in the documentation, but there are recommendations for welding in more severe conditions.

C.C. Baniotopoulos and F. Wald (eds.), The Paramount Role of Joints into the Reliable Response of Structures, 139–146.
© 2000 *Kluwer Academic Publishers.*

The performance of welding connections between aluminium profiles in protective atmosphere is a very expensive manufacturing operation. This imposed carrying out experimental investigation to determine the rigidity and load carrying capacity of mechanical connections to the basic ribs of the dome in the conditions of maximum loading. With a view to compare the results, additional experimental investigation was completed with welded joints.

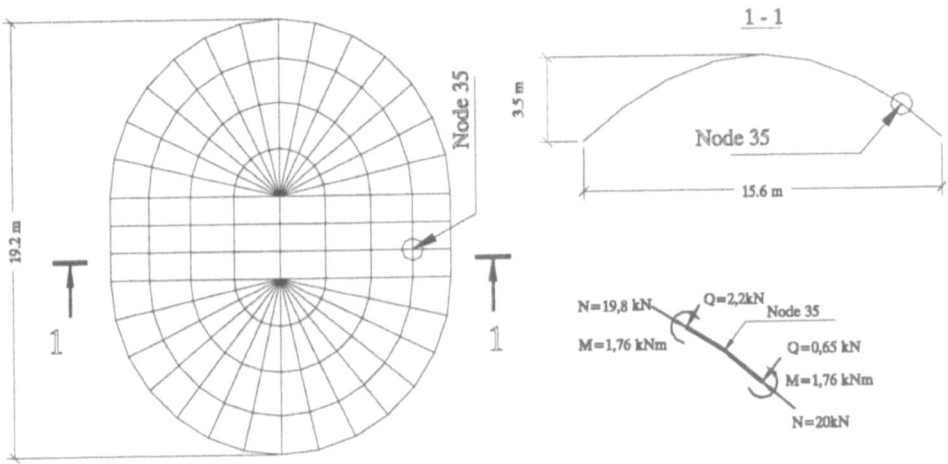

Figure 1

2. Testing and measurements

For simulating real loading a special test platform was designed (Figure 2).

Figure 2

The test stand consisted of two-hinged systems with a tension member. For the tension member a steel band was used, which gave the opportunity to avoid rigidity at the supports of the stand.

The loading of the connections was imposed using a hydraulic jack. The loading system enabled achieving simultaneous action of the bending moment M and the normal force N in the angular part, which was similar to that in the real structure. Two types of connections were investigated (Figure 3) :

– mechanical connections – type A
– mechanical connections and additional welding – type B.

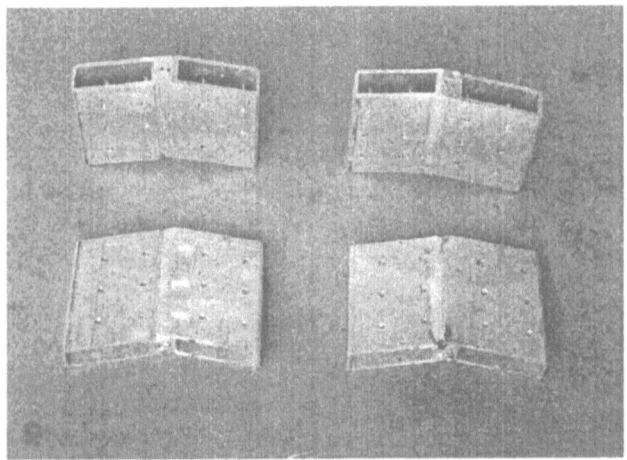

Figure 3

By the time of experimental investigation the displacement of vertical direction, horizontal displacement and deflections in the upper and bottom edge of the joints were measured. The displacement was measured using inductive transducers and the deflections – with strain gauges. The measurements were made using strain gauges system UPH (Hottinger) which was connected with a computer. For every degree of loading the system registered the indication of the measuring devices and saved the data into the computer. The scheme of experimental investigation and disposition of devices for measurement are shown on Figure 4.

3. Experimental results

The basic results from testing the connections type A (mechanical connection) are shown on Figure 5a. The results from testing the connections type B (mechanical connection and additional welding) under the same conditions of loading are presented on Figure 5b.

142

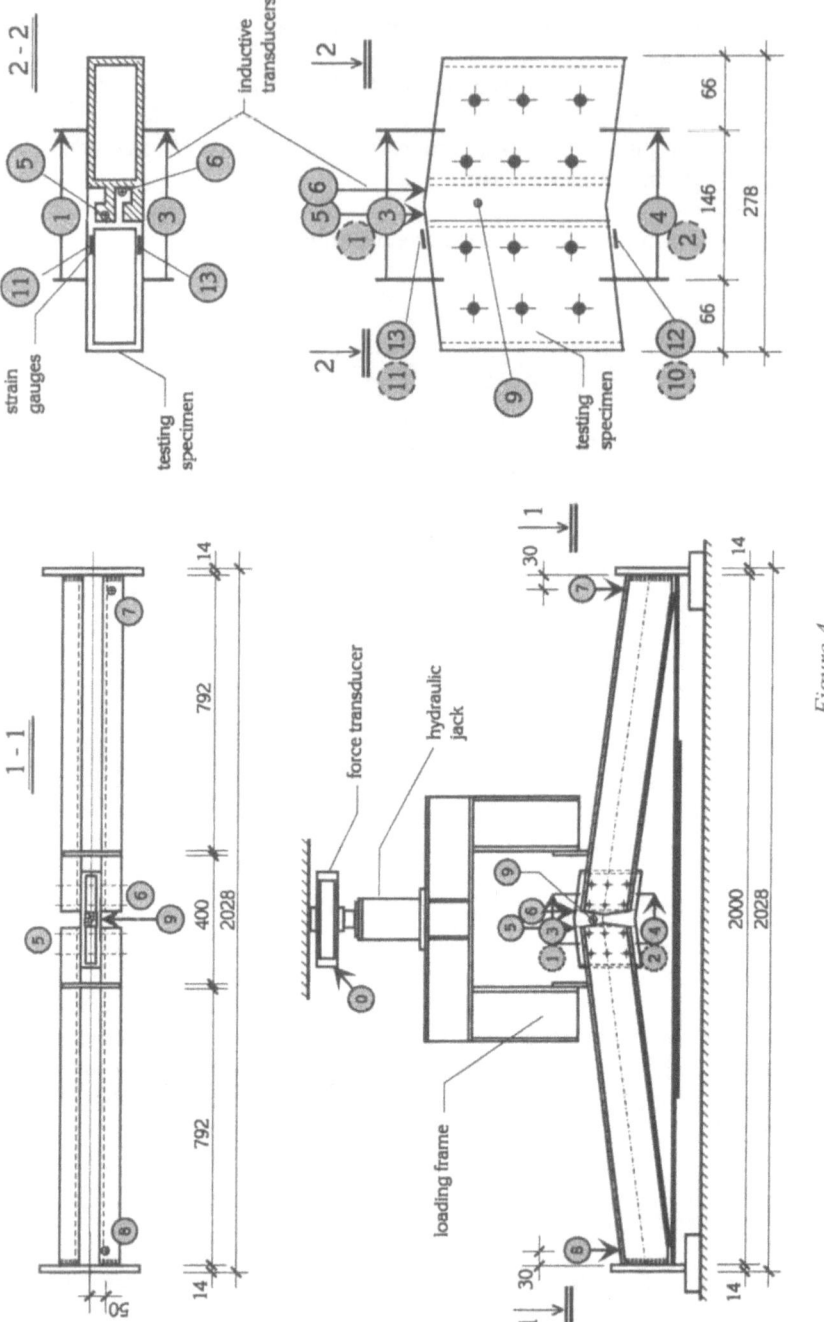

Figure 4.

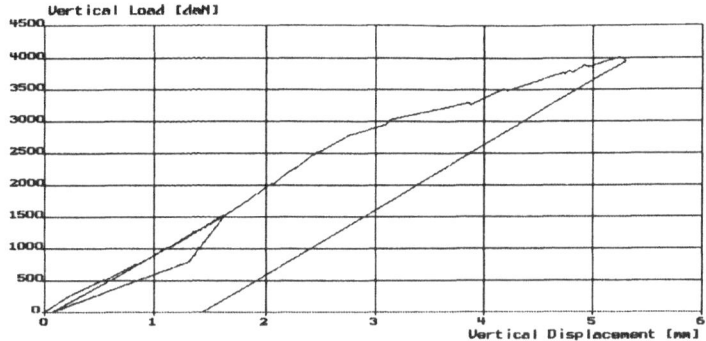

Specimen A-1

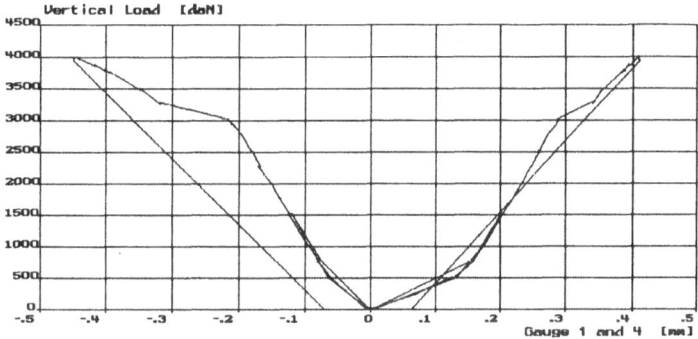

Specimen A-1

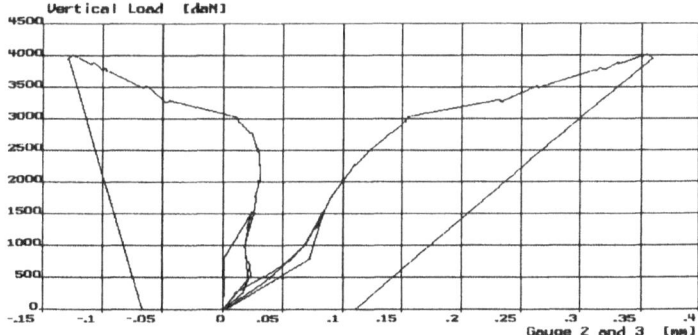

Specimen A-1

Figure 5(a)

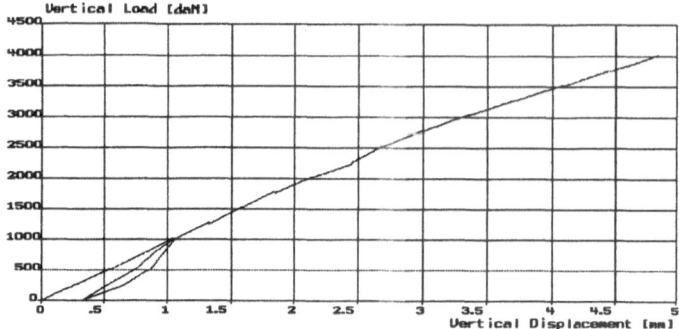

Specimen B-1

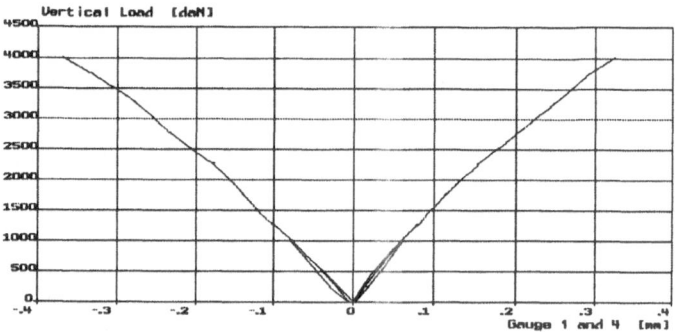

Specimen B-1

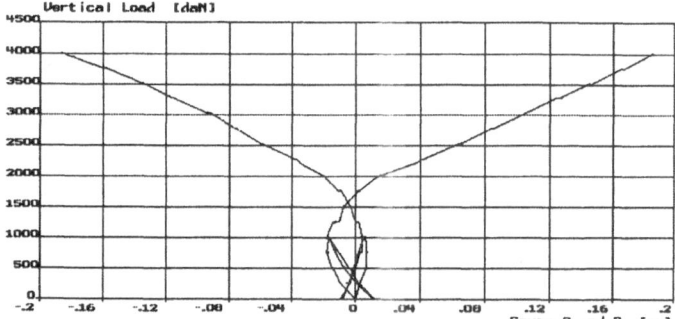

Specimen B-1

Figure 5(b)

The ultimate load P_u reached into the elastic stage of behaviour of the connection was accepted as a criterion for load carrying capacity. The load on the joint corresponding with the condition of the serviceability limit state of the structure was accepted as a characteristic load P_n. The coefficient of safety V_n was determined as a ratio between the ultimate load P_n reached into the elastic stage and the corresponding characteristic load P_n.

At the Table 1 the received experimental data for the coefficient of safety V_n and the corresponding form of rupture are presented.

TABLE 1.

Sample	Ultimate loading at the elastic stage P_n,kN	$V_n = \dfrac{P_u}{P_n}$	Remarks
A-1	25,2	2,1	squash of hole
A-2	22,67	1,89	squash of hole
B-1	30,24	2,52	deformation of bolts
B-2	25,15	2,09	–
B-3	22,78	1,90	squash of hole

At the Figure 6 initial rotational stiffness of the joints type A and type B were registered. The criterion for the rotational stiffness is the ratio between the imposed load at the connection and the angle of rotation ψ.

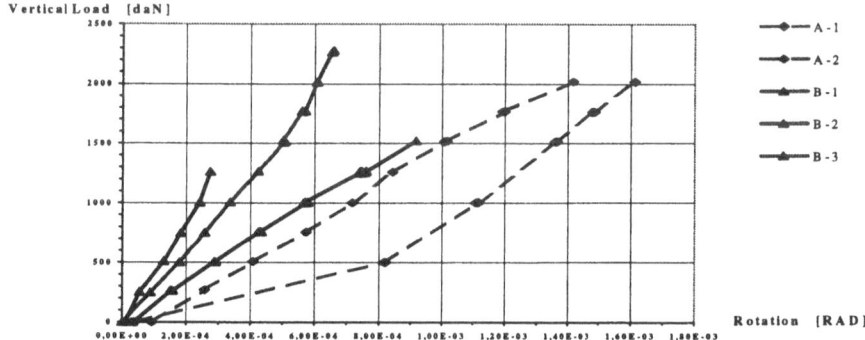

Figure 6

4. Analyses of the results

The results of the experimental investigation demonstrated that up to the characteristic load (P_n=12,00 kN) all the connections work at the elastic stage. After unloading and reloading up to the designed load (P=15,00 kN) considerable plastic deformation are not observed.

The edge deformations measured in the joints with mechanical connection (type A) are relatively larger than those in the joints with mechanical connection and additional welding. Nevertheless when using designed load the edge deformation of the joint type A do not exceed 0,55 mm, which is not dangerous for the structure of the dome.

The measured maximum normal stresses in the joints at designed load are considerably under the characteristic stress of the aluminium alloy.

The results of the experimental investigation indicate that the asymmetric connection gives rise to one side deflection out of the plane of vertical loading, when the load reaches its ultimate value. The tendency at horizontal deflection can be explained with the influence of the imperfection and the type of connection. In the real structure there is available bracing from relevant hoop which would not permit horizontal deflection of the joint.

The testing specimens did not show inclination to creeping of the element even at loading higher than the designed one.

Conclusions:

1. The general conclusion from the experimental investigation is that the joints with mechanical connection have sufficient reliability and offer a possibility to save the expensive welding works.

2. The domes with mechanical connections without welding must be calculated more precisely taking into account the deformation in the connections. This is essential for the domes with big spans.

EXPERIMENTAL AND ANALYTICAL STUDIES ON THE CYCLIC BEHAVIOUR OF COLUMN-BASE JOINTS

S. ÁDÁNY, L. DUNAI
Budapest University of Technology and Economics
1111 Budapest, Műegyetem rkp. 3., Hungary

L.CALADO
Instituto Superior Técnico
1049-001Lisbon, Av. Rovisco Pais, Portugal

1. Introduction

End-plate-type joints are widely used in steel frame structures, connecting either two steel elements (like beam-to-column, beam-to-beam or column-to-column joints) or a steel and a concrete/reinforced concrete element (like column-base joints or joints of a steel beam and a reinforced concrete column). Although these joints have numerous practical advantages, their application results in a more complicated structural behaviour which must be considered in the design.

In the recent decade lots of experimental, analytical and numerical investigations have been performed to analyse the behaviour of the various kinds of end-plate joints. The research has been focused mainly on the monotonic behaviour, while to the cyclic behaviour much less efforts have been devoted. Nevertheless, certain number of experimental programs have been performed ([3], [4], [5], [6], [7], [8]).

The complete understanding of the cyclic behaviour of end-plate joints is essential, especially in the seismic design. The importance of the problem was clearly justified during the recent earthquake events, where significant structural damage of steel frames took place in the connection zones in several cases. Thus, it is important to understand and simply but reliably asses the behaviour of the joints in case of seismic actions, in order to satisfy the required resistance, rigidity, ductility and energy absorption demands.

In this paper the first phase of a research is reported. The aim of the research is the systematic study of the monotonic and cyclic behaviour modes of steel-to-concrete column-base joints. However, in the present report the research concentrates on the steel components only, while the effect of the behaviour of concrete foundation block is not taken into consideration. (Practically it means that a rigid foundation is assumed.) In Section 2 a simple calculation method is presented, for the behaviour mode prediction. In Section 3 the calculation method is applied in designing an experimental program,

C.C. Baniotopoulos and F. Wald (eds.), The Paramount Role of Joints into the Reliable Response of Structures, 147–158.

while Section 4 presents some basic results of the tests. Finally, some basic conclusions are drawn.

2. Prediction of Behaviour Modes of End-Plate Joints

A simple calculation method is developed to predict the behaviour mode and the moment resistance of end-plate joints under monotonic loading. As it has been already mentioned, the method takes only the steel components into consideration. The calculation is based on the possible failure modes of the joint. The moment resistance is determined as the minimum of resistances belonging to the possible failure modes.

Assuming a rigid foundation, and neglecting transverse effects, the resistance of the connection (end-plate + bolts) can be determined on a simple two-dimensional connection model. Four modes of failure can be defined as illustrated in Figure 1. Mode 1 represents the pure end-plate failure without failure of bolts (3-hinge mode). Mode 4 corresponds to the pure bolt failure, without any failure of the end-plate (0-hinge mode). Mode 2 and 3 are two cases of combined bolt and end-plate failure, with two plastic hinges in the end-plate in case of Mode 2 (2-hinge mode), and one plastic hinge for Mode 3 (1-hinge mode).

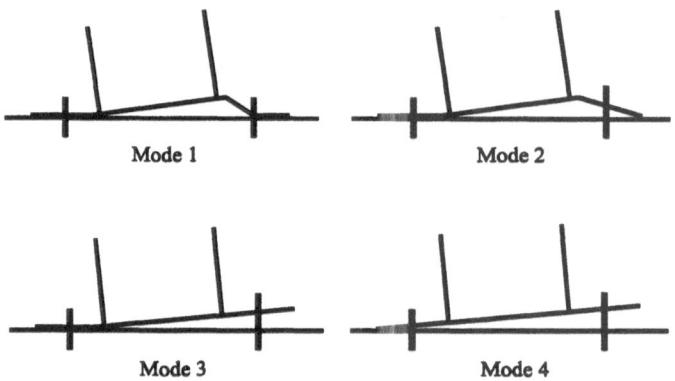

Figure 1. Failure modes of the connection

The connection resistance belonging to the various failure modes can be expressed by the following formulae, for Modes 1 to 4, respectively.

$$M_{Rd,1} = M_{ep,Rd} \cdot 2 \cdot \left(1 + \frac{h'}{m'}\right) \tag{1}$$

$$M_{Rd,2} = M_{ep,Rd} \cdot \left(2 + \frac{h'}{m'+n}\right) + F_{b,Rd} \cdot \frac{h' \cdot n}{m'+n} \tag{2}$$

$$M_{Rd,3} = M_{ep,Rd} + F_{b,Rd} \cdot (h'+m') \tag{3}$$

$$M_{Rd,4} = F_{b,Rd} \cdot b \qquad (4)$$

The notations are given as follows. $M_{ep,Rd}$ denotes the plastic resistance of the end-plate, calculated as:

$$M_{ep,Rd} = \frac{a \cdot t^2}{4} \cdot f_y \qquad (5)$$

$F_{b,Rd}$ denotes the plastic resistance of the bolts (two bolts), calculated as:

$$F_{b,Rd} = 2 \cdot A_b \cdot f_{yb} \qquad (6)$$

h' and m' can be calculated as:

$$h' = h + t \qquad (7)$$

$$m' = m - \frac{t}{2} \qquad (8)$$

A_b is the sectional area of one bolt, a, b, h, m, n, and t are geometrical dimensions of the joint, presented in Figure 2, while f_y and f_{yb} the yield stress of the end-plate material and bolt material, respectively.

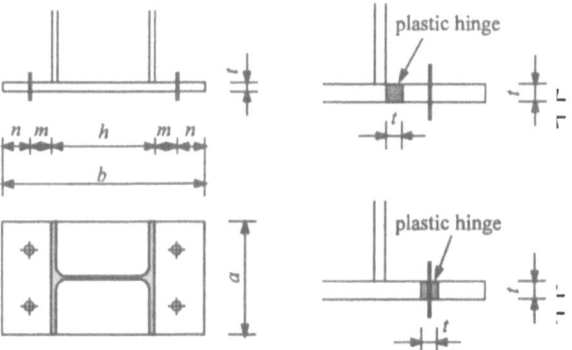

Figure 2. Notations and assumed position of plastic hinges

Moreover, the resistance of the column section is also to be considered. The corresponding resistance can be calculated as the minimum of section plastic resistance and buckling resistance, depending on the cross-section parameters. Appropriate formulae can be found in Eurocode 3 [10]. The calculation depends on the classification of the cross-section, considering plastic reserve or local buckling of the section.

Theoretically, this calculation method, presented above, is valid only in case of monotonic loading and for elastic-perfectly plastic material. However, it can also be applied for cyclic loading and for real mild steel as an approximate method, since it can be assumed that the application of cyclic loads, which may result in developing significant hardening, does not modify the behaviour mode basically, even if certain modification of the behaviour definitely takes place during the subsequent cycles. In the

150

following Section the method is applied for the prediction of the moment resistance, as well as to asses the cyclic behaviour mode in designing an experimental program.

3. Experimental Program

3.1. Test Equipment

The test equipment, on which the present experimental program is carried out, is developed to test beam-to-column joints of steel frames. The global arrangement is illustrated in Figure 3. In addition, there is a lateral frame to make possible the lateral support of the specimen, avoiding its lateral movement or twisting. The whole testing process is managed by a personal computer, by means of a data acquisition unit which commands the actuator and reads the data from the load cell and displacement transducers. More information about the test equipment can be found in [5].

In designing the test, the specimen characteristics are determined in accordance with the parameters of the existing test set-up, by considering the geometrical properties, the load capacity of the actuator and load cell, as well as the displacement capacity of the inductive displacement transducers. The main geometrical dimensions of the specimens are presented in Figure 3. The arrangement represents a column base joint, with an H-shaped column and a practically rigid base. The top part of the specimen has the role to ensure the restraint against lateral movement and twisting of the column.

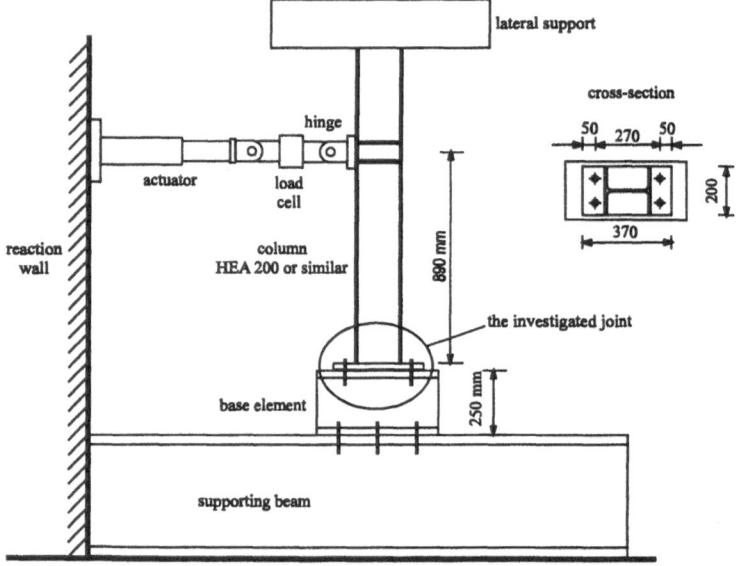

Figure 3. Global arrangement with the main dimensions of the specimens

3.2 Preliminary Calculations

In designing the specimens the calculation method presented in Section 2 is applied. The resistance of the connection part is calculated for various end-plate thicknesses.

The main geometrical dimensions are given in Figure 3. The column section is assumed to be a HEA 200 profile, or similar, the height of which is $h = 190$ mm. For the bolt position and extension of end-plate $m = 40$ mm and $n = 50$ mm are applied. The thickness of the end-plate is treated as a parameter, varying between 10 and 50 mm. The bolt diameter is 16 mm, which gives approximately $A_b = 200$ mm^2 for the area of one bolt.

The applied bolt is of grade 8.8, while the material is S235, which means that the characteristic value of the yield strength is equal to 640 MPa and 235 MPa for the bolt and the base material, respectively, according to the Eurocode 3 [10]. Generally, these characteristic values are adopted as the basis of the calculations. According to previous experiences, however, a higher value is considered for the base material (270 MPa).

The calculations are summarised in Table 1, showing the resistances of the various failure modes for the various plate thickness values. The most probable failure mode is the one to which the minimal resistance belongs.

TABLE 1. Moment resistance calculation of the joints

t [mm]	Plate res. [kNm]	Mode 1 [kNm]	Mode 2 [kNm]	Mode 3 [kNm]	Mode 4 [kNm]	Resistance [kNm]
10	1.35	_18.1_	33.0	55.5	85.2	18.1
12	1.944	_27.0_	36.3	56.3	85.2	27.0
16	3.456	51.4	_44.5_	58.3	85.2	44.5
20	5.4	86.4	_55.2_	60.7	85.2	55.2
22	6.534	108.6	_61.5_	62.1	85.2	61.5
25	8.4375	148.8	72.2	_64.3_	85.2	64.3
30	12.15	238.1	93.7	_68.6_	85.2	68.8
40	21.6	540.0	152.0	_79.2_	85.2	79.2
50	33.75	1147.5	234.7	92.5	_85.2_	85.2

It can be seen from Table 1 that pure bolt failure is not realistic to achieve since it occurs only in case of extremely thick end-plate. For that reason it was decided to eliminate the pure bolt failure from the study and, finally, three pieces of end-plate thickness were chosen, according to failure modes 1, 2 and 3.

Two types of column cross-section are designed. One is a HEA 200 hot-rolled profile, which stocky enough to avoid local buckling. Its resistance is 116.0 kNm, more than any of the resistance values tabulated in Table 1. This means that in all these cases the column failure cannot be expected.

The other section, presented in Figure 4, is designed to study the effect of local buckling, by applying a slender welded column profile with less resistance than that of the connection itself. It is to be noted that the welded section is designed so that its system lines (mid-lines of the plane elements) would be identical with those of HEA 200 section. The cross-section belongs to Class 4 with a capacity equal to 56.5 kNm according to Eurocode 3 [10]. It means that the column local buckling can be expected as governing phenomenon if the end-plate thickness is more than 20 mm, since in this case the resistance of the column section is less than the capacity of the bolts or end-plate.

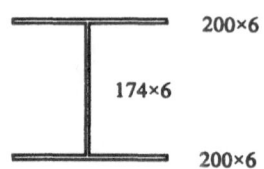

Figure 4. Welded column cross-section

3.3 The Specimens

Altogether five specimens have been designed. The main characteristics of the specimens are summarised in Table 2. Four characteristic behaviour types are anticipated, as well as the effect of bolt pre-tensioning is also studied. Note that CB1 and CB4 are identical, the only difference is in the bolt tightening.

TABLE 2. Specimens main characteristics

Specimen	Column section	End-plate thickness	Bolt tightening	Anticipated behaviour
CB1 (CB1R)	HEA200	25 mm	hand-tightened	Mode 3
CB2	HEA200	16 mm	hand-tightened	Mode 2
CB3	welded	25 mm	hand-tightened	local buckling
CB4	HEA200	25 mm	pre-tensioned	Mode 3
CB5	HEA200	12 mm	hand-tightened	Mode 1

3.4. Loading

The specimens are tested under cyclic loading, by adopting the loading history proposed in the ECCS Recommendations [9]. However, some minor modifications are introduced, according to previous experiments of the authors. The loading process is controlled by the displacement that belongs to the limit of elasticity (e_y). This yielding displacement has been determined by finite element calculation for each specimen. More details about the calculations can be found in [1]. The applied loading history is summarised in Table 3.

TABLE 3. Loading history

Cycle nr.	1	2	3	4	5	6	7	8	9	10	11	
Displ. amplitude	$\frac{1}{4}e_y$	$\frac{1}{2}e_y$	$\frac{3}{4}e_y$	e_y	$2e_y$	$2e_y$	$3e_y$	$3e_y$	$4e_y$	$4e_y$	$5e_y$	etc.

4. Results

4.1. Behaviour Modes

Figures 5 to 8 present the typical deformation patterns experienced during the tests. In case of CB1 and CB4 the behaviour follows Mode 3, with two plastic hinge-lines in the end-plate, and without considerable column deformations, as it clearly can be observed in Figure 5. CB2 specimen shows Mode 2 behaviour, with one hinge-line in the end-plate (Figure 6). In case of CB3 the governing phenomenon is definitely the local buckling of the column cross-section, as it is demonstrated in Figure 7, with practically no end-plate and bolt deformations. For CB4 specimen (Figure 8) the behaviour is between Mode 1 and Mode 2, since the end-plate deformation is similar to that of Mode 1, but certain plasticity in the bolts is also developed.

Figure 5. End-plate deformation in CB1 and CB4 tests

Figure 6. End-plate deformation in CB2 test

154

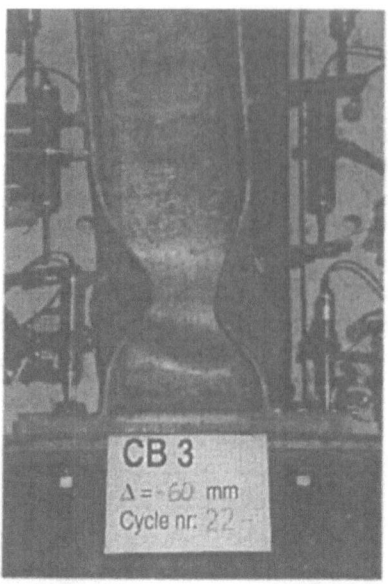

Figure 7. Column local buckling in CB3 test

Figure 8. End-plate deformation in CB5 test

4.2. Moment-Rotation Curves

The moment-rotation relationship of the joint is established on the basis of the measured forces and displacements. To be able to show the results in a unified way, a "joint reference section" is introduced, which is used to calculate the moments and rotations. This section is defined at a distance of twice the column section depth from the base-plate in order to be not disturbed by the intensive deformation due to local buckling.

The moment-rotation curves are presented in Figure 9. Six curves are plotted, because CB1 test was repeated due to some technical problem. The repeated test is referred as CB1R.

Figure 9 demonstrates the typical moment-rotation characteristics for the typical behaviour modes. A general observation that whenever bolt elongation takes place, a considerable rigidity degradation occurs, together with the degradation of energy absorption capacity. At the same time, end-plate deformations and, especially, column local buckling lead to more stable hysteresis behaviour. More detailed evaluation of the results can be found in [1] and [2].

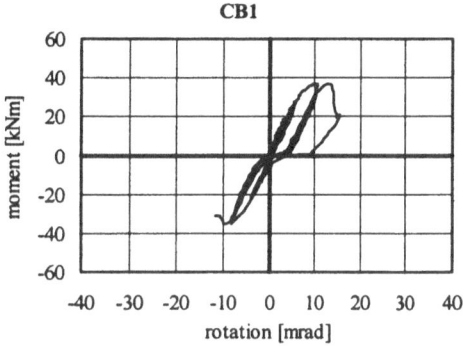

Figure 9a Moment-rotation curve of test CB1

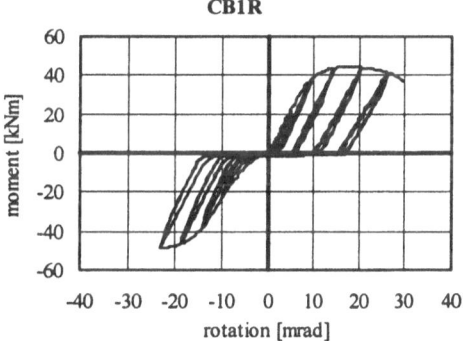

Figure 9b Moment-rotation curve of test CBR1R

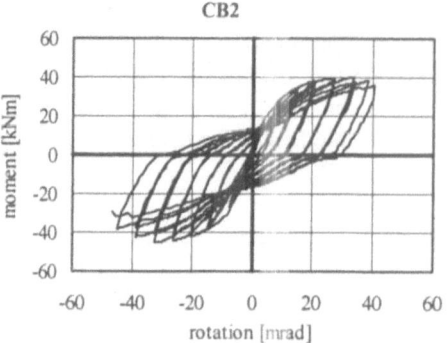

Figure 9c Moment-rotation curves of test CB2

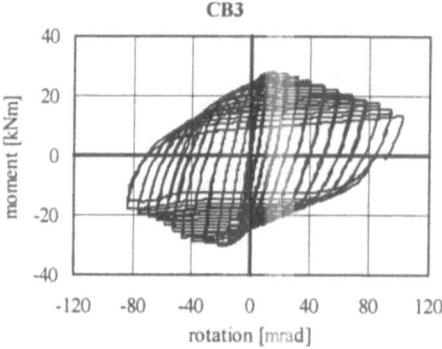

Figure 9d Moment-rotation curves of test CB3

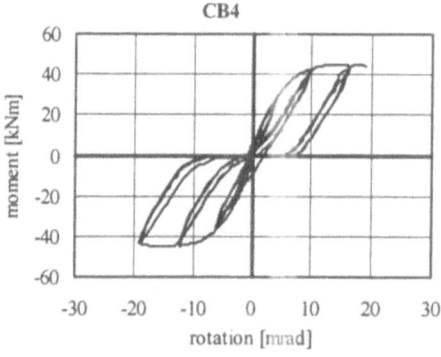

Figure 9e Moment-rotation curves of test CB4

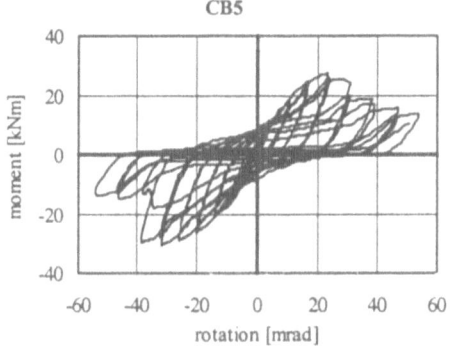

Figure 9f Moment-rotation curves of test CB5

5. Conclusion

In this paper a research on steel bolted end-plate joints is presented with the primary aim of providing information on the typical behaviour modes. Here, some general conclusions are drawn.

It can be stated that the experienced behaviour of each specimen is in accordance with the expected behaviour. Thus, the applied calculation method for the prediction of behaviour mode is justified.

The tests justified the existence of the pre-defined types of behaviour. However, the important effect of weld cracks is also highlighted. Whenever there is intensive end-plate deformation the failure is caused by the cracks occurred at the flange to end-plate welds. The cracks also influence the cyclic characteristics causing significant degradation of the moment resistance and the energy absorption capacity.

The obtained results are applicable for the verification and calibration of numerical models. Detailed experimental data are provided for various behaviour types corresponding to the same joint topology. It is important, however, to study the concrete behaviour, which can be the topic of further investigations.

5   Acknowledgements

The research work has been conducted under the financial support of the following projects: OTKA T029635, PBICT/P/CEG/2359/95, TÉT P-4/99.

158

7. References

1. Adany, S. 2000. "Numerical and Experimental Analysis of Bolted End-Plate Joints under Monotonic and Cyclic Loading", *PhD Dissertation*, Budapest University of Technology and Economics.
2. Adany, S., Calado, L. & Dunai, L. 1999. "Experimental Program on Bolted End-Plate Joints", *ICIST Internal Report*, Lisbon, Portugal.
3. Ballio, G., Calado, L. & Castiglioni, C. A. 1997. "Low Cycle Fatigue Behaviour of Structural Steel Members and Connections", *Fatigue & Fracture of Engineering Materials & Structures*, Vol. 20, No. 8, pp. 1129-1146.
4. Calado, L., Bernuzzi, C. & Castiglioni, C. A. 1998. "Structural Steel Components under Low-cycle Fatigue: Design Assisted by Testing", Structural Engineering World Congress, SEWC, San Francisco.
5. Calado, L. & Lamas, A. 1998. "Seismic Modelling and Behaviour of Steel Beam-to-Column Connections", 2nd World Conference on Steel Construction, San Sebastian, Spain.
6. Calado, L. & Mele, E. 1999. "Experimental Research Program on Steel Beam-to-Column Connections ", Report ICIST, DT no 1/99, ISSN:0871-7869
7. Calado, L., Mele, E. & De Luca, A. 1999. "Cyclic Behaviour of Steel Semirigid Beam-to-Column Connections", *to be published in ASCE.*
8. Dunai, L., Fukumoto, Y. & Ohtani, Y. 1996. "Behaviour of Steel-to-Concrete Connections under Combined Axial Force and Cyclic Bending", *Journal of Constructional Steel Research*, Vol. 36, No. 2, pp. 121-147.
9. ECCS 1986. "Recommended Testing Procedure for Assessing the Behaviour of Structural Steel Elements under Cyclic Loads", Technical Committee 1, TWG 1.3, No. 45.
10. ENV 1993-1-1: Eurocode 3 1993. "Design Rules for Steel Structures", Part 1-1, "General Rules and Rules for Buildings."

BOLTED JOINTS IN THIN WALLED CROSS-SECTIONS OF TYPE Z MADE FROM COLD FORMED STEEL

M. D. ZYGOMALAS
Aristotle University, Department of Civil Engineering
GR-54006, Thessaloniki, Greece

1. Introduction

The following Figure 1 and Table 1 show all the kinds of type Z cross-sections used for the construction of purlines in metallic roofs.

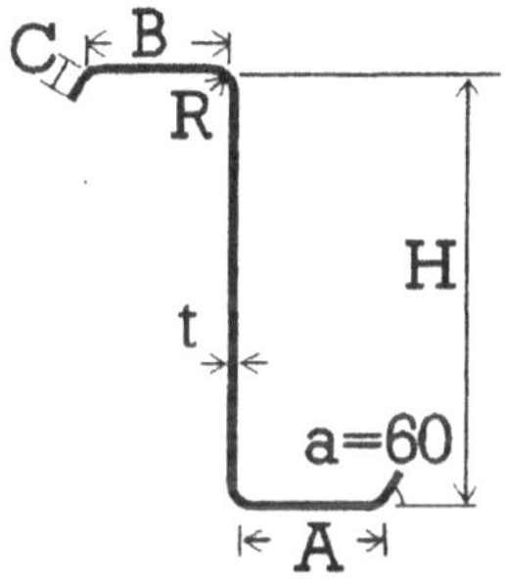

TABLE 1. Geometrical values of cross sections used (in mm)

	t	H	A	B	C	R
Z140.15	1.46	140	65	60	20	4
Z140.20	1.96	140	65	60	20	4
Z180.15	1.46	180	65	60	20	4
Z180.20	1.96	180	65	60	20	4
Z180.25	2.46	180	65	60	20	4
Z210.15	1.46	210	65	60	20	4
Z210.20	1.96	210	65	60	20	4
Z210.25	2.46	210	65	60	20	4

Figure 1. (cross section)

Table 1 covers all cross sections of type Z available in market where the material used is steel type GC-280 Mpa and all the pieces are galvanized.

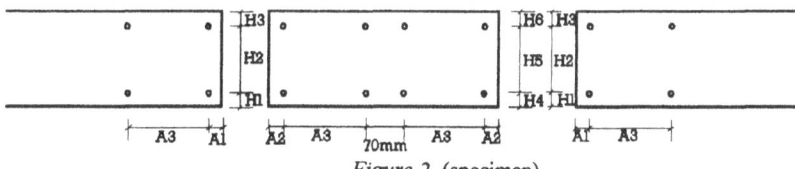

Figure 2. (specimen)

Having as scope reliable experimental results, for each one of the existing cross-sections three same specimens have been tested.

An additional piece of the same material used for the beams, has been used to connect the two pieces. The diameters of the holes were 2 mm bigger than the diameters of the bolts. For the cross-sections Z140, bolts of 12 mm diameter were used, while 16 mm were used for all the other cross-sections. Figure 2 shows the details of the

159

C.C. Baniotopoulos and F. Wald (eds.), The Paramount Role of Joints into the Reliable Response of Structures, 159–168.

respective joints, the position of the holes used for the bolted connection on the knot, which depends of the size of the profile.

TABLE 2 Distances for Fig. 2

A/A	A1	A2	A3	H1	H2	H3	H4	H5	H6
Z140	32	35	230	34	70	36	36	70	34
Z180	32	35	300	44	90	46	46	90	44
Z210	32	35	360	59	90	61	61	90	59

The dimensions shown in Table 2 describe the location of the centres of the holes in the connected pieces of the beam with the additional metallic piece, depending on the height of the cross section as defined in Eurocode.[1] Diameters of holes for profiles of 140 mm height have been chosen equal to 14 mm, while for all the other types of profiles equal to 18 mm.

2. Theoretical Approach

Parts of the theoretical solution of Bryan [2] have been used. The mathematical solution for the semi-rigid connections at hand is:

$$c = 5n(\frac{10}{t_1} + \frac{10}{t_2} - 2)10^{-3} \text{ mm/kN} \tag{1}$$

where c is the flexibility of the joint (in mm/kN), t_1 and t_2 are the thicknesses of the two connected pieces (in mm and $t_1 < 8mm$ $t_2 < 8mm$) and n is a factor which has to do with the number of bolts used in the connection, the nests or interlock and whether the shear plane is on plane shank or on thread. The values of the factor n are given in the Tables 3 and 4

TABLE 3 Values of n in eqn(2.1)for bolted joints in tension	One bolt connection	Two or more bolt connection
Shear plane on plain shank	3.0	1.8
Shear plane on thread	5.0	3.0

TABLE 4 Values of n in eqn(2.1) for bolted joints under moment	Simple bolted joints	Joint which nest or interlock	Joint which nest and interlock
Shear plane on plain shank	1.8	1.4	1.2
Shear plane on thread	3.0	2.4	2.0

3. Experiments

For the experiments performed the horizontal symmetrical simply supported beams have been used. The static load was placed in the middle of the span, was vertical. The specimens used are presented in figure 4. They were prepared in such a way so as to have the same operation as in real-life constructions. The load had to be parallel with the vertical piece of the cross-section and pass through the middle of the thickness of the joint. Because of the risk of the specinan being turned over during the experiment, special mechanical devices have been constructed to prevent the movement or rotation in bearings and in the middle of length, where the load acted. The devices shown in Figure 4 have been constructed for the bearings of the beam. They had a cylinder to slot and rotate and two plates to keep the cross-section vertical and undeformed. In the position where the load acted, a reproduction of real construction has been made. So, two 8mm thick metal plates were used in perpendicular position, as a bearing of the joint to the truss of the roof. This part of the construction is immovable and rigid. There were 4 holes on it to realize the bolted connection with the beam. On the upper horizontal plate, a cylindrical grove was made to stabilize the position of the load in a place where it was vertical passing through the middle of the thickness of the vertical part of the beam.

For every experiment the displacement of 7 different point have been measured and stored (Figure 4). Two of the measured points corresponded to the ends and one of the middle of the additional metallic part of the beam (Points 3,4,1 Fig. 4). Two points were near the end of the additional part on the beams, (Points 2,5 Fig.4) and the last two points were at the end of two beamson the joint (Points 6,7 Fig.4). The sensors used to measure the displacements were high quality linear extensiometers. A linear extensiometer, connected to the hydraulic press was used to measure the acting load on the specimen.

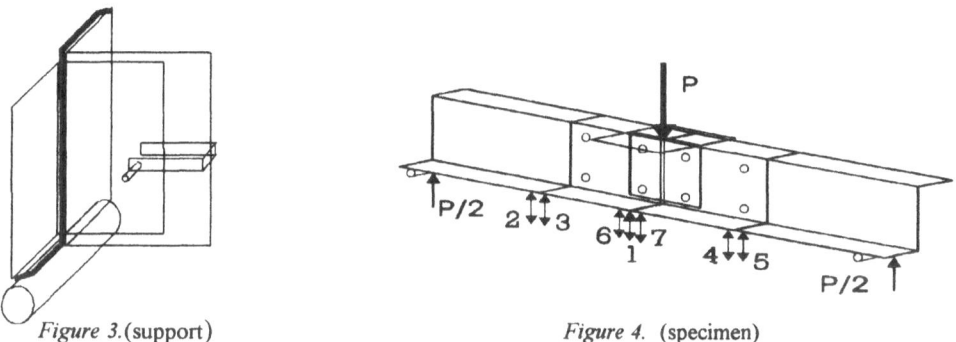

Figure 3.(support)   *Figure 4.* (specimen)

The recordings were made using a computer every 500 msec for 7 values of displacement and one value for load. 500 recordings were made for each one of the experiments.

The transformation of the linear electric signals to digital ones have been made by using an A/D, D/A card with an accuracy of 14 bits with 16 inputs and 2 outputs.

In real constructions, the creation of the joint starts with the connection of the additional metallic beam to the truss through a 8 mm thick type L metallic piece with 4 holes for the bolts of the connection. This additional part is the bearing for the two beams. For each one of the joints, a total number of 8 bolts was used as shown in Fig. 4. 4 out of the 8 bolts connect two thicknesses of cross-sections of type Z at the end of the additional part. Besides the two thicknesses, the 4 central bolts connect the 8 mm type L piece, which is the bearing between the joint and truss.

Each experiment was carried out till the full collapse of the specimen. At the end of each experiment, the specimen was dismounted and the traces due to the plastic deformation in the area of the holes were examined, in order to determine the position of the centre of the real elastic rotation for each metallic beam.

4. Static Solution for the Beam

The composite beam used was a simply supported beam supported by two rolling bearings, subjected to concentrated static load P in the middle of its length. Obviously,

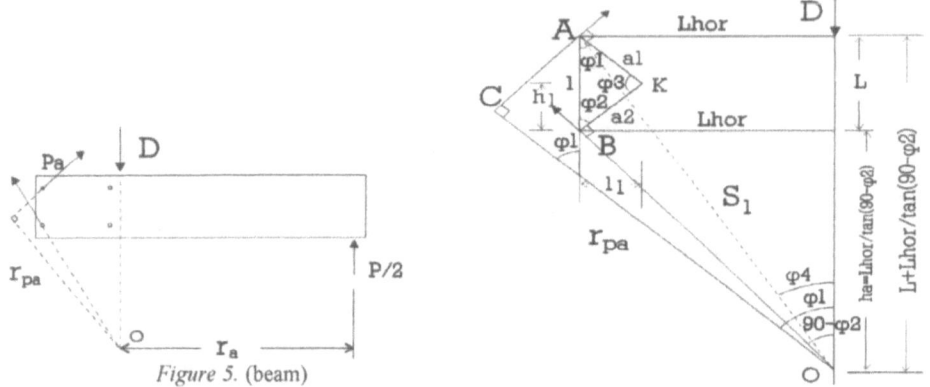

Figure 5. (beam)

Figure 6. (geometrical data)

the reactions were symmetrical and equal to the half of the load (P/2). They were vertical because the bearings, as they were described before, can only receive and distribute vertical loads.

Examining the loads acting on the structure, we found out that they were acting isostaticaly and the calculation of all of them was an easy task.

Fig. 5 shows the loads acting on the part of the beam at the right side of the joint. The definition of the direction of the two loads on the left edge of the beam, was done after the examination of the traces of the plastic deformation in the holes done by the bolts. Load D is the reaction at the end of the additional part and is vertical, because there is a slipping contact in the region between the two metallic pieces and any pass of horizontal load is impossible. This contact makes the two bolts under it inactive, because the differential movement of the holes is impossible and the bolts passing through them are not able to carry loads because of the 2 mm difference in diameter between bolts and holes. We conclude to the same result after the experiment because there is no deformation around the holes in the area of these bolted connections. The calculation of the load Pa (Fig. 5) derives from the moment balance around point O. This equation is:

$$Pa x r_{pa} = (P/2) x r_a \tag{2}$$

Equation 2 gives the value of the load **Pa** because all the other values are known.

Fig. 6 presents the distances used for the geometrical calculation of the length r_{pa} which is the distance between load **Pa** and the section of load **D** and the load of the bolt in hole B.

After the detailed measurement on the specimens, the values of the angles φ_1 and φ_2, and distance AB=1 between the centres of the holes A and B are known. The lengths L and L_{hor} from the specimen geometrical data are also known. Angle φ_3, is calculated using the geometrical equation:

$$\varphi_3 = 180 - \varphi_1 - \varphi_2. \tag{3}$$

In triangle ABK point K, which is the centre of the plastic rotation, is calculated using the equations:

$$a_2 / \sin(\varphi_1) = a_1 / \sin(\varphi_2) = 1 / \sin(\varphi_3) \tag{4}$$

$$\text{and } h_1 = a_2 \times \cos(\varphi_2) \quad l_1 = a_2 \times \sin(\varphi_2) \tag{5}$$

The distance of centre O, used for the moments balance and the centre of hole A is:

$$S_1 = (h_a + L)/\cos(\varphi_4) \tag{6}$$

Finally the distance between O and load **Pa** is:

$$r_{pa} = S_1 \times \cos(\varphi_1 - \varphi_4) \tag{7}$$

The distance between load P/2 and point B (L_{hor} in Fig.6) is equal to 615 mm for cross sections Z140, 765 mm for cross sections Z180 and 965 mm for cross sections Z210. Finally:

$$r_a = L_{hor} - A_2 - A_3 \tag{8}$$

5. Results of Experimental Measurements.

As a continuation of the previous calculations, the description of the rotation on the joints is theoretically possible. The flexibility of the joint, the position of the centre of the plastic rotation and the load which the bolt passes to the beam are known. After the calculations and the detailed examination of the experimental results, the following empirical equations were found:

$$l_1 = 0.12 \times L_{hor}$$
$$h_1 = 0.63 \times L \tag{9}$$

These equations are only for beams consisting of two long pieces and one additional piece in the centre, all of them with cross-section of type Z (as shown in Table 1) using 8 bolts for the joint (as in Fig.2 and Table 2).

164

After the processing of the experimental measurements, the diagrams shown in section 6 have been created. In all these cases the horizontal axis represents the moment in the joint in kNcm and the vertical axis represents the rotation between the two long pieces of the joint in degrees.

Each one of the diagrams contains the results of three same specimens. Having in mind to simulate the joint similarly to the real situation in steel roofs, the position of the centre of the bolts in the holes were randomly put in the connection area.

6. Conclusions

All the diagrams start with a branch where the elevation of the line is parallel to the calculated semi-rigid connection, or they are equal to zero for a small moment. This means that the composite beam deforms at its ends, where the bearings are, without rotating the joint or it contains a part of the slipping due to the difference in bolt and hole diameters.

In all the diagrams, a branch follows wherever slipping of the joints takes place. The elevation of this part of the diagram is not vertical but it grows up with the moment.

Another branch follows wherever the moment and torsion of the joint verify the equation 1. In some cases the next branch is the breakdown of the beam. In some others, the elevation changes and represents the new situation after the conduct of the upper parts of cross-sections Z, and the plastic deformation of the additional part to the joint. For this case the plastic rotation centre moves towards the contact point on the axis of symmetry. Therefore, the calculation in section 4 is not real. The above takes place just before the breakdown of the beam and after the semi-rigid deformation and being not important, it is not examined.

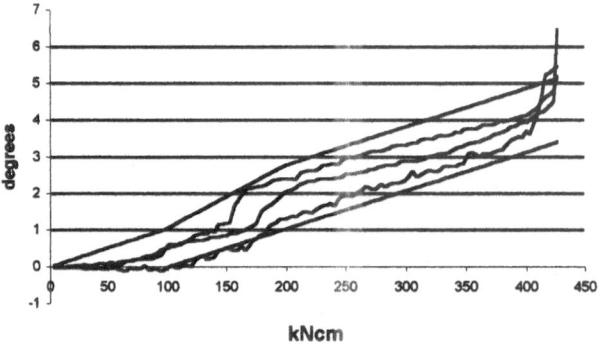

Figure 7. (Z 140/15)

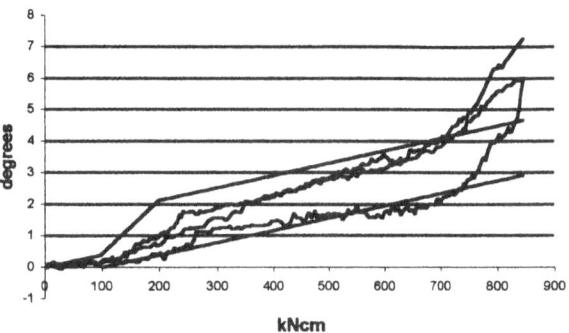

Figure 8. (Z 140/20)

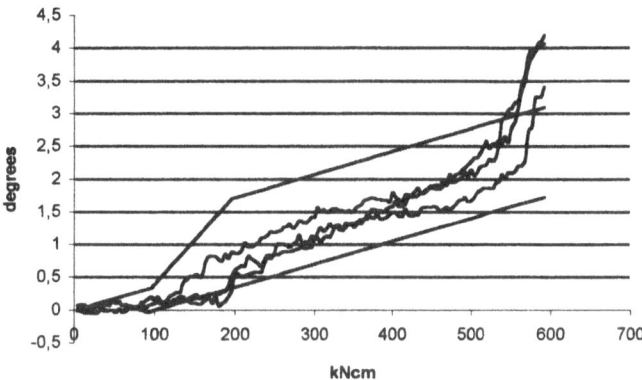

Figure 9. (Z 180/15)

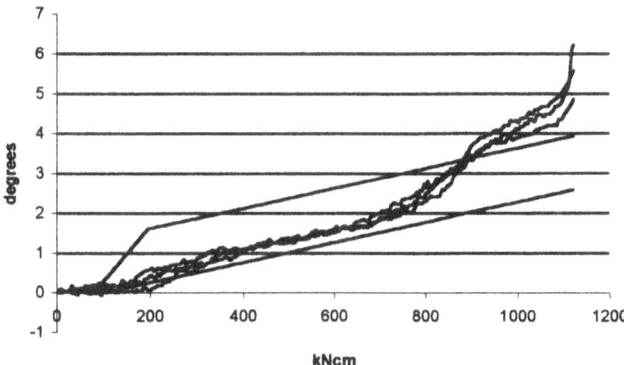

Figure 10. (Z180/20)

166

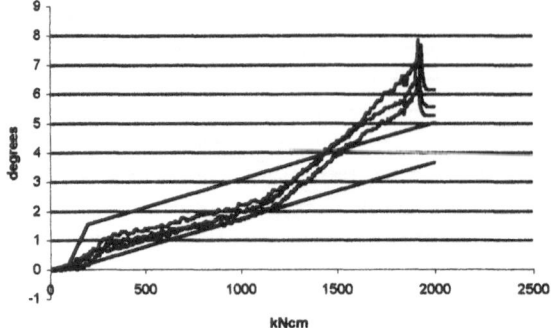

Figure 11. (Z 180/25)

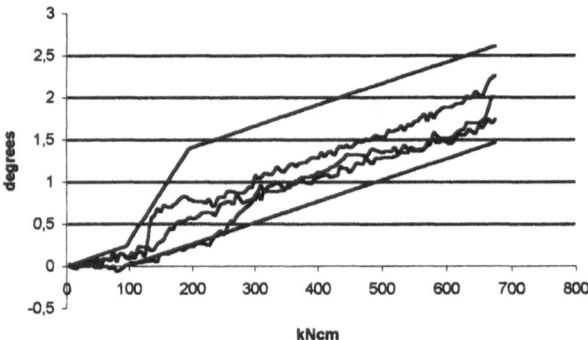

Figure 12. (Z 210/15)

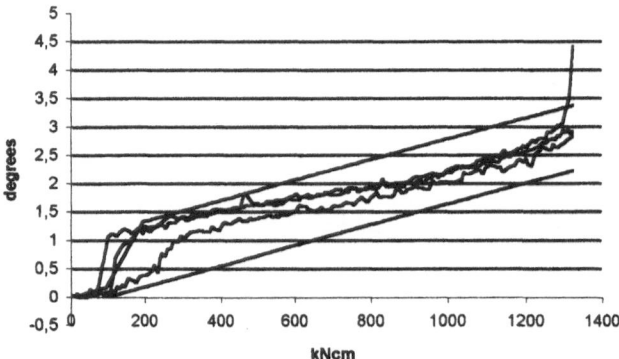

Figure 13. (Z 210/20)

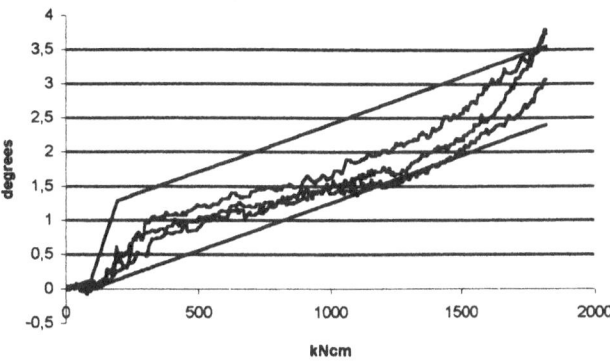

Figure 14. (Z 210/25)

It must be noticed that out of three iron plates in every bolt only two deformed, carrying loads exhibiting a semi-rigid behaviour. For the cross sections Z140 the two thin walled place just before the breakdown of the beam and after the semi-rigid deformation and being not important, it is not examined.

It must be noticed that out of three iron plates in every bolt only two deformed, carrying loads exhibiting a semi-rigid behaviour. For the cross sections Z140 the two thin walled plates deformed, when for all the other cross-sections the thin-walled additional part of the joint didn't deform. But the beams had a semi rigid connection with the 8 mm thick iron plate.

Based on the previous remarks, the following conclusive results have been obtained:

1. With the assembling of the joint in such a position that the whole slipping of the joint happened before the torque of the bolts, there is a moment area from 0 to 100 kNcm where possibly no rotation of the joint takes place. After that moment, the joint works as a semi-rigid connection exhibiting a response described by the equation 1. This line is the lower boundary of the values for each one of the cross sections.

2. With the assembling of the joint in such a position where the rotation of the joint is in the maximum position in the opposite direction of the real rotation, there is an area from 0 to 100 kNcm where a plastic deformation, (following the equation 1) can take place due to the partial slipping of the bolts that touch the borders of the holes. A new area follows from 100 to 200 kNcm where the slipping of plates stops, because of the difference of the 2 mm in bolt and hole diameters for every plate. After 200 kNcm, the curve follows the equation 1.

3. After the above, it is not possible to represent the relation between torsion of the joint and rotation by a single curve. However an area with a maximum border and a minimum border (as in Figures 7 to 14) can be represented. The average of the two

borders can be a good approach to the semi-rigid behaviour but this can not be the theoretical solution of the semi-rigid connections.

Acknowledgements

The continuous support of the late Prof. P.D.Panagiotopoulos (1950-1998) to the author during the performance of the present research work is gratefully acknowledged.

7. References

1. EUROCODE 3, CEN, Brussels
2. E.R.Bryan (1993) *The design of Bolted Joints in Cold-Formed Steel Sections.* Thin-Walled Structures 16 239-262
3. ANSE STANDARD (1991) *Specification for the Design of Cold-Formed Stainless Steel Structural Members*
4. WEI-WEN YU,Ph. D (1985) *Cold-Formed Steel* Design Wiley-Interscience Publication John Wiley & Sons
5. J.W.B.Stark -.W.Toma (1991) J. RHODES A.C. WALKER *DEVELOPMENTS IN THIN-WALLED STRUCTURES-1* Applied Science Publishers- LONDON

REVIEW AND CLASSIFICATION OF SEMI-RIGID CONNECTIONS

K. M. ABDALLA
Assistant Professor
Jordan University of Science and Technology
Civil Engineering Department
Irbid - Jordan

1. Introduction

Most design engineers assume the behavior of their building connections either as perfectly pinned or as completely fixed elements. This simplification results in an inaccurate prediction of frame behavior. Full-scale experiments are generally necessary to describe actual behavior of these connections. At the University of Illinois, Young [1], (1917) Wilson, and Moore [2], (1917), performed the first experiment to assess the rigidity of steel frame connections. Since then, experimental testing has been continued.

The recent AISC design code [3], referred to as the Load and Resistance Factor Design (LRFD) specification (1986), designates two types in its provision: Type FR (Full Restrained) and Type PR (Partially Restrained). The primary distortion of steel beam-to-column connections is their rotational deformation, Q_r, caused by the in - plain bending, M, Fig. 1. This connection deformation has a destabilizing effect on frame stability since it adds additional drift to the frame and results in a decrease in its effective stiffness of the member to which the connections are attached. An increased frame drift will intensify the P-D effect and hence the overall stability of the frame will be affected. Thus, the nonlinear characteristics of beam-to-column connections play a very important role in the structural design.

Prior to 1950, most connection test was focussed on riveted joint (Batho and coworkers 1931, 1934, 1936 [4,5,6], Young and coworkers, 1928, 1934, Rathbun, 1936) [7]. After 1950, high strength bolts have been used extensively in steel construction. A large number of tests have been made and reported. Jones, et al. (1980, 1983) [8], reviewed and collected a total of 323 tests from 29 separate studies. Nethercot (1984) [9] examined and evaluated more than 800 individual tests from open literature. Goverdhan (1984) [10] collected a total of 230 experimental moment - rotation curves and digitizes them to form the database of connection behavior. Kishi and Chen (1986a,1986b), [11,12]

C.C. Baniotopoulos and F. Wald (eds.), The Paramount Role of Joints into the Reliable Response of Structures, 169–178.
© 2000 *Kluwer Academic*

extended Goverdhan's collection (1984) to a total of 303 tests and crated a computerized data bank system together with modified exponential curve fitting program. Abdalla, Chen and Kishi expand this database by adding additional 46 experimental test data of steel - to - column connections that have been collected and analyzed up to date. The types of semi-rigid connections collected are given in the SCDB program. For each experimental datum, moment - rotation characteristics together with all the parameters used in prediction equation are included.

2. Modeling of Connection

There are several connection models reported in open literature moment rotation curves. These are:

2.1. LINEAR MODEL

a) The linear models were proposed by Batho [4,5,6], and Baker [13].
b) The bi-linear models were proposed by Melchers and Kaur [14], Romstad and Subramanian [15], and Lui and Chen [16].
c) The piecewise linear models were proposed by Razzaq [17].

2.2 POLYNOMIAL MODEL

Frye and Morris used an odd-power polynomial to represent the moment - rotation curve [18].

2.3. CUBIC B - SPLINE MODEL
This model can fit test data well [19,20,21].

2.4 POWER MODEL

a) The power model proposed by Batho and Lash [5]. and, Krishnamurthy et al. [22].
b) The power models proposed by Colson and Louveau [23], Goldberg and Richard [24].
c) The three parameter power model by Kishi and Chen [25,26].
d) The Ang and Morris model [30].

2.5 EXPONENTIAL MODEL

a) Chen and Lui parameter model.
b) Kishi and Chen extend Chen - Lui model.

c) Yee and Melchers [32] four parameter exponential model.

3. Classification of Connections

Connections can be classified according to the technique of fastening, rigidity of connection, and type of force action.

3.1. CLASSIFICATION ACCORDING TO FASTENING

Bolts were used first, followed by pinned, riveted, and welded connections. At the present time, most connections are either welded or bolted.

3.2. CLASSIFICATION ACCORDING TO CONNECTION RIGIDITY

According the rigidity of connections, they can be divided into three types [AISC specification (3.35) section 1.2, p. 5-14]:

Type 1. Rigid frame connections.
Type 2. Simple connection.
Type 3. Semi-rigid connections.

3.3. CLASSIFICATION ACCORDING TO THE TYPE OF FORCE ACTION

Depending on the position of the connected members and the corresponding force action, connections can be exposed to:

a) Axial shear,
b) Torsion and eccentric shear,
c) Tensions in bolted connections,
d) Combined axial tension and shear.

3.4. TYPE OF SEMI-RIGID CONNECTIONS

Semi-rigid connection in steel can be divided into the following eleven basic types:

3.4.1. *Single Web Angle*

Comparison results between actual experimental test and theoretical prediction of moment rotation curves are given in figure 1.

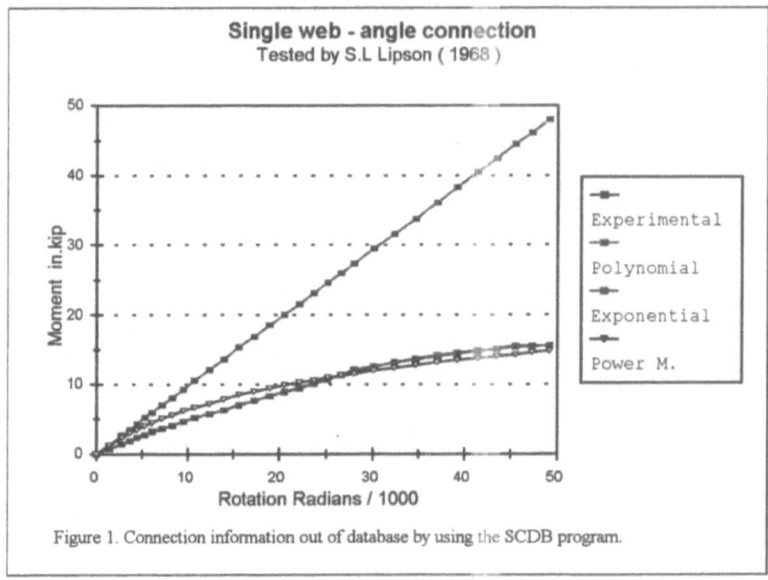

Figure 1. Connection information out of database by using the SCDB program.

3.4.2. *Double Web Angle*

Comparison results between actual experimental test and theoretical prediction of moment rotation curves are given in figure 2.

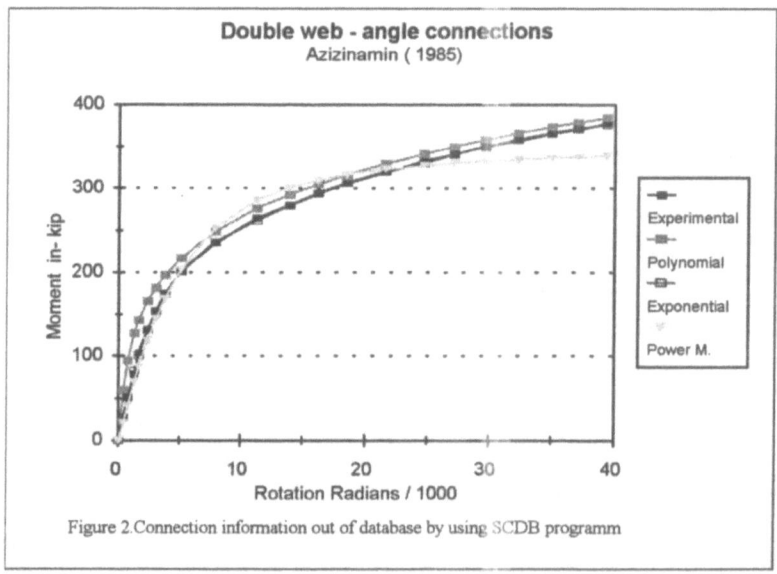

Figure 2.Connection information out of database by using SCDB programm

173

3.4.3. *Top – and Seat –Angle with Double Web Angle*

Comparison results between actual experimental test and theoretical prediction of moment rotation curves are given in figure 3.

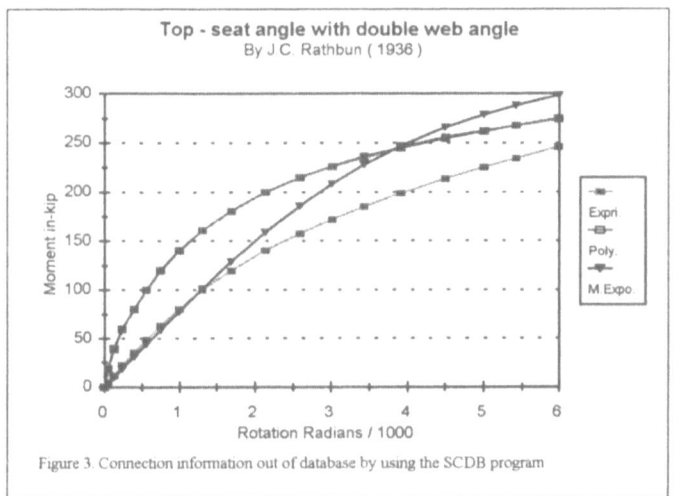

Figure 3. Connection information out of database by using the SCDB program

3.4.4. *Top and Seat-Angle Connections*

Fifty-three sets of moment-rotation curves have been obtained. Comparison results between actual experimental test and theoretical prediction of moment rotation curves are shown in figure 4.

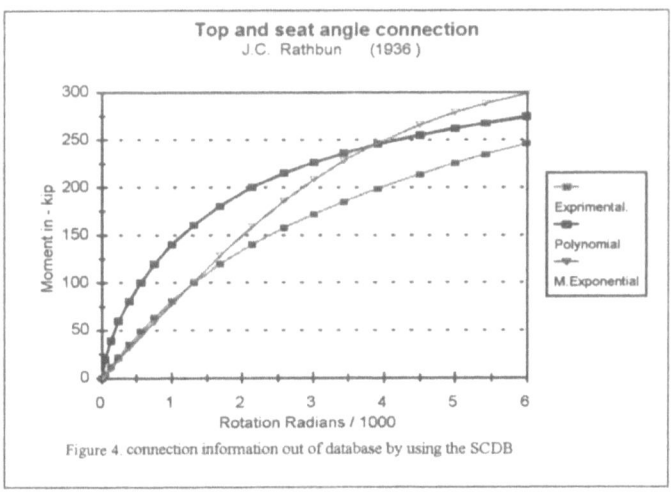

Figure 4. connection information out of database by using the SCDB

3.4.5. *Extended end - plate connections*

One hundred sets of moment – rotation curves are available in Table 5. Ffigure 5 shows typical moment-rotation curves.

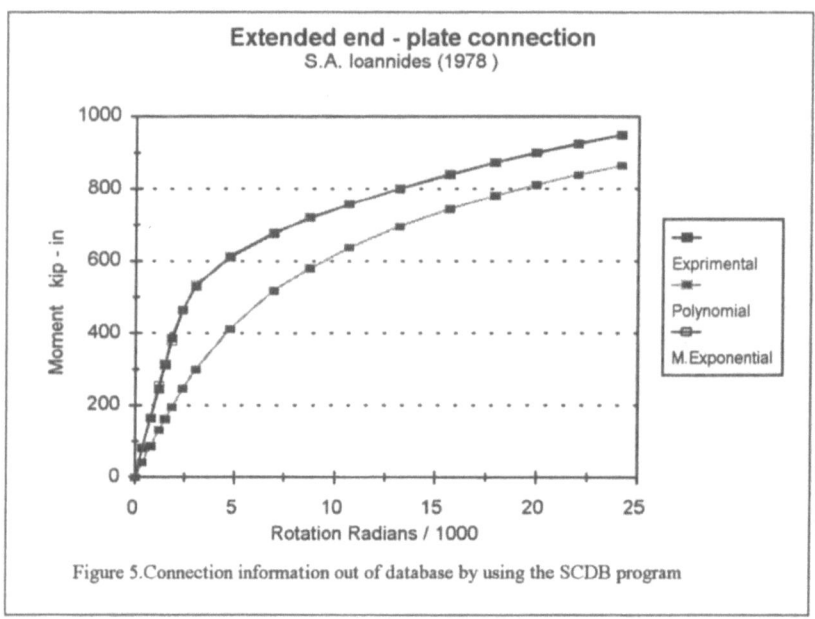

Figure 5.Connection information out of database by using the SCDB program

3.4.6. *Flush End – Plate Connections*

Data on typical moment-rotation curve is given in table 1. Comparison results between actual experimental test and theoretical prediction of moment rotation curves are shown in figure 6.

TABLE 1. FLUSH END – PLATE CONNECTION

No	Author	Test Id.	Beam	Column	li(in)	g(in)	tp(in)	db
1	J.R.Ostrander (1970)	TEST 1	W10X21	W8X28	5	3.5	0.5	¾"
2		TEST 3	W10X21	W8X28	5	3.5	0.375	¾"
3		TEST 4	W10X21	W8X28	5	3.5	0.25	¾"
4		TEST 9	W10X21	W8X28	5	3.5	0.75	¾"
5		TEST 11	W12X27	W8X40	7	4	0.375	¾"
6		TEST 12	W12X27	W8X40	7	4	0.5	¾"
7		TEST 13	W12X27	W8X40	7	4	0.625	¾"
8		TEST 17	W12X27	W8X24	7	4	0.375	¾"
9		TEST 18	W12X27	W8X24	7	4	0.5	¾"

10		TEST 19	W12X27	W8X24	7	4	0.625	¾"
11		TEST 23	W12X27	W8X48	7	4	0.625	¾"
12	J.R.Ostrander (1970)	TEST 2	W10X21	W8X28	5	3.5	0.5	¾"
13		TEST 5	W10X21	W8X28	5	3.5	0.5	¾"
14		TEST 6	W10X21	W8X28	5	3.5	0.375	¾"
15		TEST 7	W10X21	W8X28	5	3.5	0.25	¾"
16		TEST 8	W10X21	W8X28	5	3.5	0.25	¾"
17		TEST 10	W10X21	W8X28	5	3.5	0.75	¾"
18		TEST 14	W12X27	W8X40	7	4	0.375	¾"
19		TEST 15	W12X27	W8X40	7	4	0.5	¾"
20		TEST 16	W12X27	W8X40	7	4	0.625	¾"
21		TEST 20	W12X27	W8X24	7	4	0.375	¾"
22		TEST 21	W12X27	W8X24	7	4	0.5	¾"
23		TEST 22	W12X27	W8X24	7	4	0.625	¾"
24		TEST 24	W12X27	W8X48	7	4	0.625	¾"

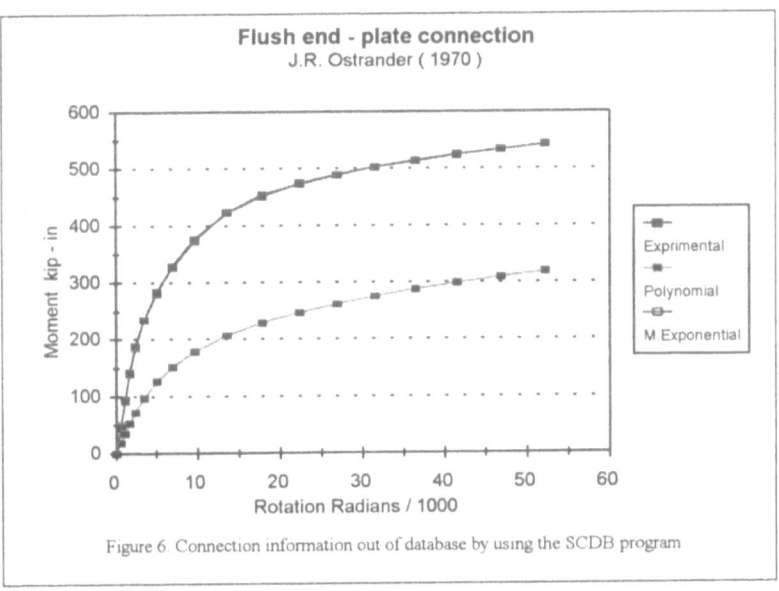

Figure 6. Connection information out of database by using the SCDB program

3.4.7. *Header Plate Connections*

Twenty set of moment-rotation curves are available and a typical moment-rotation is given in table 2. Comparison results between actual experimental test and theoretical prediction of moment rotation curves are shown in figure 7.

176

TABLE 2. HEADER PLATE CONNECTION

No	Author	Test Id.	Beam	Column	lp(in)	g(in)	tp(in)	db
1	W.H.Sommer (1969)	TEST 5	W 18X45	--	15	4	0.25	3/4"
2		TEST 6	W 24X76	--	9	4	0.25	3/4"
3		TEST 7	W 24X76	--	12	4	0.25	3/4"
4		TEST 8	W 24X76	--	15	4	0.25	3/4"
5		TEST 9	W 24X76	--	18	4	0.25	3/4"
6		TEST 10	W 18X45	--	9	4	0.375	3/4"
7		TEST 11	W 18X45	--	12	4	0.375	3/4"
8		TEST 12	W 18X45	--	15	4	0.375	3/4"
9		TEST 13	W 24X76	--	9	4	0.375	3/4"
10		TEST 14	W 24X76	--	12	4	0.375	3/4"
11		TEST 15	W 24X76	--	15	5.5	0.375	3/4"
12		TEST 16	W 24X76	--	18	5.5	0.375	3/4"
13		TEST 17	W 24X76	--	18	5.5	0.25	3/4"
14		TEST 18	W 24X76	--	15	5.5	0.25	3/4"
15		TEST 19	W 24X76	--	12	5.5	0.5	3/4"
16		TEST 20	W 24X76	--	15	5.5	0.5	3/4"
17		TEST 25	W 18X45	--	12	4	0.25	3/4"
18		TEST 26	W 18X45	--	9	4	0.25	3/4"
19		TEST 27	W 12X27	--	9	4	0.25	3/4"
20		TEST 28	W 12X27	--	6	4	0.25	3/4"

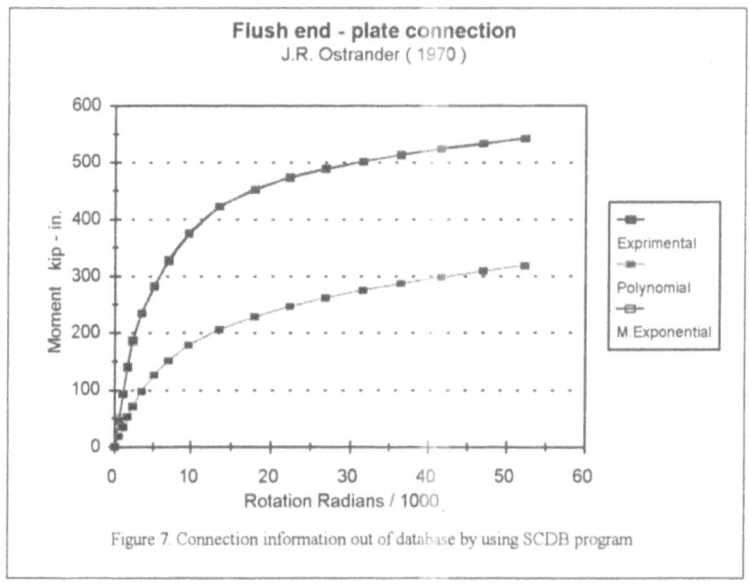

Figure 7. Connection information out of database by using SCDB program

3.4.8. *Web Side Plate*
Eighty fife set of moment-rotation curves by Lipson (1968, Canada), and seven set by Richard et al. (1980, U.S.A.).

3.4.9. *Tee Studs*
Eight set by Bannister (1966, U.K.) and eight by Zotenmijer (1974, Netherlands) of moment rotation curves.

3.4.10. *Top plate and Seat Angle*
Three set of moment rotation curves by Van Dalen and Godoy (1982,Canada).

3.4.11. *Tee Studs and Web*
Therteen set of moment rotation curves by Zoetemejir (1974, Metherlands).

4. References

1. Young, C.R. Bulletin no.4, (1917) Engineering Experiment station, University of Illinois.
2. Wilson, W.M. and. Moore, H.F (1917) Tests to determine the rigidity of riveted joints in steel structures. *Bulletin no.104, Engineering Experiment station, University of Illinois, Urbane.*
3. AISC (1986), Load and resistance factor design specification for buildings, pp. 5.130-5.145. American Institute of steel Construction, Chicago, IL.
4. Batho, C. (1931) Investigations on beam and stanchion connections. *1st Report, steel structures Research Committee, Department of Scientific and Industrial Research, Vol. 1-2*, p.p. 61-137. *HMSO, London*
5. Batho, C. and. Lash, S. D. (1936) Further investigations on beam and stanchion connections encased in concrete. Together with laboratory Investigation on a full-scale steel frame, *Final Report, steel structures Research Committee, Department of Scientific and industrial Research*, pp. 276-363. *HMSO, London.*
6. Batho, C. and. Rowan, H. C. (1944) Investigations on beam and stanchion connections. *2nd report, Steel Structures Committee. Department of Scientific and Industrial Research.* p.92. *HMSO. London.*
7. Young, C.R. and Dunbar, W.B. (1928) Permissible stresses on rivets in tension. *Bulletin, no. 8, Section no. 16, School of Engineering Research, University of Toronto.*
8. Jones S.W, Kirby, P.A. and Nethercot, D.A. (1980) Effect of semi-rigid connections and steel columns strength, *J. Construt. Steel Res. 1*, 38-46.
9. Nethercot, D.A. (1985) Steel beams to column connections a review of test data and their applications to the evaluation of joint behaviour on the performance of steel frames. *CIRIA Project Study, London*, p. 338.
10. Goverdhan A.V. (1984) A collection of experimental moment - rotation curves and evaluation of prediction equations for semi - rigid concertinos Ph. D. Thesis, Vanderbilt University, Nashvill, TN.

11. Kishi, N. and Chen, W.F. (1986) Steel connection data bank program. *Structural Engineering Report no. CE-STR-86-18*, Purdue University, W. Lafayette, IN.

12. Kishi, N. and Chen, W. F. (1986) Database of steel beam - to- column connections, Vols I and II. *Structural Engineering Report no. CE-STR-86-26*, School of Civil Engineering, Purdue University, W. Lafayette, IN.

13. Baker, J. F. (1934) A note on the effective length of a pillar forming. *2nd Report, Steel structures research Committee*, Department of Scientific and Industrial Research of Great Britain, HMSO, London, pp. 13-34.

14. Mechers, R. E. and Kaur, D. (1982) Behaviour of frames with flexible joints. *In Proc. 8th Australian Conf. Mech. of Structural Materials, Newcastle, Australia*, pp. 271-275.

15. Romstad, K. M. and Subramanian, C. V. (1970) Analysis of frames with partial connection rigidity. *J. Struct. Div. ASCE 100, (ST6), Proceeding paper 7664*, pp. 2283-2300.

16. Lui, E. M. and Chen, W. F. (1983) Strength of columns with small end restraints. *J. Inst. Struct. Engrs, 61B*, pp. 17-26.

17. Razzaq, Z. (1975) End restraint effect on steel column strength. *ASEC J. struct. Div. 109 (ST2)*, pp. 314-334.

18. Frye, M. J. and Morris, G. A. (1975) Analysis of flexibly connected steel frames. *Can. J. civil engng 2*, pp. 280-291.

19. Cox, M. G. The numerical evaluation of B-splines. *J. Inst. Math. Applic. 10*, pp. 134-149.

20. Jones, S. W. Kirby, P. A. and Nethercot, D. A. (1981) Modelling of semi-rigid connection behaviour and its influence on steel column behaviour. In Joints in Structural Steelwork 9 Edited by Howlett, J. H. Jenkins, W. M. and Stainsby, R., Pentch Press, London.

21. Jones, S. W. Kirby, P. A. and Nethercot, D. A. (1982) Columns with semi-rigid joints. *J. struct. Div. ASCE. 108(ST2)*, pp. 361-372.

22. Krishnamurthy, H. T. Hung, P. K. Jeffrey, P. K. and Avery, L. K. (1979) Analitical moment - rotation curves for end-plate connections. *J. struct. Div. ASCE. 105 (ST1)*, pp. 133-145. *Proc. paper 14294*.

23. Colson, A. and Louveau, J. M. (1983) Connections incidence on the behaviour of steel structures. *Euromech. Colloq. 174*.

24. Goldberg, J. E. and Richard, R. M. (1936) Analysis of nonlinear structures. *J. struct. Div. ASCE, 89 (ST4)*, pp. 333-351.

25. Kishi, N. and Chén W. F. (1987) Moment-rotation relation of Top-and seat- angle connections. *Structural Engineering report no. CE-STR-87-4*. School of Civil Engineering, Purdue University, W. Lafayette, IN.

26. Kishi, N. and Chen W. F. (1987) Moment - rotation relation of semi-rigid connections. *Structural Engineering report no. CE-STR-87-29. School of Civil Engineering*, Purdue University, W. Lafayette, IN.

27. Ang, K. M. and Moriss G. A. (1984) Analysis of three-dimensional frames with flexible beam-column connections. *Can. J. Civil Engrs 11.*, pp. 245-254.

III. BEHAVIOUR OF EARTHQUAKE RESISTANT STRUCTURES INCLUDING JOINT BEHAVIOUR

BEHAVIOUR OF CONNECTIONS OF SEISMIC RESISTANT STEEL FRAMES

F. M. MAZZOLANI
University of Naples "Federico II"
Department of Structural Analysis and Design
80, P.le Tecchio, 80125 Naples, Italy

1. Introduction

An unexpected brittle failure of connections and, in same cases, of members occurred during the last earthquakes of Northridge (1994) and Kobe (1995). Immediately after, the international scientific community and, in particular, the earthquake prone Countries of the Mediterranean area and of the Eastern Europe were aware of the urgent need to investigate new topics and to improve the current seismic provisions consequently. In addition, it was observed that the whole background of the modern seismic codes deserves to be completely reviewed in order to grasp the design rules which failed during the last above mentioned seismic events. This revision must be aimed at the updating of seismic codes and in particular at the improvement of Eurocode 8, whose application will be widespread in the next years after the so-called conversion phase, which is now in course.

In this scenario the European research project dealing with the "Reliability of moment resistant connections of steel building frames in seismic areas" (RECOS) has been worked out with the sponsorship of the European Community within the INCO-Copernicus joint research projects (Mazzolani, 1999 [14]). The aim of this project has been to examine the influence of joints on the seismic behaviour of steel frames, bringing together knowledge and experience of specialists coming from different Countries.

In particular, it has been developed through the co-operation of thirteen Universities and Institutions of eight European Countries (Belgium, Bulgaria, France, Greece, Italy, Portugal, Romania, Slovenia).

The aim of the joint research program has been to provide an answer to the above questions and this goal has been accomplished through the following objectives:

a) analysis and synthesis of research results, including code provisions, in relation 'with the evidence coming from the earthquakes;

b) identification and evaluation by experimental tests of the structural performance of beam-to-column connections under cyclic loading and under extreme conditions;

c) setting up of sophisticated models for interpreting the connection response under seismic actions;

C.C. Baniotopoulos and F. Wald (eds.), The Paramount Role of Joints into the Reliable Response of Structures, 181–196.
© 2000 *Kluwer Academic Publishers.*

182

d) numerical study of the connection influence on the seismic response of steel building frames;

e) assessment of new criteria for selecting the behaviour factor for different structural schemes and definition of the corresponding range of validity in relation to the connection typologies.

The RECOS project has been completed in November 1999 and the main results are illustrated in this paper. The whole out-put of the research is collected into a volume (Mazzolani, 2000 [16]).

2. Seismic Input and Codification

The codification of building structural analysis under seismic loading represents a very important and difficult matter. Recently, important progress in this field has been reached, allowing for designing structures by accounting for the inelastic deformation capability of structural components. Nevertheless, the actual behaviour of steel buildings, as evidenced by recent earthquakes occurred all around the World, has been demonstrated to be not always correctly interpreted by present codes. Therefore, the need of further important revisions of existing provisions has been identified.

The analysis of several existing codes belonging to different geographical areas (Europe, USA and Japan) has allowed a comparison among the most important codified rules to be carried out. American code (UBC) is certainly the most up-to-dated (1997), including important aspects arisen from the recent Northridge earthquake. In all codes the method to be chosen for performing global analysis is based on structural regularity and, sometimes, on the importance of building. A specific characteristic of the Japanese code is that the ultimate limit state is investigated through a plastic analysis rather than an elastic one.

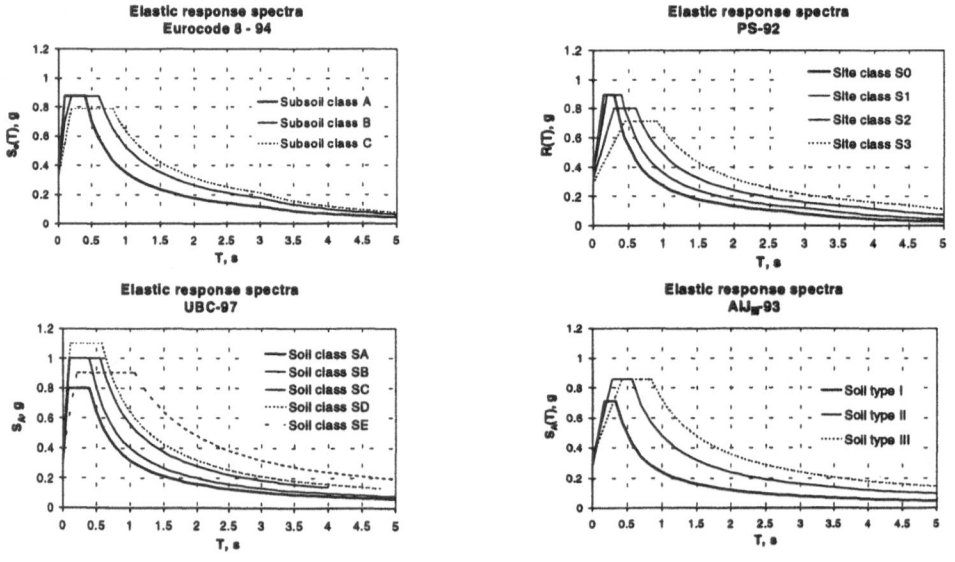

Figure 1. Comparison among the design spectra

As far as the definition of seismic force is concerned, analytical expressions of design spectra appear to be quite different to each other, but the shape of the curve is very similar. Seismic risk is essentially based upon the seismicity of the zone, which is defined by peak ground acceleration and type of soil. The comparison among the design spectra of the examined codes (Eurocode 8-94, french code PS-92, american UBC-97 and japanise AIJ.93) is given in Figure 1.

The influence of the seismic ground motion types on the structure response has been deeply investigated, considering the effect provided by both near-source and far-source, artificial and recorded, earthquake types on the response of MR frames. Due to large rapid pulses characterising ground motions generated in the vicinity of the source, the structural response of frames may be very different in relation to the peculiarities of the seismic input and the code provisions must predict the corresponding behaviour, accordingly. The main parameters of influence due to the type of ground motions are illustrated in Figure 2.

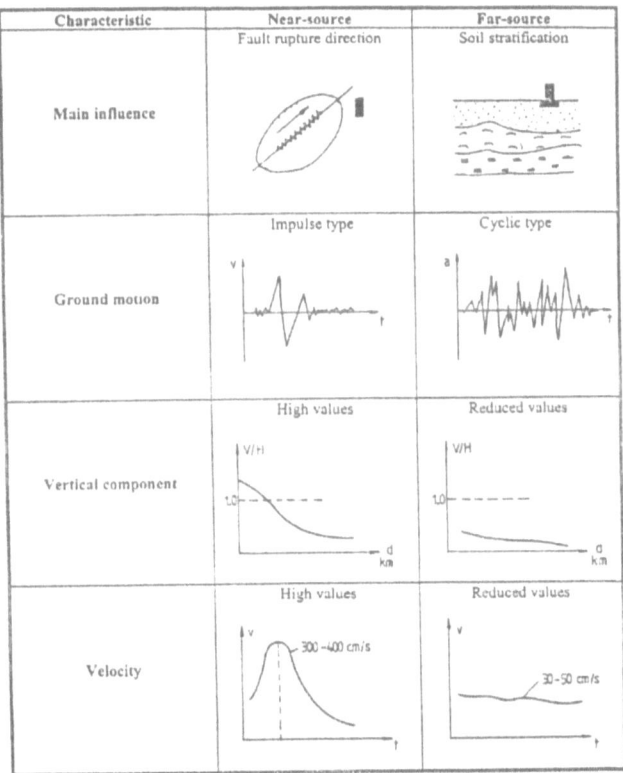

Figure 2. Main differences between near-source and far-source earthquakes

From the present investigation it seems that additional great efforts in this direction are required, aiming at providing safer and more detailed approaches for assessing the seismic risk.

184

3. Ductility of Members and Connections

The evaluation of available ductility of members and connections is a very important topic for the seismic analysis of structures and it becomes essential when sophisticated inelastic time-history analyses are performed. In fact, in such a case, the collapse of structures may be easily predicted by comparing the developed inelastic deformations in all structural components to the corresponding available ones. In the past, several studies have been carried out with reference to the ductility of members under monotonic loading. More recently, the main efforts have been addressed in general to connections under cyclic loading. Two different approaches have been developed to determine the available ductility: (1) using the collapse plastic mechanism method and (2) using the component method. In the first case, a general computer program "DUCTROT" has been set up, it being very versatile and accounting for several influencing phenomena (strain-rate, cyclic loading, tensile failure), allowing steel members and several joint typologies to be analysed. The main features of the DUCTROCT computer program are shown in Figure 3 (Gioncu & Pectu, 1995 [9]).

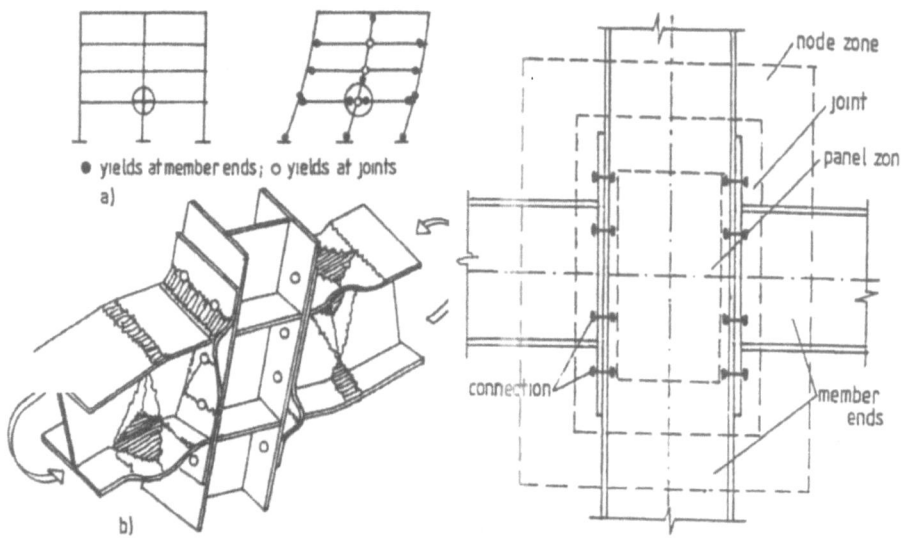

Figure 3. The collapse plastic mechanism of a beam-to-column connection according to the DUCTROCT method

In the second case, the approach is rather difficult but very attractive. In this project a consistent theoretical study has been developed with reference to the T-stub model, which is the main component in all bolted joints.
Figure 4 shows as the T-stub model can interpret the plastic deformation capacity of bolted beam-to-column connections. Results are very satisfactory and they are available to be extended to other components and joint typologies (Faella et al., 1997 [6]).

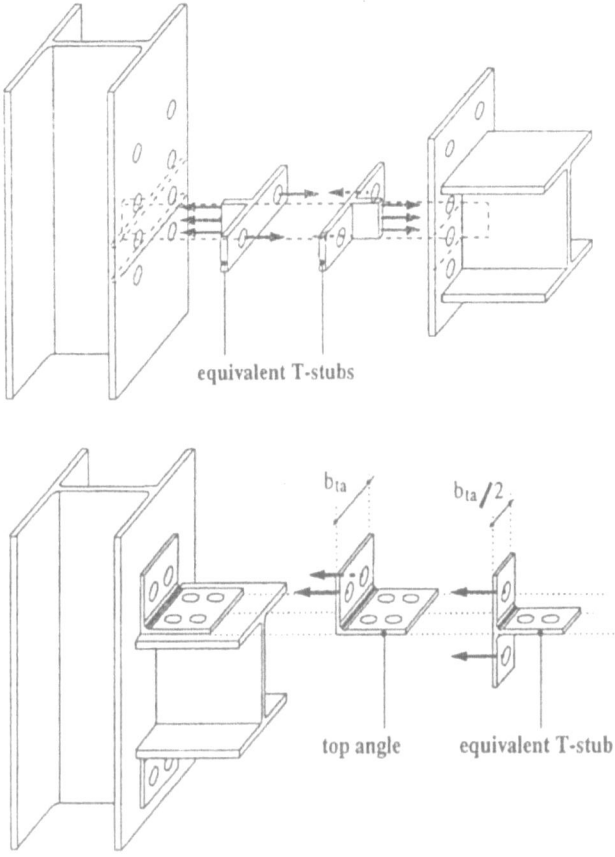

Figure 4. The T-stub as a basic component of end-plate and top and seat-angle beam-to-column connections

4. Testing Activity on Cyclic Behaviour of Beam-to-Column Connections

4.1 BARE STEEL

The most appropriate and reliable method to analyse the cyclic behaviour of connections is still represented by the direct experimentation. Several full-scale tests on a number of different joint typologies have been executed within the current project. Attention has been paid to many influencing factors, trying to cover lacks in the existing literature. In particular, influence of: (a) strain-rate, (b) loading asymmetry, (c) beam haunching, (d) column size, (e) static pre-loading and (f) partial strength have been analysed according to Table 1, where the involved Laboratories are listed.

TABLE 1
DISTRIBUTION OF EXPERIMENTAL ACTIVITIES AMONG PARTNERS

Type of connection	Laboratory	Type of influence					
		a	b	c	d	e	f
Welded	Lisbon				▨		
	Ljubljana	▨					
	Liège	▨					
	Sofia			▨			
	Timisoara		▨				
Bolted	Lisbon				▨		
	Ljubljana	▨				▨	
	Liège	▨					▨
	Sofia						
	Timisoara		▨				

The influence of strain-rate on mechanical response of structures is a present matter, it being worthy of a special regard. In fact, in some case, especially related to near-source earthquakes, due to the great velocity of the seismic action, loading rate on structural members may be much higher than the one commonly applied in laboratory quasi-static tests. On the other side, it is well known that materials exhibit higher yielding and ultimate strength as far as strain-rate increases, but such increments are not homogeneous and, therefore, the mechanical response of the components can undergo some variation. The experimental evidence is essential to understand the potentiality of such an effect on the poor brittle behaviour that some steel structures exhibited during recent earthquakes of Northridge and Kobe. But, it is undeniable that the available experimental results appear questionable, showing different influences due to strain-rate. Contradictory results have been also obtained from tests performed within this project at the Laboratories of the Universities of Liège and Ljubljana, where rigid full-strength extended end-plate connections revealed a different influence of high loading velocity on the cycle number to failure in relation to the specific strength of the adopted steel. The topic is very interesting and attractive, but it must be deepened in further research studies.

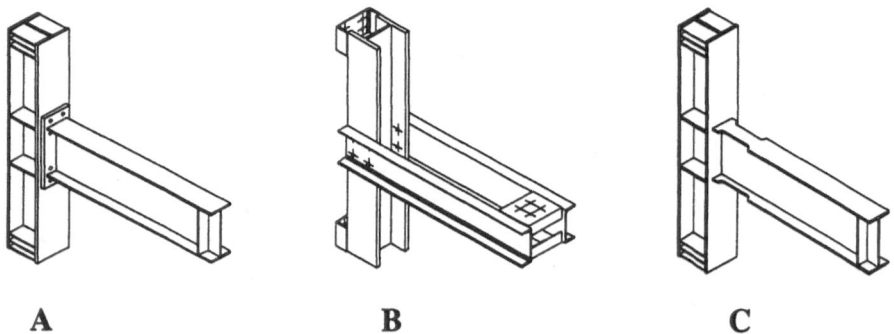

A   **B**   **C**

Figure 5. Three types of beam-to-column connections tested at the Laboratory of Liège

The effects of partial strength (type A and B) and of the "dog-bone" (type C) configurations have been examined at the University of Liège (see Figure 5), confirming the good performance of the latter.

The influence of static pre-loading has been examined at the University of Ljublijana on both symmetrical and unsymmetrical end-plate bolted connections (see Figure 6), showing a very different hysteresis loop with a loss of energy absorption capacity in the unsymmetrical case.

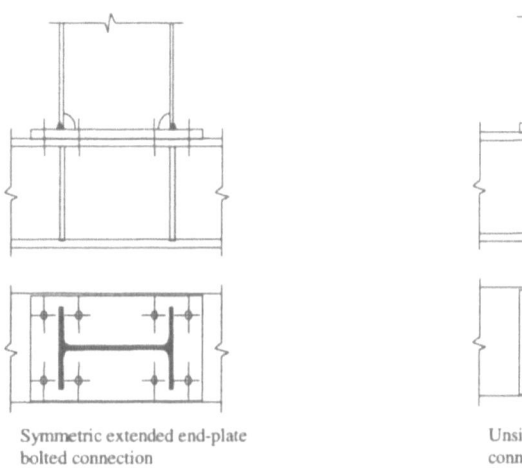

Symmetric extended end-plate bolted connection

Unsiymmetric end-plate bolted connection

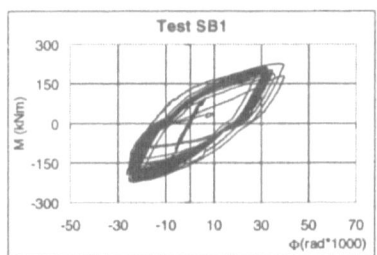

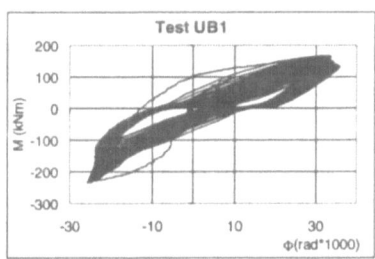

Figure 6. Symmetrical and unsymmetrical bolted end-plate connections tested at the Laboratory of Ljubljana and the corresponding hysteresis loops

The influence of loading asymmetry has been investigated as well. For this scope, a number of different connection typologies (welded and end-plate bolted) have been considered at the Laboratory of the University of Timisoara (see Figure 7). It has been observed that the panel zone in shear is the component that potentially may induce a strong variation of connection behaviour. As a consequence, loading asymmetry affects some response parameters of beam-to-column joints and this difference must be duly accounted for in design procedures. This experimental activity identified such an influence, essentially from a qualitative point of view.

188

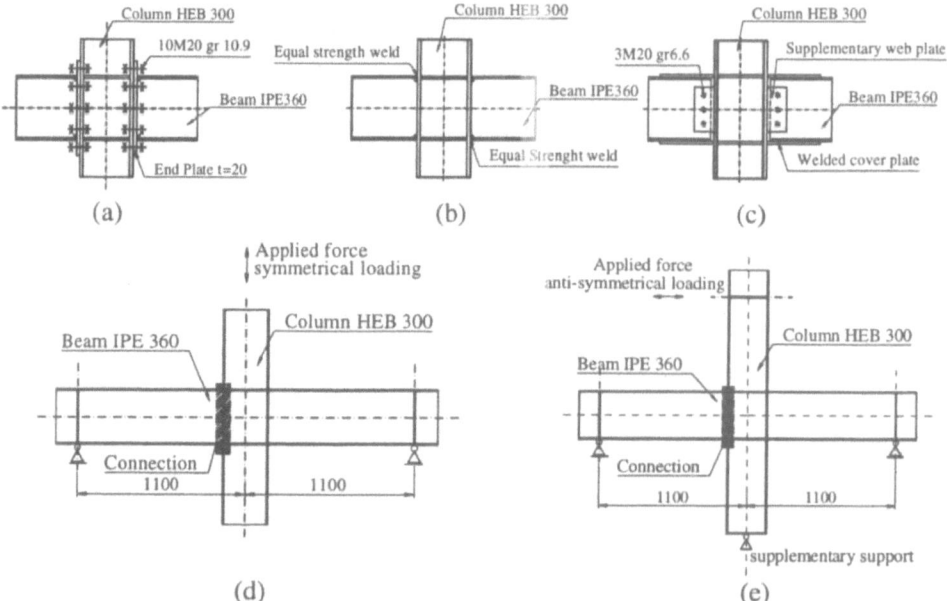

Figure 7. Specimens (a, b, c) tested at the Laboratory of Timisoara under symmetrical (d) and anti-symmetrical (e) loading

The influence of haunching has been analysed at the Laboratory of the University of Sofia, where the reliability of new connection details and their effect on the dissipative capacity of beam-to-column connections has been tested. This is a new trend in steel constructions, essentially promoted in U.S.A. and Japan after recent earthquakes in Northridge and in Kobe. Strengthening and weakening strategies are now being widespread elsewhere, because simple details may produce an important performance improvement in the behaviour of the whole structures. This advantage has been confirmed by these test results, which undeniably evidence that the beam haunching is able to move the plastic hinge away from the column face, therefore conferring to the joint a high ductility and very stable hysteresis cycles. Both the two tested typologies, like tapered flange and radius cut flange, exhibited satisfactory cyclic behaviour. In addition, the latter also assures some advantage connected to the possibility to correctly predict the actual behaviour of the connection and therefore it appears to be more convenient for design practice.

An important concern affecting the response of moment resisting steel frames is the ratio between beam and column strength. The influence of column size, with a fixed beam, on both welded and bolted connections has been studied at the University of Lisbon (Figure 8). Results show that the control of such a ratio may be somewhat important in case of welded connections, whose behaviour is conditioned by the panel zone. These connections may behave as semi-rigid or rigid as a function of the column size. Also, their strength is affected by the same factor. On the contrary, the influence of column size seems to be not important in case of bolted top and seat angle connections, where the connecting elements themselves usually represent the weakest components of the whole joint.

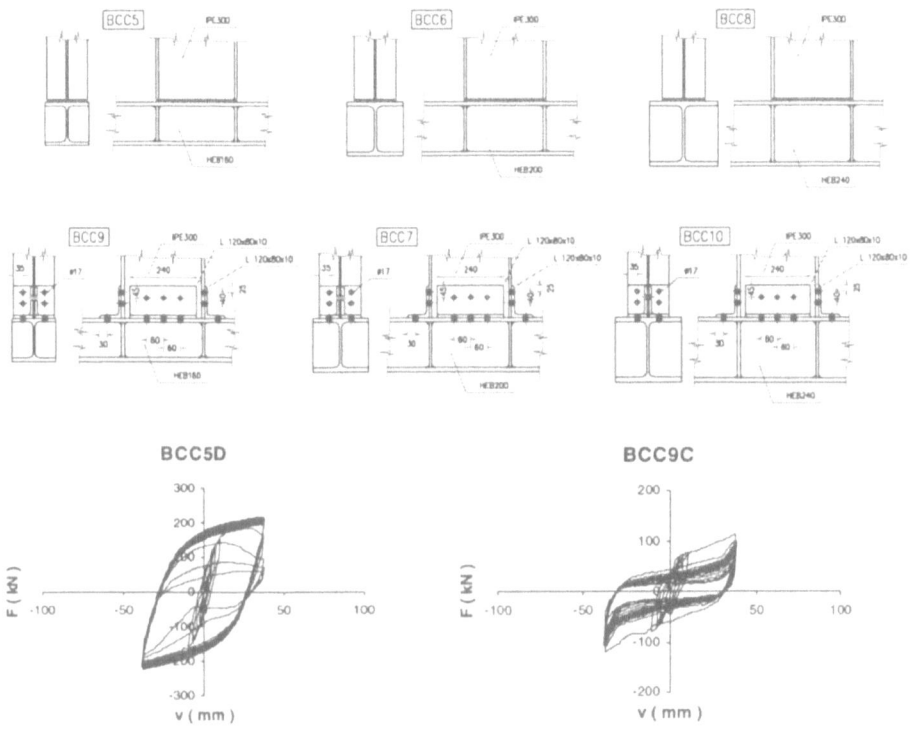

Figure 8. Specimens with the corresponding hysteresis loops tested at the Laboratory of Lisbon

4.2 STEEL-CONCRETE

Steel-concrete composite connections have been investigated at INSA - Rennes Laboratory. Nowadays, composite constructions are becoming more and more popular, especially due to the improved mechanical and fire resistance respect to bare steel elements. But, the knowledge concerning the actual behaviour of beam-to-column composite joints is still incomplete. The main aspect analysed in this study is the risk of degradation of the slab and the shear connectors under cyclic loading. Tests have concerned with monotonic and cyclic response of different slab and connector typologies, currently used in buildings. Preliminarily, push-out and push-pull tests on connectors evidenced a peculiar cyclic behaviour for such type of components, which requires the control of their ductility and fatigue resistance. Tests on the whole end plate composite joints (see Figure 9) have shown that the rotation capacity may be detrimentally affected by partial shear connectors under seismic actions as well as by other factors such as the number of ribs in the slab and the mutual position of shear connectors and reinforcement. A theoretical determination of the collapse based on the

190

usual fatigue formulation provided by EC3 has been also proposed and applied, providing interesting results. Further activities should be devoted to analyse different types of joints in order to better state which factors can improve the rotation capacity of composite joints.

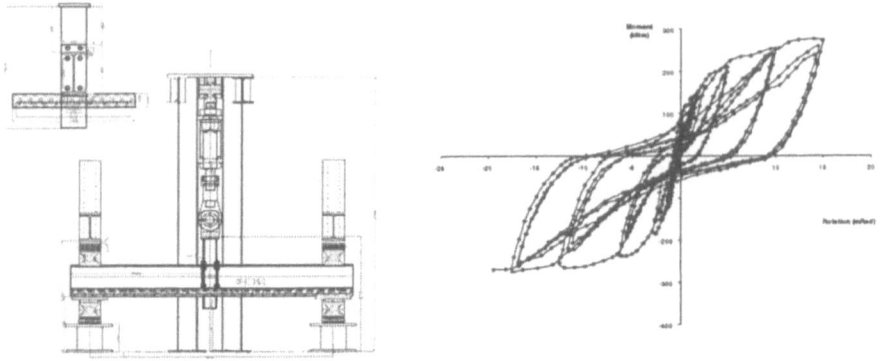

Figure 9. Composite beam-to-column connection tested at INSA - Rennes

4.3 RE-ELABORATION OF EXPERIMENTAL RESULTS

Even though tests have been performed in different Laboratories, a special attention has been paid to the methodology of analysis and the adopted measuring convention. As a consequence, test results are perfectly homogeneous and they may be directly compared to each other. Since the assessment of the number of cycles to failure under cyclic loading is one of the most important objective to be got by experimental tests, an unified re-elaboration of all results has been provided as well. This allowed the fatigue life endurance of tested specimens to be determined by the same procedure, which is based on a fixed level of probability of failure. Results show that the S-N lines are sensitive to specimen typology, but they may be predicted with an acceptable level of reliability (Castiglioni & Calado, 1996 [3]).

5. Numerical Analysis

5.1 EVALUATION OF GLOBAL SEISMIC PERFORMANCE

The evaluation of seismic resistance of steel frame buildings has been analysed though a number of numerical studies. Aiming at assessing the ductility demand for semi-rigid joint frames, several analyses have been carried out at the University of Timisoara, where some frames with different column cross-section sizes and types of connections have been investigated (Figure 10). As it was expected, the connection typology affects the whole performance of the frame (welded joints generally behave better than the bolted ones), but an important influence is also given by the considered acceleration record. Besides, the adopted collapse criterion may influence the judgement of reliability that may be assigned to the frame. Several factors affect the numerical response of the analysed structures and they must be correctly accounted for in comparing structure performances of different building configurations (Dubina et al., 1999 [5]).

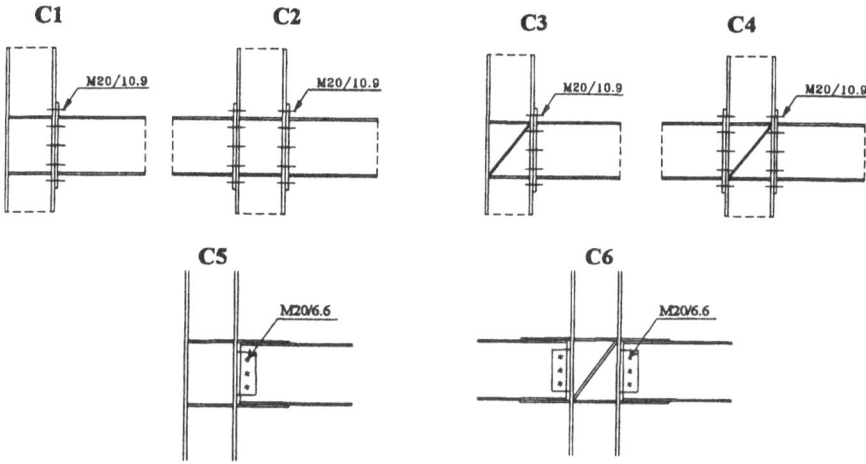

Figure 10. Types of connections considered in the global analysis

In the same direction, the activity developed at the University of Athens has been related to the interaction between local and global ductility properties for frame structures. The analysed frames are shown in Figure 11.

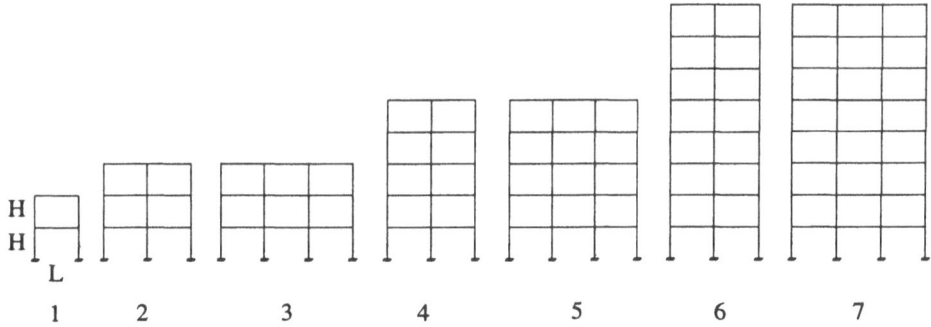

Figure 11. Types of frames considered in the study on the local to global ductility interaction (Mazzolani & Piluso, 1996 [10])

In principle, the task is very interesting because the demand in terms of global ductility by numerical analyses must be compared with the available ones, which is related to the level of local ductility of members and connections. Dynamic analyses have shown strong differences for different types of accelerograms, thus confirming the important rule of the acceleration record. The main trend of the performed analyses show that higher local ductility improves the structural response, but special attention should be paid to structural irregularities, especially the ones due to weak storeys, which may penalise the frame performance. Eventually, serviceability criteria may prevail over the

192

resistance ones, restricting the possibility to exploit large inelastic deformations. As a whole, the study provides useful information and criteria on how to perform numerical analyses, but also shows that the prediction of the seismic behaviour of steel moment resisting frames is almost complicated and the application of design criteria as well as the interpretation of the corresponding results is not immediate.

5.2 FAILURE MODE AND DUCTILITY CONTROL

The ductility demand control of moment resisting frames has been deepened at the University of Naples, by analysing some important aspects connected with both design methodology and numerical modelling.
It is commonly accepted that frames should be designed so to promote global type collapse mechanisms, in order to assure the maximum energy dissipation capacity and global ductility (see Figure 12).

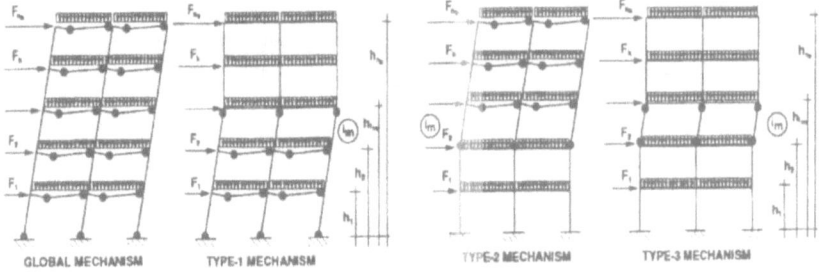

Figure 12. Types of collapse mechanism

Design procedures exist in case of steel frames with rigid full-strength joints (Mazzolani & Piluso, 1997 [12]). The first attempt to extend such a procedure to partial-strength semi-rigid frames has been also proposed by Mazzolani & Piluso [11]. Results are encouraging, especially because they show that such joint typologies, in case of long span and heavily loaded beams, may be economically advantageous, without penalising the performance of the structure in terms of ductility demand.
The influence of the hysteretic behaviour of beam-to-column connection has been analysed as well. Special cyclic models have been developed and calibrated on the basis of available experimental results for typical connection typologies (Figure 13).

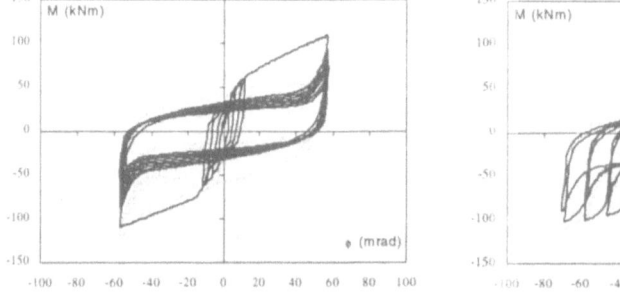
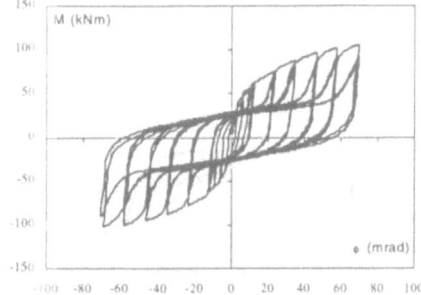

Figure 13. Comparison between experimental and numerical results

In case of frame structures, the effect of joint on ductility demand has been deeply investigated, aiming at identifying the main factors affecting the performance of the structure. Obviously, the results may be interpreted twofold: on one side, they emphasise the connection typologies allowing for a better seismic response of MR frames; on the other side, they are useful for assessing the susceptibility of numerical analysis results to suitably model all types of connections and, in particular, the possibility to adopt simplified hysteretic rules (Figure 14) (Della Corte et al. 1999 [4]).

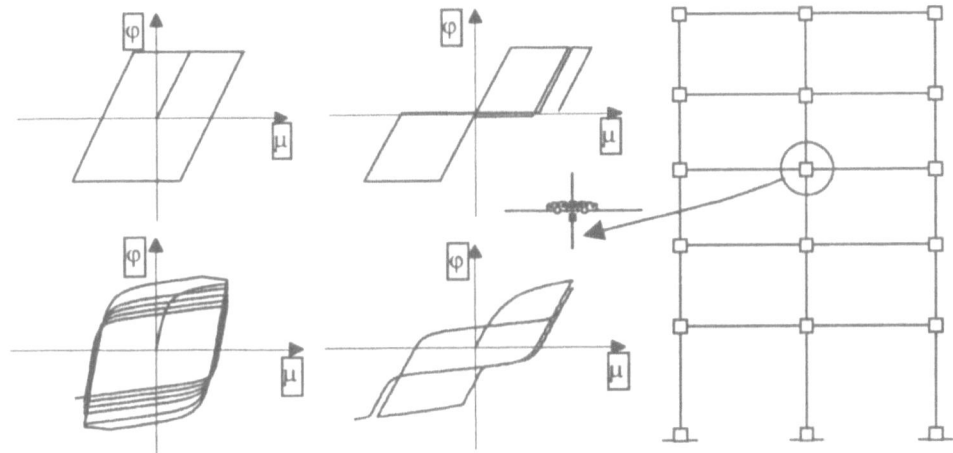

Figure 14. Influence of connection model on the frame global behaviour

Influence of structural typology has been focussed into different ways (Mazzolani & Piluso, 1997 [13]). The first is dealing with the performance of dual frame buildings, where completely pinned or semi-rigid partial strength connections are employed in either interior or exterior bays of the frame. The second is concerned with the effect of building asymmetry (Figure 15).

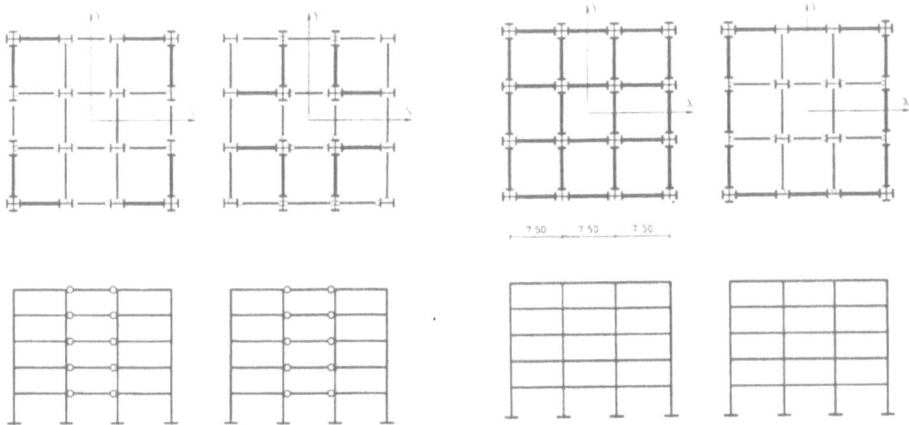

Figure 15. Different structural configurations

In both cases the developed tasks contribute to the understanding of the non-linear seismic response of frame buildings in case of non-ordinary conditions, providing the main factors having the major impact within the above influences. Besides, the way how these studies have been conducted allows some interests in the development of simplified non-linear analyses to be stated, as the one proposed Fajfar & Gaspersic [7]. This kind of simplified analyses can be very helpful to correctly and reliably assess the actual behaviour of the building frame.

5.3 DESIGN METHODOLOGY

All the above tasks dealing with numerical analyses have shown that the interpretation of results is a difficult matter, but it is the most important aspect for correctly assessing the structural behaviour and the effect of assumed design conditions. In particular, it is essential to be consistent in applying code provisions, assuming design criteria and evaluating structural performance (Fajfar & Krawinkeer, 1997 [8]). It seemed to be useful to analyse most of the issues connected with design methodology and evaluation of frame response, firstly providing basic definitions and relations of the most important mechanical factors and general methods used to characterise the building performance. Then, the evaluation of global seismic response has been performed by means of several approaches for determining the so-called q-factor (behaviour or reduction factor). The investigation has been done both by comparing the existing typical methods and by proposing and developing new "ad hoc" methods, like the one of Aribert and Grecea, so-called "base shear force approach", and the modification of the Kato-Akiyama energy approach [1].

As a whole, the present study shows that for moment resisting frames, with both rigid and semi-rigid joints (Figure 16), even if the definition and evaluation of behaviour factor is taken for granted by codes and researchers, this problem is worthy of further investigation and remark (Figure 17).

6. Conclusive Remarks

The RECOS Copernicus project sponsored by the European Commission is basically dealing with the influence of joints on the seismic response of steel moment resisting frames. It has been the largest European project in this field.

Many experimental tests, theoretical activities and numerical studies have been performed. The obtained results are very helpful for design, research and codification, because they give an answer to many important questions recently risen from the last earthquakes.

A similar activity has been contemporary carried out in USA by the SAC Steel project, in order to analyse the unexpected damage in steel structures during the Northridge earthquake.

The out-put of both these two projects will be presented during the next STESSA 2000 Conference (Montreal, August 21st - 24th) and the main results will be compared in a special session. New possible developments in this field will be stated on the basis of round-table discussion among international experts.

In addition, this topic will be of strong interest in next years within the activity of revision of the national seismic codes. In particular, EUROCODE 8 will take important advantages during is on-going conversion phase from the output of RECOS project.

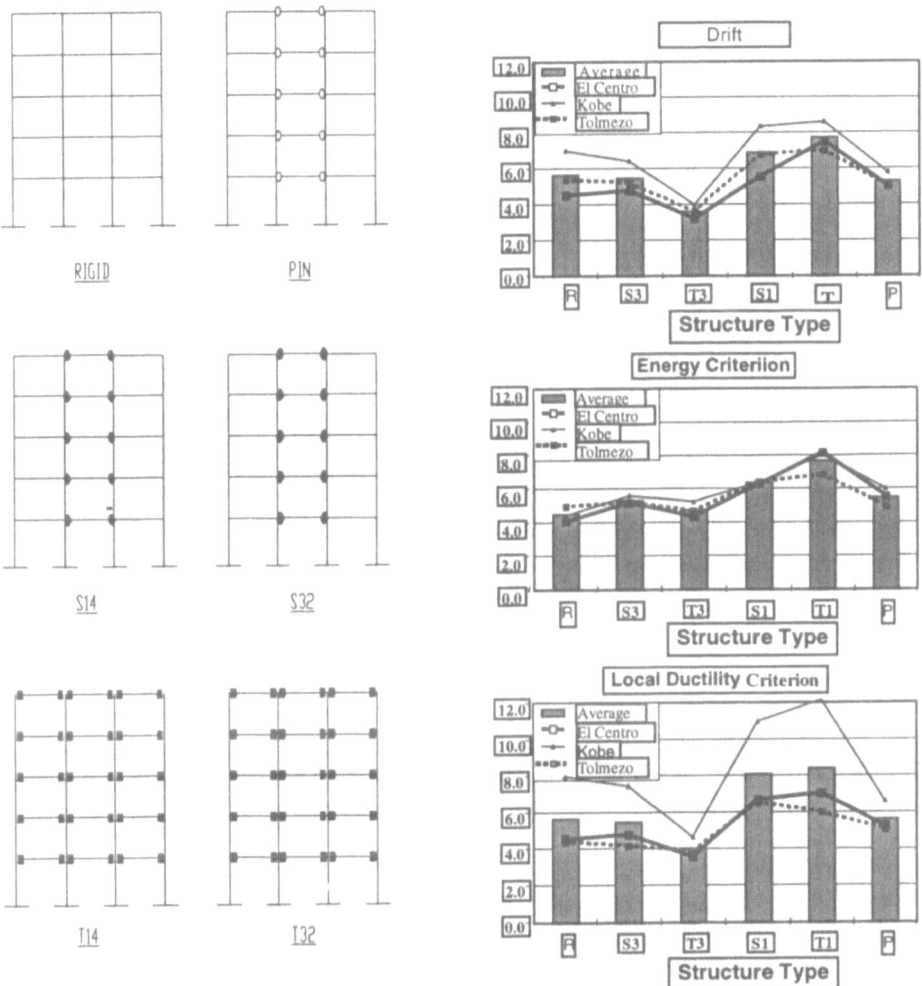

Figure 16. Frames with rigid and semi-rigid joints

Figure17. Values of q-factor for frames of Figure 16 by using different design criteria

7. References

1. Akiyama H. (1988) "Earthquake Resistant Design Based on the Energy Concept" *Proceedings of the 9th World Conference on Earthquake Engineering*, Tokyo, Kyoto.
2. Aribert J. M. and Grecea D. (1997) A new method to evaluate the q-factor from elastic dynamic analysis and its application to steel frames. *Proceedings of the International Conference on Behaviour of Steel Structures in Seismic Areas,*

196

STESSA '97 (edited by F. M. Mazzolani & H. Akiyama), Kyoto, Japan, 4-7 August.

3. Castiglioni C. A. and Calado L. (1996) Seismic Damage Assessment and Failure Criteria for Steel Members and Connections, *Proceedings of the International Conference on Advances in Steel Structures*, Hong Kong, pp 1021-1026.

4. Della Corte G., De Matteis G. and Landolfo R. (1999) A mathematical model interpreting the cyclic behaviour of steel beam-to-column joints. *Proceedings of XVII CTA Congress*, Naples, 3-5 October.

5. Dubina D., Ciutina A., Stratan A. and Dino F. (1999) Global Performances of Steel Moment Resisting Frames with Semi-rigid Joints. *Proceedings of the 6th international colloquium on Stability and Ductility of Steel Structures*, Timisoara, Romania, 367-377.

6. Faella C., Piluso V. and Rizzano G. (1997) Plastic deformation capacity of bolted T-stubs, *Proceedings of the International Conference on Behaviour of Steel Structures in Seismic Areas*, STESSA '97, (edited by F. M. Mazzolani & H. Akiyama), Kyoto, Japan, 4-7 August.

7. Fajfar P. and Gaspersic P. (1996) The N2 Method for the Seismic Damage Analysis of RC Buildings. *Earthquake Engineering and Structural Dynamics*, 25(1), 31-46.

8. Fajfar P. and Krawinkler H. edrs (1997) *Seismic design methodologies for the next generation of codes*, Balkema, Rotterdam.

9. Gioncu V. and Petcu D. (1995) Numerical investigations on the rotation capacity of beams and beam-columns. *"Stability of Steel Structures"* (ed. M. Ivanyi), 21-23 September, Budapest, Vol. 1, 129-140.

10. Mazzolani F. M. and Piluso V. (1996) *Theory and Design of Seismic Resistant Steel Frames*, Chapman & Hall, London,

11. Mazzolani F. M. and Piluso V. (1996) Plastic Design of Semirigid Frames for Failure Mode Control. *Proceedings of the IABSE Colloquium on Semi-Rigid Structural Connections*, Instanbul, September 1996.

12. Mazzolani F. M. and Piluso V. (1997) Plastic Design of Seismic Resistant Steel Frames. *Earthquake Engineering and Structural Dynamics*, Vol. 26, pp. 167-191.

13. Mazzolani F. M. and Piluso V. (1997) The influence of the design configuration on the seismic response of moment-resisting frames. *Proceedings of the International Conference on Behaviour of steel Structures in Seismic Areas*, STESSA '97, (edited by F. M. Mazzolani & H. Akiyama), Kyoto, Japan, 4-7 August.

14. Mazzolani F. M. (1999) Reliability of moment resistant connections of steel building frames in seismic areas: the first year of activity of the RECOS project. In *Proceedings of the 2nd European Conference on Steel Structures, Eurosteel 1999*, Praha, Czech Republic. May 26-29.

15. Mazzolani F. M. (1999) Reliability of moment resistant connections in steel building frames in seismic areas. *Proceedings of the International Seminar on Seismic Engineering for Tomorrow*, in honor of professor Hiroshi Akiyama, Tokyo, Japan, November 26.

16. Mazzolani F.M. (2000) *Moment resisting connections of steel frames in seismic areas: design and reliability*, published by E & FN SPON, London (now in press).

OBSERVED BEAM-COLUMN JOINT FAILURES DURING 17 AUGUST 1999 KOCAELI AND 12 NOVEMBER 1999 DUZCE, TURKEY EARTHQUAKES

O.C. ÇELİK
Istanbul Technical University, Faculty of Architecture, Department of
Theory of Structures
Taşkışla, Taksim, 80191, İstanbul, Turkey

1. Introduction

The earthquakes which occurred in Kocaeli (İzmit) and in Düzce affected a wide area in the Marmara region of Turkey, leading to huge loss of life and structural damage. Kocaeli and Düzce and their environs are on the Northern Anatolian Fault Line, and include Turkey's most earthquake prone areas. Records of past earthquakes in Turkey reveal that numerous earthquakes of $M_s \geq 7.0$ occurred in the region. The epicentre of the Kocaeli earthquake, Richter magnitude $M_s = 7.4$, was located at lattitude $40.70°$ N and longitude $29.91°$ E, [1]. According to Boğaziçi University Kandilli Observatory's Earthquake Research Institute, its focal depth was $H = 10 \sim 15$ km and the intensity at the epicentre was $I_o = X\text{-}XI$ MSK. In terms of the devastation it caused, the 1999 Kocaeli Earthquake was the most serious one to hit a Turkish city, and is the second largest earthquake of the century after the 1939 Erzincan Earthquake. For the same reasons, it is one of the most serious earthquakes to have occurred worldwide. The epicentre of the Düzce earthquake having a magnitude of $M_s = 7.2$, was located at lattitude $40.77°$ N and longitude $31.15°$ E. According to the evaluations of USGS, its focal depth was $H = 10$km.

After the 17 August 1999 Kocaeli and 12 November 1999 Düzce, Turkey earthquakes, detailed site investigations have shown that the main causes of the earthquake hazard and loss of human lives cannot only be explained by the inadequate quality and the quantity of the material used in construction. There exist some fatal architectural and engineering design and construction mistakes, [2,3,4,5]. In this respect, the role of beam-to-column joints in framed buildings and the effect of joint behaviour on the dynamic response and performance were remarkable. Observed damages on buildings proved that buildings having irregular structural systems did not behave well. Here, common types of irregular buildings are summarized and their effect on the seismic response of framed buildings is given in detail. The importance of capacity design and the effect of the strong column-weak beam design philosophy on the seismic behaviour of framed buildings are re-evaluated based on the available data. Many buildings within the earthquake region were heavily damaged or collapsed due to inadequate joint resistance during the recent earthquakes, [7]. The role of beam-to-

C.C. Baniotopoulos and F. Wald (eds.), The Paramount Role of Joints into the Reliable Response of Structures, 197–206.
© 2000 *Kluwer Academic Publishers.*

column joints on the seismic response of framed buildings is discussed and possible suitable solutions to prevent/minimize joint failures are proposed with reference to the new Turkish Earthquake Code.

2. Observed Damage

Reinforced concrete buildings have been widely used in the earthquake region and in other parts of Turkey. Office and residential buildings with 2~6 stories are generally designed and constructed with moment resisting frames with masonry infill walls. Infill walls are mostly used as nonstructural partition walls. Both positive and negative effects of infill walls were observed during the recent earthquakes. In the region, steel framed buildings are generally used as one story, single or multiple bay industrial buildings. No important damage was observed on steel structures. After the development of prefabrication technology in Turkey, the number of reinforced or prestressed concrete prefabricated buildings have been increased rapidly. As a result, prefabricated buildings have had an important market in the industrial type of buildings as well as in multi-story office and residential buildings. The performance of prefabricated buildings was not good during these last earthquakes. Most of the prefabricated buildings had severe damage or totally collapsed. The reasons of very common types of damages encountered during the site investigation may be summarized below:

a) Short column (Photo 1,2), short beam
b) Poor material quality (concrete (Photo 3), reinforcing bars)
c) Soft/weak story (Photo 4)
d) Poor joint detailing (especially beam-to-column joints, (Photo 5,6)
e) Irregular structural systems (Photo 6)
f) Inadequate element dimensions
g) Inadequate lateral stiffness (P-Δ effect)
h) Pounding between adjacent buildings having different dynamic characteristics
i) Damage to nonstructural elements
j) Local soil conditions
k) Earthquake parameters

Most of the structural damage could be attributed to poor beam-to-column joint behaviour. Formation of the plastic hinges at the end points of the columns, near the joint, were widely observed. From the investigations, rather thin plain bars of normal yield strength were used as transverse reinforcement in the beam-to-column joint regions. They had 90-degree hooks and spacings between 20cm to 40~50cm. Besides this, many detailing deficiencies were observed in damaged or collapsed structures. A common problem was the lack of appropriate anchorage of the beam reinforcement into the beam column joints. In many damaged buildings, the column reinforcement within the joint was not supported by transverse reinforcement. Buckling of column longitudinal reinforcement near the joint was frequently encountered. On the other hand, insufficient splice length for the longitudinal reinforcement was widely observed. It was also observed that the compressive strength of concrete in the structural elements was about 8~12 N/mm^2. These values are rather below the minimum values recommended by the old and new Turkish Earthquake Codes. Since many reinforced concrete buildings had no shear walls, excessive story drifts resulted in second order effects which produced plastic hinges in beam-to-column joints.

Photo 1. Short column

Photo 2. Short column.

Photo 3. Poor concrete quality.

Photo 4. Soft story, irregular building.

Photo 5. Beam-to-column joint failure.

Photo 6. Compression failure in RC column. irregular Building.

3. Code Requirements

The last earthquake code of Turkey named "Specification for Structures to be Built in Disaster Areas" have been effective since 01.01.1998, [6]. The general principle of earthquake resistant design to this Specification is to prevent structural and non-structural elements of buildings from any damage in low-intensity earthquakes; to limit the damage in structural and non-structural elements to repairable levels in medium-intensity earthquakes, and to prevent the overall or partial collapse of buildings in high-intensity earthquakes in order to avoid the loss of life. Requirements of this Specification shall be applicable to newly constructed buildings as well as to buildings to be modified, enlarged and to be repaired or strengthened prior to or following the earthquake. It is proposed in the code that design and construction of irregular buildings should be avoided. Structural system should be arranged symmetrical or nearly symmetrical in plan and torsional irregularity should preferably be avoided. In this respect, it is essential that stiff structural elements such as structural walls should be placed so as to increase the torsional stiffness of the building. On the other hand, vertical irregularities leading to weak/soft story at any story should be avoided. In this respect, appropriate measures should be taken to avoid the negative effects of abrupt decreases in stiffness and strength due to removal of infill walls from some of the stories and in particular from the first storey of buildings which may possess considerable stiffness in their own planes, even though they are not taken into account in the analysis.

According to the code, structures are classified as structural systems of high ductility level and structural systems of nominal ductility level. Different structural behaviour factors are given for reinforced concrete buildings and for structural steel buildings. In structural systems denoted as being high ductility level, ductility levels shall be high in both lateral earthquake directions. Systems of high ductility level in one earthquake direction and of nominal ductility level in the perpendicular earthquake direction shall be deemed to be structural systems of nominal ductility level in both directions.

3.1. REINFORCED CONCRETE STRUCTURES

3.1.1. *Special Seismic Hoops and Crossties*

Hoops and crossties used in columns, beam-column joints, wall end zones and beam confinement zones of all reinforced concrete systems of high ductility level or nominal ductility level in all seismic zones shall be special seismic hoops and special seismic crossties, Fig.1. Special seismic hoops shall always have 135 degree hooks at both ends. However, 90 degree hook may be made at one end of the special seismic crossties. In this case, crossties with 135 degree and 90 degree hooks shall be placed on one face of a column or wall in a staggered form in both horizontal and vertical directions. Special seismic hoops shall engage the longitudinal reinforcement from outside with hooks closed around the same rebar. Diameter and spacing of special seismic crossties shall be the same as those of hoops. Crossties shall be connected to longitudinal reinforcement always at both ends. Hoops and crossties shall be firmly tied such that they shall not move during concrete pouring.

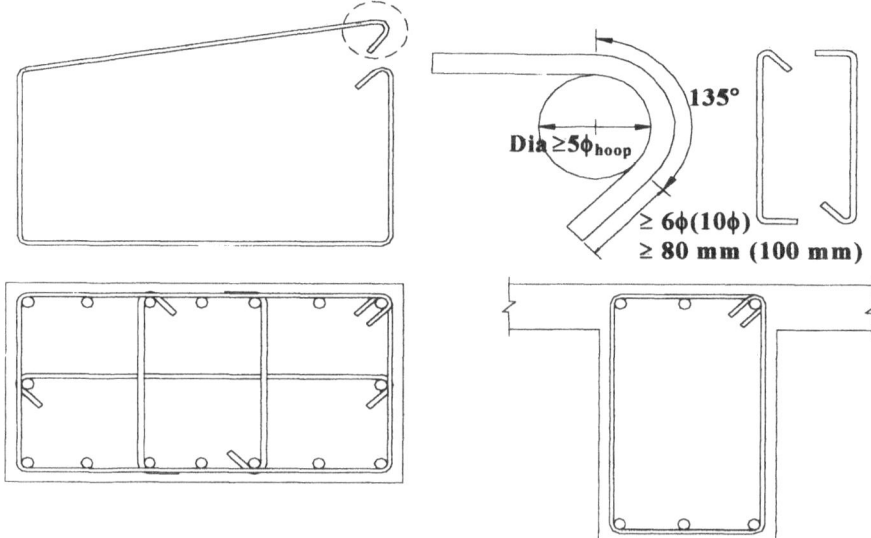

Figure 1. Special seismic hoops and crossties in RC elements.

In the case where the column cross-section changes between consecutive stories, slope of the longitudinal reinforcement within the beam-column joint shall not be more than 1/6 with respect to the vertical. Fig. 2.

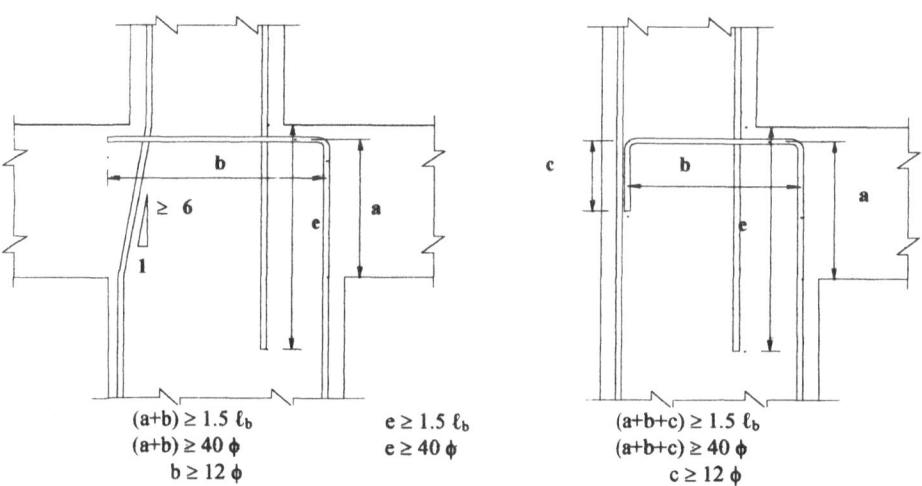

Figure 2. Reinforcement detailing in sectional changes.

3.1.2. *Transverse Reinforcement Requirements*

Special confinement zones shall be arranged at the bottom and top ends of each column. Length of each of the confinement zones shall not be less than smaller of column cross section dimensions (diameter in circular columns), 1/6 the clear height of column (measured upward from floor level or downward from the bottom face of the deepest beam framing into the column), and 500 mm. Requirements for transverse reinforcement to be used in confinement zones are given below. Such reinforcement shall be extended into the foundation for a length equal to at least twice the smaller of column cross section dimensions.

3.1.3. *Requirement of Having Columns Stronger Than Beams*

In structural systems comprised of frames only or of combination of frames and walls, sum of ultimate moment resistances of columns framing into a beam-column joint shall be at least 20% more than the sum of ultimate moment resistances of beams framing into the same join, Fig. 3.

$$(M_{ra} + M_{r\ddot{u}}) \geq 1.2 (M_{ri} + M_{rj}) \tag{1}$$

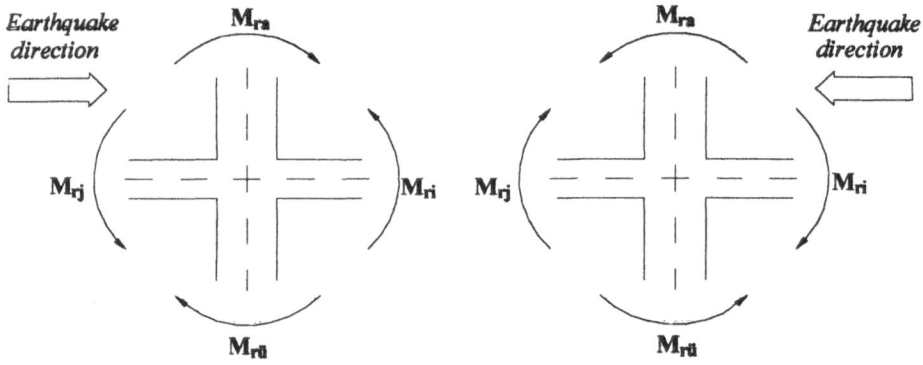

Figure 3. Strong Colomn-Weak Beam

3.1.4. *Beam-column joints of frame systems of high ductility level-Confined and unconfined joints*

In the case where beams frame into all four sides of a column and where the width of each beam is not less than 3/4 the adjoining column width, such a beam-column joint shall be defined as a confined joint. All joints not satisfying the given conditions shall be defined as unconfined joint.

In confined joints, at least 40 % of the amount of transverse reinforcement existing in the confinement zone of the column below shall be provided along the height of the joint. However, diameter of transverse reinforcement shall not be less than 8 mm and its spacing shall not exceed 150 mm. In unconfined joints, at least 60 % of the amount of transverse reinforcement existing in the confinement zone of the column below shall be provided along the height of the joint. However in this case, diameter of transverse reinforcement shall not be less than 8 mm and its spacing shall not exceed 100 mm.

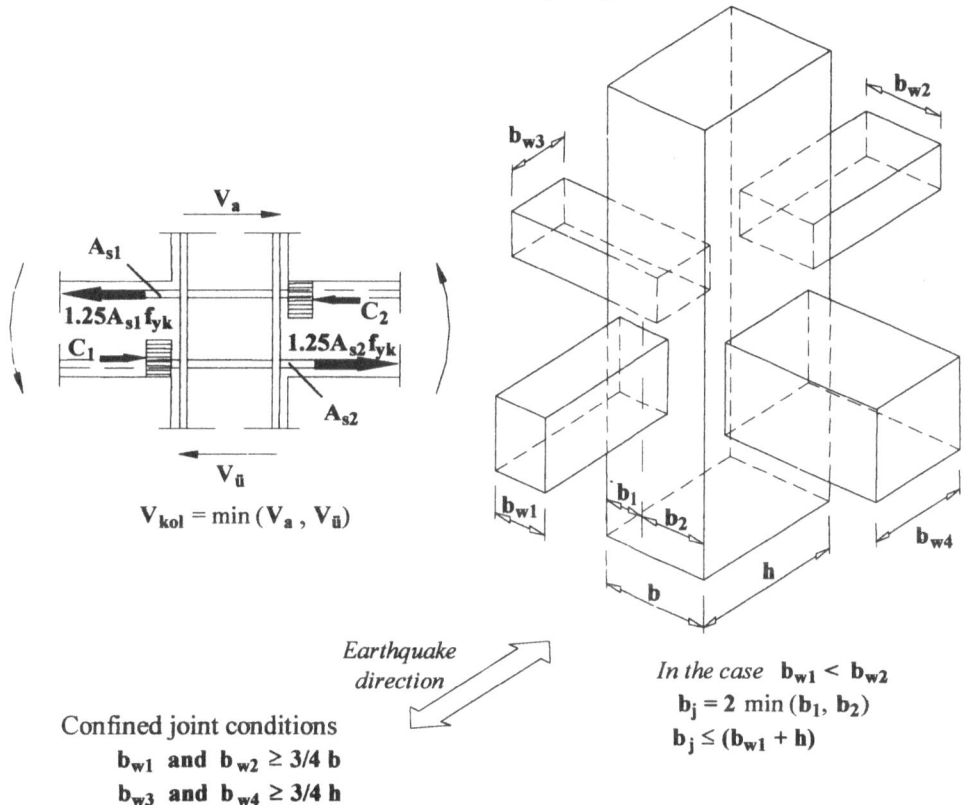

$V_{kol} = \min(V_a, V_{\ddot{u}})$

In the case $b_{w1} < b_{w2}$
$$b_j = 2 \min(b_1, b_2)$$
$$b_j \leq (b_{w1} + h)$$

Confined joint conditions
$$b_{w1} \text{ and } b_{w2} \geq 3/4 \, b$$
$$b_{w3} \text{ and } b_{w4} \geq 3/4 \, h$$

Figure 4. Confined joint conditions

3.2. SPECIAL REQUIREMENTS FOR PREFABRICATED BUILDINGS

3.2.1. *Frames with Hinged Connections*

With the exception of single storey industrial-type buildings, prefabricated frame type structural systems with hinge connections (which are unable to resist moments) may be permitted provided that reinforced concrete cast-in-situ structural walls are constructed in both directions to fully resist against the seismic loads. Welded hinge connections and other hinge connections shall possess sufficient strength to resist 1.5 times and 1.2 times the seismic connection forces, respectively.

3.2.2. *Moment Resisting Frames*

Connections shall possess sufficient strength to transfer moments, shear forces and axial forces to be developed at the ultimate strength level without any reduction in strength and ductility. In welded connections and other type of connections, 1.5 times and 1.2 times the seismic connection forces, respectively. Connections must be arranged in sufficient distance from the potential plastic hinges that can develop within the elements connected.

3.3. STEEL STRUCTURES

Lateral load carrying systems of structural steel buildings may be comprised of steel frames only, of steel braced frames only or of combination of frames with steel braced frames or reinforced concrete structural walls. Lateral load carrying systems of structural steel buildings shall be classified with respect to their seismic behaviour into two classes as structural steel systems of high ductility level and structural steel systems of nominal ductility level. In all seismic zones, allowable stress or ultimate strength of welding shall be decreased by 25%.

3.3.1. *Requirement of Having Columns Stronger Than Beams*

In frame systems or in the frames of frame-wall (braced frame) systems, sum of the plastic moments of columns framing into a beam-column joint in the earthquake direction considered shall be more than the sum of plastic moments of beams framing into the same joint, Fig.5.

$$(M_{pa}+M_{pü}) \geq (M_{pi} + M_{pj}) \tag{2}$$

shall be applied separately for both senses of earthquake direction to yield the most unfavourable result. In calculating the column plastic moments, axial forces shall be considered to yield the minimum moments consistent with the sense of earthquake direction.

Eq.(2) need not to be checked in single storey buildings and in joints of topmost storey of multi-storey buildings. The column shall be continuous in beam-column joints of frames. In the case where the beam is connected to the column flange, web of the column shall be strengtened at the beam flange level by stiffening plates. In the first and second seismic zones, common bolts shall not be used in connections and splices transferring moments. However prestressed high strength bolts and anchor bolts are exempted from this restriction. High strength bolts shall be of ISO 8.8 or 10.9 quality. Column splices shall be made away from beam-column joint by at least 1/4 the storey height. In the case of splices with butt welds, edge preparation and deep penetration welding shall be applied. Load transfer strength of beam-column joints with fillet welds or non-prestressed bolts shall not be less than 1.2 times the strength of element connected to the joint. Load transfer strength of other types of beam-column joints shall not be less than the strength of element connected to the joint.

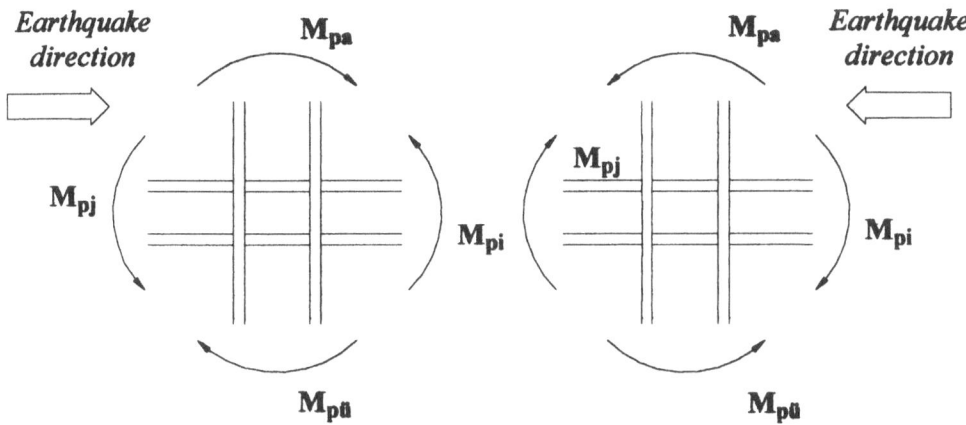

Figure 5. Strong Colomn-Weak Beam

Beam splices shall be made away from the beam-column connection by a distance at least equal to beam height. In braced frames where braces are connected to columns, connections shall be made at the column flange. Connection shall not be made to web of the column. Slenderness ratio of braces designed to resist compressive forces too shall not be more than 100. In the case where common bolts are used for brace connections, allowable stresses of bolts shall be reduced by 33%.In design calculations of splices and connections of frames of nominal ductility level, twice the internal forces obtained. In the case where braces are designed to resist tension only, slenderness ratio of braces shall not exceed 250.

4. Conclusions

The 17 August 1999 Kocaeli (İzmit) and 12 November 1999 Düzce earthquakes resulted in loss of many human lives and yielded economic crises within the region. The effect of the soil conditions on the dynamic response of low-rise reinforced concrete framed buildings was not negligible. Buildings having architectural based irregular structural systems were heavily damaged or collapsed during the earthquake. Beam-to-column joint behaviour mainly affected the performance of framed buildings. The earthquake performance of prefabricated buildings was poor due to inadequate joint detailing and weak lateral stiffness. Many buildings having regular structural system but roughly designed performed well with minor damage. Excessive use of soft stories, short columns, lack of column confinement and the use of framed systems having strong beam-weak column joints are the reasons of the catastrophic damage. Use of adequate amount of reinforced concrete shear walls in reinforced concrete buildings and use of steel bracing systems in steel framed buildings are strongly recommended to control story drifts.

5. References

1. http://ciragan.cc.itu.edu.tr/deprem/
2. Çelik,O.C., Çılı,F., and Özgen,K. (2000) Observations on the Kocaeli (İzmit) Earthquake of 17 August 1999, *Yapı*, 218, 65-76.
3. Çılı,F., Çelik,O.C., and Sesigür,H. (1999) Seismic Damage Evaluation in Reinforced Concrete Buildings After 17 August 1999 Kocaeli (İzmit) Earthquake, *Proceedings of ITU-IAHS International Conference on the Kocaeli Earthquake*, 02-05 December,313.
4. Çelik,O.C., Çılı,F., and Özgen,K. (1998) Cantilevers in Reinforced Concrete Structures, *Proceedings of 11th European Conference on Earthquake Engineering*, 06-11 September, 1-10.
5. Çılı,F., Özgen,K., and Çelik,O.C. (1994) Evaluation of Seismic Damages to RC Buildings During 1992 Erzincan Earthquake, *Proceedings of 10th European Conference on Earthquake Engineering*, 28 August-2 September, Vol.1, 773-778.
6. Ministry of Public Works and Settlement (1997) Specification for Structures to be Built in Disaster Areas.
7. Saatcioglu,M. and Bruneau, M. (1994) Performance of R/C Structures during the 1992 Erzincan Earthquake, *Proceedings of 10th European Conference on Earthquake Engineering*, 28 August-2 September, Vol.1, 805-811.

INFLUENCE OF SEMI-RIGID AND/OR PARTIAL-STRENGTH JOINTS ON THE SEISMIC PERFORMANCES OF STEEL MRF

D. GRECEA
"Politehnica" University of Timisoara, Department of Steel Structures
and Structural Mechanics
RO-1900, Timisoara, Romania

1. Introduction

Generally European codes are based on elastic static global analysis when designing a steel structure subject to seismic actions. Advantage of the very significant dissipative phenomena in the structure is taken by means of the q-factor reducing the seismic forces which would be obtained assuming a perfectly elastic behaviour of structural steel. As reminder, a steel frame with elements (beams and columns) of high rotation capacity (for example when the cross-sections are Class 1) can reveal values of q-factor greater than 6 (CEN 1995) provided that their joints in dissipative zones are rigid and show sufficient overstrength. In fact, Eurocode 8 does not provide any application rule for the use of partial-strength joints, so that the procedure to evaluate the effect of the behaviour of partial-strength joints on the q-factor remains to be established totally.

A parametrical investigation on different types of steel MRF with semi-rigid and/or partial-strength joints subject to seismic motions is presented in comparison with the case of the same structures but with rigid and full-strength joints. Seismic behaviour of the analysed frames is evaluated by the behaviour q-factor, related essentially to the maximum inelastic base shear force of the structure deduced from inelastic dynamic analyses.

Dynamic calculations of several types of multi-bays multi-storeys steel frames are performed using two severe accelerograms (Bucharest 1977 and Kobe 1995). Three values of rotation capacity have been selected a priori, namely 0.015, 0.030 and 0.045 radians, which are considered significant for practical design. The other investigated parameters are the initial rotational stiffness of the joints (less than the classification value between rigid and semi-rigid behaviours) and their resistance moment (less than the plastic resistance moment of the connected beams). But these parameters do not appear determinant of the q-factor values contrary to the rotation capacity. Additional comments are given briefly about 2nd order geometrical effects and possible occurrence of plastic hinges in a few columns.

Finally, a conservative table is proposed for steel frames in low seismicity zones, providing the q-factor as a function of the joint rotation capacity and the nominal ground acceleration together.

C.C. Baniotopoulos and F. Wald (eds.), The Paramount Role of Joints into the Reliable Response of Structures, 207–216.
© *2000 Kluwer Academic Publishers.*

2. Determination of the q-factor

As clearly illustrated in a chapter of Mazzolani and Piluso's book [12], comparisons of different methods existing in the scientific literature to evaluate the q-factor show a large scattering of results. That may be explained by the conventional nature of the adopted definitions, which generally are not consistent with the "directions for use" of the q-factor.

2.1. DEFINITION OF q-FACTOR

Contrary to the choice of the literature (Ballio & Setti [5], Sedlacek & Kuck [14], Mazzolani & Piluso [12]), this paper is using a new method of Aribert & Grecea [1], [2], [3] which is characterising the maximum response of the structure, not by a displacement (generally the upper horizontal deflection δ is considered significant), but by the horizontal base shear force V of the structure. Increasing λ step-by-step, such a determination should be repeated systematically up to a certain ultimate value λ_u which corresponds to the attainment of the rotation capacity of beam and column elements or a particular joint. Curve OEU in Figure 1 represents the variation of V versus the multiplier λ where the shear-force values $V^{(e)}$ and $V^{(inel)}$ correspond to the first yielding state and the ultimate limit state, respectively. In the same figure, straight line OEU* would be the curve (V, λ) assuming an ideal elastic behaviour of the structure up to λ_u.

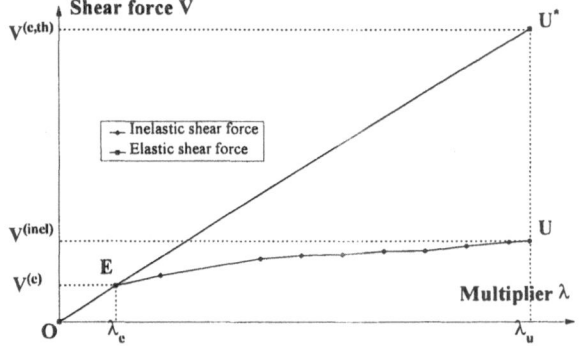

Figure 1. Maximum base shear force versus accelerogram multiplier

Whereas the q-factor in the literature is generally defined by the ratio λ_u/λ_e, this new definition given by Aribert and Grecea [1] is the ratio between the elastic theoretical base shear force, $V^{(e,th)}$, and the real inelastic base shear force, $V^{(inel)}$:

$$q = V^{(e,th)}/V^{(inel)} = \left(V^{(e)}/\lambda_e\right)/\left(V^{(inel)}/\lambda_u\right) \tag{1}$$

It is important to mention a complementary aspect of definition (1) which has been neglected likely in the literature; as an explicit result of the determination procedure itself, the q-factor should be associated with a precise value of the maximum ground acceleration, namely:

$$a_N^{(u)} = \lambda_u \, max|a(t)| \qquad (2)$$

Practically, $a_N^{(u)}$ may be considered as a nominal acceleration.

2.2. SOME RESULTS FROM A PARAMETRICAL INVESTIGATION ON STEEL FRAMES WITH RIGID CONNECTIONS

A few results extracted from a parametric study carried out by Grecea [11], dealing with the behaviour q-factor of frames with different configurations and with rigid and full strength joints are examined in this paragraph.

The main objective of the study was to dispose of significant values of q-factor, and to establish a sort of definition of q as a function of nominal acceleration a_N, parameter of plastic redistribution capacity α_u/α_y, fundamental period T_1, and interstorey drift sensitivity coefficient θ_j :

$$q = q\left(a_N^{(u)}, \alpha_u/\alpha_y, T_1, \theta_j\right) \qquad (3)$$

α_u and α_y are, respectively, the maximum and first yielding values of the horizontal seismic forces multiplier determined by a static elastic-plastic analysis ; θ_j is defined in EC3 (5.2.5.2) and EC8 (4.2.2).

The frames adopted in the parametric study are shown in Figure 2 with the associated characteristics given in Table 1. In Table 1, w means the storey load.

The concerned frames were subject to two different accelerograms namely these of Vrancea – 1977and Kobe - 1995 (Figure 3). Inelastic dynamic analyses of the frames were performed using the DRAIN-2DX computer code, developed at Berkeley University by Prakash & all. [13].

The obtained results from the parametrical study are presented in Table 2, for all the frames and for the three accelerograms.

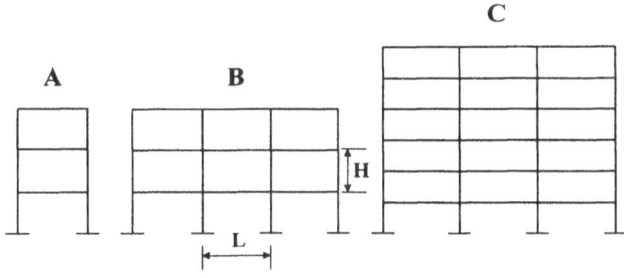

Figure 2. Investigated frames

Table 1. Characteristics of analysed frames

Frame	L[m]	H[m]	w[kN/m]	m_j[kg]	Beams	Columns
A	5.0	4.0	22	11000	IPE300	HEB240
B	4.0	4.0	32	38400	IPE330	HEB240
C	4.5	3.0	35	47250	IPE330	HEB360 (1,2) HEB300 (3-5) HEB260 (6)

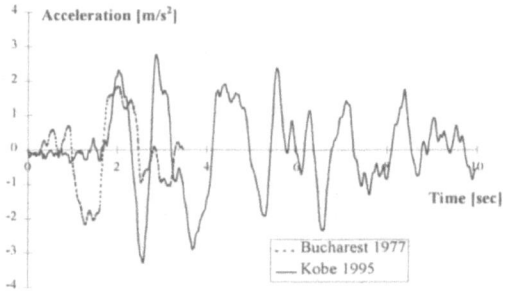

Figure 3. Accelerograms of Bucharest (1977) and Kobe (1995)

Table 2. Results of q-factor from parametric investigation

Accelerogram	Frame	A	B	C
of :	T_1 (sec)	0.883	0.917	1.330
	α_u/α_y	1.38	1.27	1.41
	q	1.6	1.5	3.4
BUCHAREST	$a_N^{(u)}$	3.1	2.6	4.8
	θ_j	0.020	0.030	0.130
	q	3.1	3.8	6.7
KOBE	$a_N^{(u)}$	4.6	5.6	7.9
	θ_j	0.070	0.160	0.170

Analysing these results, the following remarks may be underlined :
• Values of q-factor determined according to the new method are smaller than those given usually by codes and by other methods of the literature.
• Values of q-factor are clearly influenced by the type of frame. For Vrancea accelerogram, q-factor is varying from 1.5 or 1.6 for a simple frame like A or B until 3.4 for a dissipative frame like C.
• But values of q-factor are also strongly influenced by the shape of the accelerogram and of its associated spectrum, with regard to the fundamental period of the structure. So, for Kobe accelerogram, q-factor is varying from 3.1 for frame A until 6.7 for frame C.

- The reference value q should be associated with the nominal ground acceleration $a_N^{(u)}$ which can be regarded as the consequence of only q (the dissipation level depending eventually on the spectrum shape and the fundamental period of the frame).
- The plastic redistribution parameter α_u/α_y and the interstorey drift sensitivity coefficient θ_j have a slight influence on the values of the q-factor, but not always evident due to the prevailing influence of the spectrum shape and the fundamental period of structure.
- According to Table 2, for q values higher than 2.5 or 3.0, P-Δ effects cannot be neglected because of $\theta_j > 0.10$ (for example, frames C subject to accelerogram of Vrancea, and frames B, subject to accelerogram of Kobe).

3. Parametric investigation for partially dissipative structures

As already mentioned, the structure should be considered partially dissipative when the rotation capacity of dissipative members is limited (due to cross-sections in Class 2 or 3) or partial strength joints are used in dissipative zones.

In this case, the maximum values (q, $a_N^{(u)}$) should be determined as a function of the allowable rotation capacity directly from dynamic analyses.

3.1. DATA

A parametric investigation [11] was carried out on the structures presented in Figure 2 (structures A, B, C with characteristics of Table 1), using partial strength joints with different properties and considering the two accelerograms of Figure 3 in order to evaluate their influence on the q-factor. The moment resistance $M_{j,R}$ and initial rotational stiffness $S_{j,ini}$ of the studied joints are given in Table 3 where $M_{b,pl,R}$ is the plastic resistance moment of the connected beam to the joint and K_{sup} is the limit rotational stiffness corresponding to the distinction between rigid joint behaviour and semi-rigid one according to Annex J of Eurocode 3. Figure 4 shows the skeleton moment-rotation curves of the joints (for a quarter space) assuming here perfect parallelograms without stiffness degradation when repeated cyclic bending moments are applied. This type of joint behaviour was introduced in the inelastic dynamic analyses performed using DRAIN-2DX computer code (Prakash 1993). As often as not, partial strength joints have got a limited rotation capacity ϕ_u. Here three values of rotation capacity were selected, namely 0.015 ; 0.030 and 0.045 radians which seem realistic to cover most of applications.

This type of joint moment-rotation cyclic curve based on an elastic-perfectly plastic model, but with limited values of the rotation capacity ϕ_u can be considered significant as proved by a research program developed at INSA Rennes (Aribert and Grecea [2]).

212

Table 3. Joint characteristics of analysed frames

Moment resistance $M_{j,R}$	Rotational stiffness $S_{j,ini}$
$1.0M_{b,pl,R}$	K_{sup}
	$0.8K_{sup}$
	$0.6\,K_{sup}$
$0.8M_{b,pl,R}$	K_{sup}
	$0.8K_{sup}$
	$0.6\,K_{sup}$
$0.6M_{b,pl,R}$	K_{sup}
	$0.8K_{sup}$
	$0.6\,K_{sup}$

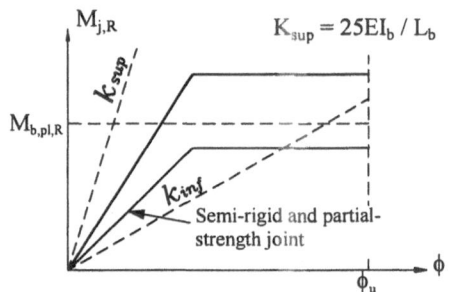

Figure 4. Skeleton moment-rotation curves

3.2. NUMERICAL RESULTS

Tables 4 and 5 collect, for example the results of the q-factor for structure C subject to the accelerogram of Vrancea (1977) and Kobe (1995), respectively.

3.3. INTERPRETATION OF THE PARAMETRIC INVESTIGATION

From examination of Tables 4 and 5 it is clear that the initial rotational stiffness $S_{j,ini}$, even for the lowest value $0.6K_{sup}$, has no influence practically on the q-factor. Moreover, decrease in the joint moment resistance tends to increase slightly the q-factor, maybe favouring the occurrence of global dissipative mechanism. So, these results incite to present the average values of the q-factor for the two accelerograms, as a function only of the joint resistance moment and of the rotation capacity. These average results are collected in Tables 6 and 7 and represented in Figure 5.

Table 4. Values of q-factor for accelerogram of Bucharest (1977)

$M_{b,pl,R}$	Criterion	K_{sup} q	a_N	θ_j	$0.8K_{sup}$ a_N	q	θ_j	$0.6K_{sup}$ q	a_N	θ_j
1.0	$\phi_{0.015}$	1.5	1.30	0.042	1.30	1.6	0.039	1.6	1.30	0.043
	$\phi_{0.030}$	2.0	2.20	0.061	2.20	2.0	0.061	2.1	2.20	0.065
	$\phi_{0.045}$	2.8	3.00	0.090	3.00	2.9	0.094	3.0	3.00	0.096
0.8	$\phi_{0.015}$	1.8	1.20	0.044	1.20	1.8	0.049	1.8	1.20	0.049
	$\phi_{0.030}$	2.3	2.10	0.074	2.00	2.3	0.074	2.4	2.00	0.075
	$\phi_{0.045}$	2.9	2.90	0.097	2.80	2.9	0.101	3.0	2.80	0.102
0.6	$\phi_{0.015}$	2.0	1.10	0.060	1.10	2.0	0.060	2.0	1.10	0.058
	$\phi_{0.030}$	2.4	2.00	0.086	2.00	2.4	0.087	2.6	2.00	0.090
	$\phi_{0.045}$	2.9	2.60	0.111	2.60	3.0	0.112	3.0	2.60	0.115

Table 5. Values of q-factor for accelerogram of Kobe (1995)

$M_{b,pl,R}$	Criterion	K_{sup} q	a_N	θ_j	$0.8K_{sup}$ a_N	q	θ_j	$0.6K_{sup}$ q	a_N	θ_j
1.0	$\phi_{0.015}$	1.5	0.50	0.044	0.60	1.6	0.046	1.8	0.80	0.054
	$\phi_{0.030}$	2.3	1.00	0.054	1.20	2.8	0.066	2.7	1.20	0.071
	$\phi_{0.045}$	3.4	1.60	0.069	1.60	3.5	0.073	3.2	1.60	0.079
0.8	$\phi_{0.015}$	1.5	0.50	0.043	0.50	1.7	0.047	1.9	0.70	0.052
	$\phi_{0.030}$	2.8	1.00	0.068	1.00	2.9	0.070	2.8	1.10	0.074
	$\phi_{0.045}$	3.5	1.40	0.069	1.50	3.8	0.072	3.5	1.60	0.078
0.6	$\phi_{0.015}$	1.9	0.50	0.042	0.50	2.3	0.049	2.3	0.60	0.056
	$\phi_{0.030}$	3.2	1.00	0.065	1.00	3.5	0.070	3.2	1.10	0.075
	$\phi_{0.045}$	4.2	1.40	0.075	1.50	4.6	0.079	4.0	1.70	0.082

Table 6. Average values of q-factor for accelerogram of Bucharest (1977)

$M_{j,R}$	$1.0M_{b,pl,R}$			$0.8M_{b,pl,R}$			$0.6M_{b,pl,R}$		
ϕ	$\phi_{0.015}$	$\phi_{0.030}$	$\phi_{0.045}$	$\phi_{0.015}$	$\phi_{0.030}$	$\phi_{0.045}$	$\phi_{0.015}$	$\phi_{0.030}$	$\phi_{0.045}$
q	1.6	2.0	2.9	1.8	2.3	2.9	2.0	2.4	3.0
a_N	1.30	2.20	3.00	1.20	2.00	2.80	1.10	2.00	2.60
θ_j	0.04	0.06	0.09	0.05	0.07	0.10	0.06	0.09	0.11

Table 7. Average values of q-factor for accelerogram of Kobe (1995)

$M_{j,R}$	$1.0M_{b,pl,R}$			$0.8M_{b,pl,R}$			$0.6M_{b,pl,R}$		
ϕ	$\phi_{0.015}$	$\phi_{0.030}$	$\phi_{0.045}$	$\phi_{0.015}$	$\phi_{0.030}$	$\phi_{0.045}$	$\phi_{0.015}$	$\phi_{0.030}$	$\phi_{0.045}$
q	1.6	2.6	3.4	1.4	2.8	3.6	2.2	3.3	4.3
a_N	0.60	1.20	1.60	0.60	1.00	1.50	0.50	1.00	1.50
θ_j	0.05	0.06	0.07	0.05	0.07	0.07	0.05	0.07	0.08

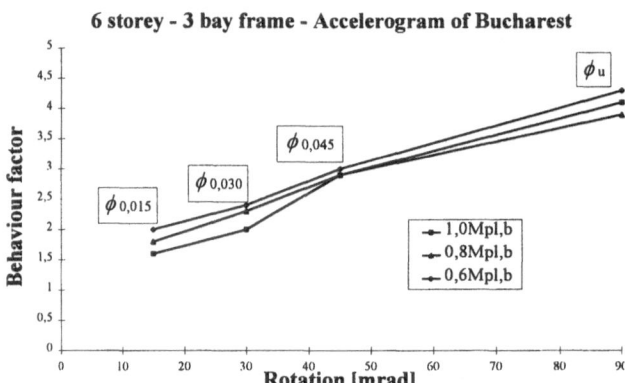

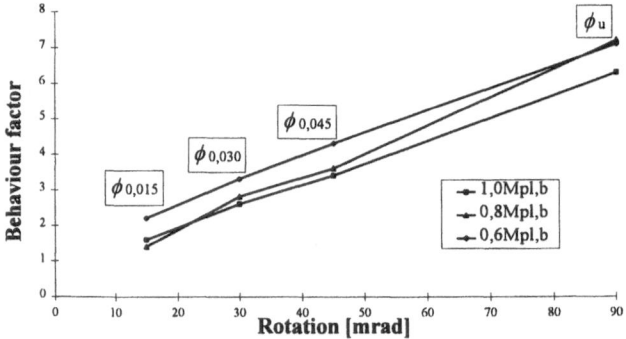

Figure 5. Average of q-factor as a function of the joint rotation capacity for accelerograms of Bucharest (1977) and Kobe (1995)

Average values of q for $\phi_{0.015}$ can be deduced from Tables 6 and 7, applicable to the three values of resistance moment of the joints $1.0M_{b,pl,R}$, $0.8M_{b,pl,R}$ and $0.6\ M_{b,pl,R}$. For Bucharest accelerogram, q can be taken equal to 1.8 provided that a_N and θ_j do not exceed $1.2m/s^2$ and 0.050, respectively. Similarly for Kobe accelerogram, the values become q=1.7, with $a_N=0.6m/s^2$ and $\theta_j=0.050$.

For the criterion $\phi_{0.030}$ and Bucharest accelerogram, it is obtained q=2.2 with $a_N=2.1m/s^2$ and $\theta_j=0.075$, while for Kobe accelerogram, q=2.7, with $a_N=1.0m/s^2$ and $\theta_j=0.070$.

Lastly for the criterion $\phi_{0.045}$ and Bucharest accelerogram, it is obtained q=2.9 with $a_N=2.8m/s^2$ and $\theta_j=0.100$, while for Kobe accelerogram, q=3.8, with $a_N=1.5m/s^2$ and $\theta_j=0.075$.

Following to the previous comments, some more general conclusions can be expressed.

For q-values higher than q=3.0 it appears that the geometrical second order effects (or P-Δ effects) begin to influence the structural behaviour. So, for q-values higher than this value P-Δ effects cannot be neglected.

In accordance with all performed analyses, even for low q-values, most of the plastic hinges are developed in beams or joints but also someones in columns. Consequently, care must be taken to control the rotation capacity at the column ends with respect to the risk of local mechanism (due to occurrence of an intermediate plastic hinge).

For the chosen three rotation capacities, the joint moment resistance seems also not to influence the q-factor, noting that with the reduction of the moment resistance, the q-factor has a little tendency to grow up a little bit. In practice, it should be preferable to keep constant the value of q-factor till the value $0.6M_{b,pl,R}$ (Figure 5).

3.4. PROPOSAL OF CONSERVATIVE q-FACTORS

According to the numerical study, an indicative proposal of q-factors could be promote for steel structures with semi-rigid and partial-resistant joints, for different categories of joint rotation capacity (ϕ), acceleration (a_N) and type of accelerogram, as in Table 8. Evidently, these values are imbued with the results deduced from the accelerograms of Bucharest and Kobe, which are a quite severe in comparison with those given by most of the seismic codes, specially in Europe. Proposed values of Table 8 could be generalized and improved by further researches ; they are already giving a first answer to the design problem of steel frames with partial-strength dissipative joints.

Table 8. Conservative values of q-factor for steel frames with partial-strength joints

	Accelerogram duration			
	≤ 4 sec.		≤ 20 sec.	
	Q	$a_N\ [m/s^2]$	q	$a_N\ [m/s^2]$
$\phi_{0.015}$	1.3 – 1.8	1.20 – 2.20	1.7 – 1.8	0.60 – 1.70
$\phi_{0.030}$	1.7 – 2.2	2.10 – 2.90	2.0 – 2.9	1.10 – 2.20
$\phi_{0.045}$	2.0 – 2.9	2.80 – 3.70	2.3 – 3.8	1.50 – 2.70

4. References

1 Aribert, J.M. and Grecea, D. (1997) A new method to evaluate the q-factor from elastic-plastic dynamic analysis and its application to steel frames. Proceedings of STESSA'97, Kyoto, 3-8 August 1997.

2 Aribert, J.M. and Grecea, D. (1998) Experimental behaviour of partial-resistant beam-to-column joints and their influence on the q-factor of steel frames. The 11th European Conference of Earthquake Engineering, Paris, 6-11 September 1998.

3 Aribert, J.M. and Grecea, D. (1999) Dynamic behaviour control of steel frames in seismic areas by equivalent static approaches. Proceedings of SDSS'99, Timisoara, 9-11 September 1999.

4 AFNOR-Règles PS 92 appliquables aux bâtiments, NFP 06.013, Décembre 1995.

5 Ballio, G. (1985) ECCS approach for the design of steel structures against earthquakes. Symposium on Steel in Buildings, IABSE-AIPC-IVBH Report, Vol.48, pp. 373-380, Luxembourg, September 1985.

6 CEN (1994) Eurocode 8 (ENV): Design provisions for earthquake resistance of structures. Part 1-2: General rules for buildings.

7 CEN (1995) Eurocode 8 (ENV): Design provisions for earthquake resistance of structures. Part 1-3, Section 3: Specific rules for steel buildings.

8 ECCS TWG1.3. (1985) Recomanded testing procedure for assesing the behaviour of structural steel elements under cyclic loads.

9 Eurocode 3. Part 1.1. Revised annex J : Joints in building frames. ECCS Committee TC 10-Structural Connections. April 1996.

10 Galea Y. and Bureau A. (1995) PEP-micro. Analyse plastique au second ordre de structures planes à barres. Manuel d'utilisation, Version 2b, CTICM, France.

11 Grecea D. (1999) Caractérisation du comportement sismique des ossatures métalliques - Utilisation d'assemblages à résistance partielle. Thèse de Doctorat. INSA de Rennes, France.

12 Mazzolani, F.M. and Piluso, V. (1996) Theory and design of seismic resistant steel frames. E & FN Spon, London

13 Prakash, V., Powell, G.H. and Campbell, S. (1993) DRAIN-2DX, base program description and user guide. Version 1.10.

14 Sedlacek, G and Kuck, J. (1993) Determination of q-factors for Eurocode 8, Aachen, 31 August 1993.

15 Setti, P. 1985. Un metodo per la determinazione del coefficiente di strutura per la construzioni metalliche in zona sismica. Construzioni Metaliche n°3.

CYCLIC BEHAVIOUR OF END-PLATE BEAM-TO-COLUMN COMPOSITE JOINTS

RUI SIMÕES & LUÍS SIMÕES DA SILVA
University of Coimbra, Department of Civil Engineering
Polo II, Pinhal de Marrocos, 3030-290 Coimbra, Portugal

PAULO CRUZ
University of Minho, Department of Civil Engineering
Campo de Azurém, 4600 Guimarães, Portugal

1. Introduction

The behaviour of joints under cyclic loading, when compared to the corresponding static monotonic response, presents the added difficulty of degradation of strength and stiffness in successive loading cycles. Composite joints in seismic regions must provide adequate performance under load reversal, with good energy dissipation. To try to provide some additional insight into this problem, an experimental research program on end-plate beam-to-column composite joints under cyclic loading carried out at the University of Coimbra is described in this paper.

The cyclic behaviour of a joint is always unstable, exhibiting a progressive degradation of its mechanical properties (strength, stiffness and energy dissipation capacity), as shown in Fig. 1 [1]. In seismic areas, characterised by repeated load

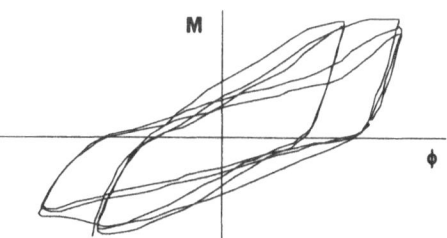

Figure 1. Cyclic behaviour of joints

reversal, the resulting joint response should remain as symmetrical as possible, an imposition much harder to ensure in common beam-to-column joints, given the asymmetry of the joint with respect to the centroidal axis.

From the experimental results of tests on internal and external node configurations, with and without composite columns [2], it was possible to identify the various failure modes and to fit the corresponding hysteretic curves to the Richard-Abbott and

C.C. Baniotopoulos and F. Wald (eds.), The Paramount Role of Joints into the Reliable Response of Structures, 217–226.
© 2000 *Kluwer Academic Publishers.*

Mazzolani models. These curve-fitting exercises highlighted the need to adapt both models, either for improved ease of application, or to deal with some aspects previously not covered by those models, as described next.

2. Analytical evaluation of the dynamic behaviour of composite joints

2.1 RICHARD-ABBOTT MODEL

The Richard-Abbott model is based on a formula developed in 1975 [3] to reproduce the elastic-plastic behaviour of several materials and was initially used to simulate the static monotonic response of joints and later applied to cyclic situations [4]. According

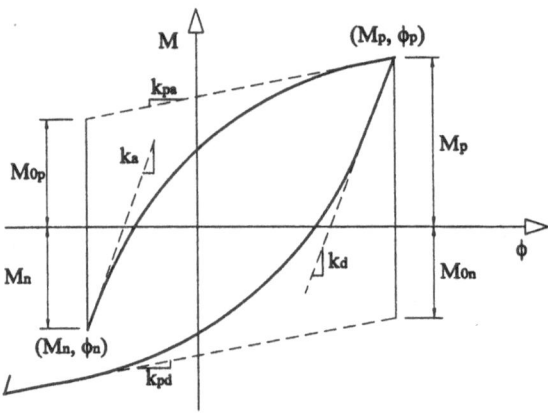

Figure 2. Richard-Abbott model, adapted to deal with different behaviour under positive and negative bending

to this model, the loading branch of the moment-rotation curve of a joint is described by the following equation, here presented in a modified form to deal with unsymmetrical joints with respect to the centroidal axis, as is usually the case for composite joints,

$$M = M_n - \frac{(k_a - k_{pa}) \cdot (\phi_n - \phi)}{\left[1 + \left|\frac{(k_a - k_{pa}) \cdot (\phi_n - \phi)}{M_{0a}}\right|^N\right]^{1/N}} - k_{pa} \cdot (\phi_n - \phi) \tag{1}$$

where $M_{oa} = M_n + M_{0p}$. Parameters k_a and k_{pa} (and, for the unloading branch, k_d, k_{pd}), that depend on the mechanical properties of the joint, are defined in Fig. 2, while N allows for the adjustment of the curvature, further details being found in [2]. The unloading branch of the curve is described by a similar equation, by replacing point (M_n, ϕ_n) by (M_p, ϕ_p) and parameters M_{0a}, k_a and k_{pa} by the corresponding values evaluated at unloading, M_{0d}, k_d and k_{pd}.

In general, whenever a joint is subjected to successive loading cycles in plastic regime, parameters k, k_p, M_0 and N (either for the loading or unloading branches) do not remain constant. In particular, stiffness k and moment M_0 exhibit a tendency to reduce, corresponding to the degradation of the mechanical properties of the joint.

2.2 MODIFIED MAZZOLANI MODEL

The model proposed by Mazzolani [5,6], based on the Ramberg-Osgood model, allows the mathematical simulation of hysteretic behaviour with slipage, where the cycles have the shape shown in Fig. 3. As originally proposed, each complete cycle was divided in four branches (I, II, III e IV), the definition of branches I and II being similar to branches III and IV. However, in unsymmetrical joints, as is the present case of composite joints, all parameters must be defined separately for the positive (branch I and II) and negative (branch III and IV) zones.

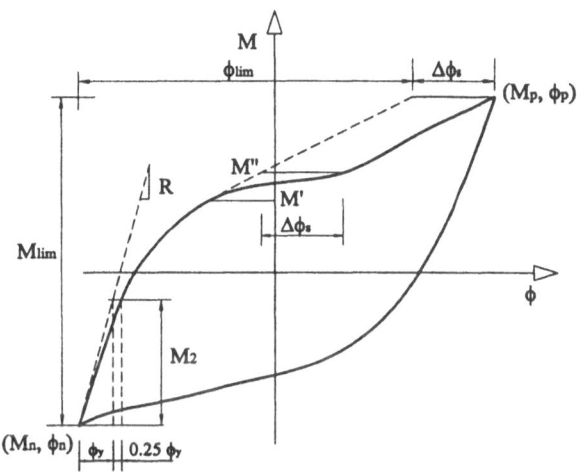
Figure 3. Definition of a complete cycle

Given that, in joints with slipage, the corresponding branch may start in the unloading zone, thus preventing the application of the model, a modified version is proposed in this paper. It consists of the definition of each cycle with two single branches, ascending and descending, as described in Fig. 3, thus eliminating the limitation of slipage not being able to occur in the unloading branch.

The mathematical description of the ascending branch ($M_n \leq M \leq M_p$) is given by equations (2a-c). The initial rotation (ϕ_n) and initial stiffness (R) are evaluated at point (M_n, ϕ_n):

$$\phi = \phi_n + \frac{M - M_n}{R} + c_1 \cdot \left(\frac{M - M_n}{M_2}\right)^{c_2} \qquad \text{if} \quad M_n \leq M \leq M' \qquad (2a)$$

$$\phi = \phi_n + \frac{M - M_n}{R} + c_1 \cdot \left(\frac{M - M_n}{M_2}\right)^{c_2} + \frac{\Delta\phi_s}{2} + \left(\frac{\Delta\phi_s}{2} - K_m\right) \cdot \rho \cdot |\rho|^{s-1} + K_m \cdot \rho$$

$$\text{if} \quad M' < M \leq M'' \tag{2b}$$

$$\phi = \phi_n + \frac{M - M_n}{R} + c_1 \cdot \left(\frac{M - M_n}{M_2}\right)^{c_2} + \Delta\phi_s \qquad \text{if} \quad M'' \leq M \leq M_p \tag{2c}$$

Moment M_2 is used to constrain the curve to an intermediate point, defined in the context of the present work to correspond to a rotation of the order of 1.25 the elastic rotation, so that this point lies outside the slipage branch. Parameters c_1 and c_2 are obtained from eqs. (3) and (4):

$$c_1 = 0.25 \cdot \phi_y = 0.25 \cdot \frac{M_2}{R} \tag{3}$$

$$c_2 = \frac{\ln\left[\left(\phi_{\lim} - \frac{M_{\lim}}{R}\right) \Big/ c_1\right]}{\ln\left(M_{\lim} / M_2\right)} \tag{4}$$

where $M_{\lim} = |M_n| + |M_p|$ and $\phi_{\lim} = |\phi_n| + |\phi_p| - \Delta\phi_s$.

The stiffness at the start of each cycle (R) is obtained as a function of the accumulated energy of the previous cycle (Ω) as described in [3]. Equations (2b-c) reproduce the slipage and post-slipage branches, the corresponding parameters $(M'$, M'', $\Delta\phi_s$, ρ, s and $K_m)$ being also defined in [5].

Starting from the positive extremum of the previous half-cycle, the descending branch is defined in similar fashion.

To reproduce the degradation of strength in the current model, a degradation curve is proposed for $M_{\lim}$ similar to the one considered for stiffness (R). Starting from an initial value, the current value for a given cycle is given by:

$$M_{\lim} = M_{\lim 0} \cdot \left[1 - \frac{\Delta M_{\lim}}{M_{\lim 0}} \cdot \left(\frac{\Omega}{\Omega_{máx}}\right)^p\right] \tag{5}$$

where:

$M_{\lim 0}$ - Initial $M_{\lim}$ obtained from the static monotonic moment-rotation results

$\Delta M_{\lim}$ - Difference between $M_{\lim}$ evaluated at the first and last cycle before collapse (Figure 4).

Ω - Accumulated energy at the end of the previous cycle

$\Omega_{máx}$ - Accumulated energy at collapse

p - Parameter defined according to Figure 4.

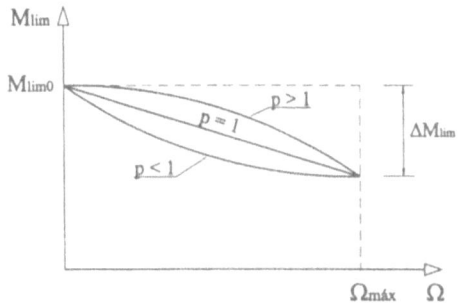

Figure 4. Relation between M_{lim} and the accumulated energy

In contrast with stiffness, M_{lim} does not decrease for cycles with amplitudes little greater than the elastic amplitude; on the contrary, from the elastic amplitude ($\Omega = 0$) and up to a certain value, an increase of M_{lim} is noted. Consequently, eq. (5) should only be used for cycles of amplitude equal or larger than the maximum attained moment. In the context of the present work, this corresponds to 4 times the elastic amplitude, M_{lim} being obtained directly from the static monotonic or cyclic envelope of the moment-rotation curve for smaller amplitudes.

3. Experimental behaviour

The test program performed at the Civil Engineering Department of the University of Coimbra included 4 prototypes, being 2 in internal nodes and 2 in external nodes, a thorough description being found elsewhere [2]. The description of each model includes the geometric definition, the material properties and the testing and instrumentation procedures. The prototypes, covering internal and external nodes, were defined such

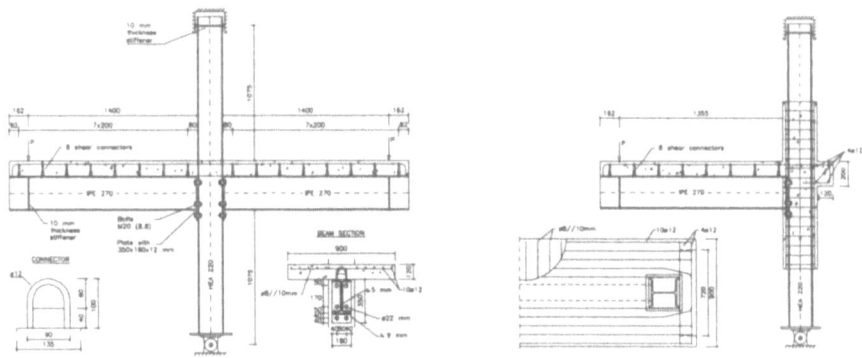

Figure 5. Experimental models for internal and external nodes

that they could reproduce the connections in a common framed structure, with spans of about 7m, 4m spacing between frames, live loads up to 4 kN/m² and a high energy dissipation capacity and a good fire resistance [7,8]. According to the objectives of this study, the steel connection is the same in all prototypes, corresponding to a beam connected to the column by one end plate, welded to the beam and bolted to the column.

In all cases, the beams consist of an IPE 270, rigidly connected to a reinforced concrete slab (full interaction) by 8 shear connectors. The slab, 900 mm wide and 120 mm thick, is reinforced with 10ϕ12 longitudinal bars and 10ϕ8 transversal bars per meter, with 20 mm cover. The steel connection consists of a 12 mm thick end plate, welded to the beam and bolted to the column flange through 6 M20 bolts (class 8.8). The end-plate is flushed at the top and extended at the bottom, in order to achieve similar behaviour under positive and negative moments. The steel column is the same in all the tests (HEA 220), being envolved by concrete (300 × 300 mm) in tests E10 and E12, with longitudinal reinforcement of 4ϕ12, with one bar in each corner of the section and stirrups consisting of ϕ6 bars 0.08 m apart. The following materials were chosen: S235 in the steel components, steel class 8.8 in the bolts, steel A400 NR in the reinforcing bars.

Two tests were performed in internal nodes, test E11 corresponding to the prototype arrangement between composite beams and a steel column shown in Figure 6a) and test E12 between composite beams and a composite column. The loads were applied to the beams 1.40 m from the steel column face with two dynamic actuators with a capacity of 200 kN and 600 kN, and maximum displacement of 20 cm and 10 cm, respectively. In both tests, loading was applied according to the methodology proposed by the ECCS [9], with both joints (left and right) being equally loaded but out of phase by one half cycle. Each cyclic test comprised 4 elastic cycles followed by 15 cycles in plastic

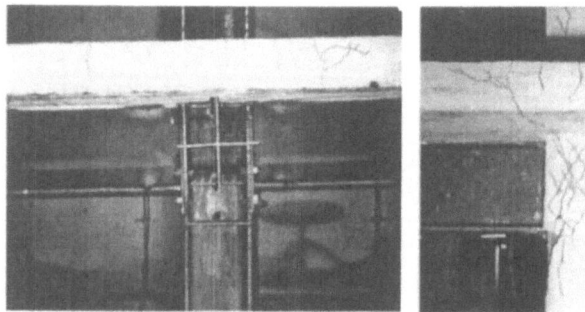

Figure 6. Photos of prototypes after testing

regime (3 x ±2e_y, 12 x ±4e_y), the elastic displacement, e_y = 12 mm, being evaluated from the results of earlier equivalent static tests [10].

As for the internal nodes, two tests were performed in external nodes, test E9 corresponding to a steel column, and test E10 corresponding to a composite column and illustrated in Figure 5b). Loading was applied similarly to the internal node tests but with a larger number of amplitudes.

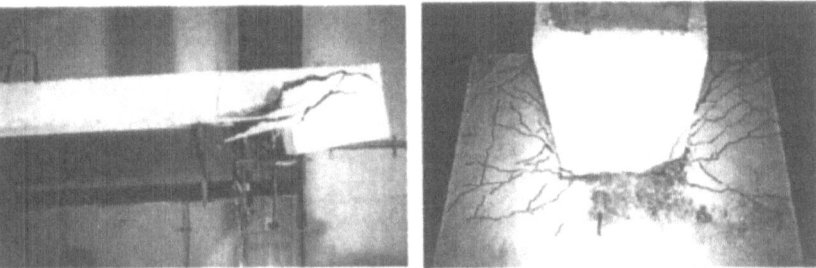

Figure 7. Photos of prototypes after testing

The evaluation of results was based on the moment rotation response of the joint, the various elastic parameters being summarized in Table 3, evaluated according to [9], from the envelope of the cyclic tests:

TABLE 3. Elastic parameters for tests E9 to E12.

Elastic Parameters	Test E9		Test E10		Test E11		Test E12	
	Env. +	Env. -	Env. +	Env. -	Env. +	Env. -	Env. +	Env. -
K_y (kNm/mrad)	24.57	26.81	36.95	40.83	16.50	18.83	34.44	36.22
ϕ_y (mrad)	4.64	4.29	4.60	3.89	5.05	4.28	3.39	3.25
M_y (kNm)	114.00	114.99	169.81	158.92	83.41	80.68	116.83	117.78

For the tests on internal nodes (E11 and E12), the joints presented high ductility (ductility ratio close to unity), with similar response for hogging and sagging moment. Because the maximum amplitude was not very high, the strength degradation was low.

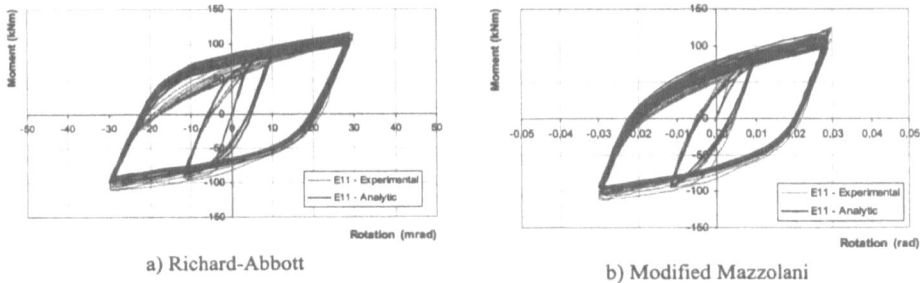

a) Richard-Abbott   b) Modified Mazzolani

Figure 8. Moment-rotation response for test E11

224

However, for the stiffness and energy degradation ratios, strong reductions were noted [2].

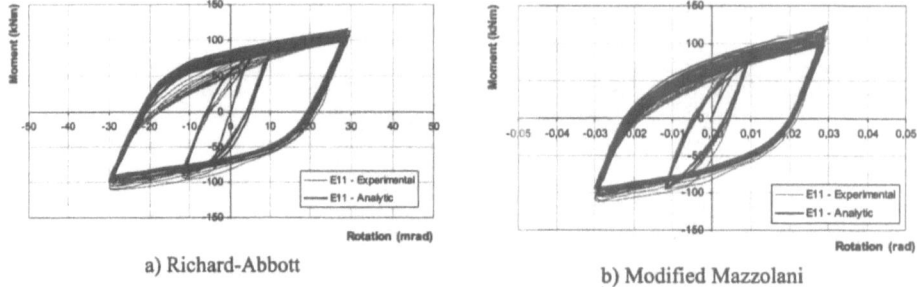

a) Richard-Abbott

b) Modified Mazzolani

Figure 9. Moment-rotation response for test E12.

Figures 8 and 9 illustrate the experimental results for the internal node tests, shown superimposed with results from the (a) Richard-Abbott analytical model and (b) modified Mazzolani model.

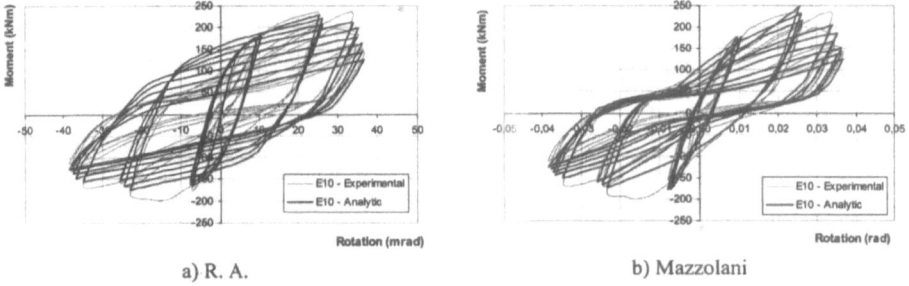

a) R. A.

b) Mazzolani

Figure 10. Moment-rotation response for test E9

For the tests on external nodes (E9 and E10), the ductility ratios remained high (above 0.8), except for test E10 under negative hogging moment. The strength degradation was severe, while the energy degradation ratios have shown that the joints reached collapse. Figures 10 and 11 illustrate the moment-rotation response for these two tests, again superimposed with the (a) Richard-Abbott analytical model and (b) modified Mazzolani model.

The analytical predictions based on the Richard-Abbott model for the cyclic response of the composite joints presented in Figs. 8a to 11a were directly derived from the static monotonic response. Parameter k_p was taken constant and equal to 5 % of the initial elastic stiffness k (equal to K_y^+ or K_y^-). Parameter N was also kept constant for all cycles and obtained as the average of a previous adjustment cycle by cycle directly from the experimental results (simulation I). A decaying law based on the accumulated dissipated energy (obtained analytically) was assumed for parameters k and M_0, based on the values obtained for each cycle in the previous section; these curves were

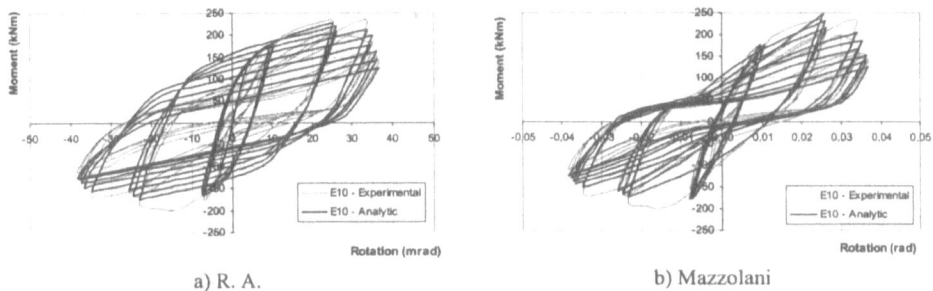

a) R. A.   b) Mazzolani

Figure 11. Moment-rotation response for test E10

The analytical predictions based on the modified Mazzolani model also rely on decaying laws for the (i) tangent (stiffness) at the origin of the ascending and descending branches (R), (ii) bending moment (M_{lim}), and (iii) slipage $(\Delta\phi_s)$. These decaying laws are defined from initial values as a function of the accumulated dissipated energy. The initial values of R and M_{lim} may be obtained from the static monotonic results; for the remaining parameters, representative values are chosen based on available experimental data.

4. Conclusions

The developments presented in this paper were directed at the daunting task of predicting the dynamic behavior of steel and composite joints.

The Richard-Abbott model was adjusted so that the error on the maxima moments reached at the end of each half-cycle was minimised. Consequently, agreement between analytical and experimental results for this criterium was good, the maximum difference being obtained for test E9 with 8% difference for positive bending and 10% for negative bending. As for the energy dissipation, the Richard-Abbott model is only able to reproduce hysteretic curves whenever slipage does not occur. For tests E11 and E12 on internal nodes where slipage hardly occurred, the model showed good agreement, with maximum error of +33% and -1% for the positive and negative zones, respectively.

The modified Mazzolani model, with more parameters and able to simulate slipage, gave a much better agreement between experimental and analytical results, particularly with respect to dissipated energy, a crucial aspect in terms of seismic behaviour. Although all parameters were evaluated directly from the static monotonic results, several coefficients being calibrated from the available test results, the error on the evaluation of total dissipated energy was only of –7%. For tests E11 and E12, slipage was not modelled, resulting in an error of +28% for the latter, since some slip was observed on the negative zone. In terms of bending moment, the maximum error occurred for test E10, with an average error of +6%.

Finally, it is noted that the Mazzolani model, with the modifications introduced in this work, constitutes a good tool to predict the cyclic behaviour of joints, particularly beam-to-column composite joints. Its application, however, requires the calibration of a certain number of coefficients that must be done experimentally. Given that the number of tests used in this work was only 4, all presenting different aspects, its validity should be checked against a wider base of experimental data.

5. Acknowledgments

Financial support from "Ministério da Ciência e Tecnologia" - PRAXIS XXI research project PRAXIS/P/ECM/13153/1998 is acknowledged.

6. References

1. ECCS (1994) ECCS Manual on Design of Steel Structures in Seismic Zones, by F. M. Mazzolani and V. Piluso, TC 13 Seismic Design, N° 76, Napoli, Italia.
2. Simões, R.A.D. (2000) Behaviour of beam-to-column composite joints under static and cyclic loading (in portuguese), PhD Thesis, Civil Engineering Department, Universidade de Coimbra, Coimbra, Portugal.
3. Richard, R. M. and Abbott, B. J. (1975) Versatile Elasto-Plastic Stress-Strain Formula. *Journal of the Engineering Mechanics Division*, ASCE, **101**, EM4, 511-515.
4. Elsati, M. K. and Richard, R. M. (1996) Derived Moment Rotation Curves for Partially Restrained Connections. *Structural Engineering Review*, **8** (2/3), 151-158.
5. De Martino, A., Faella, C. and Mazzolani, F. M. (1984) Simulation of Beam-to-Column Joint Behaviour Under Cyclic Loads. *Construzioni Metalliche*, **6**, 346-356.
6. Mazzolani, F.M. (1988) Mathematical Model for Semi-Rigid Joints Under Cyclic Loads, in R. Bjorhovde et al. (eds) *Connections in Steel Structures: Behaviour, Strength and Design*, Elsevier Applied Science Publishers, London, 112-120.
7. Eurocode 3, ENV 1993-1-1 (1992) Design of Steel Structures, CEN, European Committee for Standardization, Ref. No. ENV 1993-1-1: 1992, Brussels, Belgium.
8. Eurocode 4, ENV 1994-1-1 (1996) Proposed Annex J for EN 1994-1-1, Composite joints in building frames, CEN, European Committee for Standardization, Draft for meeting of CEN/TC 250/SC 4, Paper AN/57, Brussels, Belgium.
9. ECCS (1986) Recommended Testing Procedure for Assessing the Behaviour of Structural Steel Elements under Cyclic Loads - N°45.
10. Simões, R. A. D., Simões da Silva, L. A. P., and Cruz, P. J. S. (1999) Experimental models of end-plate beam-to-column composite connections, *2nd European Conference on Steel Structures, EUROSTEEL '99*, 26-29 May, Praha, Czech Republic, 625-629.

A NONCONVEX OPTIMIZATION APPROACH FOR THE DETERMINATION OF THE CAPACITY CURVE AND THE PERFORMANCE POINT OF MR STEEL FRAMES EXHIBITING SOFTENING UNDER SEISMIC LOADING

E.S. MISTAKIDIS
Department of Civil Engineering, University of Thessaly,
GR-38334, Volos, Greece

1. Introduction

After the last catastrophic earthquake in Greece a vast need arized for the assessment of the structural response of existing buildings under seismic loading. To this end the elastic analysis methods which include code static and dynamic lateral force procedures and elastic procedures for the capacity design are insufficient. Although an elastic analysis indicates where first yielding will occur, it cannot predict failure mechanisms and account for the redistribution of forces during progressive yielding. This problem is bypassed in the phase of design of a new structure by the introduction of the behavioral factor of the structure (the well known q-factor) which is selected according to the specific structural typology of a new building. This factor is a divider of the lateral forces applied on the structures and accounts for the elastoplastic behaviour of the structure under seismic loading. At a later stage of the design, the value of the q-factor and the achievement of the predefined elastoplastic behaviour are ensured through the capacity checks of the seismic codes.

However, coming to the seismic assessment of existing structures, one has to cope with an existing structural system with an unknown elastoplastic behaviour and therefore the previous procedure cannot be applied. The most accurate inelastic analysis procedure is the complete nonlinear time history analysis which at this time is considered extremely complex and impractical for general use. To face the problem the following simplified nonlinear analysis methods have been proposed in the literature [1], [2]:

- the Capacity Spectrum Method (CSM) which uses the intersection of the capacity curve with the reduced response spectrum to estimate the maximum displacement of the structure under seismic loading.

- The Displacement Coefficient Method (DCM) that uses the capacity curve and a modified version of the "equal displacement approximation" to estimate the maximum displacement.

C.C. Baniotopoulos and F. Wald (eds.), The Paramount Role of Joints into the Reliable Response of Structures, 227–236.
© 2000 *Kluwer Academic Publishers.*

Inelastic analysis procedures help demonstrate how buildings really work by identifying modes of failure and the potential for progressive collapse. However, both
methods require the determination of the capacity curve of the structure which
depends on the strength and deformation capacities of the individual components
of the structure. At this point it becomes apparent the necessity of the accurate
description of the nonlinear response of the individual components as e.g. of the
connections.

In the classical approach, a bilinear form of the moment rotation relationship is
employed without or with hardening (Fig. 1a). However, steel members exhibit
softening, right after having reached the maximum resistance moment (Fig. 1b).
This form of the $M - \varphi$ curve (which has also been confirmed by experiments [3],
[4] is due to a combination of material nonlinearity with severe local buckling and,
if torsional restraints are not provided, with the occurrence of lateral-torsional
buckling. Notice that more complex moment-rotation diagrams have also been
proposed in the literature [2] (Fig. 1c). Such diagrams may reduce considerably
the rotational capacity of the corresponding element.

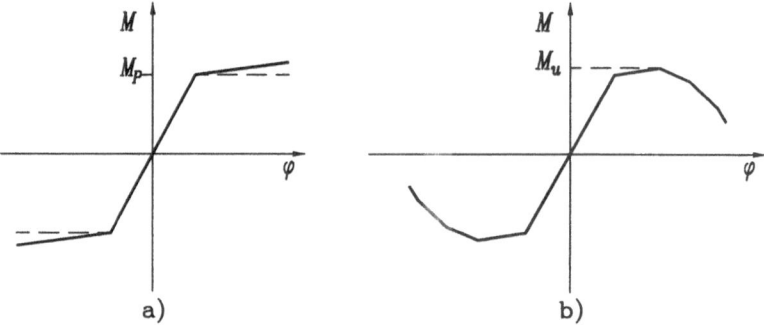

Figure 1: Monotone and nonmonotone moment - rotation laws

Here the empirical method of Kato-Akiyama [5] is used. According to this method,
the rotational capacity of the members can be described by means of the general
moment-rotation curve of Fig. 2. In the diagram of Fig. 2, the parameters m, φ_y,
φ_u, K_h, K_d are calculated taking into account the axial loading of the member,
the slenderness of the web and the flange and the slenderness of the member with
respect to the weak axis of the section. The method can also take into account the
flexural-torsional buckling phenomena. In this case the coupling of the buckling
modes leads to a higher slope of the softening branch.

In this paper the solution of the nonconvex optimization problem which gives the
equilibrium configuration of the structure under the above mentioned exceptional
conditions (non-monotone moment-rotation law) is obtained by the application
of the heuristic nonconvex optimization algorithm described in detail in [6]. The
method is based on the theoretical developments of [7] and [8].

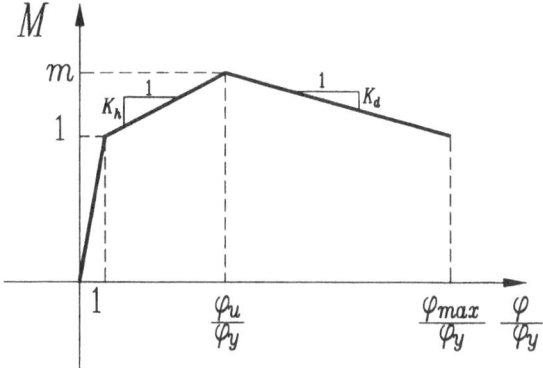

Figure 2: General nonmonotone moment-rotation curve

2. Basic aspects of the CSM and DCM

In the following the basic aspects of CSM and DSM are presented. The key elements of both the design procedures are demand and capacity. Demand is a representation of the earthquake ground motion. Capacity is the ability of the structure to resist the seismic demand. The performance is dependent on the manner that the capacity is able to handle the demand. In other words, the structure must have the capacity to resist the demands of the earthquake such that the performance of the structure is compatible with the objectives of the design.

Both methods require the determination of the capacity curve of the structure. A lateral force distribution is applied on the structure and the capacity curve is constructed by plotting the roof displacement versus the base shear. The analysis method followed in this paper for the determination of the capacity curve is presented in Section 3.

The capacity spectrum method (Fig. 3) is based on finding a point on the capacity spectrum that also lies on the appropriate response spectrum which is reduced from the elastic design spectrum. For this methodology, spectral reduction factors are calculated (in terms of effective damping) based on the shape of the capacity curve, the estimated displacement demand and the resulting hysteresis loop. The intersection of the two curves (the capacity curve and the demand curve) determines the performance point of the structure. This point represents the condition for which the seismic capacity of the structure is equal to the seismic demand imposed on the structure by the specified ground motion. Notice that for the application of the method, it is necessary to convert the capacity curve, which is in terms of base shear and roof displacement to what is called a capacity spectrum, which is a representation of the capacity curve in Acceleration - Displacement Response Spectra (ADRS) format [1].

The displacement coefficient method is based on statistical analysis of the results of time history analysis of single degree of freedom models of different types. The demand displacement is called the target displacement and is calculated by

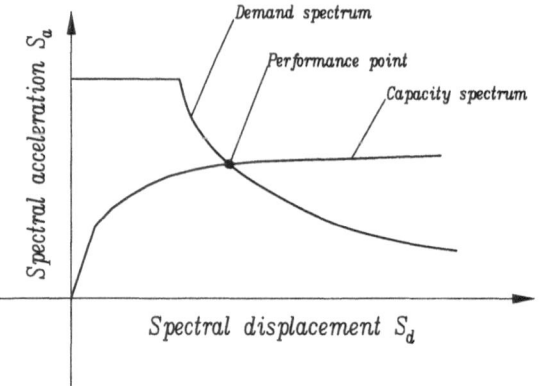

Figure 3: Graphical representation of the CSM

modifying the displacement obtained applying the "equal displacement approach" (which is based on the assumption that the inelastic spectral displacement is the same as that which would occur if the structure remained perfectly elastic) by various coefficients. These coefficients relate the spectral displacement and the likely building displacements, the expected maximum inelastic displacements to the displacements calculated for linear elastic response and also take into account the effect of the hysteresis loop on the maximum displacement response and the increased displacements due to second order or softening effects.

After the determination of the performance point or the target displacement, a performance check can verify that the structural or non-structural components are not damaged beyond the acceptable limits of the performance objective for the forces and displacements implied by the displacement demand.

3. Approximation of nonconvex potentials using convex potentials

Here the method applied for the solution of the arising nonconvex optimization problem is described leaving out any points that would cause unnecessary difficulty to the practicing engineer. The theoretically interested reader is referred to [6]. The method is presented by means of Fig. 4. Let us assume that the moment - rotation diagram OAB of Fig. 4 describes the behaviour of the member under consideration. Moreover let us assume for simplicity that the structure is made up of a single element, and we want to calculate the response of the structure under certain loading. We assume first that instead of the initial nonmonotone law, the simple, monotone law $h^{(1)}$ holds. This is an elastic - plastic law and the solution of this problem is trivial and can be obtained using a simple Quadratic Programming (QP) algorithm. The solution of the problem gives as a result a certain value for the plastic rotation φ let us say $\varphi^{(1)}$. We recognize that this is not a solution of the initial problem, as the point $(M^{(1)}, \varphi^{(1)})$ doesn't lie on the nonmonotone law. We continue the approximations of the nonmonotone law by making this time the assumption that the law $h^{(2)}$ holds. Again this is a plasticity problem that gives as a result the rotation $\varphi^{(2)}$. Again, the point $(M^{(2)}, \varphi^{(2)})$ is not a solution of the initial problem. Then we solve a new "ideal plasticity" problem with the assumption that the law $h^{(3)}$ holds, and so on until the solution point $(M^{(n)}, \varphi^{(n)})$ lies on the nonmonotone law.

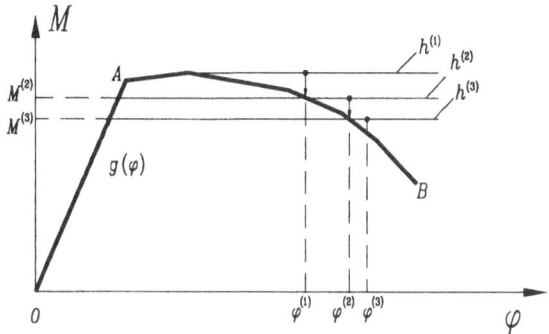

Figure 4: Graphical representation of the algorithm

For this simple and engineering oriented algorithm one can find a robust mathematical explanation [6]. The approximation of the nonmonotone law problem by simple monotone law problems is equivalent to the substitution of the nonconvex optimization problem by a sequence of convex optimization problems and more specifically by a sequence of quadratic optimization problems. These problems are treated efficiently by robust numerical algorithms able to handle hundreds or even thousands of unknowns. In this way we manage to extend the advantages of the methods used to solve the classical plasticity problems to nonmonotone problems.

4. Numerical applications

The method presented in the previous sections will be now applied for the analysis of two simple examples. In both examples treated here, the steel grade is Fe360 with a yield stress of 235 N/mm². The units are m and kN.

4.1 SIMPLE PITCHED ROOF FRAME

As a first example the simple structure of Fig. 5a is considered. The structure is a moment resisting frame consisting of HEA450 columns and a IPE450 beam. The structure is loaded with the vertical loads $P_V = 13kN$ which are considered constant and with the horizontal load P_H which increases until the collapse of the structure. The moment-rotation curve for the beam was calculated according to [5]. It is assumed that the beam is laterally restraint. The moment rotation diagram is presented in Fig. 5c. In order to investigate the influence of the moment-rotation law to the total ductility of the structure and to compare the results against the results of more classic approaches, two cases are considered.

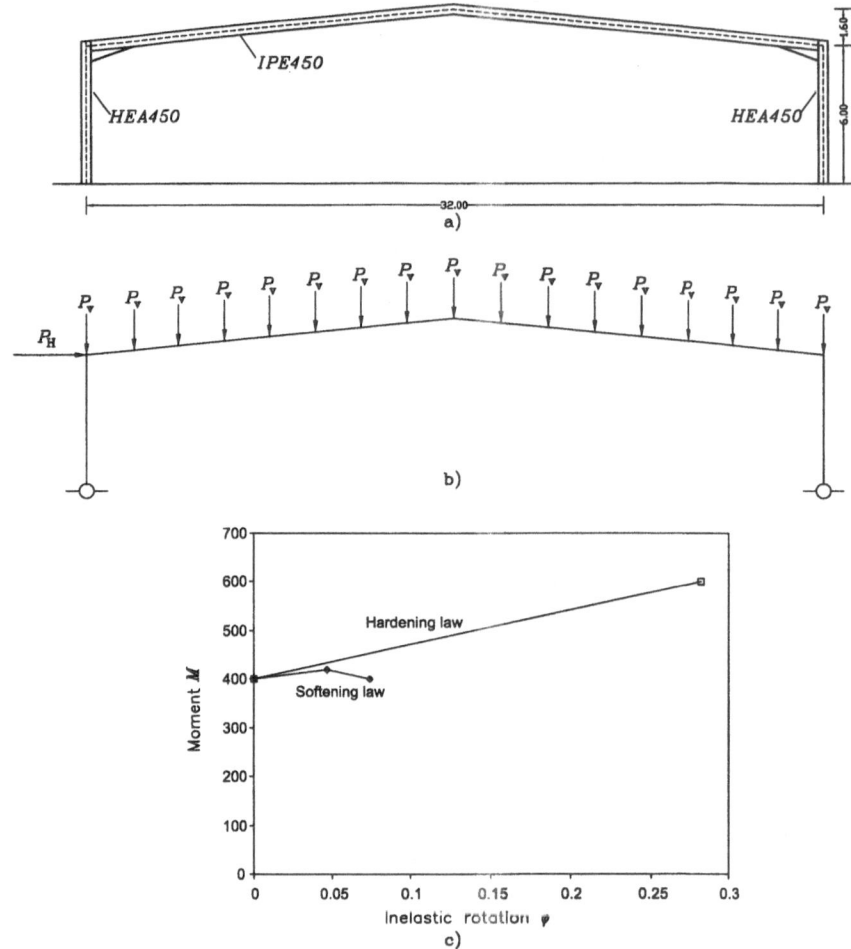

Figure 5: Pitched roof structure and the corresponding moment-rotation diagram for the beam

- A softening moment-rotation law

- An elastic-plastic moment-rotation law with hardening

The application of the algorithm of the previous section to the above structure gave the capacity curves presented in Fig. 6 in ADRS format. Then, the elastic response spectrum was constructed according to the Greek seismic design code [9] by using the following data:

- Spectral acceleration: 0.24

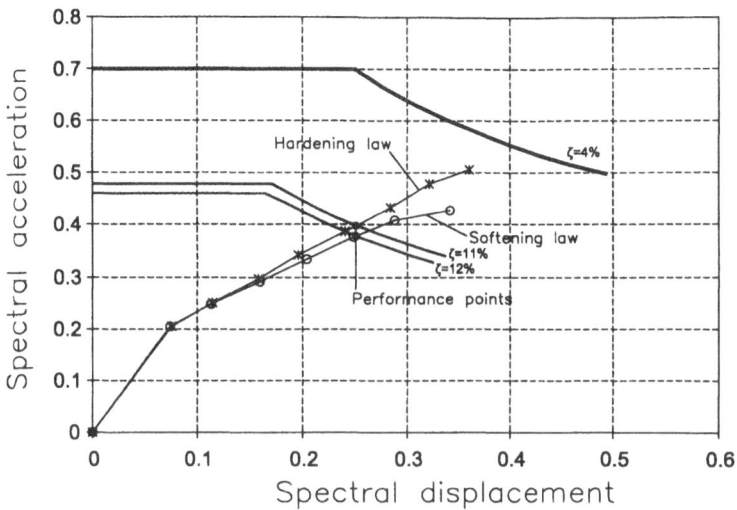

Figure 6: Results for the first example (ADRS format)

- Soil type: D

- Characteristic period: 1.2 sec

- Effective damping: 4%

- Importance factor: 1.15

- Foundation factor: 0.90

Finally, the application of the capacity spectrum method gave the performance points presented in Fig. 6. As the horizontal loading increases plastic hinges are formed at the beams. Notice that although the displacement demand is similar for both cases, in the case of the hardening law, the structure is able to undertake greater horizontal forces. The ratio between the base shear that corresponds to the elastic response spectrum to the base shear that corresponds to the inelastic response spectrum gives the q factor. For the case of the hardening law a value of $q=1.75$ was obtained while the softening law gave a value of $q=1.85$. From the calculations it can easily be verified that at the performance point the plastic rotations are 8% greater in the case of the softening law.

4.2 APPLICATION TO MULTISTOREY FRAMES

The multistorey plane frame of Fig. 7 is now considered. The structure consists of HEB240 columns and IPE300 beams and is loaded with the vertical loads $P_V = 40kN$ assumed as constant and with a pattern of horizontal loads which increase until the collapse of the structure. For the moment-rotation law of the

234

beams two cases are considered, as in the previous examples, a softening one and an elastic-plastic one with hardening. The adopted moment-rotation laws for the beams are presented in Fig. 7b. For the columns an elastic-plastic law with hardening was considered for both cases and is presented in Fig. 7c.

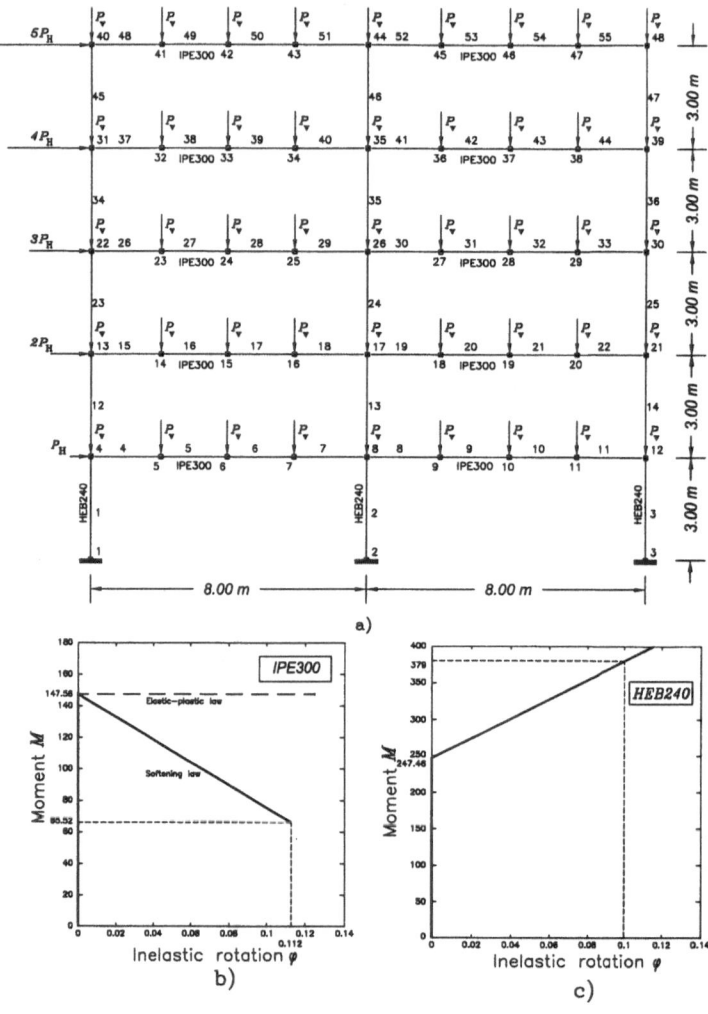

Figure 7: Multistorey structure and the corresponding moment-rotation diagrams

The structure is analyzed using the algorithm presented in Section 3. Fig. 8, presents the capacity curves obtained for the two cases. The elastic response spectrum was calculated according to [9] for the following data:

• Spectral acceleration: 0.12

- Soil type: B

- Characteristic period: 0.8 sec

- Effective damping: 4%

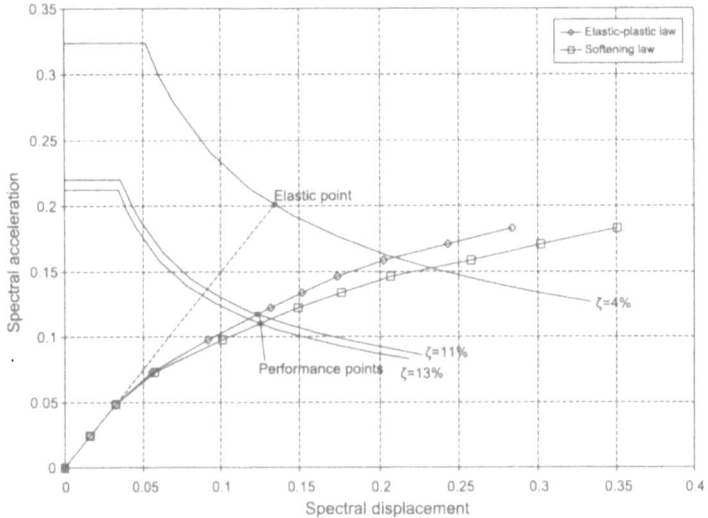

Figure 8: Results for the second example (ADRS format)

As the horizontal loading increases, more and more beams enter into the softening region of the moment-rotation curve. This phenomenon has as a result the redistribution of the stresses of the whole frame, and the increase of the moments at the columns. Further increase of the horizontal loading has as a result the plastification of the lower sections of the columns and the frame collapses.

The performance points for the two cases were calculated by applying the capacity spectrum method. The displacement demand is slightly larger for the case of the softening law, while the forces applied on the structure for the displacement demand are lower in the case of the softening law. For the case of the hardening law a value of $q=1.71$ was obtained while the softening law gave a value of $q=1.98$. But the most important fact is that at the performance point the plastic rotations are 17% greater in the case of the softening law.

5. Conclusions

In this paper, a new numerical method has been proposed which allows the investigation of the influence of the softening branch of the moment-rotation law to the seismic behaviour of steel frames.

From the numerical examples treated here it was concluded that the softening branch plays a secondary role to the overall seismic behaviour of the structure.

Due to the greater plastic rotations that result in the case of the softening law, the modeling of the softening branch is very important because the increased rotations that correspond to the softening law might be beyond the acceptability limits required by the performance objective. This is also an important difference between displacement based design and force based design.

6. References

1. ATC 40 (1996) Seismic evaluation and retrofit of concrete buildings, Vol. 1, *Applied Technology Council, Seismic Safety Institute*, Report No. SSC 96-01, California.

2. FEMA 273 (1997) NEHRP Guidelines for the seismic rehabilitation of buildings, *Building Seismic Safety Council*, Washington.

3. Kecman, D. (1983) Bending collapse of rectangular and square tubes, *Int. J. of Mechanical Sciences*, **13(9-10)**, 623–636.

4. Mitani, I. and Mahino, M. (1980) Post local buckling behaviour and plastic rotation capacity of steel beam-columns, In *7th World Conference on Earthquake Engineering*, Istanbul, 1980.

5. Kato, B. and Akiyama, H. (1981) Ductility of members and frames subject to buckling, *ASCE Convention*.

6. Mistakidis, E.S. and Stavroulakis, G.E. (1997) *Nonconvex optimization in Mechanics. Algorithms, heuristics and engineering application by the F.E.M.*, Kluwer, Boston.

7. Panagiotopoulos, P.D. (1985) *Inequality problems in mechanics and applications. Convex and nonconvex energy functions*, Birkhäuser, Basel - Boston - Stuttgart. (Russian translation, MIR Publ., Moscow 1988).

8. Panagiotopoulos, P.D. (1993) *Hemivariational inequalities. Applications in mechanics and engineering*, Springer, Berlin - Heidelberg - New York.

9. EAK2000, (2000) Greek Seismic Design Code, *Greek Organization for Seismic Planing and Protection (OASP)*, Athens.

BI-DIRECTIONAL PSEUDODYNAMIC TEST OF A FULL-SIZE THREE STOREY STEEL CONCRETE BUILDING WITH RIGID AND SEMI-RIGID JOINTS

M. GERADIN and F. J. MOLINA
European Commission, Joint Research Centre
Institute for Systems, Informatics and Safety
Safety in Structural Mechanics Unit
21020 Ispra (Varese), Italy

1   Introduction

The current lack of knowledge on the behaviour of structures under seismic loading justifies nowadays the execution of large-deformation tests on civil-engineering specimens. Most of them consist of quasi-static cyclic tests or shaking table tests. However, PsD tests [13] may seriously compete with these traditional techniques since they are, in principle, able to combine the advantages of both quasi static testing, i.e., large specimens, accurate control and measurement of displacements and forces, easy observation of damage progression and possibility of halting the test procedure, sub-structuring and shaking table testing, i.e. reproduction of dynamic response for specified ground motion. In practice, since PsD testing is performed quasi-statically with on-line numerical time integration of a discrete system of motion equations, it also has its own specific limitations:

1. The stiffness of the structure should not be large compared to the stiffness of the actuators. A large mass of the specimen may also induce difficulties in the control as it happens even more drastically for a shaking table test. In general, systematic experimental errors (especially control errors) need to be strictly limited in order not to significantly alter the structural response [12].
2. The discrete model should reproduce properly the response of the distributed-mass structure.
3. Specimens made of strain-rate sensitive materials may only be tested if the effect of load application rate is susceptible of a calibrated compensation [8].

In many real cases, the control system [6, 7] developed at ELSA (European Laboratory for Structural Assessment) has allowed successful PsD tests on large specimens which have contributed to a better understanding of the seismic behaviour of such structures [3]. However, as a novelty, the test described in this paper represents the first PsD test

C.C. Baniotopoulos and F. Wald (eds.), The Paramount Role of Joints into the Reliable Response of Structures, 237–256.

performed at the ELSA using 3 DoFs per floor. Earlier performed bi-directional PsD tests have already been described in the literature. In particular, Thewalt and Mahin [14] have reported a PsD test at reduced-scale on a 1-storey model which successfully reproduced the results of a previous shaking table test. Unfortunately, it was not clear whether their testing method could be applied to large size real structures. In fact, they report that, in order to get acceptable accuracy in the control of the PsD test, it was necessary to uncouple from the specimen most of the real inertial mass (nearly 5 tons).

The general concept of co-ordinates transformation between actuators and floor generalised DoFs on which our PsD test has been based is essentially the same as described by Thewalt and Mahin [14]. However, it differs on the one hand by a rigorous handling of geometric non-linearity in the geometric transformation of forces and displacements and, on the other hand, by introducing the possibility of having more than 3 actuators and control displacement transducers per floor. Both contributions, together with the unique characteristics of the used control set-up, have allowed to perform, to our knowledge, the first 3-DoF-per-floor PsD test on a large-size building for strong motion earthquake.

Fig. 1. Photograph of the test structure during a PsD test in the ELSA laboratory

The test structure was a three-storey composite frame with 3-bay by 3-bay column layout and overall dimensions of 16 x 12 m in plan and 9.5 m in height (Figure 1). The seismic PsD test as well as the quasi-static cyclic tests performed on the structure at ELSA served for the evaluation of the influence of the concrete slab on the seismic moment capacity for typical beam-column connections. The information obtained is currently being used for the improvement of Eurocode 8 [4, 1, 11]. Apart from the PsD test performed on that specimen for a major earthquake, a small amplitude random burst test was also conducted, both dynamically and pseudodynamically, in order to validate the developed testing methodology.

2 Test Modelling for 3-DoF floors

The PsD integration of the horizontal response of the structure is performed in terms of 3 generalised DoFs at each floor consisting of the in-plane displacements (d_x, d_y) and the rotation d_θ at the centre of mass (CoM) of the floor with respect to the ground (figure 2).

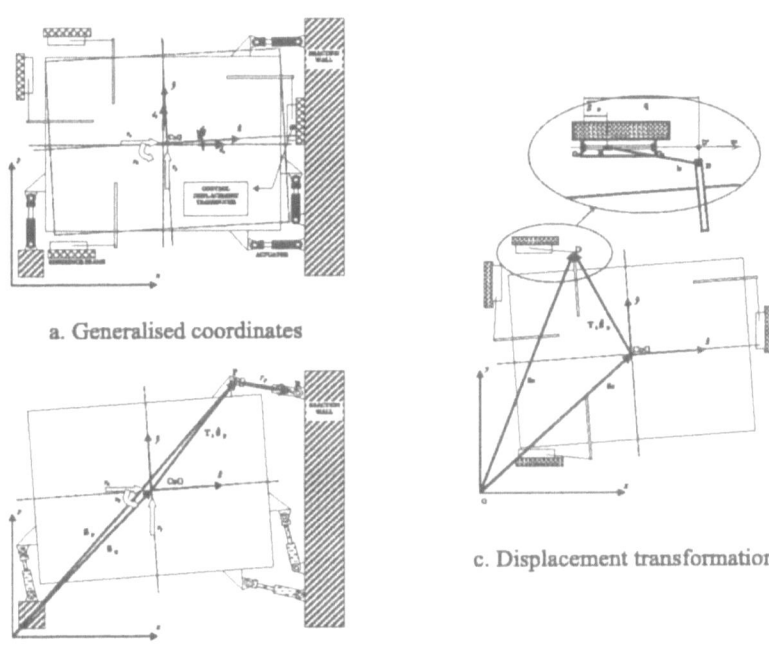

a. Generalised coordinates

c. Displacement transformation

b. Force transformation

Fig. 2. Generalised coordinates, force and displacement transformations

They are collected in the vector of floor generalised displacements

$$d = [d_x \quad d_y \quad d_\theta]^T \tag{1}$$

The in-plane resultant forces (r_x, r_y) and the moment r_θ at the CoM are contained in the vector of conjugated generalised restoring forces

$$d = [r_x \quad r_y \quad r_\theta]^T \tag{2}$$

2.1 Equations of motion

Assuming for each floor of the structure a rigid body behavior, its horizontal motion is fully described by the generalised displacements (1) and its equations of motion result from the application of D'Alembert's Principle. In matrix form

$$m + r = -mja_g \tag{3}$$

Assuming that the entire structural mass m and moment of inertia I are concentrated at the floor CoM, the floor mass matrix takes the form

$$m = \begin{bmatrix} m & 0 & 0 \\ 0 & m & 0 \\ 0 & 0 & I \end{bmatrix} \tag{4}$$

with the vector of relative accelerations at the CoM

$$a = \begin{bmatrix} a_x & a_y & a_\theta \end{bmatrix}^T = \frac{d^2 d}{dt^2} \tag{5}$$

The vector appearing at the right-hand side of equation (3)

$$a = \begin{bmatrix} a_{gx} & a_{gy} & a_{g\theta} & a_{g\alpha} & a_{g\beta} \end{bmatrix}^T \tag{6}$$

contains the ground accelerations in translations x and y, twist θ, roll α and pitch β and the matrix

$$j = \begin{bmatrix} 1 & 0 & -y_C & 0 & z_C \\ 0 & 1 & x_C & -z_C & 0 \\ 0 & 0 & 1 & 0 & 0 \end{bmatrix} \tag{7}$$

is a geometric influence matrix obtained from the current position (x_C, y_C, z_C) of the CoM.

Thus, the system of equations of motion for the multi-storey structure can be written as

$$MA + CV + R = -MJa_g \tag{8}$$

where the matrices M, A, R and J collect the different floors contribution of the building. Here, an assumed viscous damping term CV has been added for the sake of generality. Usually, introducing such a damping term is not necessary since, for classical materials, most of the dissipation is of hysteretic type and is thus implicitly taken into account in the restoring forces r.

Equations (8) are expressed in terms of generalised forces and displacements. However, the control system used for the test is based on a set of linear actuators and displacement transducers attached at prescribed locations on each floor of the structure. The necessary transformations between both systems of co-ordinates are developed in the following subsections.

2.2 Transformation from generalised to transducer displacements

The measurement of the floor displacements for control purposes is achieved using $n_D \geq 3$ high-resolution linear displacement transducers attached to each floor. As shown in Figure 1.c each one of these control displacement transducers consists of a slider $G_1 - G_2$ on which a body M translates and gives a measure of its relative position on the slider. This slider is attached to a fixed reference frame while the mobile body M is connected to the measuring point D on the structure through a pin-jointed rod.

During the test, the computed generalised displacement of the floor is imposed through the actuators with feedback from the displacement transducers. Thus, in order to determine the target displacements at transducer level, a geometric transformation is needed. Starting from a known set of generalised displacements (1), the transformation is obtained through the following steps:

Step 1) The position of the floor CoM is updated as

$$S_C = S_C^0 + d \tag{9}$$

where $S_C = [x_C \quad y_C]^T$ are its global co-ordinates and S_C^0 their reference value for zero displacement. Note that letter S will in general denote a position in the following formulae.

Step 2) The position of the measuring point D is likewise updated as

$$S_D = S_C + T_{\theta_C} \hat{S}_D \tag{10}$$

Where $\hat{S}_D = [\hat{x}_D \quad \hat{y}_D]^T$ are the co-ordinates of point D in a reference system local to the floor, centred at its CoM, and

$$T_{\theta_C} = \begin{bmatrix} \cos\theta_C & -\sin\theta_C \\ \cos\theta_C & \sin\theta_C \end{bmatrix} \tag{11}$$

is a rotation matrix. Assuming an infinitely rigid floor, the local co-ordinates S_D are constant.

Step 3) In order to express the position of body M (see figure 2) of the transducer along its slider $G_1 - G_2$, let us compute the relative position of the measuring point D with respect to the slider origin G_1

$$\tilde{S}_D = \overrightarrow{G_1 D} = S_D - S_{G_1} \tag{12}$$

Then, the projection of vector (12) along the slider is computed as

$$q = \overline{G_1 D'} = \tilde{S}_D^T e_G \tag{13}$$

where e_G is the constant unit vector defined in the direction of positive measurement along the slider. The position of body M along the slider is then given by

$$\overline{\overline{S}}_D = \overline{G_1 M} = \overline{G_1 D'} - \overline{M D'} = q - \text{sign}(q)\sqrt{\ell_D^2 - \tilde{S}_D^T \tilde{S}_D + q^2} \tag{14}$$

where ℓ_D is the constant length of the rod.

The corresponding measure at the transducer is $d_D = \overline{\overline{S}}_D - \overline{\overline{S}}_D^0$ where $\overline{\overline{S}}_D^0$ is a reference position for zero displacement of the transducer.

2.3 Transformation from actuator to generalised forces

Generally, every displacement transducer has associated a piston acting approximately along the same axis, but preferably at a point not too close to the measuring point. This is done to avoid the influence on the measurements of local deformations generated by the concentrated load of the piston.

Once the prescribed displacements are achieved, the acting axial force at every actuator is measured by its load cell. However, in order to express these forces as resultant generalised forces at the CoM of the floor, a static transformation is needed. Since the ends of the actuator are pin-jointed, it is assumed that it produces a purely translational force along the line PR connecting the ends of the actuator. Starting from the load-cell measure r_P of this force and assuming that the current position of the floor CoM is known, the transformation results from the following steps:

Step 1) The global position of the loading point P is obtained as

$$S_P = S_C + T_{\theta_C} \hat{S}_P \tag{15}$$

where T_{θ_C} is given by equation (9) and, similarly to S_D , $\hat{S}_P$ contains the co-ordinates of point P in the local reference system to the floor.

Step 2) The global components of the piston force are then computed as

$$p_P = r_P e_P \tag{16}$$

where e_P is a unit vector in the direction of $\overrightarrow{PR}$.

Step 3) The floor generalised restoring forces (2) are obtained by summing up the effects of all pistons acting on the floor:

$$r = \sum T_P D_P \quad \text{with} \quad T_P = \begin{bmatrix} 1 & 0 \\ 0 & 1 \\ -y_P & x_P \end{bmatrix} \tag{17}$$

2.4 Estimation of generalised displacements from measured displacements

Due to geometry and control errors and to the flexibility of the floor, the actual displacements and rotation at the CoM differ from the prescribed input. In order to get an estimate of the actual generalised displacements, the measures given by all the control displacement transducers may be exploited as follows:

Step 1) Since the relations between generalised and transducer displacements (Section 2.2) are non-linear and thus cannot easily be inverted in closed form, the solution is achieved through a Newton-Raphson iteration process starting from a first estimate of the generalised displacements, e.g.,

$$d^{EST} = 0 \tag{18}$$

Step 2) The non-linear equations giving the associated transducer displacements d_D are obtained by substituting d by d^{EST} into equation (9) and applying formulae

$$d^{EST} = d(d^{EST}) \tag{19}$$

Step 3) The difference between the estimate (19) and the measured transducer displacements d is computed to provide an new estimate of the generalised displacements

$$d^{EST} \longleftarrow J^{-1}(d - d^{EST}) + d^{EST}) \tag{20}$$

where $J = \frac{\partial d_D}{\partial d}|_{d^{EST}}$ is the Jacobian matrix computed at d^{EST}.

If the number of control transducers on the floor exceeds 3, equation (19) must be solved in a least-squares sense, in which case the inverse of J in (20) is replaced by the pseudo-inverse

$$psinv(J) = \left[J^T J\right]^{-1} J^T \tag{21}$$

Step 4) Step 4) Steps 2) and 3) are iteratively repeated until a specified tolerance is reached.

2.5 Optimal distribution of piston loads

When more than three pistons act on a rigid floor, the use of the individual displacement transducers on the structure as feedback signals for each piston may lead to an unstable control. Using structural displacements as feedback signals in equal number to the number of DoFs is then necessary, while the redundant pistons are controlled by other means in order to maintain an acceptable distribution of loads among all the pistons. This can be done by implementing an algorithm capable of optimising the distribution of piston loads for a known set of generalised floor loads. Even distribution of forces is also desirable because it leads to a better approximation to the distributed inertial forces of a real dynamic event.

This section describes an algorithm to compute an 'optimal' distribution of piston loads compatible with a known set of floor generalised loads. The piston forces determined are assumed to be statically equivalent to the generalised loads while minimising a penalty function, which becomes infinite when a piston force reaches its working limit.

A suitable expression for the penalty function is

$$f(r_P) = \sum_P \frac{1}{M_P^2 - r_P^2} \tag{22}$$

where M_P is the working limit of the absolute value of the piston load r_P. Clearly, minimising the function (22) will guarantee that all piston loads are kept far from their limit. Using expressions (16) and (17), the conditions of static equivalence of the piston loads with the known set of generalised loads give a constraint on the minimisation problem which can be written in the vector form

$$g(r_P) = \sum_P T_P e_P r_P - r = 0 \tag{23}$$

where r_P is the known set of generalised forces (2). The following constrained minimisation problem results

$$
\begin{cases}
min \ f(r_P) \\
\\
g(r_P) = 0
\end{cases}
\tag{24}
$$

It can be solved by the Lagrange multipliers method in terms of an augmented functional

$$
h(x) = f(r_P) + \lambda^T g(r_P) \ \Rightarrow \ \frac{\partial h}{\partial x} = 0
\tag{25}
$$

The solution vector

$$
x^T = [\, r_P^T \quad \lambda^T \,]
\tag{26}
$$

contains as unknowns the piston loads r_P and Lagrange multipliers λ. The non-linear equation (18) may be solved by a Newton-Raphson iteration procedure as done previously to obtain the generalised displacements.

3 Testing method

This section describes the time stepping algorithm used for the time integration of the equation of motion and its application to 3-DoF-per-floor structures using the model described in section 2. It also describes the implementation of an appropriate control strategy and the characteristics of the software in charge of the test execution.

3.1 PSD time integration algorithm

This subsection describes the numerical time integration of an equation of motion of the type

$$
MA + CV + R = F.
\tag{27}
$$

Equation (27) is a generalisation of equation (8) in which F is a general external force vector including either seismic equivalent forces or directly applied forces. Two different algorithms are proposed here for its step-by-step integration:

- The Explicit Newmark method, which is equivalent to the Central Difference algorithm, and
- The α Operator Splitting method, which is a hybrid explicit-implicit method. Both algorithms can in fact be regarded as particular cases of the α-generalised method, an extension of the Newmark scheme, which can be stated as [5]:

$$
\begin{aligned}
&MA_{n+1} + (1+\alpha)CV_{n+1}^\star - \alpha CV_n^\star + (1+\alpha)R_{n+1}^\star - \alpha R_n^\star = (1+\alpha)F_{n+1}^\star - \alpha F_n^\star \\
&D_{n+1}^\star = D_n^\star + \Delta t V_n^\star + \Delta t^2 \left[(\tfrac{1}{2} - \beta)A_n + \beta A_{n+1} \right] \\
&V_{n+1}^\star = V_n^\star + \Delta t \left[(1 - \gamma)A_n + \gamma A_{n+1} \right]
\end{aligned}
\tag{28}
$$

Since the basis for the formulation is available elsewhere [10] [2], the present description will focus only on programming aspects. It will be shown that, operationally, the explicit method can be dealt as a particular case of the hybrid one. We start thus with the latter.

The α Operator Splitting algorithm is characterised by the coefficients

$$-\frac{1}{3} \le \alpha \le 0 \qquad \beta = \frac{1-\alpha^2}{4} \qquad \gamma = \frac{1-2\alpha}{2} \tag{29}$$

and defines the implicit approximations to displacements and velocities at time n as

$$D_n^\star = D_n + \Delta t^2 \beta A_n \qquad V_n^\star = V_n + \Delta t \gamma A_n \tag{30}$$

Then, by comparison with (28), the explicit predictions of displacements and velocities at time n+1 can be computed as:

$$D_{n+1} = D_n^\star + \Delta t V_n^\star + \Delta t^2 (\frac{1}{2} - \beta) A_n \qquad V_{n+1} = V_n^\star + \Delta t(1-\gamma)A_n \tag{31}$$

They are explicit in the sense that they are computed solely from the information at the previous step. They are also the ones that are imposed to the structure and for which the restoring forces

$$R_{n+1}^\star = R(D_{n+1}) \tag{32}$$

are measured. The method assumes that the difference between explicit and implicit forces is small and can be approximated by a linear model

$$R_{n+1} - R_{n+1} = R(D_{n+1}^\star) - R(D_{n+1}) \simeq K_I(D_{n+1}^\star - D_{n+1}) \tag{33}$$

in which K_I is an implicit stiffness matrix which preferably overestimates the initial stiffness of the structure.

Starting from the known initial values D_1, V_1, A_1 at time step $n = 1$ ($t = 0$), the successive computation stages of the method at every time step are:

Stage 1) Compute at time n the implicit displacements and velocities $D_n^\star$, $V_n^\star$ using equations (30).

Stage 2) Compute next at time $n + 1$ the explicit values of displacements and velocities $D_{n+1}^\star, V_{n+1}^\star$ using equations (31).

Stage 3) Impose to the structure the new explicit displacement D_{n+1} and measure the associated restoring force 32.

Stage 4) Increment the step counter $n \longleftarrow n + 1$.

Stage 5) Compute the acceleration at new time n from the equilibrium equation

$$A_n = \tilde{M}^{-1}[(1+\alpha)(F_n - R_n - CV_n) \\ -\alpha(F_{n-1} - R_{n-1} - CV_{n-1}) + \alpha(\gamma\Delta tC + \beta\Delta t^2 K_I)A_{n-1})] \tag{34}$$

with

$$\tilde{M} = M + (1+\alpha)(\gamma\Delta tC + \beta\Delta t^2 K_I) \tag{35}$$

Stage 6) Go back to stage 1) until the final time is reached.

Similarly, the Explicit Newmark algorithm is defined by the coefficients:

$$\alpha = 0 \qquad \beta = 0 \qquad \gamma = \frac{1}{2} \tag{36}$$

The same stages 1) to 6) can then be performed as for the Operator-Spitting algorithm. The advantage in this case is that no implicit stiffness matrix is required. The disadvantage is that the algorithm is only conditionally stable, imposing thus small time steps compared to the minimum period of the structure.

3.2 Marching procedure

The application of the time integration method to the 3-DoF-per-floor model described in Section 2 yields the following experimental step-by-step procedure. The algorithm starts from known initial values D_1, V_1, A_1 and the computations are organised in the following stages:

Stage 0) Let $n = 0$.

Stage 1) Transform the generalised displacements into target displacements d_D at the control transducers computed from the geometric model described in Section 2.2:

$$(d_D)^{TARGET}_{n+1} = d_D(D_{n+1}) \tag{37}$$

Stage 2) Send these target displacements (37) to the controllers which impose them to the specimen by performing a ramp from the previous position.

Stage 3) Let $n \longleftarrow n+1$.

Stage 4) Measure the current displacements $(d_D)^{MEAS}_n$ and restoring loads $(r_P)_n$ at the controllers, r_P being the forces applied by the pistons.

Stage 5) From the measured control displacements, estimate the current generalised displacements on each floor using the least-squares solution described in Section 2.4

$$D^{MEAS}_n = D\left((d_D)^{MEAS}_n\right) \tag{38}$$

Stage 6) From the measured piston forces, compute the current generalised restoring forces using the transformation formulae of Section 2.2:

$$R_n = R\left((r_P)_n, D^{MEAS}_n\right) \tag{39}$$

Stage 7) If $n \geq 1$, compute the new accelerations using expression (34):

$$A_n = A_n\left(F_n, R_n, V_n, F_{n-1}, R_{n-1}, V_{n-1}, A_{n-1}\right) \tag{40}$$

Stage 8) Predict the generalised displacement and velocity at the next time increment using the finite-difference approximation (30), (31) of the integration algorithm:

$$(D_{n+1}, V_{n+1}) = f(D_n, V_n, A_n) \tag{41}$$

Stage 9) Go back to stage 1) until reaching final time.

3.3 Control strategy

As already mentioned in Section 2.5 the fact of using more than three actuators per floor in order to have a better distribution of forces may lead to difficulties in the control strategy. Let us assume that PID controllers are used and that three actuators are controlled using as feedback one displacement transducer on the structure. In that case, there are, in principle, three options in the choice of the feedback transducer to control the redundant actuators: first, the "closest" displacement transducer on the structure as for the other three actuators, second, the actuator load cell or, third, the actuator internal displacement transducer.

The first option is not practicable because the control system becomes unstable for acceptable controller gains. In fact, if the system turned out to be stable and accurate for this configuration, it would mean that the floor is relatively flexible and that more than three degrees of freedom could be taken into account. The second option may lead to a stable control system, but, usually, force control strategies do not produce the best accuracy. In fact, its accuracy is much lower than the one that would be obtained with only three pistons and using the structural displacements as feedbacks. This second option could be acceptable for a cyclic test, but not for a PsD test in which relatively small control errors may result in a large distortion of the integrated response [12].

The third option may give at the same time accurate and stable control system thanks of being a displacement control strategy, as in the first option, but associated to a transducer that sees a more flexible subsystem than the structure itself. In fact, the displacements measured by the internal transducer of the actuator are considerably larger than the structural ones since they comprise the deformation of the reaction wall and actuator attachments. Furthermore, in this case, feedback between co-located quantities is achieved.

Thus, we have adopted the actuator internal transducers as feedbacks for the redundant pistons. The only potential problem with this option is that the load distribution remains undetermined. To solve it, the computed target for the redundant pistons is slightly adjusted at every integration step in order to control the floor distribution of loads. Instead of using expression (37), their target is computed as

$$(d_P)_{n+1}^{TARGET} = d_P(D_{n+1}) + \overline{(d_P)}_{n+1} \tag{42}$$

where the first term on the right-hand side is the theoretical elongation of the piston and the second one is a correction introduced in order to modify the force. It is updated at every time step in the form

$$\overline{(d_P)}_{n+1} = + \overline{(d_P)}_n + \frac{r_P^{OPTIMUM} - r_P^{MEASURED}}{K_P} \tag{43}$$

where the term added to the correction is the difference between the computed optimum force of the piston (Section 2.5) and the measured one at the former step, divided by a stiffness parameter K_P empirically selected. In general terms, the smaller is this parameter, the faster is the convergence of the force to the optimum value, but using a too small value may result in instability which would make the force to oscillate out of control in very few steps.

subsectionHardware and software set-up The servo-control units used for these tests were MOOG actuators with (0.5 m stroke and load capacity of 0.5 MN, except for the third floor of the specimen where the three actuators closer to the main reaction wall had a capacity of 1.0 MN. The control displacement transducers on the structure were optical HEIDENHEIN sensors with a stroke of ± 0.5 m and a 2 μm resolution. Every actuator was equipped with a strain-gage load cell and a TEMPOSONIC internal displacement transducer. Each actuator had its own PID controller based on a 486DX4 processor. The characteristics of this hardware have been fully described by Magonette [6, 7].

All the controllers are connected to a master unit by means of an EFIWAY net. For the previous PsD tests performed at the ELSA, this master unit assumed two tasks:

Task 1) sending the targets to the controllers and receiving and displaying the associated measurements.

Task 2) performing all computations for the PsD integration.

The corresponding software (based on C language) had to combine real-time capabilities with high-level logic and algebraic computation. However, for this occasion, due to the significant complexity added by the 3-DoF-per-floor model, this master unit has been divided in two units: a communication unit in charge of task 1) and a computation unit in charge of task 2).

These two units are separated processors communicating at a very high speed thanks to a dual-RAM interface module. The computation unit consists of an NT workstation running the testing procedure in MATLAB interpreted language, while the communication unit is a DOS PC running the real-time application in C. The advantage of this task splitting is that the interpreted language offers much more flexibility (and simplicity of programming and debugging) for high level operations and user interface. On the other hand, the interpreted language is not able to perform the real-time tasks still reserved to the communication unit working in C.

This configuration allowed developing a software which implemented the described model and marching procedure with the following capabilities:

- PsD integration either by Explicit Newmark or α-Operator-Splitting algorithms.
- Pause and Continue buttons.
- Possibility of on-line change of parameters such as excitation spans, step minimum duration and ramp speed, alarm limits, control-strategy parameters such as K_P (see Section 3.3).
- Possibility to restart from any previous time step or initial conditions.
- Possibility of substituting the real specimen by a linear model for checking purposes.
- Graphic monitoring of controller co-ordinates, generalised co-ordinates, energy variables, deformed shape or any other programmed function.
- Centralised alarm protection based on actuator and transducer limits, which is also applied to the target before it is sent.

4 Description of test campaign

4.1 Test specimen

The test structure was a 3-storey steel-concrete composite building constructed within the ELSA laboratory. Its overall dimensions were 16 x 12 m in plan and 9.5 m in height. It had four frames (1,2,3,4) in the x (EW) direction, four frames (A,B,C,D) in the y (NS) direction and three floors (see Figure 3). Frames 2 and 3 in the x direction and frames A and B in the y direction were moment-resisting frames. The frames were made of standard Fe360 rolled sections, while the floors consisted of a 15-cm slab of reinforced C30/37 concrete poured on corrugated metal decking. The structure is representative of current European construction methods and the design has been made according to Eurocode 8 but introducing at different joints different kinds of connections between the slab and the beams and columns in order to study different composite behaviours. More details on the design can be found in the references [4, 1, 11].

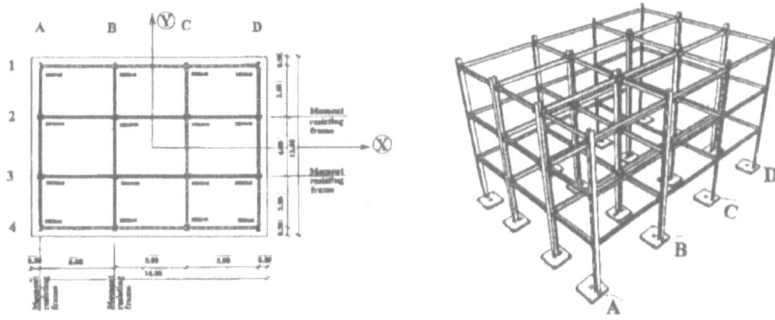

Fig. 3. Views of the test structure

4.2 Instrumentation

In addition to the load cells and control displacement transducers of the four actuator used for each floor, resistive strain gages and potentiometer displacement transducers were used, yielding 550 channels of continuous measuring in total, each one taking one averaged measure at every test step. The strain gages were glued on the columns in order to measure the bending moment in x and y directions at two sections and the axial load at one section. The displacement transducers were used to measure at every beam the vertical deflection at three points as well as the relative horizontal displacements between the columns in x and y directions. These data have been used in the post test analysis either directly, in the form of moment-curvature diagrams for every joint, or indirectly, by means of the identification of a numerical model which follows the plastic rotation at the beams.

4.3 Test programme

Initially, three cyclic tests have been performed on the specimen, each one with increasing amplitude up to a nominal global drift of 2% (180 mm of displacement at the third floor). The first quasi-static cyclic test was executed in the x direction, the second one in the y direction and the third one in a combination of both directions. Afterwards, a PsD seismic test described below was performed. The intensity of this test was chosen so that the response would keep every drift under the values of the previous cyclic tests. Then a major quasi-static cyclic test has been done in the x direction with amplitude up to 4.7% of global drift. After this test, the global strength had fallen by 30% and the lower beam flange at many of the joints in the x direction were torn up, apart from other types of observed damage. Finally, a small-amplitude dynamic random -excitation test has been performed followed by its PsD reproduction in order to check the validity of the PsD modelling and testing methodology. The results of the cyclic tests have been described and analysed by Bouwkamp et al [1]. The seismic and random tests are described in the next two subsections.

4.4 Psd seismic test

The steel-concrete composite structure had been designed according to Eurocode 8 with a design spectrum characterised by 5% of damping ratio, a behaviour factor q = 6 and a soil profile B. For the PsD seismic test, two independent artificial accelerograms with duration of 10 seconds were used for the x and y directions. The intensity of the normalised spectra was multiplied by 2.25 so that the peak ground acceleration was 2.25 m/s2 in both directions (Figure 4). All the rotational components of the ground acceleration vector 6 were considered zero.

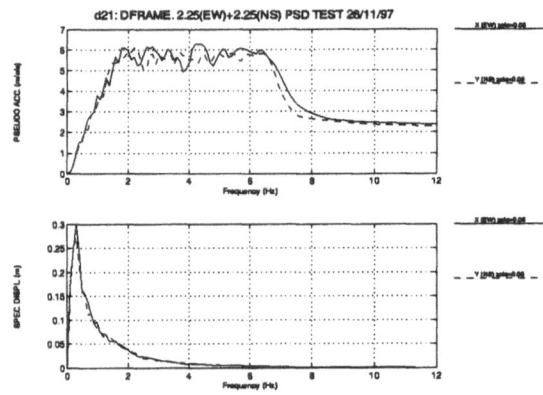

Fig. 4. Response spectra of the ground accelerograms

The 9-DoF equation of motion for the PsD test has the form described in Section 2.1 with translational and rotational masses of $m = 124.10^3 \ kg$ and $I = 4464.10^3 \ kg \ m^2$ for every floor. These masses do not correspond to the mass of the specimen but to the design mass relating to the earthquake action. The viscous damping has been considered as negligible, all the dissipation being thus introduced hysteretically by the experimental restoring forces.

The algorithm used for the integration was the Explicit Newmark, as described in Section 3.1, with a time increment of 0.005 seconds. Since the maximum frequency was estimated to be under 10 Hz, such time increment is small enough to guarantee stability and accuracy without an excessive number of time steps. The control strategy used for the imposition of the displacements at every floor was the one described in Section 3.3. With two actuators acting in the x direction and the other two in the y direction, one of the actuators in the x direction was indirectly controlled in order to optimise the force distribution, while the other three were directly fed back by their control displacement transducers on the structure.

Some of the results of this PsD seismic test are shown in Figure 5 to Figure 7. In Figure 5, the three curves represent the displacements at the three floors in x direction (upper graph), y direction (middle) and torsion θ (lower). The maximum generalised displacement was 120 mm and was recorded at the CoM of the third floor in the x direction.

However, the maximum displacement at the control transducers (not plotted in the figure) was of 200 mm. It was recorded in the East transducer at the third floor (the torsion centre of the structure being closer to its West end). Thus, although no rotational ground motion was introduced, an important torsional response took place due to the lack of symmetry of the structure, especially in the y direction.

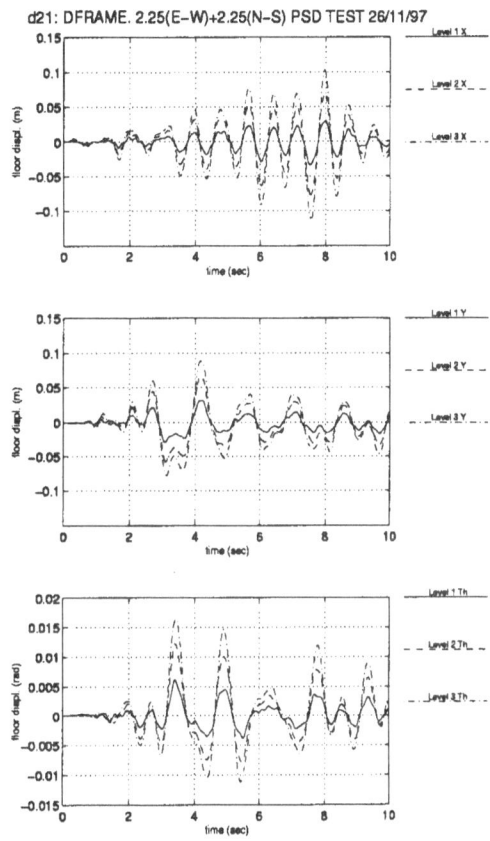

Fig. 5. Seismic test: Floor generalised displacements in x (upper), y (middle) and θ (lower graph) directions

Figure 6 shows the base shear versus top displacement cycles obtained in the three directions. Although the test did not produce severe damage, some plastic dissipation is observed from these curves. On the other hand, each one of these graphs contains in fact two curves. The solid line corresponds to the base shear computed from the piston forces, while the dashed line corresponds to base shear computed from the strain-gage bridges attached to the columns. The vicinity of both curves provides a check of the reliability of these strain-gage measurements.

252

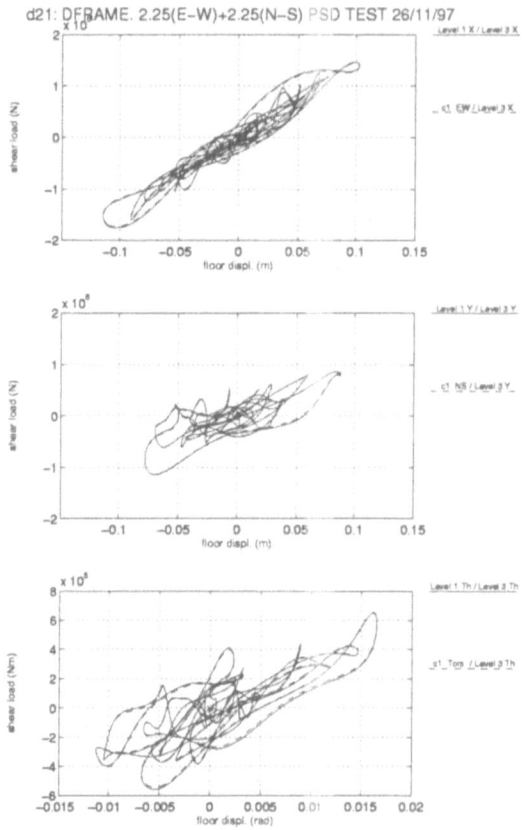

Fig. 6. Seismic test: Base shear-top displacement cycles in x (upper), y (middle) and θ (lower graph) directions

The effect of hysteretic dissipation can be better quantified in Figure 7, where the absorbed energy

$$E_a = \int r^T dd \qquad (44)$$

is plotted as a function of time, distinguishing the three directions (x, y and θ) and the total. The absorbed energy (44) may be calculated either from the measured displacements or from the computed ones. In the first case it represents the energy really dissipated by the specimen-apart from measurement errors-while in the second case it represents the energy seen by the integration algorithm which computes the response. If the difference, or error energy

$$E_d = \int r^T (dd^{MEASURED} - dd^{ALGORITHM}) = \int r^T d(\epsilon) \qquad (45)$$

is large in relation to the total (44), the PsD response obtained should not be accepted as representative of that specimen. This could occur due to inadequate quality of the control system or to an excessive testing velocity. The error energy (45) obtained within this PsD test is plotted at the lower part of Figure 7; it is negative and amounts to 6% of the total dissipated energy, which means that the specimen has an effective damping of about 0.94 times the damping shown in the PsD integration.

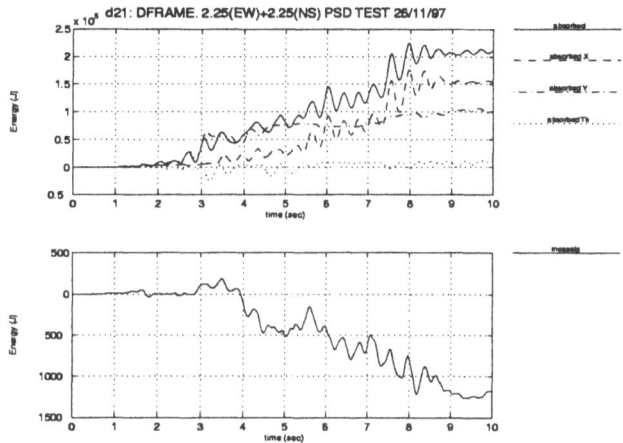

Fig. 7. Seismic test: Absorbed energy (upper) and error energy (lower)

Another interesting result of this seismic test is shown in Figure 8. It consists of the eigen-frequencies and modes as obtained by a transfer function method applied to the input accelerograms and output displacements. The method is based on a time-domain identification of a filter model, which has already shown to be well adapted for PsD results [9].

Using the response during the whole earthquake (2000 discrete time instants), a linear 9-DoF model was adjusted by a minimisation of the square error. Playing with different orders (2,4,6, ...) for the model, a good convergence was found to the values shown in the figure. Since the real structure behaves non-linearly, these linear-equivalent results depend on the amplitude of the response as well as on the time interval; this last effect is due to the damage accumulation. Consequently, the values shown should be considered as an average during the whole earthquake and may characterise the response of the structure, in its current degradation state, to earthquakes with similar spectrum and intensity as the one used for the input. Looking at the form of the displacement response spectrum (Figure 5) a significant response was to be expected for the first three modes. The predominance of the lower frequencies is also clear in Figure 5.

4.5 Random burst test

After all the main tests had been run on the building specimen, all the fragile instrumentation could be removed and a real dynamic random burst test was performed on it. This

254

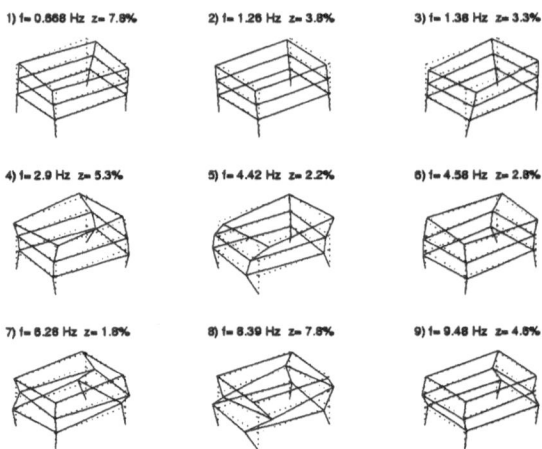

1) f= 0.868 Hz z= 7.8%   2) f= 1.26 Hz z= 3.8%   3) f= 1.36 Hz z= 3.3%

4) f= 2.9 Hz z= 5.3%   5) f= 4.42 Hz z= 2.2%   6) f= 4.56 Hz z= 2.8%

7) f= 6.28 Hz z= 1.8%   8) f= 8.39 Hz z= 7.8%   9) f= 9.48 Hz z= 4.6%

Fig. 8. Seismic test: Identified linear-equivalent natural frequencies, damping ratios and mode shapes

small-amplitude dynamic test was reproduced afterwards by a PsD test in order to verify the validity of the 3-DoF-per-floor model and the testing methodology developed. The advantage of this type of dynamic check test is that it uses the same loading set-up as the PsD test.

For the dynamic test, 9 statistically-independent reference signals were generated for 9 pistons acting on the structure, using the actuator internal displacement transducer as feedback. The reference signals had a significant content up to 12 Hz, but, due to the limitations of the loading devices at high frequencies, the measured applied forces were intense only for frequencies up to 10 Hz. The response at the control displacement transducers on the structure was measured at a sampling period of 0.005 sec. The generalised displacements obtained at the third floor are the ones shown in Figure 9. The maximum displacement in x direction at the CoM of the floor was of about 6 mm, while in y direction it was of about 12 mm.

For the PsD test, the same methodology as for the seismic test was applied, except for the seismic equivalent forces which were substituted by the external forces F in equation (25) which are in fact the ones measured during the dynamic test. The masses and moments of inertia introduced for the different floors were the ones estimated for the specimen: $m_1 = 68.10^3 kg$, $m_2 = 65.10^3 kg$, $m_3 = 72.10^3 kg$, $I_1 = 2680.10^3 kg\ m^2$, $I_2 = 2620.10^3 kg\ m^2$, $I_3 = 2710.10^3 kg\ m^2$.

Under these conditions, the PsD test was performed twice with the results shown in Figure 9, where a comparison is made with the dynamic test. One can see in this figure to which extent the PsD tests are able to acceptably reproduce the shape and especially the maximum of the dynamic response. The discrepancies observed with respect to the dynamic curves may be mainly attributed to the damage accumulation in the specimen at every repetition of the test. This could occur even for such small amplitudes because, unfortunately, at this stage the specimen showed already major cracks at critical points.

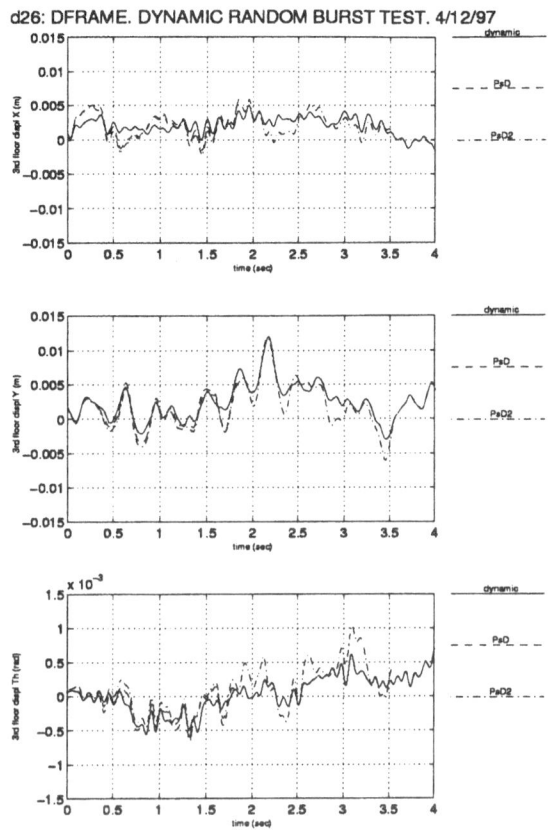

Fig. 9. Random burst test: Dynamic and PsD displacements at the third floor in x (upper), y (middle) and θ (lower graph) directions

5 Conclusions

The PsD testing technology has been successfully extended to bi-directional excitation of real-size buildings by implementing the test modelling and procedure described in the paper. Some important features of the methodology are:

- The capability of imposing large displacements by means of a rigorous geometrically non-linear transformation between actuator co-ordinates and floor generalised co-ordinates.
- The ability to use more than 3 actuators and control displacement transducers per floor by applying a stable and accurate control strategy which guarantees suitable distribution of the forces among the actuators.
- The development of a new, flexible software structure through separation of high-level tasks and real-time tasks which communicate through a dual-RAM interface module. The high-level tasks, comprising the geometric transformation and the time

integration, are written in MATLAB (interpreted language) while the real-time tasks are programmed and compiled in C.

The performed seismic test on a 3-storey steel-concrete composite building gave useful information about the seismic behaviour of that type of structure and served to show the applicability of the proposed testing technology for large-size specimens. The random burst test executed dynamically and pseudodynamically on that specimen gave information on the reliability of the PsD methodology.

References

1. J. G. Bouwkamp, H. Parung, and A. Plumier. Bi-directional cyclic response study of 3-d composite frame. Proceedings of the XIth European Conference on Earthquake Engineering, 1998.
2. D. Combescure and P. Pegon. Alpha-operator splitting time integration technique for pseudodynamic testing. error propagation analysis. Technical Report Special Publication No. I.94.65, European Commission, JRC, Ispra, 1994.
3. J. Donea, G. Magonette, P. Negro, P. Pegon, A. Pinto, and G. Verzeletti. Pseudo-dynamic capabilities of the elsa laboratory for earthquake testing of large structures. *Earthquake Spectra*, 12(1):163–180, 1996.
4. J. G.Bouwkamp, H. Parun, S. Qorraj, A. Plumier, and C. Doneux. Research on the energy dissipation capacity of composite steel/concrete structures. Final report of contracts Nos. 11708-96-03 ED ISP D and 11708-96-03 ED ISP B, European Commission, JRC, Ispra, 1998.
5. H.M. Hilber, T.J.R. Hughes, and R.L. Taylor. Improved numerical dissipation for time integration algorithms in structural dynamics. *Earthquake Engineering and Structural Dynamics*, 5:283–292, 1977.
6. G. Magonette. Digital control of pseudo-dynamic tests. in: Experimental and Numerical Methods in Earthquake Engineering (Donea, J. and Jones, P. M. ed.), Kluwer Academic Publishers, 1991.
7. G. Magonette, P. Pegon, F. J. Molina, and Ph. Buchet. Development of fast continuous substructuring tests. Proceedings of the Second Word Conference on Structural Control, 1998.
8. F. J. Molina, E. Gutierrez, G. Magonette, V. Renda, D. Tirelli, and G. Verzeletti. Pseudodynamic simulation of base isolation on a reinforced concrete building by means of substructuring. Proceedings of the First European Conference on Structural Control, 1996.
9. F. J. Molina, P. Pegon, and G. Verzeletti. Time-domain identification from seismic pseudodynamic test results on civil engineering specimens. Proceedings of the 2nd International Conference on Identification in Engineering Systems, The Cromwell Press Ltd, 1999.
10. M. Nakashima, T. Akawaza, and O. Sakaguchi. Integration method capable of controlling experimental error growth in substructure pseudo-dynamic test. *J. of Struct. Constr. Engng. AIJ (in Japanese)*, (454):61–71, 1993.
11. A. Plumier, C. Doneux, J. G. Bouwkamp, and C. Plumier. Slab design in connection zone of composite frames. Proceedings of the XIth European Conference on Earthquake Engineering, 1998.
12. P. B. Shing and S. A. Mahin. Cumulative experimental errors in pseudo-dynamic tests. *Earthquake Eng. Struct. Dyn.*, 15:409–424, 1987.
13. K. Takanashi and M. Nakashima. A state of the art: Japanese activities on on-line computer test control method. Report of the Institute of Industrial Science **32**, 3, University of Tokyo, 1986.
14. C. R. Thewalt and S. A. Mahin. Non-planar pseudo-dynamic testing. *Earthquake Eng. Struct. Dyn.*, 24:733–746, 1995.

INFLUENCE OF JOINT FLEXIBILITY IN THE SEISMIC PERFORMANCE OF MOMENT RESISTING STEEL FRAMES

I.VAYAS
National Technical University of Athens
Patission 42, 10682 Athens, Greece

F.DINU
Romanian Academy, Timisoara Branch
Mihai Viteazul 24, 1900 Timisoara, Romania

1. Introduction

Semi-rigidity in beam-to-column joints of steel frames has increasingly attained attention during the last years. For braced frames it was repeatedly demonstrated that semi-rigid joints may lead to remarkable material savings compared to simple joints, mainly due to a beneficial moment distribution along the members and provision of higher rigidity to satisfy serviceability requirements. Such an evidence is still lacking for moment resisting frames, where joint flexibility may lead to higher lateral deformations and increased 2nd order effects.

The effects of recent strong earthquakes in different parts of the world have shown that, unlike what was widely believed, steel buildings may also be vulnerable to them. The most widely observed type of failure was cracking in the region of beam-to-column joints of moment resisting frames. As such joints were formed according to different design and construction practices, it is considered worthwhile to examine types of joints other than rigid, where the strength demands are more evenly distributed along the girders.

The present work refers to the study of the behaviour to earthquakes of moment resisting frames with rigid and semi-rigid joints. This behaviour is examined by means of usual performance criteria of stiffness, strength and ductility, as well as criteria referring to the low-cycle fatigue strength of joints.

Several frames with different characteristics in respect to geometry, vertical loading and joint flexibility are subjected to different earthquake motions and investigated by means of non-linear dynamic analysis. Interesting conclusions are drawn.

2. Performance criteria

Frames designed against earthquakes have to comply with specific criteria such a stiffness, strength and ductility. Such criteria are either introduced in seismic Codes or used during building inspections after an earthquake event. In the present work, four

C.C. Baniotopoulos and F. Wald (eds.), The Paramount Role of Joints into the Reliable Response of Structures, 257–266.
© 2000 *Kluwer Academic Publishers.*

performance criteria associated with relevant limit states are used for the evaluation [6], as following:

- Drift limitation (A)

Seismic design Codes require a limitation of inter-storey drifts in the event of moderate earthquakes in order to limit damage in nonstructural elements. In Eurocode 8 [1] the relevant limits range between 0,6% and 0,4% in dependence on whether those elements are flexible or not. Taking into account a ratio of 3 between peak ground accelerations of strong to moderate earthquakes, limit drifts around 1,5% result. In the present work a limit drift of 2%, more liberal than the Code prescriptions, was considered as a stiffness criterion.

- Residual drift limitation (B)

Such a criterion is considered as an indication of the building condition during inspections after a strong earthquake. For residual drift exceeding 3% the building should be demolished [3]. In the present work a residual limit drift of 1%, corresponding to a moderately damaged building was adopted as a criterion. This is characterized as a strength criterion due to the fact that in case of low structural strength, large inelastic deformations will occur potentially leading to unacceptably high residual drifts.

- Available Rotational Ductility (C)

It is well known that overall ductility is directly associated to the rotation capacities at local zones, where plastic hinges develop. A lot of studies exist for the definition of the required rotation capacity of members and connections for steel frames. For frames subjected to earthquakes, the cyclic character of the response must be taken into account. The evidence is not conclusive yet. However, in this work, a rotation capacity of 0.03 rad, proposed by the recent AISC-Code (1997) for special moment frames was adopted as a criterion.

- Low-cycle fatigue strength (D)

The development of cracks in the beam-to-column joint regions may be associated to the exhaustion of the low-cycle-fatigue strength. This strength may be expressed in terms of plastic rotation. It may be shown, that the number of cycles for a certain range of plastic rotation is given by:

$$N = \frac{1}{2}(\frac{\varphi_{mon}}{\Delta\varphi_p})^m \tag{1}$$

where φ_{mon} corresponds to the rotation capacity under monotonic loading.

For variable ranges of plastic rotation, the damage assessment is performed in accordance to the linear Palmgren-Miner cumulative law in accordance to:

$$D = \Sigma \frac{n_i}{N_i} \tag{2}$$

where n_i = number of cycles of deformation range $\Delta\varphi_i$, and N_i = number of cycles of the same deformation range that cause failure .

For the rotation capacity under monotonic loading φ_{mon}, a value equal to 0.05 radians, higher as the corresponding value for cyclic loading, was adopted. For the determination of the design spectrum in the fatigue assessment, the *rainflow* or *reservoir* method for counting the cycles for a certain deformation history has been employed [5].

3. Parametric studies

The parametric studies refer to frames with different geometric and loading conditions. The frames under consideration are shown in Table 1. The loading conditions are characterized by the amount of beam strength that is required for the support of vertical loading. Here two levels 40% and 60%, were adopted. The level of vertical loading corresponds to the seismicity of the region. Obviously, the higher the level the lower the seismic region. Two types of behaviour for the beam-to-column joints were examined, rigid joints and semi-rigid joints, with rigidity, according to the definitions of Eurocode 3, 0.4K. The parameter $(K = 25 \cdot EI_b / L_b)$ expresses the stiffness of the beam connected at the joint.

TABLE 1. Geometric properties of the frames under consideration

Frame	L(m)	H(m)	T (sec) 40%	T (sec) 60%	Beam	Column
1	5	3	0.62	0.76	IPE300	HEB180
2	4	4	0,99	1.21	IPE330	HEB240
3	4	4	1.12	1.37	IPE330	HEB240
4	4	3	1.14	1.39	IPE360	HEB280
5	4	3	1.15	1.42	IPE360	HEB280
6	4	3	1.41	1.73	IPE450	HEB360

The frames were subjected to three seismic records, two records from Thessaloniki (1978) and Aigion (1985), Greece and one from Kobe (1995), Japan. The spectra of these records are shown in Figure 1.

In order to investigate the influence of the type of seismic record on the structural response, the maximal accelerations α_u of each record, for which a specific frame meets at the limit the specified performance criteria, were determined by appropriate scaling. Besides, the yield acceleration, at which the first plastic hinge develops, was calculated. The analysis was performed by means of the general purpose DRAIN-2DX software package [2].

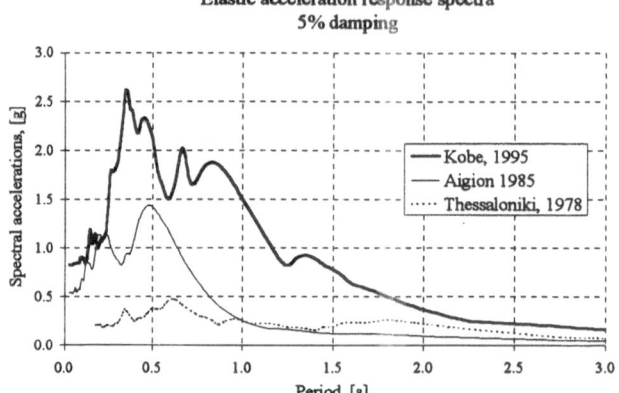

Figure 1. Acceleration response spectra of records

4. Results

The results of the parametric studies are shown in Table 2.

TABLE 2. Yield and ultimate accelerations

Frame	Level of vertical loading	Stiffness	Ultimate Criteria	Aigion earthquake Accel. [g]		Thessaloniki earthquake Accel. [g]		Kobe earthquake Accel. [g]	
				Elastic	Ultimate	Elastic	Ultimate	Elastic	Ultimate
1	40%	04Ksup	A	0.24	0.67	0.23	0.59	0.13	0.28
			B		2.42		0.74		0.35
			C		1.35		0.82		0.44
			D		1.28		0.65		0.46
		Rigid	A	0.15	0.65	0.21	0.53	0.09	0.42
			B		1.84		1.39		0.45
			C		1.17		1.05		0.51
			D		1.25		0.84		0.50
	60%	04Ksup	A	0.12	0.62	0.09	0.30	0.03	0.27
			B		1.35		0.25		0.33
			C		0.57		0.24		0.17
			D		0.94		0.33		0.25
		Rigid	A	0.11	0.83	0.08	0.42	0.06	0.24
			B		1.50		0.50		0.27
			C		1.14		0.58		0.42
			D		0.96		0.36		0.24
2	40%	04Ksup	A	0.42	1.10	0.21	0.65	0.12	0.31
			B		1.95		0.92		0.93
			C		1.55		0.73		0.36
			D		1.98		0.93		0.60
		Rigid	A	0.32	1.13	0.10	0.66	0.09	0.51
			B		2.21		0.85		1.02
			C		1.86		0.93		1.05
			D		1.99		0.61		0.46
	60%	04Ksup	A	0.23	1.04	0.20	0.75	0.10	0.57
			B		1.40		0.76		0.70
			C		1.41		0.76		0.42
			D		1.59		0.98		0.54

TABLE 2. (Continued)

		Rigid	A		1.12		0.64		0.51
			B	0.18	1.51	0.12	0.68	0.05	0.58
			C		1.50		0.83		0.79
			D		1.69		0.80		0.48
3	40%	04Ksup	A		1.03		0.61		0.41
			B	0.35	1.97	0.20	0.77	0.12	0.74
			C		1.81		1.10		0.81
			D		1.87		1.21		0.70
		Rigid	A		1.12		0.59		0.48
			B	0.36	1.79	0.10	0.75	0.11	0.94
			C		2.01		0.92		1.02
			D		1.98		0.93		0.78
	60%	04Ksup	A		0.85		0.70		0.41
			B	0.17	1.10	0.17	0.72	0.10	0.44
			C		1.15		0.85		0.45
			D		1.34		1.08		0.61
		Rigid	A		1.00		0.62		0.46
			B	0.16	1.36	0.10	0.61	0.06	0.83
			C		1.34		1.14		0.56
			D		1.45		0.94		0.58
4	40%	04Ksup	A		0.84		0.22		0.15
			B	0.21	1.56	0.14	0.50	0.09	0.54
			C		1.68		1.24		0.81
			D		1.41		1.08		0.80
		Rigid	A		1.12		0.62		0.77
			B	0.26	1.78	0.08	0.73	0.08	0.78
			C		1.56		0.85		0.81
			D		1.64		0.91		0.71
	60%	04Ksup	A		0.39		0.21		0.11
			B	0.10	0.88	0.10	0.48	0.06	0.38
			C		1.02		0.64		0.48
			D		1.62		0.85		0.59
		Rigid	A		0.93		0.62		0.48
			B	0.10	1.13	0.08	0.50	0.06	0.51
			C		1.19		1.06		0.73
			D		1.05		-		0.64
5	40%	04Ksup	A		0.72		0.62		0.39
			B	0.24	0.85	0.30	0.74	0.15	0.75
			C		1.23		1.35		0.88
			D		-		-		-
		Rigid	A		1.13		0.60		0.67
			B	0.19	1.60	0.21	0.68	0.10	0.70
			C		1.71		0.89		0.86
			D		1.62		0.95		0.60
	60%	04Ksup	A		0.67		0.72		0.60
			B	0.10	0.82	0.14	0.63	0.06	0.62
			C		0.91		0.90		0.57
			D		1.05		0.96		0.43
		Rigid	A		0.83		0.74		0.43
			B	0.09	1.13	0.09	0.68	0.05	0.54
			C		1.04		0.99		0.64
			D		0.93		0.88		-
6	40%	04Ksup	A		0.84		0.64		0.45
			B	0.22	1.60	0.21	1.12	0.14	1.16
			C		1.28		0.93		1.08
			D		-		-		-
		Rigid	A		1.12		1.05		0.60
			B	0.21	1.66	0.22	0.90	0.14	0.71
			C		1.16		1.25		0.54
			D		1.51		1.31		0.67

TABLE 2. (Continued)

60%	04Ksup	A		0.04	0.61	0.06	0.69	0.02	0.34
		B			0.65		0.70		0.38
		C			0.70		1.12		0.41
		D			-		-		-
	Rigid	A		0.03	0.69	0.07	0.98	0.02	0.58
		B			0.76		0.93		0.64
		C			1.02		1.33		1.00
		D			0.63		-		-

The yield acceleration α_y for two levels of vertical loading, rigid and semi-rigid joints and the Thessaloniki record is presented in Figure 2. It may be observed that in most cases yielding occurs at higher accelerations for semi-rigid joints. This is due to moment shedding from the, more flexible, joint to the beam span. The initiation of yielding is also a function of the level of vertical loading, and happens, as expected, earlier at higher levels of such a loading.

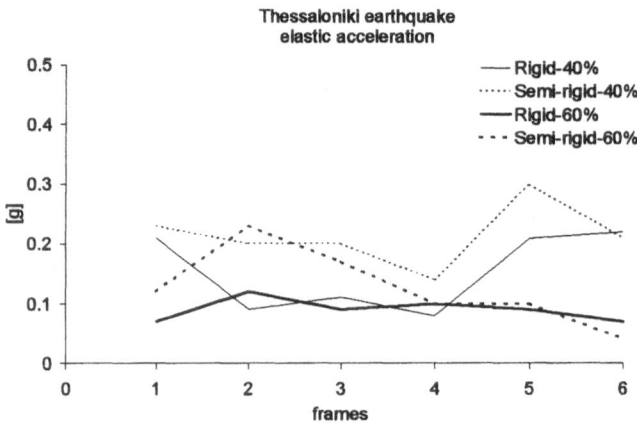

Figure 2 Yield accelerations for the Thessaloniki record for two levels of vertical loading

The same effects may be observed for the other examined records, the main difference being on the level of yield acceleration. The Kobe record provided, due to its high energy-content, the lowest values, the Aigion record the highest values.

The maximal accelerations, α_u for the Thessaloniki record, for all criteria and rigid and semi-rigid joints are shown in Figure 3 and Figure 4.

Figure 3 shows that for 40% loading the limit accelerations are similar, while for 60% loading, higher frames can sustain, due to a better redistribution, higher accelerations. As expected, the limit accelerations are for all cases lower for 60% than for 40% loading. For frames designed for stronger earthquakes, 40% loading, the critical criterion is associated to the drift limitation, and therefore to their stiffness, while for those designed for lower earthquakes the critical is the residual drift, and therefore the strength. Similar observations may be done for semi-rigid joints, except that stiffness becomes more critical. Comparing frames with rigid and semi-rigid joints, it may be observe that, with exception frame 4, the limit accelerations are similar, indicating that semi-rigidity is not disadvantageous.

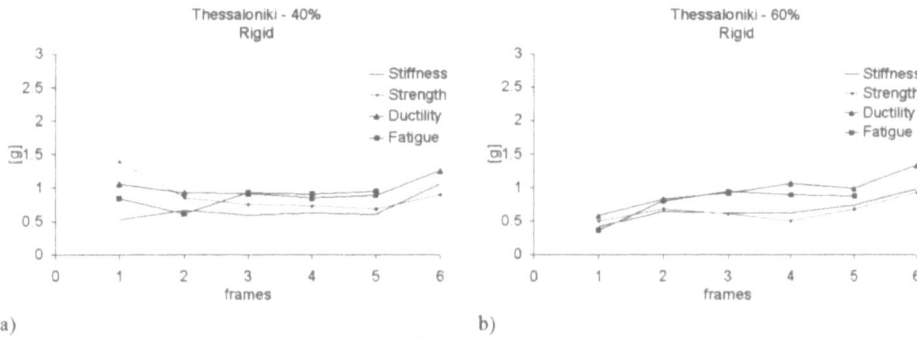

Figure 3. Limit accelerations for the Thessaloniki record and rigid joints;
a) 40% vertical loading; b) 60% vertical loading

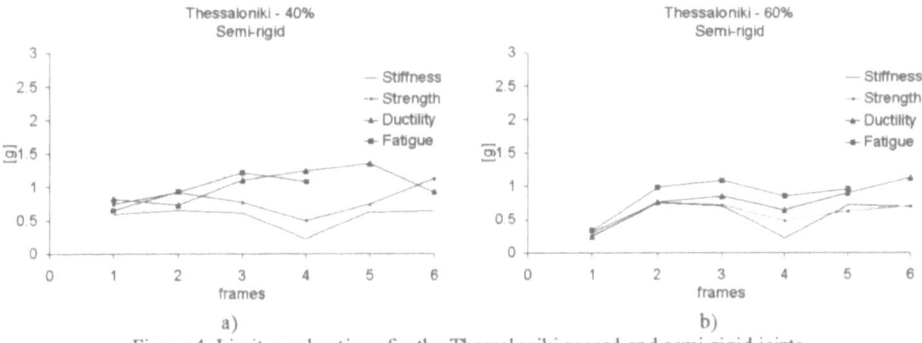

Figure 4. Limit accelerations for the Thessaloniki record and semi-rigid joints
a) 40% vertical loading; b) 60% vertical loading

Figure 5 and Figure 6 show the limit accelerations for the Aigion and the Kobe record for 40 % loading and rigid and semi-rigid joints. For the Aigion record, which was of a shock type, stiffness seems to be the most critical. Semi-rigidity does not considerable influence the response. However, for the Kobe record, low-cycle fatigue seems to govern the response when the joints are rigid .

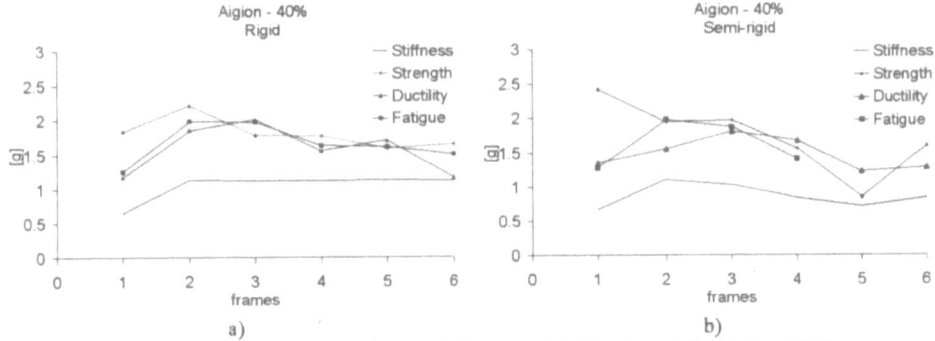

Figure 5. Limit accelerations for the Aigion record rigid and semi-rigid joints (40%)
a) rigid joints; b) semi-rigid joints

264

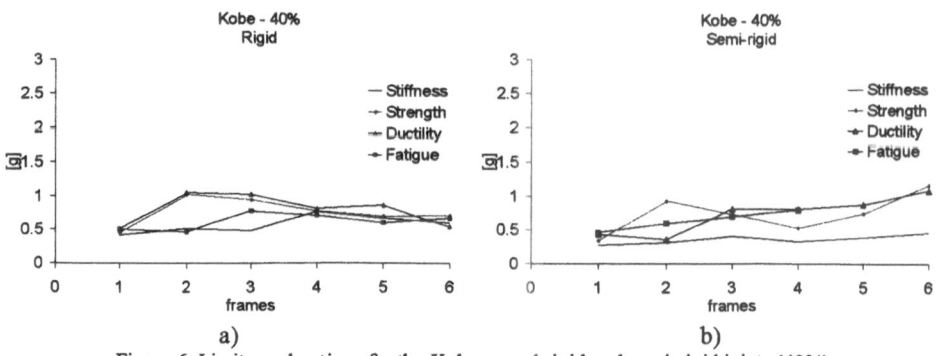

Figure 6. Limit accelerations for the Kobe record rigid and semi-rigid joints (40%)
a) rigid joints; b) semi-rigid joints

This is in line with what was really observed after the earthquake and is due to the large number of cycles during the Kobe earthquake.

The introduction of flexibility in the joints alleviates this effect, providing almost two times higher limit accelerations corresponding with the most critical ones that arise from the stiffness criterion.

Figures 7 to 10 give the ratios between the limit and yield accelerations for the different criteria as mean values for the three records. These ratios express, in accordance with Eurocode 8, the behaviour factor of the frame. The most important parameter seems to be the initial design of the frame. If the frame is primarily designed to support vertical loading, which corresponds to 60% level, the behaviour factor is low, independent on the joint flexibility. If the frame is primarily designed to support seismic loading, which corresponds to 40% level, the behaviour factor varies between 3 and 10. In almost all cases the behaviour factor is lower for semi-rigid joints, expressing the previously referred observation that yielding starts generally later for semi-rigid joints. However, the importance of this factor is limited in the present work, the more important fact being the directly evaluated limit accelerations.

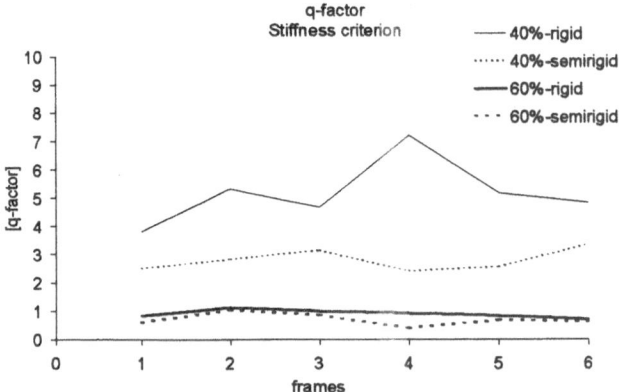

Figure 7. Behaviour factors for the stiffness criterion

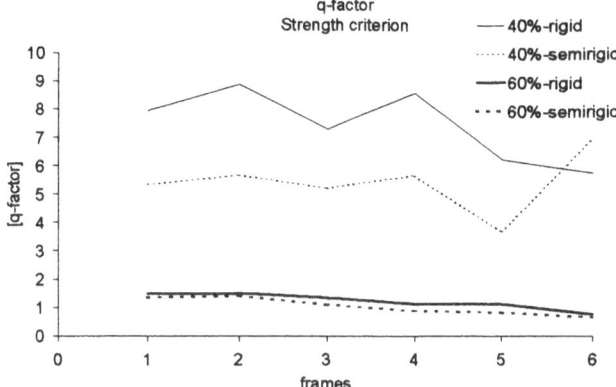

Figure 8. Behaviour factors for the strength criterion

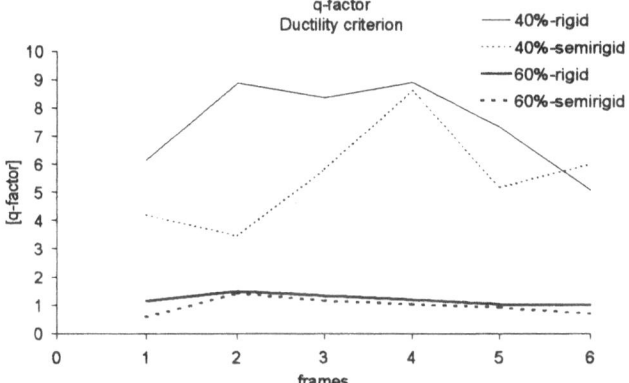

Figure 9. Behaviour factors for the ductility criterion

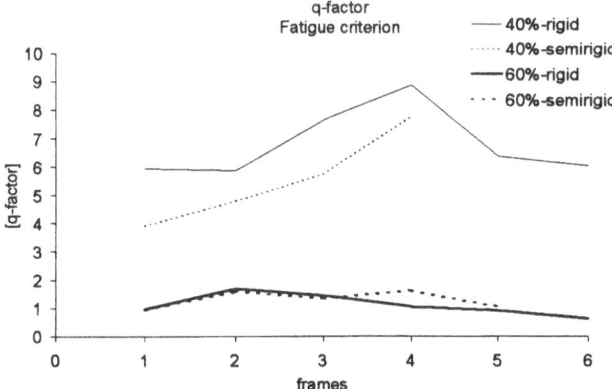

Figure 10. Behaviour factors for the fatigue criterion

5. Conclusions

Studies of a number of frames with rigid and semi-rigid beam-to-column joints subjected to different earthquakes lead to following conclusions:

- Serviceability criteria are generally the most critical for moment resisting steel frames
- For records with long duration and many deformation cycles, low-cycle fatigue may be critical.
- The introduction of flexible joints almost excludes the appearance of low-cycle fatigue problems.
- When not low-cycle fatigue but stiffness is the main design criterion, the introduction of semi-rigid joints is disadvantageous.
- For conclusive evidence, more studies are required.

6. References

1. Eurocode 8 Design provisions for earthquake resistance of structures (1994). CEN, European Committee for Standardization, ENV 1998-1-1.
2. Kannan, A. and Powel, G. DRAIN-2D. (1975) A general purpose computer program for dynamic analysis of inelastic plane structures, EERC 73-6 and EERC 73-22 reports, Berkeley, USA.
3. Ohi, K. and Takanashi, K. (1998) Seismic diagnosis for rehabilitation and upgrading of steel gymnasiums. *Engineering structures,* **20, No 4-6,** 533-539.
4. Seismic Provisions for Structural Steel Buildings. (1997). American Institute of Steel Construction, Chicago, Illinois.
5. Vayas, I. (2000) Interaction between local and global ductility. in F. Mazzolani (ed.), Moment resisting connections of steel building frames in seismic areas, E & FN SPON.
6. Vayas, I. and Dinu, F. (2000) Evaluation of the response of moment frames in respect to various performance criteria, *Third International Conference on Behaviour of Steel Structures in Seismic Areas STESSA 2000*, Montreal, Canada.

DYNAMIC ANALYSIS OF STIFFENED COUPLED SHEAR WALLS WITH FLEXIBLE CONNECTIONS UNDER EARTHQUAKE EFFECTS

O. AKSOGAN & H.M. ARSLAN
Civil Engineering Department
Cukurova University, 01330 Adana, Turkey

1. Introduction

In modern tall buildings, the lateral loads that arise from the effects of winds and earthquakes, are resisted by shear walls. However, wall openings are inevitably present due to windows, doors and service ducts. These features turn simple shear walls into coupled ones, which can be considered as two smaller walls, coupled together by a system of connecting beams. For a quick initial assessment of the foregoing type of structural component, the elegant method called Continuous Connection Method (CCM) has been widely used. In this method, the connecting beams are assumed to have the same properties and spacing along the entire height of the wall. Consequently, the discrete system of connecting beams can be assumed to be replaced by continuous laminae of equivalent stiffness capable of transmitting actions of the same type as the connecting beams. This modification enables the properties of the structure to be expressed as continuous functions of the longitudinal coordinate x (see Fig. 1).

A straightforward application of CCM for the dynamic analysis of coupled shear walls results in a sixth order differential equation, which cannot be solved in closed form. Moreover, in the case of multiple sections separated by stiffeners, as in the present work, the analysis would necessitate the treatment of a coupled combination of as many such equations as there are separate sections. Hence, a special method will be employed to solve the foregoing problem.

The special method used in this study has two steps. In the first step, the structure is considered as a discrete system of lumped masses with as many degrees of freedom as the number of lumped masses. The number of lumped masses being chosen freely, the amount of mass in each of them is found using the averaged mass per unit length in the longitudinal direction. Thus, the mass matrix is determined. The second step is the determination of the stiffness matrix of the structure for the degrees of freedom chosen during the first step. This procedure is carried out by applying a horizontal unit force at each and every height with a lumped mass. For every one of these loadings, a solution is carried out making use of CCM and writing down the compatibility equations for the midpoints, which are assumed to be the points of contraflexure, of the actual stiffeners as well as the assumed laminae (Fig. 2). During this procedure the connections of the lintel beams and stiffeners to the walls are considered to be flexible ones equivalent to linear rotational springs. Once the axial forces in the walls in different sections are

C.C. Baniotopoulos and F. Wald (eds.), The Paramount Role of Joints into the Reliable Response of Structures, 267–276.
© *2000 Kluwer Academic Publishers.*

determined, writing down the moment-curvature relation, the deflected form of the system of coupled shear walls in the horizontal direction is determined in a straightforward manner. Each unit loading gives one column of the flexibility matrix as the displacements at the heights at which the lumped masses are located. Hence, the analysis for one general unit loading case will suffice to introduce the complete solution procedure for the determination of the flexibility matrix. The stiffness matrix of the structure will be determined by taking the inverse of the flexibility matrix. Substituting the mass and stiffness matrices, thus obtained, in the equations of motion and assigning the forcing load vector on the right hand side, in the form of a seismic effect, the system of equations for the problem in hand is obtained.

In this work, the seismic analysis of flexibly connected stiffened coupled shear walls with stepwise changes in width, resting on a flexible foundation, is carried out by the foregoing method. This method comprises an elegant tool for the predesign computations related to the treatment of high-rise buildings. To verify the present analysis, the computer programs prepared in FORTRAN language and in MATHEMATICA to implement the results of the foregoing analysis has been used for the solution of two examples which were also solved by employing the SAP90 Structural Analysis Program [2]. The results of the two methods showed a perfect agreement.

2. Free Vibration Analysis

To solve the problem in the heading a joint use of the well-known CCM and a discretization of the mass of the structure has been employed. The mass matrix of the structure is taken as a diagonal matrix, employing the lumped mass assumption [3]. For this purpose the top and bottom of the system of coupled shear walls, the levels of the stiffeners and the levels of width change will be called "ends" and a part of the structure between any consecutive pair of these "ends" will be called a "section". Each "section" will be divided into a suitable number of equal parts and the total mass of this "section" divided by this number will be assigned to each internal point and half of it will be assigned to each "end" of the "section". After carrying out the foregoing procedure and adding the extra mass of a stiffener, compared to a connecting beam, to each pertinent "end" the lumped mass matrix $\underline{M}$ will be found as a diagonal matrix. Obviously, the dimension of this mass matrix will be m×m where m is the number of lumped masses assigned to discretize the total mass of the structure (see Fig. 1).

Forming the mass matrix as described in the previous paragraph and choosing the lateral displacements as the only degrees of freedom, the effects of the vertical inertia forces and the rotatory inertia effects have been automatically neglected in comparison with the effects of the horizontal inertia forces. Although a slight error may show up on the higher modes, this assumption gives perfect results for the fundamental frequency and the corresponding mode of vibration.

Despite the fact that the vertical and rotatory inertias are not considered, more important effects of the flexibility of the foundation is included in the present analysis. This is done by simulating the foundation by a combination of three springs, a horizontal, a vertical and a rotational one. Furthermore, although most problems of coupled shear walls in the literature, solved by CCM, do not allow for the effect of shear deformation, it has been taken into consideration in the present work.

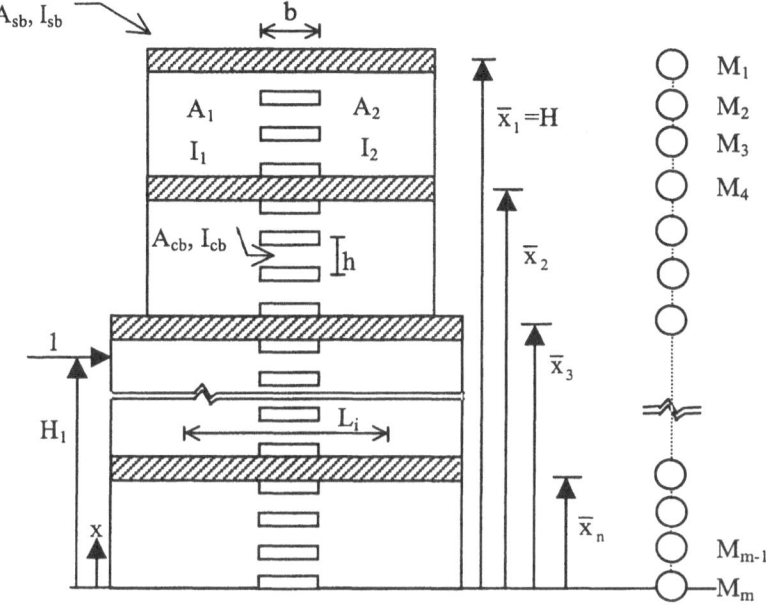

Figure 1. Lumped mass model of coupled shear wall.

In the present work, two crucial assumptions are made. One of them is about the axial deformations of the connecting beams being ignored and the other one is about the replacement of the actual discrete connecting beams and shear forces in them by a uniform (in each "section") connecting medium, or set of laminae, of equivalent flexural rigidity and shear flow function, respectively. The first assumption renders the lateral deflections of the walls equal and yields equal slopes and curvatures at the same height. This assumption can be justified taking into account the enhancement of the in-plane stiffness of the floor slabs (recall "rigid floor diaphragm" assumption). Furthermore, the connecting beams being loaded only at the ends and the end conditions being symmetrical, this assumption renders their midpoints, points of contraflexure. Another consequence is that the bending moments of the two walls at a certain height are in the same ratio as their moments of inertia. The second assumption, i.e. the replacement of the discrete set of equidistant connecting beams, each of flexural rigidity EI_c, by a set of laminae, of flexural rigidity EI_c/h per unit height, and likewise, the replacement of the discrete shear forces in the connecting beams by a shear flow function q, along the points of contraflexure, expressing the shear force per unit height render the discrete two dimensional system a continuous one dimensional one along the height (see Fig. 2).

To determine the stiffness matrix of the structure, an imaginary cut is assumed through the points of contraflexure of laminae (which are their midpoints) the compatibility condition yields

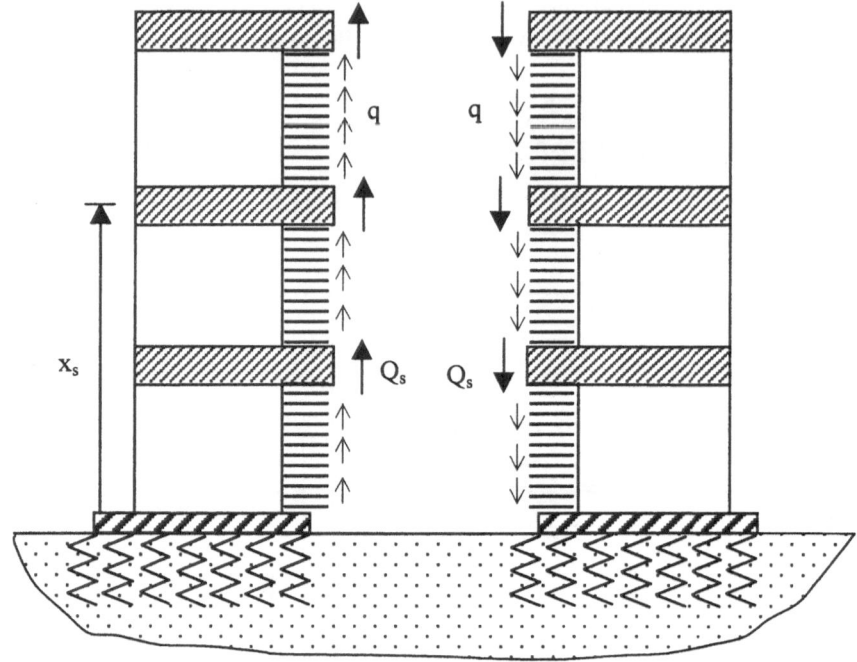

Figure 2. Equivalent structure.

$$L\frac{dy}{dx}+\frac{\langle H_1-x\rangle^0 L\mu_w}{G(A_1+A_2)}-\frac{hb^3q}{12EI_{cb}}+\frac{b\,\mu_{cb}\,hq}{G\,A_{cb}}-\frac{b^2\,hq}{2\,C_{cb}}-\frac{1}{E}\left(\frac{1}{A_1}+\frac{1}{A_2}\right)\int_0^x T\,dx-\delta_0+\delta_e=0$$

$$(1)$$

where μ_w, μ_{cb}, E, G, y, q, T, δ_0, C_{cb} and δ_e are the shear area factors for walls and the connecting beams, the elasticity and shear moduli, the lateral deflection of the walls, the shear flow in the laminae, the axial force in the walls, the relative vertical displacement of the bases of the two walls, the equivalent rotational spring constant for the ends of the connecting beams and the relative vertical displacement of the walls due to the eccentricity caused by the changes in width, respectively. δ_e is found by summing up the products of the eccentricities at the levels of change of width and the corresponding slopes for the structure below the "section" for which Equation (1) is being written. Equation (1) is written for each "section" separately. However, in this context when the unit load is applied at an internal point of a "section" it divides that "section" into two new "sections". The system of Macaulay's brackets should be understood, here and in the sequel, as

$$\begin{aligned}&<x-x'>^n=\left(x-x'\right)^n &&and &&<x-x'>^0=1 &&if &&x>x'\\&<x-x'>^n=0 &&and &&<x-x'>^0=0 &&if &&x<x'\end{aligned}$$

$$(2)$$

The successive terms in (1) represent the relative vertical displacements of the midpoint of the lamina due to the bending and shear deformations of the walls, the bending and shear deformations of the lamina, the flexibility of the connections, the axial deformations of the walls, the relative vertical settlement of the bases of the walls and the eccentricity due to the changes in width.

The moment-curvature relationship for any "section" of the structure can be given as follows,

$$EI \frac{d^2 y}{dx^2} = < H_1 - x >^1 - T L \tag{3}$$

where I expresses the sum of the moments of inertia of the two walls. On the other hand, the shear flow function q can be expressed as

$$q = -\frac{dT}{dx} \tag{4}$$

from the equilibrium of a differential element dx of one of the walls. The axial force in one of the walls can be expressed by adding, the integral of the shear flow function q from the cross-section of concern to the top, to the sum of the shear forces, Q_s (subscript s expreses the number of stiffener), in all the stiffeners in that interval, as follows:

$$T(x) = \sum_x^H Q_s + \int_x^H q \, dx \tag{5}$$

Differentiating (1) with respect to x and employing (3) and (4) to eliminate the variables y and q the governing equation for T is found as

$$\frac{d^2 T}{dx^2} - \alpha^2 T = -\gamma < H_1 - x >^1 \tag{6}$$

where

$$\gamma = \frac{1}{\dfrac{E I}{L} \left(\dfrac{b^3 h}{12 E I_{cb}} + \dfrac{b \mu_{cb} h}{G A_{cb}} + \dfrac{b^2 h}{2 C_{cb}} \right)} \quad \text{and} \quad \alpha^2 = \gamma \left(L + \frac{I A}{L A_1 A_2} \right) \tag{7}$$

in which $A = A_1 + A_2$. The general solution of (6) is

$$T = B \, Cosh \, \alpha x + C \, Sinh \, \alpha x + \frac{\gamma}{\alpha^2} < H_1 - x >^1 \tag{8}$$

The corresponding shear flow expressions can be found by using (4) as follows:

$$q = -\left(B\,\alpha\,Sinh\,\alpha x + C\,\alpha\,Cosh\,\alpha x - \frac{\gamma}{\alpha^2} < H_1 - x >^0 \right) \tag{9}$$

For the determination of the shear forces in the stiffeners, the compatibility equation for the points of contraflexure can be written as follows:

$$L\frac{dy}{dx}\bigg|_{x=x_s} + \frac{\langle H_1 - x \rangle^0 L \mu_w}{G(A_1 + A_2)} - \frac{Q_s b^3}{12 E I_{sb}} - \frac{b\,\mu_{sb}\,Q_s}{G A_{sb}} - \frac{b^2 Q_s}{2 C_{sb}} - \frac{1}{E}\left(\frac{1}{A_1} + \frac{1}{A_2}\right)\int_0^{x_s} T\,dx - \delta_0 + \delta_e = 0 \tag{10}$$

where μ_{sb}, x_s, I_{sb}, A_{sb}, C_{sb} and Q_s are the shear area factor, the height, the moment of inertia, the cross-sectional area, the rotational spring constant and the shear force of the pertinent stiffener, respectively. The shear force in a stiffener is found by comparing (10) with (1) evaluated at level $x=x_s$ as

$$Q_s = H\,\beta\,q(x_s) \tag{11}$$

where

$$\beta = \frac{\dfrac{hb^2}{12\,E\,I_{cb}} + \dfrac{hb}{2\,C_{cb}} + \dfrac{h\,\mu_{cb}}{G\,A_{cb}}}{\dfrac{b^2}{12\,E\,I_{sb}} + \dfrac{b}{2\,C_{sb}} + \dfrac{\mu_{sb}}{G\,A_{sb}}} \tag{12}$$

The integration constants B and C for all "sections" are determined from the vertical force equilibrium in the walls at the "ends" of the "sections" (excluding the bottom and including the level of the unit load) and the continuity of the slope at the "ends" (excluding the top and including the level of the unit load).

In writing these equations the axial force at the top from above should be taken to be zero and the slope and relative displacement at the bottom should be taken as

$$\theta_0 = \frac{H_1 - T(0)\times L}{K_r}, \qquad \delta_0 = \frac{T(0)}{K_v} \tag{13}$$

where K_r and K_v are the equivalent rotational and vertical rigidities of the foundations, respectively. When there is a stiffener at the bottom, the axial force, $T(0)$, must be increased by an amount equal to the shear force in that stiffener.

Substituting (8) in (3) and integrating with respect to x twice

$$y = \frac{1}{EI}\left\{\frac{1}{6}\left(1-\frac{\gamma L}{\alpha^2}\right)< H_1 - x >^3 - \frac{L}{\alpha^2}\left(B \, Cosh\,\alpha x + C \, Sinh\,\alpha x + Dx + F\right)\right\} \quad (14)$$

in which the integration constants for all "sections" are found using the continuity conditions for the horizontal displacements and the corresponding slopes at the common points of each and every pair of neighbouring "sections" and setting the horizontal displacement equal to $1/K_h$ and the rotation equal to θ_0 at the bottom of the wall. Here, θ_0 is to be taken from (13) and K_h is the equivalent horizontal rigidity of the foundation.

Having determined the lateral displacements for unit loadings at each and every one of the levels of lumped masses, the flexibility matrix, and thereby, the stiffness matrix of the structure can be found. Finally, the circular frequencies are determined from the following standard frequency equation for the lumped mass system:

$$\left| \underline{K} - \omega^2 \underline{M} \right| = 0 \quad (15)$$

where ω is the circular frequency, $\underline{M}$ is the mass matrix and $\underline{K}$ is the stiffness matrix of the structure. The respective eigenvectors, $\underline{u}_i$, are found by substituting each and every circular frequency, ω_i, in the following equation at a time:

$$\left(\underline{K} - \omega_i^2 \underline{M}\right)\underline{u}_i = 0 \qquad\qquad i = 1,2,\ldots,m \quad (16)$$

3. Forced Vibration Analysis

The equation of motion for a multi-degree of freedom system is as follows:

$$\underline{M}\,\ddot{\underline{X}} + \underline{C}\,\dot{\underline{X}} + \underline{K}\,\underline{X} = \underline{P}(t) \quad (17)$$

where $\underline{C}$ is the damping matrix and $\underline{P}(t)$ is the time dependent loading vector. This system of equations is always coupled. Hence, for shear walls with many degrees of freedom, the solution, using exact methods, will be tedious. For this reason, the solution has been carried out using the mode superposition method in the present work. To apply the foregoing method to (17), the displacement vector $\underline{X}$ is replaced by the modal displacement vector $\underline{Y}$, using the following transformation:

$$\underline{X} = \underline{\Phi}\,\underline{Y} \quad (18)$$

where $\underline{\Phi}$ is the modal matrix. Substituting expression (18) and its derivatives in (17) and premultiplying by $\underline{\Phi}^T$ the following uncoupled system of equations is obtained:

$$\widetilde{\underline{M}}\,\ddot{\underline{Y}} + \widetilde{\underline{C}}\,\dot{\underline{Y}} + \widetilde{\underline{K}}\,\underline{Y} = \widetilde{\underline{P}}(t) \quad (19)$$

The loading term on the right of the i'th equation is given as:

$$\tilde{\underline{P}}_i(t) = -\underline{\Phi}_i^T \underline{M} \underline{r} \ddot{x}_g \tag{20}$$

in which $\underline{\Phi}_i^T$ is the transpose of the i'th column of the modal matrix and $\underline{r}$ is called the pseudostatic vector. The latter vector, for the case in hand, has all terms equal to unity. Hence, (19) can be written in open form as

$$\ddot{Y}_i + 2\zeta_i\omega_i\dot{Y}_i + \omega_i^2 Y_i = -\gamma_i \ddot{x}_g \qquad\qquad i = 1,2,3...,n \tag{21}$$

where ζ_i is the damping ratio and

$$\gamma_i = \frac{\underline{\Phi}_i^T \underline{M} \underline{r}}{m_i} \tag{22}$$

and the maximum displacement for each mode can be expressed as:

$$\max|Y_i| = \gamma_i S_d(T_i, \zeta_i) \tag{23}$$

where S_d is the displacement spectrum and T_i is the period corresponding to i'th mode.

The eigenvectors are multiplied by their corresponding maximum displacements in the uncoupled system of equations to find the corresponding maximum displacement vectors. Having determined the maximum displacements for each mode, the maximum displacement for each node is found by the method of the Square Root of the Sum of the Squares (SRSS).

4. Numerical Results

Two example structures (see Fig. 3 and 4) were chosen both for verification and for application purposes. To justify the results of the present work the example structures were solved both by the present method and by the Structural Analysis Program (SAP90). The results of both methods are presented together in Figs. 5, 6 and 7. As can be seen, the results of the two methods match perfectly. The thickness of the walls is taken to be 0.2m, the modulus of elasticity 2.0×10^7 kN/m^2, the density of reinforced concrete 2400 kg/m^3 and no damping is assumed for either example. The spectral acceleration values are taken from SAP90 user manual. The results in Figs. 5 and 6 being for rigid foundation, those in Fig. 7 are for a flexible foundation for which the elastic foundation constants are, $K_r=2.72\times10^7$ kN m/rad, $K_v=8.78\times10^5$ kN/m and $K_h=7.52\times10^5$ kN/m [3].

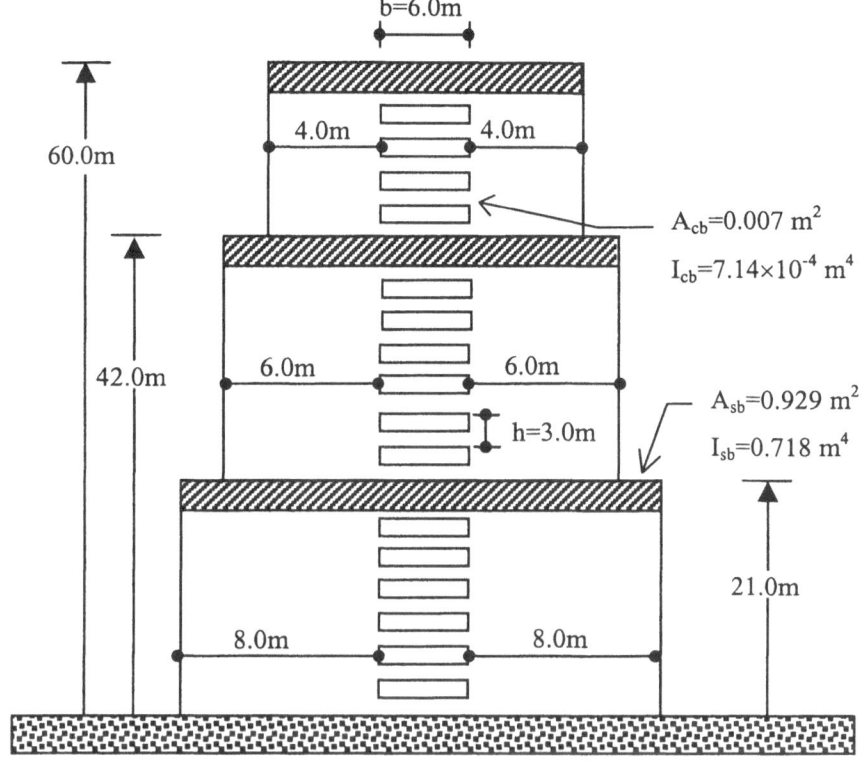

Figure 3. Coupled shear wall for example 1.

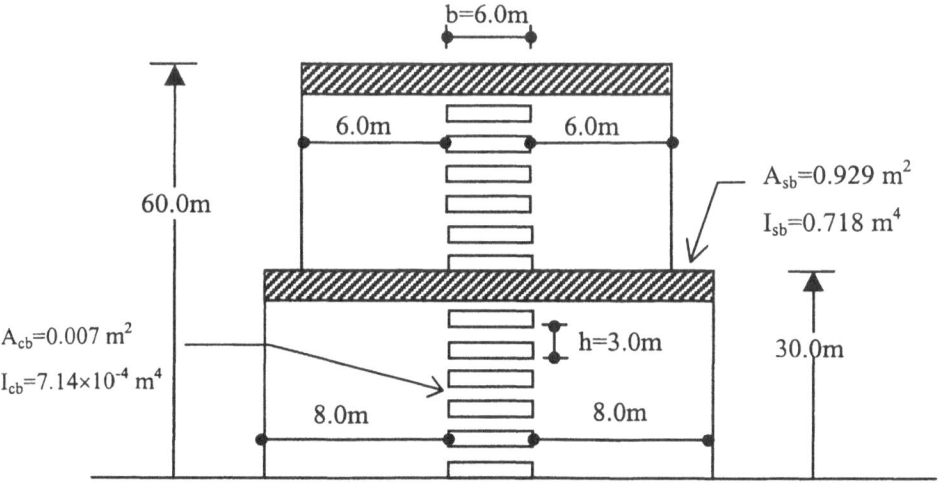

Figure 4. Coupled shear wall for example 2.

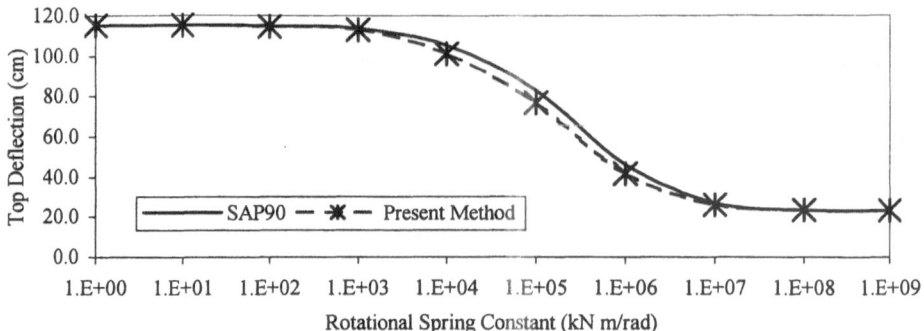

Figure 5. Variation of maximum top deflection with connection rigidity in example 1.

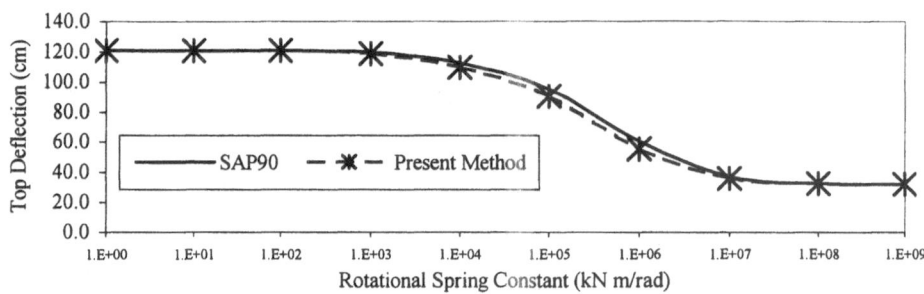

Figure 6. Variation of maximum top deflection with connection rigidity in example 2 for rigid foundation.

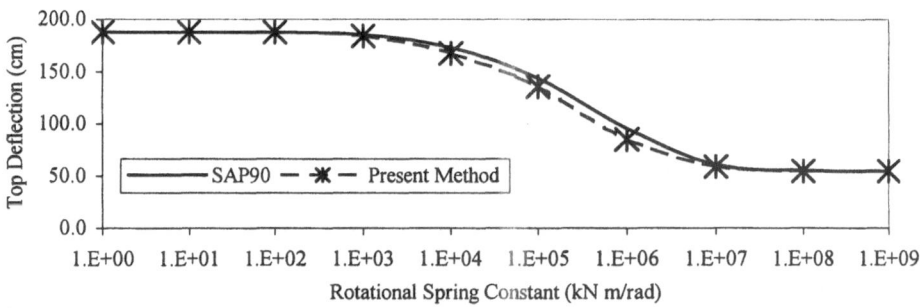

Figure 7. Variation of maximum top deflection with connection rigidity in example 2 for elastic foundation.

5. References

1. Aksogan, O., Turker, H.T., and Oskouei, A.V. (1993) Proceedings of the First Technical Congress on Advances in Civil Engineering, E.R. TUNCER and D. ALTINBILEK (eds.), *Stiffening of coupled shear walls at arbitrary number of heights*, North Cyprus, pp. 781-787.
2. Wilson, E.L. and Habibullah, A. (1992) *SAP90-Structural Analysis Users Manual*, Computers and Structures Inc., Berkeley, California.
3. Li, G.Q. and Choo, B.S. (1996) A continuous-discrete approach to the free vibration analysis of stiffened pierced walls on flexible foundations, *International Journal of Solids and Structures* **33(2)**, 249-263.

IV. NUMERICAL SIMULATION OF THE STRUCTURAL RESPONSE OF JOINTS AND FRAMES

NUMERICAL MODELLING OF THE STRUCTURAL
BEHAVIOUR OF JOINTS

N. Gebbeken
University of the Federal Armed Forces Munich
D-85577 Neubiberg, Germany

T. Wanzek
Köhler + Seitz Beraten und Planen GmbH
D-90441 Nuremberg, Germany

1 Introduction

It is the sign of the progress and it is the competition of the steel with concrete to become more and more efficient construction which needs less material and less effort in manufactoring leading to an optimal cost-benefit-ratio. Therefore new construction technologies are developed. They have to be verified and the ultimate load as well as the deformation characteristic have to be calculated. In former days it was out of question to carry out a lot of experiments to get the significant attributes. Even if a construction should be generalized for a broad range for each parameter combination a couple of experiments have to be performed because of the statistical deviation. The safety factor in the determined design criteria has to be high in order to cover uncertainties (reasons of deviation, lack of tested combinations). Nowadays the characteristic of a construction can be determined by finite element calculations. The devolopment of a new construction can be done numerically. Experiments are only needed to verify numerical results. This procedure is not as time- and cost-consuming as the pure experimental method. Numerical parametric studies are easy to conduct, the failure mechanisms are visible and therefore easy to study. So, the construction's safety factor perhaps can be reduced. But the finite element calculations have to be used carefully and their results have to be analysed critically to achieve correct qualitative as well as quantitative answers about the tested construction. In this context the main question is: How reliable are the numerical results?

The authors present in this paper the proceeding and different aspects which have to be considered to achieve reliable numerical results of the simulations of end plate connections. In addition, the experiment requirements are presented which are needed to perform a valid numerical simulation. It is shown that a fundamental numerical study prior to the actual calculation is as important as a detailed experimental study. A mesh convergence study (discretization error) and the precise investigation of the influences

279

C.C. Baniotopoulos and F. Wald (eds.), The Paramount Role of Joints into the Reliable Response of Structures, 279–292.
© 2000 *Kluwer Academic Publishers.*

of different modelling parameters is imperative in order to evaluate the reliability of the numerical results. It has to be clarified that the finite element method is an approximation method that may yield any big error. Here, the results of detailed numerical studies will be presented, which also include informations about influences of unknown experimental data (friction, imperfection). With the knowledge of these results at hand, the Munich T-stub experiments performed by Gebbeken, Wanzek and Petersen [4] will be simulated by finite element analyses. Then, the knowledge will be transferred to general bolted end plate connections. The full discussion about the experiments and the numerical studies and the background of the used brick- and contact-elements in the finite element analyses are reported in [9].

The general method and most conclusions of this paper are also transferable to other constructions. They can give a guide for performing reliable numerical calculations.

2 The Numerical Model

2.1 GENERAL

The initial step is to create a numerical model independent of any experiment results to compare with. It is very important to quantify the numerical error of the used model. This error is caused by the discretization, approximation of the material law, boundary conditions, kind of loading etc. The error of each parameter has to be calculated, in order to investigate the parameter sensitivity (but recognize, that the accumulation of small errors make a big error, too). In additon, the influence of material and geometrical deviations has to be calculated. The material deviations are yield stress, hardening function, bolt-elongation behaviour, friction coefficients etc. Geometrical uncertainties are e.g. plate thickness, bolt position, imperfection etc. From this number of parameter influences it is obvious, that the calculated joint has still some probalilistic uncertainities. But with the knowledge of the different errors of the numerical model the response range can be estimated, in which will be included the right answer.

Finally, the numerical model has to be verified by selected experiments. These experiments have to provide a lot of data required for finite element analyses (chapter 3.2).

2.2 DESCRIPTION OF THE USED MODEL

In the following some numerical studies are presented which have to be carried out in order to calculate bolted end plate connections by finite element analyses. Further information about these numerical studies can be found in Wanzek and Gebbeken [10]. The numerical studies are based on T-stub connections, because their behaviour is similar to the behaviour of the tension zone of bolted end plate connections. Therefore, the knowledge obtained from the numerical T-stub studies can be transferred to general end plate connections.

Benefitial to the finite element studies is the geometrical symmetry of T-stub connections. Therefore, only an eighth of the T-stub connection has to be modelled (Figure 1). The symmetry-plane between the two flanges is modelled by contact elements without friction at the rigid base (contact target in Figure 1). Certainly, because of the thread, the bolt elongation behaviour is not symmetric but the behaviour will be simulated as explained below. The nodes in the two remaining planes are fixed by symmetric geometrical boundary conditions (constraints to the symmetry plane). The perpendicular material of each component of the T-stub model (rolled section, bolt, washer) is simulated by its individual material behaviour. For the elementation 3d-brick-elements are used.

2.3 DESCRIPTION OF TWO EXPERIMENTS

Due to restricted space only three experiments are presented. The first two T-stub models refer to the P1K and P2K experiments, which are two of the six Munich experiment series [4] (a short presentation of the Munich experiments is given later). The third one is the Liege T-stub experiment. Usually, in this paper the Munich T-stub will be used because of its given detailed experimental data. Only the influence of bolt modelling will be shown with the Liege experiment, because in comparison to the flange the bolts were not as stiff as in the Munich experiments.

2.4 NUMERICAL BOLT MODEL

In the numerical bolt model the complicated bolt, composed of head, shank, thread and nut is simplified to a head and an "equivalent" shank. The equivalent shank is dimensioned such that it represents the half elongation behaviour of the real bolt (half

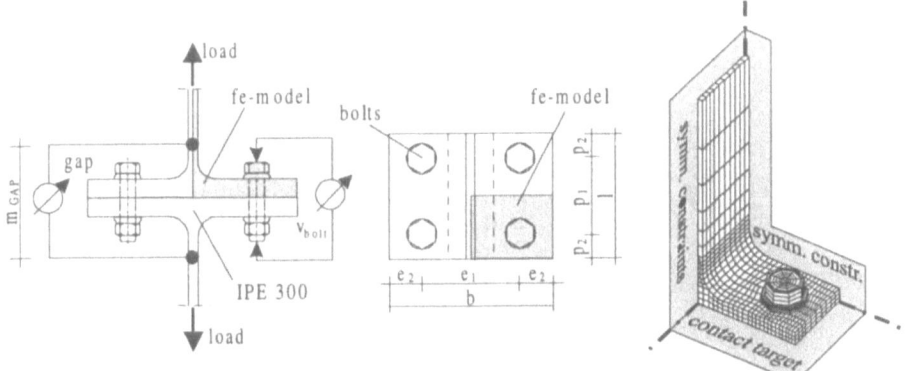

Figure 1: T-stub connection with characteristic measurements and denotation.
Marked area shows the part which is numerically modelled like on the left side.

flange because of symmetry). So the measures of the shank represent the geometrical stiffness of the bolt. It can be stated that the influence of the bolt stiffness on the T-stub deformation as well as on the bolt forces depends significantly on the relative strength of the bolt. In the Munich experiments, the bolts were very stiff compared with the flange stiffness. Therefore, the bolt stiffness nearly had no influence on the T-stub behaviour. In contrast to the Munich experiments, the bolts of the experiment from Bursi [3] were very weak compared to the flange. A bolt stiffness study (Figure 2) shows the little influence of the bolts. The curves 'Lbs1' refer to a bolt stiffness according to Agerskov [1] and the bolt stiffness of the model 'Lbs2' is only half of it.

2.5 DISCRETIZATION

The correct elementation and the selection of proper element types are, of course, main tasks in finite element modelling. The load-carrying behaviour of the flange or of the end plate, requires an element type that exhibits a good performance for bending dominated as well as plastic state problems. Because of the contact elements, the brick elements should not have midside nodes, i.e. 8-noded based brick elements will be applied here. In this work, the 8-noded 3d EAS-element (enhanced assumed strain) has been used. This finite element is based on a formulation of Simo, Amero and Taylor [8]. It has been implemented in an inhouse FE-code. In several element studies, the EAS-element has shown excellent capabilities even in the plastic regime.

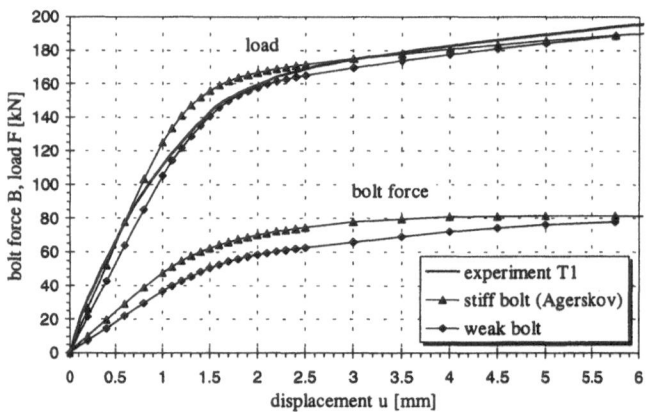

Figure 2: Study with two different bolt stiffnesses. Experiment T1 from [3].

2.5.1 Flange Elementation

Main attention should be directed towards the elementation of the flange (respectively the end plate), because its deformation is usually dominating. In addition the flange generates the most complex stress state with its bending in the plastic regime. Therefore, an element mesh as fine as necessary and as coarse as possible has to be developed. For this purpose, a discretization convergence study will be performed with respect to two parameters. The first parameter is the number of elements across the flange thickness in order to check the capability of representing the development of plastic hinges accurately. The second parameter is the element fineness in the flange (number of elements between web and hole and surroundings), in order to represent the strong bending problem of this part. The different elementations are labled with 'NxFy'. In the first part 'Nx', the number x is identical to the number of elements across the flange thickness, and 'Fy' represents the degree of discretization (Figure 3).

The deformation behaviour of the model with only one element across the thickness (N1F0) is obviously wrong (Figure 4). The model 'N2F0' is stiffer than the finer models 'N3F0' and N5F0' because the shear locking response overrules the plastic effects. If the discretization between web and bolt becomes finer the typical response of bending dominated problems in plasticity is shown. The deformation behaviour of the models with more elements across the thickness are stiffer because the development of the plastic hinges (in the center still elastic) and plastic zones is better represented (geometrical versus material stiffness). Therefore, for future studies the model with three elements across the flange thickness will be adopted. Next, a convergence study with respect to the elementation between web and bolt will be performed. The model 'N3F4' satisfies convergence requirements. The deviation from the model 'N3F3' is very small.

Remark. From the "weaker" response of the models with less elements across the flange thickness, the conclusion cannot be drawn, that the weaker plastic response can

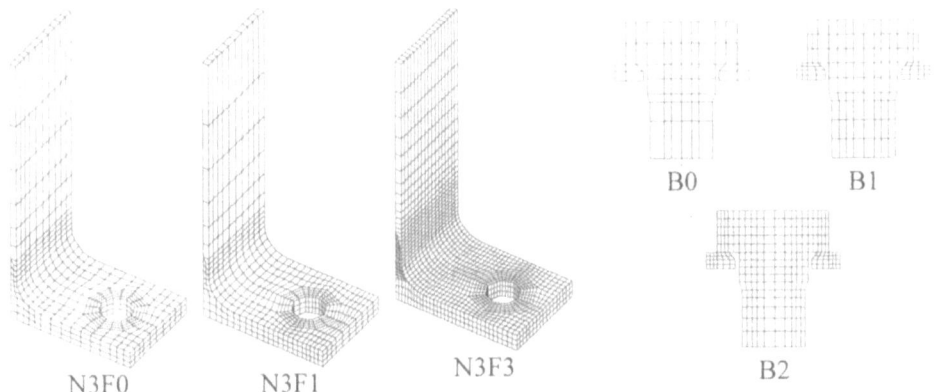

Figure 3: Discretization study. Flange elementations of the N3Fy-series. Three bolt elementation each with 8 elements around the circumferences.

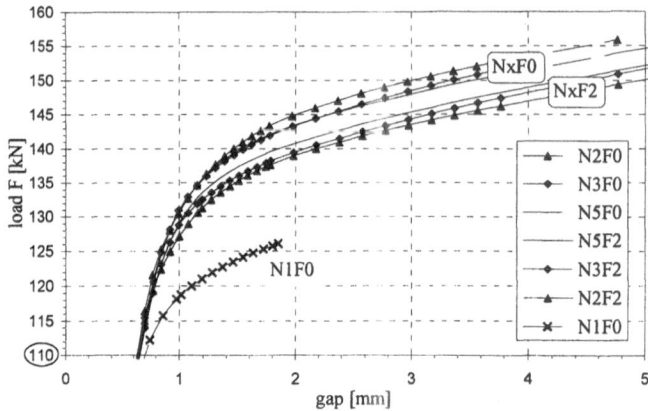

Figure 4: Discretization study. Number of Elements across flange thickness. P1K-experiment.

compensate the shear locking effect. The global deformation behaviour seems to be identical but the stress distribution is quite different. The structural analysis has to satisfy the unique physical reality, locally as well as globally.

2.5.2 Bolt Elementation

The number of elements are determined decisively by the discretization of the circumference of the bolt. For a bolt model with an area-equivalent square cross-section, only a few elements are needed. But the length of one side is less than the real diameter. In addition the square cross section leads to a direction depending model and the sharp edges produce stress concentrations. Therefore, the square bolt model will not be considered here. The load-deformation curves and bolt forces of T-stub models with 8, 12 and 16 elements in bolt circumference are nearly identical. Consequently, the bolt model with 8 elements will be used. The influence of the remaining elementation will be determined by three different degrees of discretization (Figure 3). Simulating both T-stub connections (Munich and Liege) the coarse bolt B0 overestimates the bolt forces. The bolt forces obtained by the models B1 and B2 are nearly identical.

2.6 MATERIAL DATA

The required material data are prescribed by the used material model in the finite element analysis. For all numerical calculations, a rate- and temperature independent plasticity law with hardening is applied because the studies should be carried out for an ideal static situation. That means, the material data, provided by an experiment under a certain loading velocity, overestimates the static situation. Therefore, the loading velocity was set to be zero from time to time in order to measure the real static response (Figure 5). This way we obtained the actual static stress-strain-relationship to be considered for the numerical material model. Only in this case, a valid comparison between numerical simulation and experiment is possible.

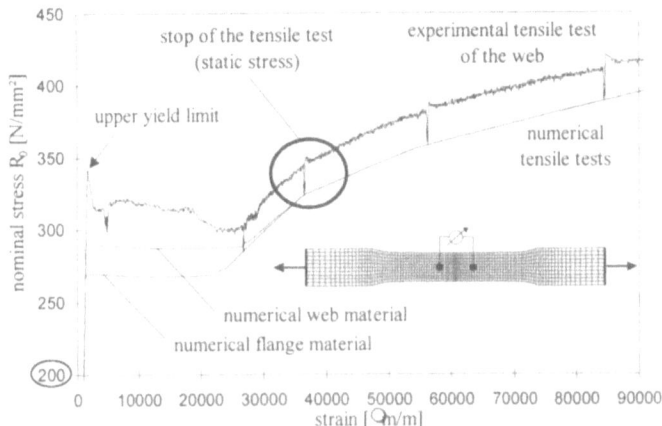

Figure 5: Comparison between the material of the experiment and of the numerical models by the tensile test. Munich experiments.

3 Benchmark Experiments

In the past numerous experiments with end plate connections were performed (e.g. Zoetemeijer [11], [12], [13], and Humer [5]), but they were not carried out as benchmarks for numerical simulation. For a reliable simulation the experiments have to provide much more than the nominal measured yield stress and a displacement-force relation. Therefore, we decided to carry out experiments in order to provide all necessary data that are required for finite element analyses. That means stress-strain relationship of the material, static as well as dynamic values, and various force-deformation-relationships. This chapter will give a brief description of these experiments. For detailed informations see Gebbeken, Wanzek and Petersen [4].

Six different types of symmetric T-stub connections with three experiments each were tested. The corresponding results of the three tests of one series deviate only a little. Because of the excellent coincidence of the deformation behaviour (gap and strain measurements) we can say that they are very reliable.

3.1 SPECIMEN

All T-stubs were made from one strand of an IPE 300 profile. One flange was cutted off at a height of *270 mm*. 24 pieces are of *l=105 mm* length and 12 pieces are of *l=210 mm* length. Types 1 and 3 are manufactured from the shorter pieces. They are distinguished by the bolt distances e_2 perpendicular to the web. Each geometric type was divided up in connections with prestressed and non-prestressed bolts.

The entire stress-strain relationship is required as input data for numerical analysis. For this sake, four standard tension test specimen are manufactured out of the steel member of the used profile. The whole testing procedure was recorded and the stress-strain curve was calculated from the strain gages and the elongation data. In order to compare different tests and for the sake of numerical simulation it was necessary to determine the static stresses. This is of importance because the stresses, especially in the plastic regime are rate-dependent. Static stresses are defined as stresses obtained during a stop of the deformation driven experiment. It was observed that the static stresses were reached after a period of approximately one minute. The total stop lasted for three minutes. It was already known that the value of stress reduction is nearly independent of the amount of strain (Petersen [6]; Scheer, Maier und Rohde [7]). Our own tests confirmed this. Consequently, it was easy to determine the static stress-strain curve (Figure 5). The stress reduction was approximately 20 N/mm^2.

All used high-strength bolts, M16, grade 10.9 according to DIN 6914, belonged to the same charge. With the help of an especially constructed testing device we determined characteristic bolt-elongation curves as well as the failure load. With the determined Young's modulus the bolt forces of the experiments can be calculated from the strain measurement at the bolts during the T-stub experiments. The bolt-elongation curve is necessary for the calibration of the bolts in a finite element model.

3.2 TESTING PROCEDURE AND MEASUREMENTS

After the specimens were totally assembled and the bolts were only tightened by hand and before the bolts were tightened to their initial force, the recording of all extensometers and strain gages started. One of the very important properties of an experiment is the static load. The static load ensures the comparison with other experiments and with numerical simulations, because it provides test speed independent results. In order to achieve the static load, the T-stub tests were interrupted temporarily similar to the static stress.

A characteristic and well defined deformation of the T-stub is the gap of the flanges measured in the centerline of the webs. This deformation can be compared with other test results which were carried out with other machines, because the gap is not affected by the elasticity of the machine, clamping conditions etc. In order to determine the bolt force, strains at the shank as well as bolt elongation were recorded. The strain measurement was the same as carried out for the bolt tests; two strain gages at opposite sides. The main interest was to compare measured strains with numerical strains obtained by the finite element simulation. This is a much more severe test than just comparing forces and deformations. So, strain gages were applied at the upper flanges of the specimen.

3.3 REMARKS

The experiments provide the necessary data and informations for finite element modelling in order to simulate these experiments. Again, we underline the importance of the determination of the static load and of the entire static stress-strain behaviour of the material. Only these informations make it possible to correctly perform numerical simulations refering to a rate-independent material law. It is self-evident, that detailed information about the measurements on the testing procedure has to be given as well.

4 Simulation of the T-stub experiments

With the experience of the numerical T-stub studies the Munich T-stub experiments [4] have been simulated by finite element analyses. The numerical material can be formulated identical to the experiment's material, because the static stress-strain behaviour of the rolled section and the bolt characteristic (force-elongation) have been determined. With this at hand, it is possible to perform a realistic detailed comparison. Besides the gap (displacement) strains and bolt forces were compared as well. Thus a valid statement about the reliability of the finite element models will be possible.

4.1 CHARACTERISTICS OF THE NUMERICAL T-STUB MODELS

The finite element analyses are performed with 3d 8-noded EAS-elements. Contact elements are generated between washer and flange with a friction coefficient $\mu = 0.5$ and between flange and symmetric ground without friction. The nodes in the interface of washer and bolt are coupled. The material of the section is calibrated with respect to the static stress-strain curve of the tension tests (Figure 5). The bolt material and the measure of the shank are calibrated with respect to the bolt elongation behaviour, which was determined in [4]. Of course the geometry is modelled with the measured data. Because of the hole clearance two simulations for each experiment will be carried out. Each simulation is generated with the extreme possible bolt position (see box in the corresponding diagram). A quadratic imperfection function represents the non-parallel flanges with an initial gap of 0.3 mm (i.e. because of symmetry 0.6 mm in reality). This initial gap was measured in the experiments. The bolt pretension at all PxK-experiments

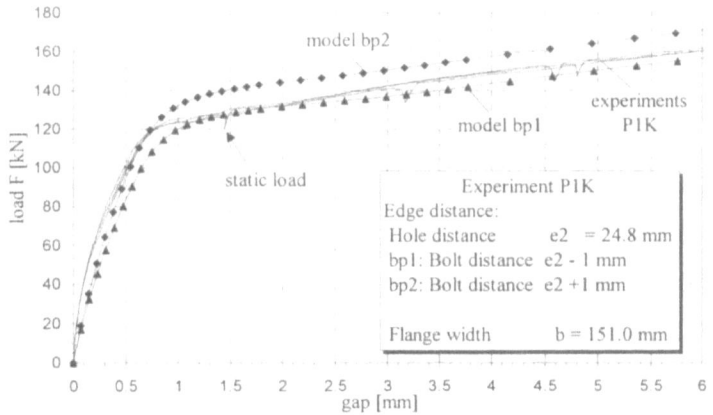

Figure 6: Global deformation behaviour. Gap-load curve of the Munich experiment P1K.

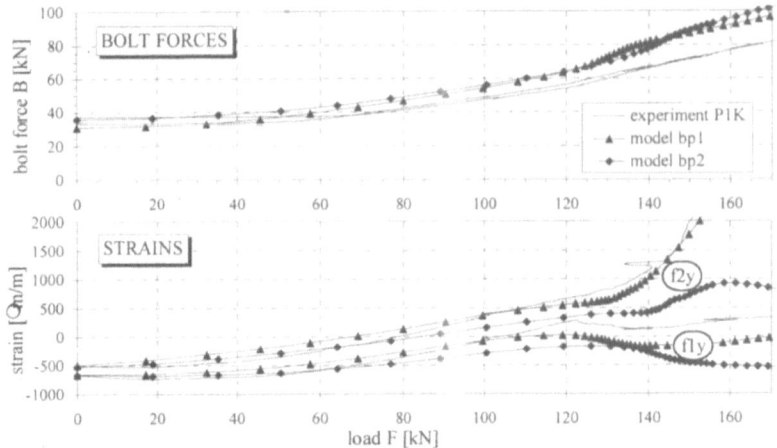

Figure 7: Bolt forces and strain behaviour. Munich experiment P1K.

is $B_0 \approx 30$ kN. The prestressing of the bolts is performed by constraints at the end of the bolt shank. The load is applied in a height of 160 mm at the web by constraints and these nodes are free during the prestressing of the bolt.

A detailed comparison between numerical simulation and experiment is performed at the Munich experiment P1K, for instance. The numerical results are compared with the experimental results in different ways. First the gap-load curve represents the global deformation behaviour. The gap is measured in a height of 40 mm at the web (Figure 1, $m_{GAP} = 80$ mm) and is set to be zero after the bolt is prestressed. Two numerical calculations with two extreme bolt positions have been carried out, because of the slighty different hole positions and the hole clearance. The significant influence of the bolt position is shown in the deformation behaviour (Figure 7). The difference of the bolt forces of these two numerical simulations (Figure 6) is not so big in contrast to the strains at certain positions, which behave quite different depending on the bolt position. Nevertheless, the qualitative and quantitative response of the numerical simulations with respect to deformation, forces, and strains correspond to that of the experiments. The numerical simulations of the other Munich experiments show the same correspondence with the experiments, e.g. the experiment P2K (Figure 8). The good correspondence with the experiment of Liège (Figure 2) is not a valid statement for the reliability of the numerical models used. Because in this case only the global deformation behaviour can be compared with the experiment. Unfortunately, some experimental information are missed.

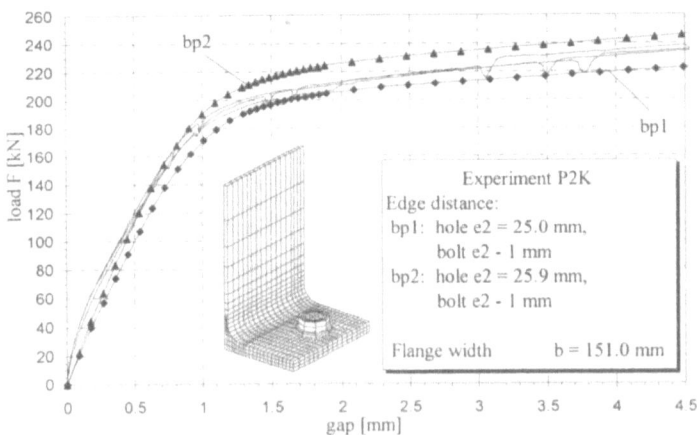

Figure 8:   Global deformation behaviour. Gap-load curve of the Munich experiment P2K.

5 Requirements for a reliable simulation

The experiences of the numerical t-stub studies will now be summarized in a list of requirements which are needed for reliable simulations. These requirements can be transferred to general bolted end plate connections. The statements about the discretization are only valid for element types similar to the EAS-element used here.

• Flange discretization between bolt and web:
  Elementation depends on bending in this part; mesh convergence study is necessary.

• Discretization in thickness direction:
  Two or three elements are minimum (depending on relative thickness, element type and integration points).

• Contact:
  Penetration controlling contact algorithm (e.g. Augmented Lagrangian); coupling of washer and bolt is possible but leads to a little stiffer deformation behaviour.

• Material data:
  Static material data for static analysis; real hardening data should be used; upper yield limit can be neglected.

• Possible imperfection has to be checked:
  Bolt position and possible hole clearance (significant influence on the T-stub); initial gap (important for detailed studies).

- Residual stresses:
  Neglectable in short pieces because the remaining stresses are small; even under full residual stresses the influence is small (e.g. column at a beam to column connection).

- Bolt:
  The connection's sensitivity with respect to the bolt stiffness should be esitmated by calculations with two extreme possible bolt characteristics.

Fulfilment of these requirements guarantees a reliable numerical investigation of end plate connections with the finite element method. Only the initial stiffness of the numerical simulations were weaker. But, usually only the very first part differs from the experimental stiffness as it is shown in the gap-load behaviour of the Munich experiments P1K and P2K (Figure 7 and 8).

6 Simulation of end plate connections

With the knowledge about the parameter that influences T-stubs end plate connections can be modelled. The requirements of the T-stub connections can be transferred to the analogous parts of the end plate connection, i.e. to the bolt region in the tension zone. In addition to these requirements the attention is focused on the consideration of the bending moment which is the load. There are two different ways to generate the moment at the end plate connection (Figure 9): a) stress distribution, b) shear force. In the first case (a) the pure moment is considered, which has the big advantage, that the contact between bolt shank and end plate or any constraints along the plate do not have to be considered. But, the distance between the stress initiation and the end plate has to be large enough, because the distribution in the beam profile near the end plate is quite different to what is usually assumed in the middle of a beam. In the other case (b), applying a shear load at the beam, the bolt shank contact has to be modelled. This needs special attention with respect to the contact condition between bolt shank and inner hole surface. Certainly, the consideration of the shear force leads to the more realistic modelling of the connection.

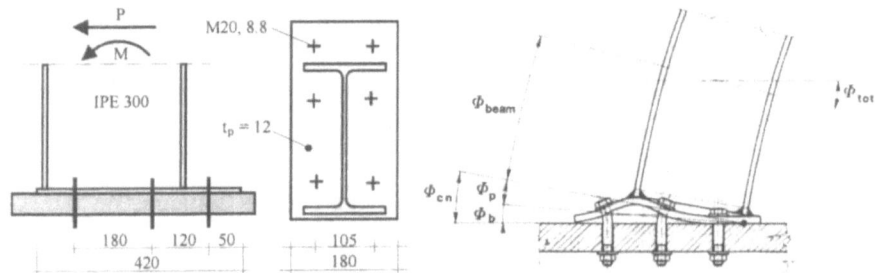

Figure 9: Experiment EP1-1 from Bernuzzi, Zandonini and Zanon [2]. Rotation definition.

But, it will be shown in an example, that the deviation in the moment-rotation behaviour of the connection between both cases does not occur if the shear stress is small compared to the normal stress. For this example we selected one of the end plate connection experiments from Bernuzzi, Zandonini and Zanon [2] (Figure 9). The discretization in the profile is coarse, because load transmission is the main task of the profile. The stress state is more complicated near the end plate. Therefore, in this part a finer discretization is necessary in axial direction as well as in thickness direction. The deformation behaviour of the end plate can be well represented with two elements in thickness direction. The deformation and the plastic zones of the finite element calculation are shown for an applied moment of 100 kNm (Figure 10). The moment of the numerical calculation 'ep-m' is applied by a moment stress distribution at the end of the beam, i.e. in a distance of 300 mm from the rigid base. In the second calculation 'ep-p' a shear force is applied at 1000 mm similar to the experiment. The moment-rotation behaviour is shown in relation to two different rotation definitions, the connection rotation Φ_{cn} and the total rotation Φ_{tot}. The total rotation, which was measured in a distance of 300 mm, relates to the kind of loading (Figure 10), because the stress state in the beam is different. On the other hand, the connection rotation Φ_{CN} as a sum of the bolt elongation and the end plate deformation is identical in both loading cases. This fact underlines that the connection rotation is the unique measurement analogously to the gap in the T-stubs.

7 Conclusions

The paper shows, that it is possible to predict a reliable behaviour of bolted end plate connection with finite element analyses, if the listed requirements are considered. Nevertheless, the final confirmation is given by the numerical simulations of

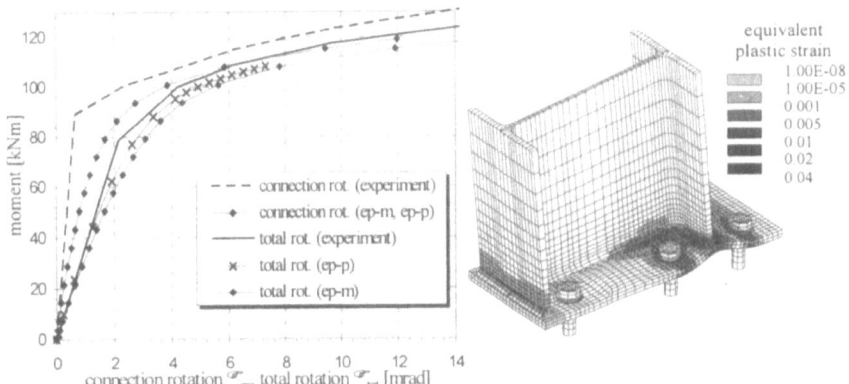

Figure 10: Finite element analyses. Moment-rotation behaviour of the experiment EP1-1. Deformation and plastic zones

experiments. These experiments have to be prepared and carried out with much more care, so that it will be really possible to perform an identical simulation of it. The comparison of experiment and finite element analyse has to include the behaviour of local zones (strain, bolt force) as well. Only if the numerical calculation can satisfy these criteria, the actual failure mechanisms of the investigated joints can be reliable simulated.

References

[1]   Agerskov, H. (1976) High-Strength Bolted Connections Subject to Prying, *Journal of Structural Division, ASCE* **102**, 161-175.

[2]   Bernuzzi, C., Zandonini, R. and Zanon, P. (1991) Rotational Behaviour of End Plate Connections, *Construzioni Metalliche* **2**, 74-103.

[3]   Bursi, O.S. (1995) A Refined Finite Element Model for Tee-stub Steel Connections. COST C1, Numerical Simulation Working Group, Doc. C1WD6/95-07.

[4]   Gebbeken, N., Wanzek, T. and Petersen, C. (1997) Semi-Rigid Connections, T-Stub Modelle – Versuchsbericht –, Report on Experimental Investigations. *Berichte aus dem Konstruktiven Ingenieurbau der Universität der Bundeswehr München.* **97/2**, ISSN 1431-1522 (*English and German*).

[5]   Humer, C. (1987) Das Momenten-Rotationsverhalten von steifenlosen Rahmenknoten mit Kopfplattenanschlüssen, Dissertation, Universität Innsbruck, Institut für Stahlbau und Holzbau.

[6]   Petersen, C. (1997) *Stahlbau*, Vieweg, Braunschweig Wiesbaden.

[7]   Scheer, J., Maier, W. and Rohde, M. (1987) Zur Qualitätssicherung mechanischer Eigenschaften von Baustahl, Bericht 6087/1, Technische Universität Braunschweig, Institut für Stahlbau.

[8]   Simo, J.C., Amero, F. and Taylor, R.L. (1993) Improved versions of assumed enhanced strain trilinear elements for 3D finite deformation problems, *Comp. Methods in Appl. Mech. and Engrg.* **110**, 359-386.

[9]   Wanzek, T. (1997) Zu Theorie, Numerik und Versuchen verformbarer Anschlußkonstruktionen, *Berichte aus dem Konstruktiven Ingenieurbau der Universität der Bundeswehr München.* **97/7**, ISSN 1431-1522.

[10] Wanzek, T. and Gebbeken, N. (1998) Versuche und numerische Berechnungen für geschraubte Kopfplattenanschlüsse, *Bauingenieur* **73**, 512-519.

[11] Zoetemeijer, P. (1974) A Design Method for the Tension Side of Statically Loaded Bolted Beam-to-Column Connection, *Heron* **20**, Technische Hogeschool Delft.

[12] Zoetemeijer, P. (1982) A Design Method for Bolted Beam-to-Column Connections with Flush End-Plates and Haunched Beams, *Stevin Report* **6-82-7**, Technische Hogeschool Delft.

[13] Zoetemeijer, P. and Munter, H. (1983) Proposal for the Standardization of Extended End-Plate Connections Based on Test Results, *Stevin Report* **6-83-23**, Technische Hogeschool Delft.

CYCLIC ANALYSIS OF STRUCTURAL JOINTS BY COMPUTATIONAL MODELS

L. DUNAI
Budapest University of Technology and Economics, Dept. of Steel Structures, H-1521 Budapest, P.O.B. 91, Hungary

S. ÁDÁNY
Budapest University of Technology and Economics, Dept. of Structural Mechanics, H-1521 Budapest, P.O.B. 91, Hungary

1. Introduction

The importance of the joint cyclic behaviour is well-known in the seismic design of dissipative steel framed structures. The local ductility of joints has significant effect on the global ductility of the structure. The local ductility of the joints can be characterised numerically by the cyclic parameters. These are the obtained from the cyclic moment – rotation relationship of the joint by standardised procedures.

Cyclic testing of joint can be done on the bases of ECCS [1] proposal and the cyclic parameters (rigidity, resistance, relative ductility and energy absorption ratios) can be determined from the test results.

Intensive experimental research work has been done on the cyclic behaviour of the joints in the last 10 – 15 years in different research institutes. Some of the most important studies are detailed in Refs. [2] - [8]. On the bases of test results empirical models are derived with the aim to extend the experimental results to more wider application in the seismic design. Some of the above mentioned models are referred in Refs. [9] - [12].

The other direction of the research activities is on the application of numerical approaches to study the joint cyclic behaviour. These numerical models are generally result in a complicated computational solution technique what is the major obstacle of the practical application. This is the reason that, despite several models are introduced for monotonic loading analysis (e.g. in Refs. [13] - [16]), only some of these models are extended and applied in the cyclic analysis of joints. In the recent years, however, the significant development in the computer technology provides with the opportunity to extend the application of the numerical models to perform parametric studies by virtual experiments.

The aim of the present paper is to give a general view about the several aspects of the cyclic modelling and analysis of structural joints. The paper concentrates mainly on the steel-to-steel bolted end-plate and steel-to-concrete base-plate type joint cyclic

C.C. Baniotopoulos and F. Wald (eds.), The Paramount Role of Joints into the Reliable Response of Structures, 293–306.
© 2000 *Kluwer Academic Publishers.*

models. Some of the modelling and computational features, however, can be extended and used for other type of joints, as well.

In the first part of the paper the requirements of the models, which can follow the cyclic loading history of a joint, are collected; the different finite element based numerical models are discussed; and the problems of non-linear solution technologies are detailed. In the second part a set of benchmark examples is proposed to verify and calibrate the models.

2. Requirements of Cyclic Joint Models

The basic requirement of the joint model is to simulate the cyclic performance on the bases of the joint geometry and material properties. To fulfil this requirement the model should have the following features:

- The model should consider the complex 2D or 3D geometry of the joint (surface or solid geometry model).
- The load model should follow the cyclic loading history.
- The material model should take into account the cyclic behaviour of the steel and concrete.
- The model should consider the local buckling of the slender plate elements under load reversal.
- The conditional connections between the joint components should be modelled under cyclic effects (contact – separation – re-contact).

The above-mentioned requirements result in a highly non-linear mathematical model. For the solution, complicated computational method should be used.

The application of numerical models calls the attention of the key question of the numerical structural analysis. This is the balance between the accuracy and efficiency. The cyclic joint model should be accurate enough to be able to use the obtained results to derive the joint cyclic characteristics for practical purposes. It can be called as „engineering accuracy". The accuracy of the model should be verified by a systematic checking, using analytical and experimental references. In this paper a multi-level testing procedure of cyclic joint models is proposed by the application of benchmark examples.

On the other hand the model should be applicable for virtual experiments on a realistic hardware/software background with realistic processing conditions. The efficiency of the model is very much dependent on the applied non-linear solution techniques. In the following Sections these topics will also be discussed.

3. Finite Element Models

3.1 FEATURES OF FINITE ELEMENT MODELS

To fulfil the above requirements of the cyclic joint model recently the only applicable solution method is the finite element method. The models which are developed by this solution technique are the finite element models.

The common features of the finite element models which can be used for cyclic joint analysis are as follows:

• Geometry model: 2D surface models (plane stress/strain, plate models) or 3D shell or solid models.
• Material non-linearity: cyclic plasticity material model.
• Geometric non-linearity: cyclic plate buckling.
• Boundary non-linearity: contact – separation – re-contact phenomena.

In the following sections some important characteristics of the finite element models are summarised.

3.2 2D FINITE ELEMENT MODELS

In these models the 3D geometry of the joint is transferred to 2D surface. The simplest 2D models concentrate on the plate bending phenomenon of a plate component of the joint. In this case the mid-surface of the plate component is modelled by plate bending finite elements. The model is used for the analysis of separated joint component (e.g. end-plate). In cyclic analysis the pure plate bending models are rarely used since the behaviour is more complex in general.

The plane stress or plane strain 2D models are used to model a plate – bolt sub-assembly, in which the 3D behaviour is reduced to 2D phenomenon. Typical applications of these models are for bolted joint components, such as T-stubs or bolted angles. Despite these models eliminate the transverse bending of the segments, the dominant behaviour can be accurately simulated. Cyclic 2D T-stub and bolted angle models are introduced and presented by the authors.

3.3 3D FINITE ELEMENT MODELS

The 3D finite element models have two major groups: shell models and solid models.

The 3D joint geometry is modelled by the mid-surface in the shell models, by shell finite elements. The shell elements are connected to each other in different planes according to the joint's geometry. The bolts are modelled as supporting springs or equivalent other finite elements (e.g. bars). The boundary non-linearity can be considered on the contact surfaces by different methods (e.g. gap elements or uni-lateral material models). In the plates the spread of the plasticity should be followed through the thickness by the application of layered shell element. The 3D shell models are used by several researchers to model joint components and complex joint, mainly for monotonic loading. The authors developed a shell model for cyclic joint analysis. The application of this model is illustrated for cyclic plate buckling and cyclic T-stub analysis in this paper as benchmark examples.

The most correct geometry model of the joint is the 3D solid model. In this case solid finite elements are used to model all parts of the joints. Very advanced models are developed and verified for monotonic loading analysis of joint components and complex joints. These studies determined the main requirements of the solid element meshing to achieve the required accuracy.

3.4 CYCLIC PLASTICITY MATERIAL MODELS

It is well-known from the experimental studies that the cyclic behaviour of steel material can not be followed by the pure elastic – plastic, isotropic or kinematic hardening models. It is experienced from the tests, that such combined isotropic and kinematic hardening models should be used which can consider both the changes of the size of the yield surface and the translation in the stress space. Basically two types of models are developed which can follow this requirement:

- Two-surface models (Dafalias – Popov type models [17]): in these models the actual and a limiting surface is defined and the movement of the actual surface is controlled by a hardening function.
- Multi-surface models (Popov – Petersson type models [18]): these models are based on a Mroz multi-surface model. The combined hardening phenomenon is controlled between two extreme stages of the material behaviour: virgin or initial stage and saturated of fully-developed stage.

Common feature of these models, that the model properties can not be derived directly from the experimental data. The properties are dependent on the applied model, they can be defined from combined testing and numerical calibrating.

In this paper the cyclic properties derived by Popov – Petersson type model are illustrated as benchmark examples.

3.5 PROBLEMS OF FINITE ELEMENT ANALYSIS

It can be summarised that in the case of finite element model based cyclic joint analysis the following problems to be solved:

- Uncertainties of input data – derivation of cyclic material properties.
- Large number of displacement degrees of freedom due to the 2D or 3D modelling.
- Combined and huge non-linearities to be considered.
- Long cyclic loading history to be followed.

These conditions result in large hardware and software background. It is underlined, that even very advanced finite element codes can not fulfil the requirements of the cyclic analysis. Beside the modelling features the applied computational solution technique has major importance on the efficiency of the analysis.

4. Non-linear Computational Methods

4.1 FEATURES OF NON-LINEAR COMPUTATION

In the non-linear solution technique the classical methods (e.g. Newton-Raphson method) can not be used purely. To follow the highly non-linear and long loading history the computational technique should have the following features:

- Iteration strategy: in the incremental-iteration technique such a method to be used, which can consider the descending branch of the load – displacement response and also the snap-back phenomenon, as it is illustrated on Fig. 1. Such methods are e.g.

the constant arc-length procedures or the Minimum Residual Displacement method (it is illustrated in Fig. 1 and used by the authors with highest efficiency).

- Convergence / divergence criterion: it is essential to determine the suitable criteria when the given equilibrium position is accepted as converged state or need to modify due to divergence. These criteria have major effect on the efficiency.
- Load step generation: the whole non-linear analysis should be automatically controlled. It requires the generation of the size of the load increments according to the level of non-linearities. The correct size of the increment reduces the required number of iterations until the converged state is reached.
- Back step definition: in interaction with the divergence criterion it is essential to define how to continue the analysis. It means that if divergence is found the load should be reduced to go back to a previous equilibrium state. From this converged reference state the solution can be continued by modified load increment.

In general it can be summarised, that the above features are in strong interaction with each other. General rules, however, can not be determined.

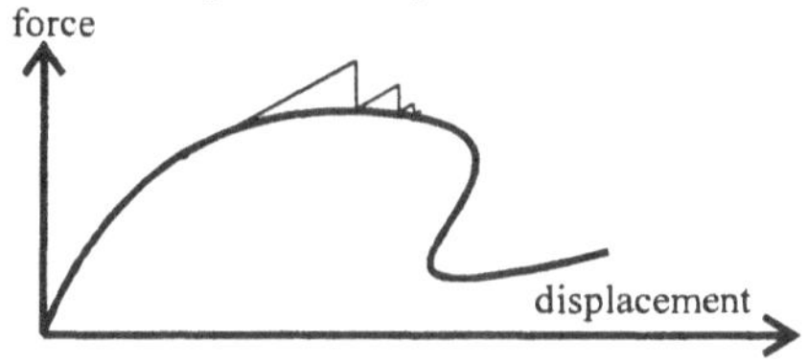

Figure 1. Illustration for the Minimum Residual Displacement iteration method

4.2 SOLUTION CONTROL PARAMETERS

The features of the non-linear solution should be expressed by numerical parameters which can be adjusted by numerical test to obtain the best efficiency of the analysis. The most required solution control parameters are as follows:

- Optimal number of iteration.
- Convergence limit.
- Divergence limit.
- Load parameter limits.
- Maximal number of iterations.
- Back-step parameter.

The background and details of the above parameters can be found in [19].

5. Testing the Cyclic Analysis

The cyclic analysis of joints should be „tested" in general sense, from two main aspects:

- Accuracy of the results.
- Efficiency of the computation.

It is evident that this testing can not be done in a simple step. It is proposed to complete multi-level testing by the application of benchmark examples. The different level of the examples:

298

1. material,
2. component (plate element, bolt),
3. sub-assembly (bolted plate element),
4. joint.

In the following Sections a set of benchmark examples is proposed on this logic. In this paper the examples are mainly illustrated; all details can be found in [19].

6. Cyclic Benchmark Examples – Material

6.1 STEEL MATERIAL

References:
- experimental cyclic tests [18]
- multi-surface model analyses [20]

Input:
- virgin and saturated (fully cycled) experimental stress-strain curves
- multi-surface model properties (Fig. 2)
- cyclic loading histories

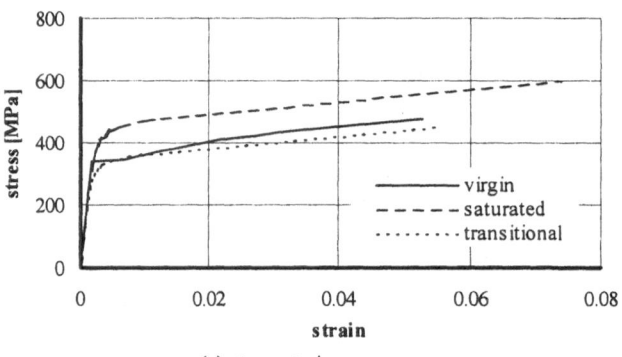

(a) stress-strain curves

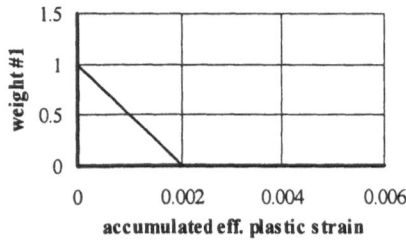 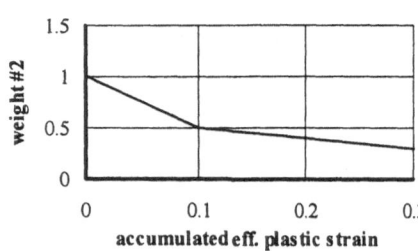

(b) weighting functions

Figure 2. Input data for the Petersson-Popov multi-surface model

Results:

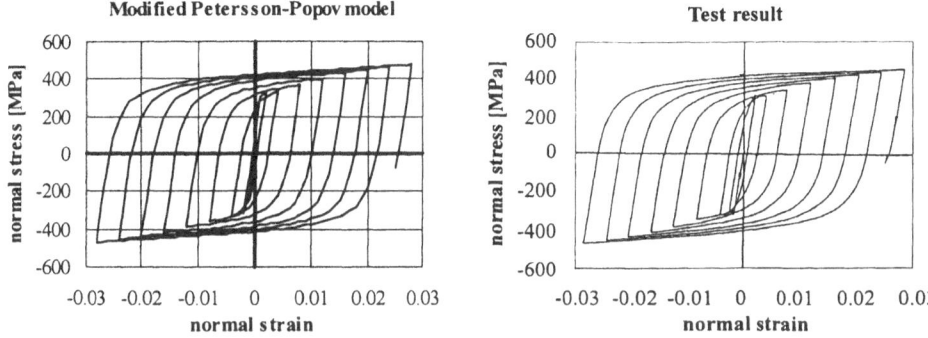

Figure 3. Experimental and numerical stress-strain curves (load history #1)

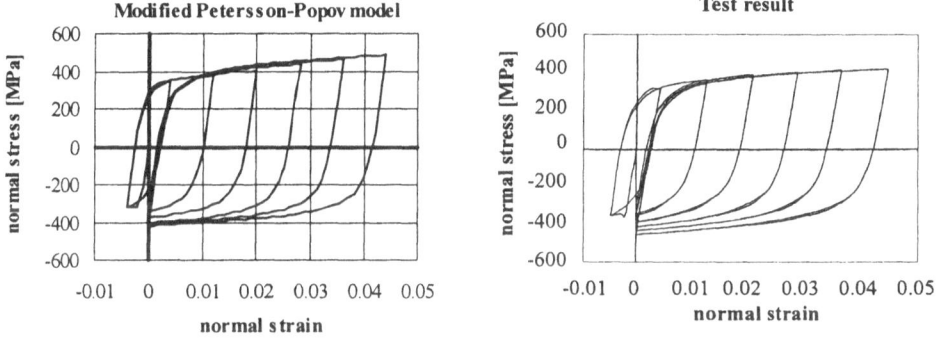

Figure 4. Experimental and numerical stress-strain curves (load history #2)

6.2 CONCRETE MATERIAL

References:
- experimental cyclic tests of authors
- uni-axial model [21]

Input:
- uni-axial stress-strain curves under cyclic compression
- uni-axial model properties (Fig. 5)
- cyclic loading histories

300

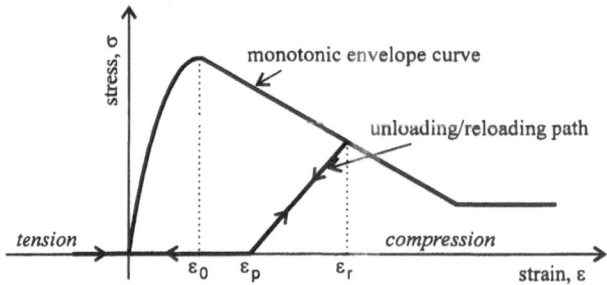

Figure 5. Uni-axial cyclic material model for concrete

Results:

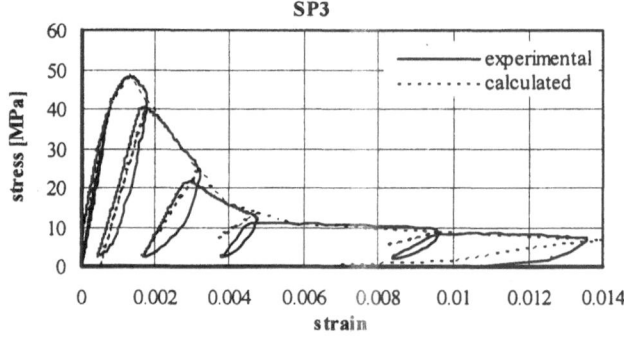

Figure 6. Experimental and numerical stress-strain curves for concrete

7. Cyclic Benchmark Examples – Component

7.1 PLATE ELEMENT

References:
- virtual cyclic tests [19], [22]

Input:
- geometric and material properties (Fig. 7)
- cyclic loading histories (Fig. 7)
- cyclic model properties [20]

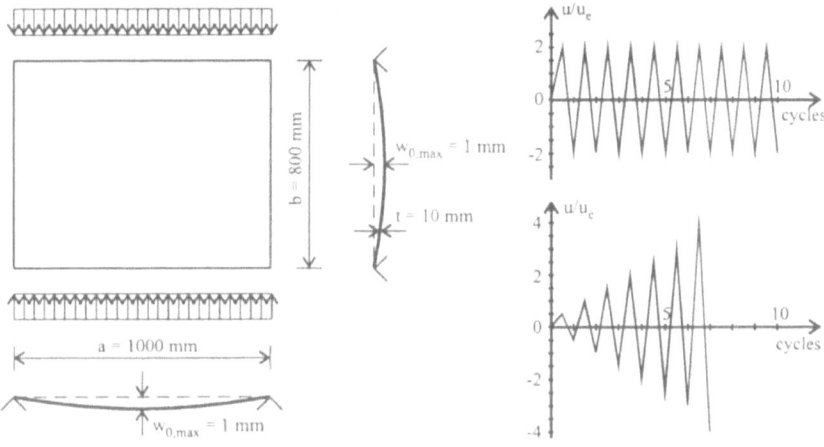

Figure 7. Input data for cyclic plate buckling example

Results:

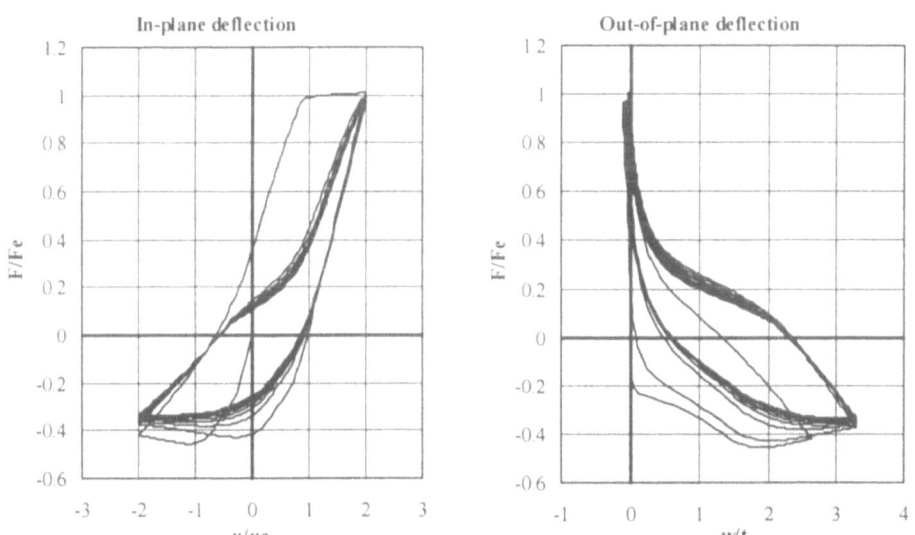

Figure 8. Load-deflection curves for loading history #1

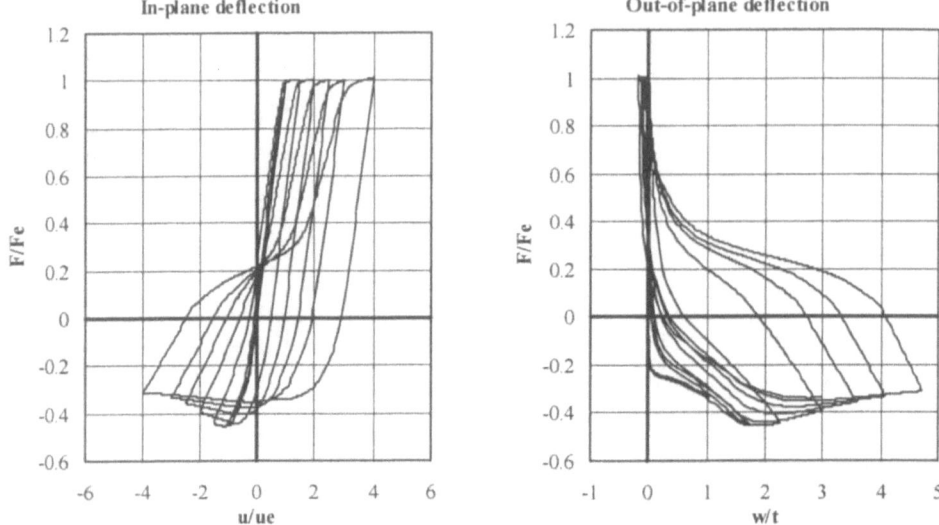

Figure 9. Load-deflection curves for loading history #2

7.2 ANCHORING ELEMENT

References:
- only a few – mainly monotonic – test result is available [19]

Input:
- geometric and material properties (Fig. 10)

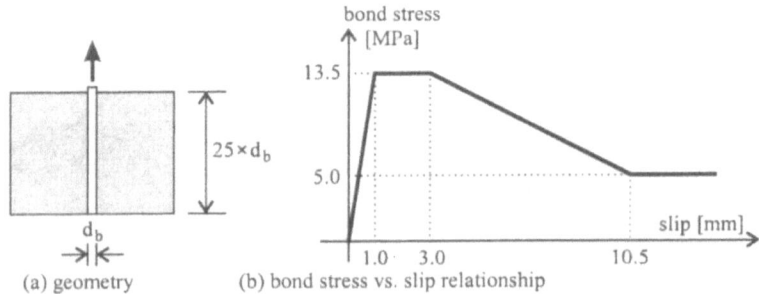

Figure 10. Input data for anchor bar example

Results:

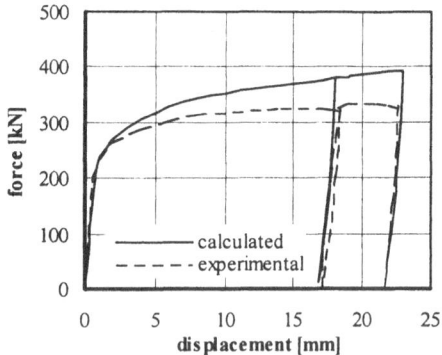

Figure 11. Experimental and numerical force-deflection curves for anchor bar

8. Cyclic Benchmark Examples – Sub-assemblage

8.1 T-STUB ELEMENT

References:
- virtual cyclic tests [23]
- experimental cyclic tests

Input:
- geometric and material properties (Fig. 12)
- cyclic loading history (Fig. 12)
- cyclic model properties [20]

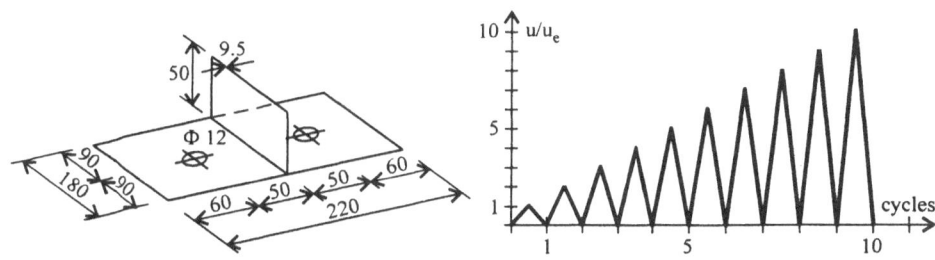

Figure 12. Input data for T-stub example

304

Results:

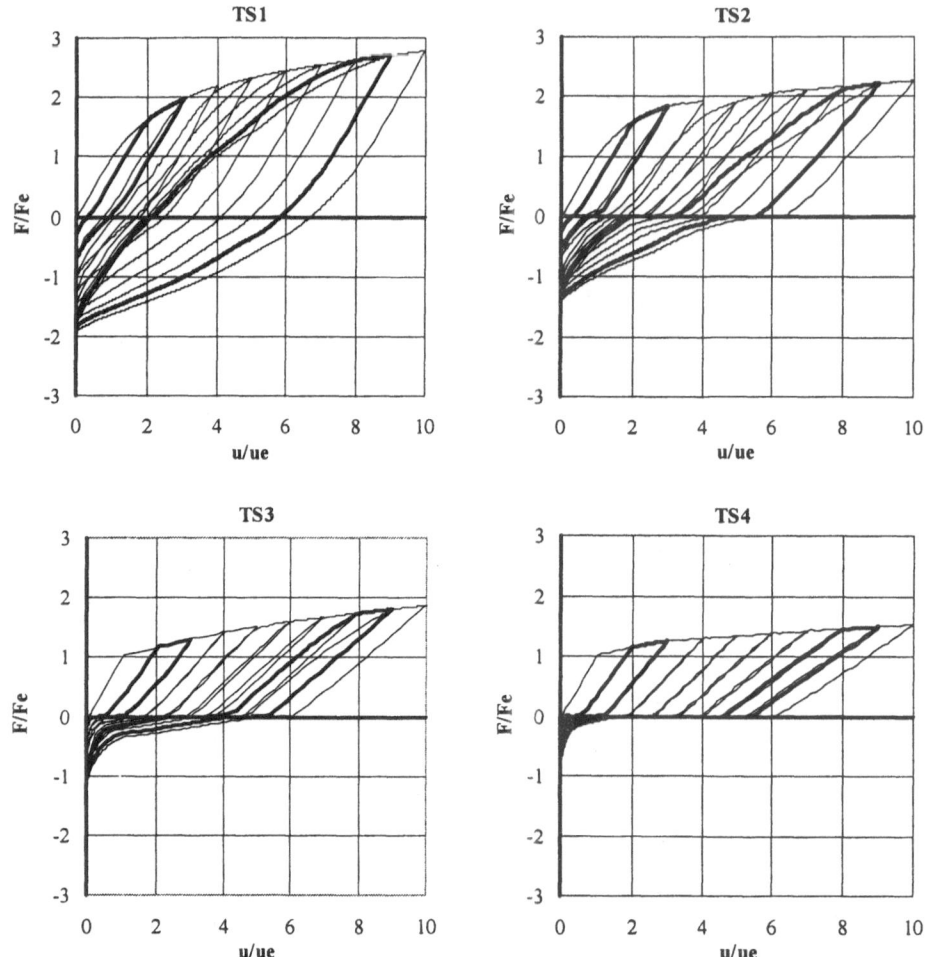

Figure 13. Load-displacement curves of T-stubs with various parameters

9. Cyclic Benchmark Examples – Joint

9.1 STEEL JOINT

References:
• experimental cyclic tests [19]
Input:
• geometric and material properties
• cyclic loading history

Results:

The referred experimental results are well documented and suitable for computer simulation; so far however, only monotonic virtual joint tests are completed.

9.2 STEEL – CONCRETE JOINT

References:
- experimental cyclic tests [6]

Input:
- geometric and material properties
- cyclic loading history

Results:

The referred experimental results are well documented and suitable for computer simulation; so far however, only monotonic virtual joint tests are completed.

10. Concluding Remarks

In the paper the computational modelling of joint under cyclic loading is discussed. It is intended to give a general view about this subject. The modelling features are discussed with the aim to obtain accurate and efficient solution for the problem. The basic requirements of the applicable models are given; these result in a highly nonlinear problem to be solved. Accordingly, the nonlinear solution techniques have major importance. The features of efficient solution methods are detailed.

In the paper there is a main focus on the verification of the applicable models. It is concluded that the models can not be tested in a single step; instead, a multi-step verification is proposed. The authors have developed a set of benchmark examples which can be used in a step-by-step checking procedure of the cyclic models and computational methods. The benchmark examples are not completed entirely. There is a need for further cyclic benchmark examples which can be done by experimental and virtual testing. It is also proposed to extend the advanced 3D solid monotonic models for cyclic testing of joints, and provide with further benchmark examples on this basis.

In the paper an important behaviour component is not discussed, it is the low-cycle fatigue. In the further step of the model development this phenomenon should be considered.

11. References

1. ECCS (1986) Recommended testing procedure for assessing the behaviour of structural steel elements under cyclic loads, *Technical Comittee 1, TWG 1.3 – Seismic Design*, No. 45.

2. Bernuzzi, C. (1992) Cyclic responses of semi-rigid steel joints, *Proceedings of the 1st COST Workshop*, Strasbourg, 194-209.

3. Astaneh, A., Bergsma, G. and Shen, J. H. (1992) Behavior and design of base plates for gravity, wind and seismic loads, *Proceedings of National Steel Construction Conference*, American Institute of Steel Constructions.

4.   Azizinamini, A. and Radziminski, J.B. (1989) Static and cyclic performance of semi-rigid steel beam-to-column connections, *J. of Struct. Eng, ASCE*, **115, 12**, 2979-2999.

5.   Chasten, C.P., Fleischman R.B., Driscoll G.C. and Lu, L-W (1989) Tope-and-seat-angle connections and end-plate connections: behaviour and strength under monotonic and cyclic loading, *Proceedings, National Steel Construction Conference, AISC*, 6-1 – 6-32.

6.   Dunai, L., Fukumoto, Y. and Ohtani, Y. (1996) Behaviour of steel-to-concrete connections under combined axial force and cyclic bending, *Journal of Constructional Steel Research*, **36, 2**, 121-147.

7.   Tsai, K.C., Shun, W. and Popov, E. P. (1995) Experimental performance of seismic steel beam-column moment joints, *J. of Struct. Eng, ASCE*, **121, 6**, 925-931.

8.   Mazzolani, F.M. (1988) Mathematical model for semi-rigid joints under cyclic loading, *Connections in Steel Structures*, Chen, W. F. ed., Elsevier, 112-120.

9.   De Stefano, M. and De Luca (1992) Mechanical models for semi-rigid connections, Constructional Steel Design, Dowling, P. et al. eds., Elsevier, 276-279.

10. Dunai, L. (1992) Modelling of cyclic behaviour of steel semi-rigid connections, *Proceedings of the 1st COST Workshop*, Strasbourg, 394-405.

11. Flejou, J.L. and Colson A. (1992) A general model for cyclic response of structural civil engineering connections, *Proceedings of the 1st COST Workshop*, Strasbourg, 419-430.

12. Bernuzzi, C. (1998) Prediction of the behaviour of top-and-seat cleated steel beam-to-column connections under cyclic reversal loading, *J. of Earthquake Eng*, **2, 1**, 25-58.

13. Bursi, O. (1992) Behaviour and modelling of semi-rigid beam-to-column joints, *SPRINT Contract RA351*

14. Gebbeken, N., Rothert, H. and Binder, B. (1994) On the numerical analysis of end-plate connections, *J. of Constructional Steel Research*, **30, 2**, 177-196.

15. Nemati, B.N. and Le Houedec (1996) The analysis and the 3D finite element simulation of steel bolted end-plate connections, *Advances in Computational Techniques for Structural Engineering,* Topping B.H.V. ed., Civil-Comp Press, Edinburgh, 179-188.

16. Dunai, L., Ádány, S., Wald, F. and Sokol, Z. (1996) Numerical modelling of column-base connections, *Advances in Computational Techniques for Structural Engineering,* Topping B.H.V. ed., Civil-Comp Press, Edinburgh, 171-178.

17. Dafalias, Y. F. and Popov, E. P. (1975) A model of nonlinear hardening materials for complex loading, *Acta Mechanica*, **21**, 173-192.

18. Popov, E. P. and Petersson, H. (1978) Cyclic metal plasticity: experiments and theory, *Journal of Engineering Mechanics Division, ASCE*, **104, 6**, 1371-1388.

19. Ádány, S. (2000) Numerical and experimental analysis of bolted end-plate joints under monotonic and cyclic loading, *PhD Thesis*, Supervisor: L. Dunai, Budapest University of Technology and Economics

20. Ádány, S. and Dunai, L. (1997) A modified multi-surface model for structural steel under cyclic loading, *Proceedings, 5th Int. Coll. on the Stability and Ductility of Steel Struct*, Nagoya, Usami, T. ed, 841-846.

21. Karsan, I. D. and Jirsa, J. O. (1969) Behaviour of concrete under compressive loading, *Journal of the Structural Division, ASCE*, **95, 2**.

22. Yao, T. and Nikolov, P.I. (1992) Numerical experiments on buckling/plastic collapse behaviour of plates under cyclic loading, *Stability and Ductility of Steel Structures under Cyclic Loading*, Y. Fukumoto, Y. Lee, G. eds., CRC Press, 203-214.

23. Ádány, S. and Dunai, L. (1998) Cyclic analysis of steel bolted components, *Journal of Constructional Steel Research*, **46**, **1-3**, 422-423 (full paper on CD-ROM).

A CONTRIBUTION TO THE FINITE ELEMENT STUDY OF PIN CONNECTIONS IN STEEL STRUCTURES

J. MADEMLIS, S. MARNOUTSIDIS & A. AVDELAS
Aristotle University, Department of Civil Engineering
GR-54006, Thessaloniki, Greece

1. Introduction

The subject of this paper is the study of pin connections in steel structures. The study of their response to loading has been carried out by the use of the finite element method. The unilateral contact between the connected members has been taken into account. The correctness of our assumptions has been verified by applying the basic concepts of the suggested simulation to a similar problem for which both computational and experimental data exist [1]. The linear response of the connected plates, under static loads, has been examined keeping under consideration the dimensions of the plates proposed by EC3 [2].

Because of their limited construction demands, pin connections are often used in steel structures. A typical example of their application is given in Figure 1. Their main advantage is the safety they provide to the engineer regarding to the prediction of their behavior, since they do not sustain moments. This leads, during the analysis of the structure, to the easy simulation of the boundary conditions of the nodes. In spite of their advantages, the theoretical and experimental study of the behavior of pin connections is not as extensive as it would be expected. A similar study, but in a different context can be found in [3].

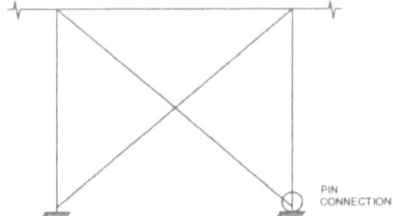

In Figure 2, the typical form of a pin connection is presented. Between the diameter of the hole d_0 and the diameter of the pin d, relation $d_0 > d$ is always valid (Figure 3).

C.C. Baniotopoulos and F. Wald (eds.), The Paramount Role of Joints into the Reliable Response of Structures, 307–316.
© 2000 *Kluwer Academic*

308

This gives to the connection the following characteristics: The contact between the pin and the connected plates is unilateral and not obstructed, the system can be freely rotated around the longitudinal axis of the pin and, finally, shear is the critical form of failure.

2. Finite Element Analysis

A two-dimensional analysis has been used for the study of the system. The graphic simulation has been carried out in the SAP2000 finite element program [4] with the aid of AutoCad R14. The unilateral contact of the pin-plate and the loading of the plates have been studied taking into account appropriate assumptions in order to simulate real conditions. It must be pointed out that the nonlinearities introduced by this unilateral contact phenomena have as a result that the contact areas between the pin and the boundary of the hole are not a priori known. The Linear Complementarity Problem that is formulated in each solution step[5], has been solved by the use of a Quadratic Programming algorithm [6].

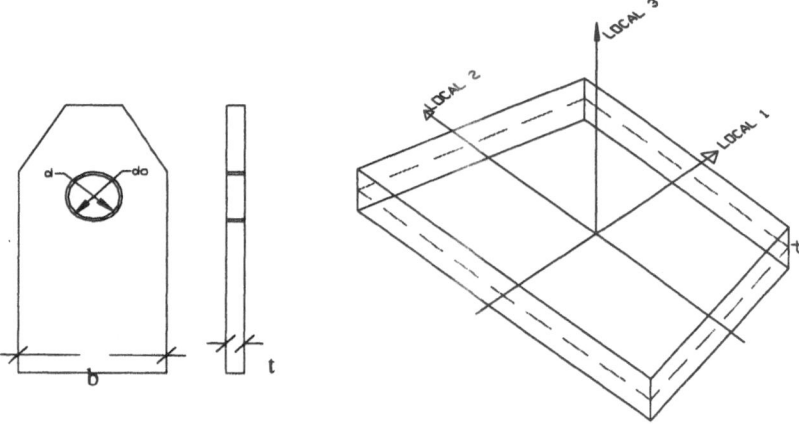

Figure 3. Thickness t and length b Figure 4. Local coordinate system of a typical shell element

The reliable study of the problem under consideration was feasible in the two-dimensional space due to the geometry and the loading conditions of the connections. The low value of the ratio thickness t over length b (Figure 3) and the symmetry of the plates contributed to the simplification of the problem [7]. Moreover, if the self-weight of the plate is not taken into account, the response of the connection under static loading can be properly described under plane stress-strain conditions. At the advantages of the two-dimensional analysis, the facilitation of the graphic simulation, the advanced supervision of the model and the limitation of the time consumed during the solving phase and the evaluation of the results can also be included.

Shell elements have been used for the analysis of the model. The shape and the local coordinate system of a typical shell element are presented in Figure 4. The z-axis has been assigned to the thickness t of the plate under consideration. The basic shape of the model has been designed using AutoCad R14 and the corresponding *.DXF file has been created. Since SAP2000 is compatible with drawing files of *.DXF format, the basic shape of the model has been easily imported. After that, the messing of the elements took place and especially at the area near to the hole.

One of the main parameters, that designate the results of the analysis, is the unilateral contact between the pin and the plates. The geometry of the contact area was determined by taking into account two basic simplifications: a) Friction has been ignored. Thus, only radial interaction between pin and plate was considered. b) Because of the crucial role of the above load condition to the response of the connection, the study has been carried out by using different values of the tension load in the Y direction.

The basic shape of the model described above has been used after further node condensing around the contour of the hole (7.5° step) [8]. Spring supports have been assigned to each of the nodes of the contour in the radial direction and equivalent uniform tension loads were enforced at the down part of the plate. The tension or compression of the spring supports determined whether there is contact or not respectively. The analysis has shown that the contact area is divided symmetrically by the Y-axis, forming an angle of 150°.

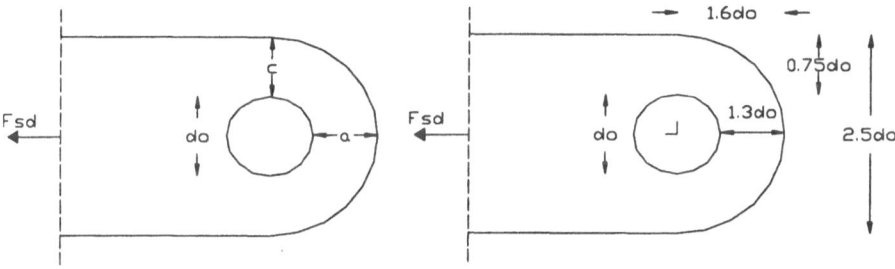

Figure 5. Geometric conditions for plates in pin connections [2]

The forces that correspond to each of the elements of the connection are transmitted from the pin to the plates and vice versa (depending on which plate is examined) through the contact area. For the study of the plates, the load has been applied to the shells included in the contact area in the form of an equivalent uniform load. The direction of the load vector coincides with the one of the local axis 2. By rotating the local axis of each shell at the radial direction, compatibility of the adopted load enforcement with the hypothesis of radial interaction between the pin and the plates has been obtained.

3. Parametric Analysis

The purpose of the parametric analysis was the examination of the behavior of several cases of pin connections. The geometric characteristics of the plates and the qualitative characteristics of the materials have been used to differentiate the different cases. The geometric characteristics of the plates have been defined according to the relevant rules of EC3.

In Table 6.5.6 of EC3 [2], the following relations are given for pin connections (Figure 5):

Type A: Constant thickness t

$$a \geq \frac{F_{sd}\gamma_{Mp}}{2tf_y} + \frac{2d_0}{3} \; : \; c \geq \frac{F_{sd}\gamma_{Mp}}{2tf_y} + \frac{d_0}{3} \tag{1}$$

Type B: Constant geometry

$$t \geq 0.7\sqrt{\frac{F_{sd}\gamma_{Mp}}{f_y}} \; : \; d_0 \leq 2.5t \tag{2}$$

Obviously, the above relations are satisfied for many values of the parameters used (a, c, d, t, F_{sd}). In the following, the critical cases will be examined in order to give an overall qualitative view. This will done in accordance with the following steps:

- The values of diameters d and d_0, that will be studied, have been specified.
- For each diameter d_0, the minimum value of the thickness t has been determined, according to the second of relations (2).
- The limit load F_{cr}, which corresponds to a plate of given thickness t, has been determined in each case from the first of relations (2).
- Cases where the pin moment M_{sd} {$f(t, F_{sd})$} is greater than the design resistance M_{rd} have been rejected.
- For the above values of d_0 and t and for $F_{sd}=F_{cr}$ the limit values of lengths a and c have been calculated from relations (1).
- For the plates, steel Fe360 and Fe510 have been used. The material of the pin has been taken as 8.8 and 10.9.

In all cases, the comparison of the yield stress f_y to the maximum Von Mises stresses has been used as a failure criterion.

4. Results

First, the above cases have been solved for the loads proposed by EC3 (F_{ec3}). Next, the limit load (F_{sap}) that every plate can resist, depending to its geometric characteristics, has been calculated. In the cases where $F_{fem}<F_{ec3}$, the modifications imposed to the geometry of the plates have been defined by the shape of the failure mode.

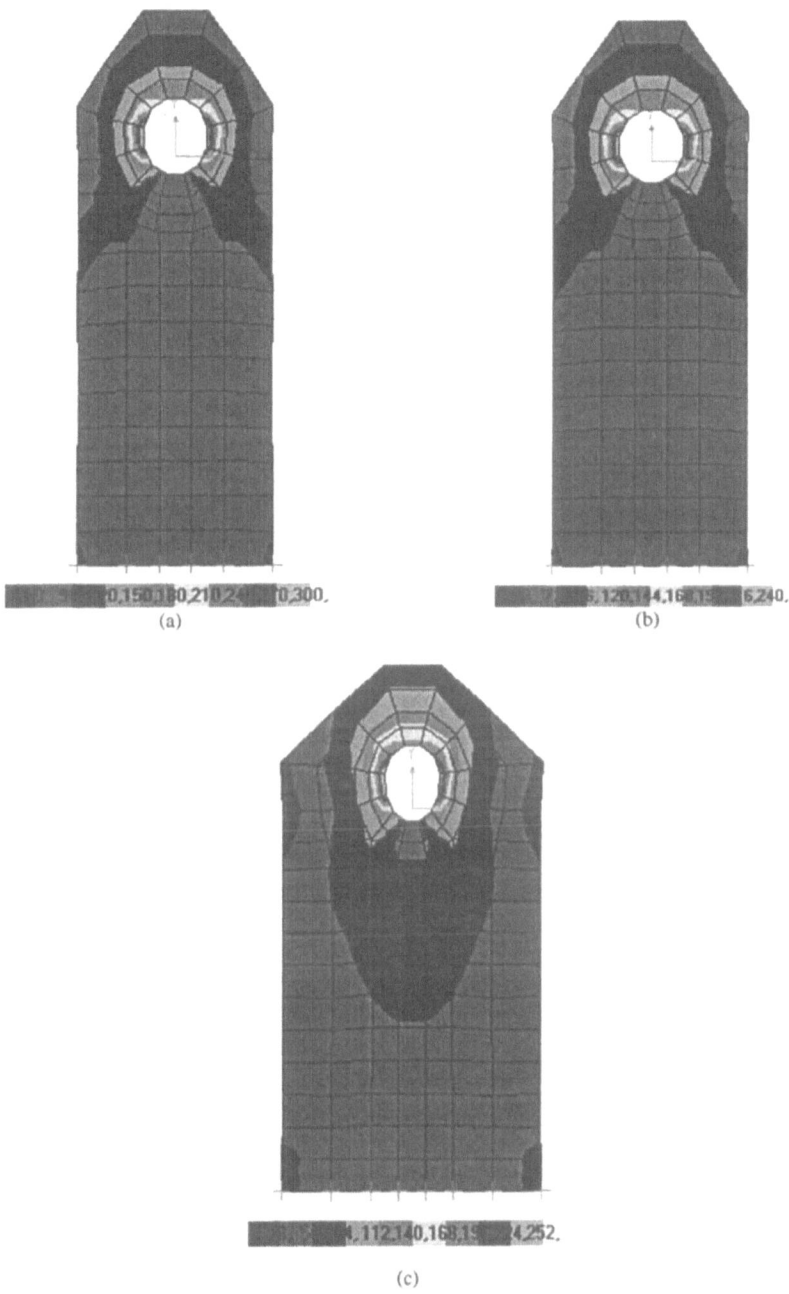

Figure 6. Stress conditions: Fe360, d_0 = 33 mm, t = 22 mm

312

Under the given loads, the shape of the hole becomes oval and the maximum stresses appear at the 0° and 180° areas – tension failure of plate [9, 10]. The parameter defining the values of the stresses at these areas, is the length c. In the case of failure, length c is modified in such a way that the plate is able to withstand the load proposed by EC3. In Figure 6 the stress conditions of the deformed shape for a typical case are presented: (a) for the loads and geometric characteristics given by EC3, (b) for the failure load according to the simulation and the geometric characteristics given by EC3 and (c) for the loading according to EC3 and for the length c modified in such a way that the plate does not fail.

In accordance with EC3, length c is given by the second of relations (1) [2]. If this relation is combined with the first of relations (2), then

$$c = t + \frac{d_0}{3}$$ (3)

is obtained.

The calculation results, derived from the study of the pin connection, show that the above relation for c is not valid in some critical cases. For this reason, a modification of the relation is proposed. It must be noted that the criterion for the modification is the diameter of the hole. In the cases where EC3 is valid, the above modification contributes to the improvement of the safety of the structure. This modification takes the form of a factor k with the values given in the table of Figure 8. This factor would multiply the second part of relation (3) that takes the form

$$c = k\left[t + \frac{d_0}{3}\right]$$ (4)

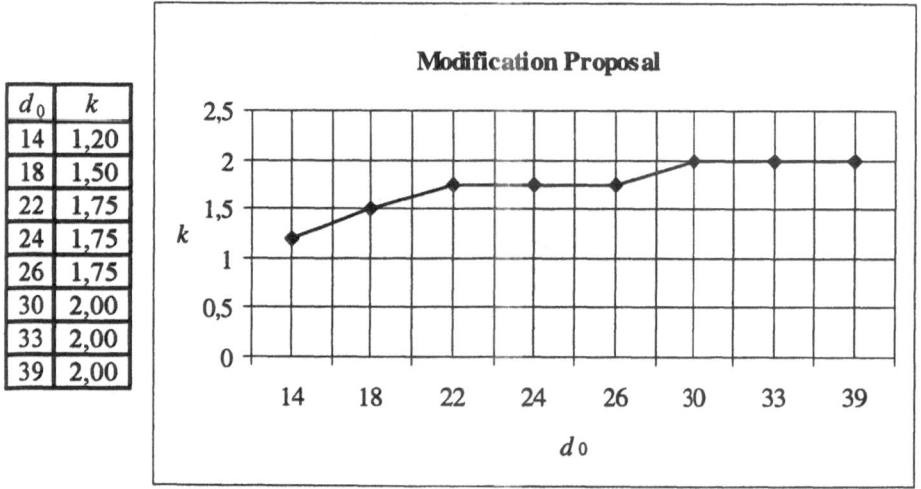

d_0	k
14	1,20
18	1,50
22	1,75
24	1,75
26	1,75
30	2,00
33	2,00
39	2,00

Figure 7. Values of the proposed factor k

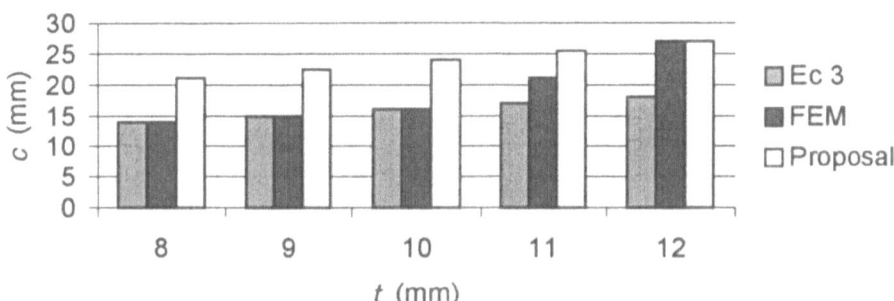

d_0=18 mm (Fe360)

Figure 8

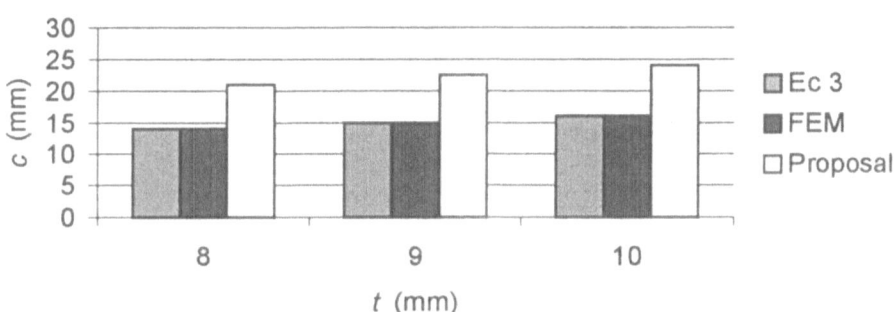

d_0=18 mm (Fe510)

Figure 9

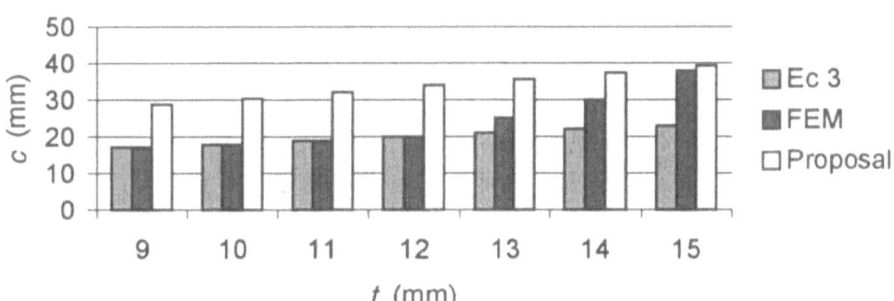

d_0=22 mm (Fe360)

Figure 10

314

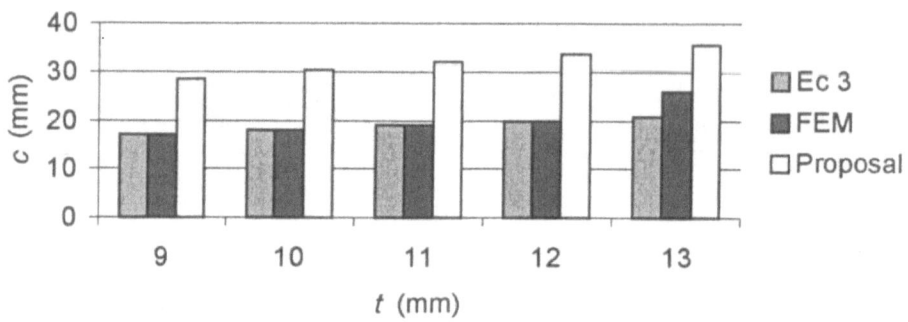

Figure 11

d_0=33 mm (Fe360)

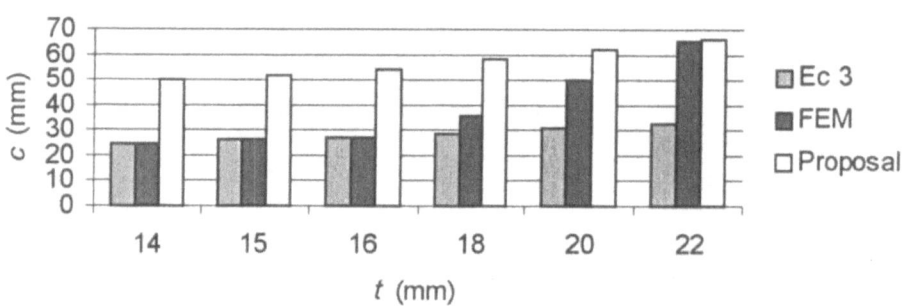

Figure 12

d_o=33 mm (Fe510)

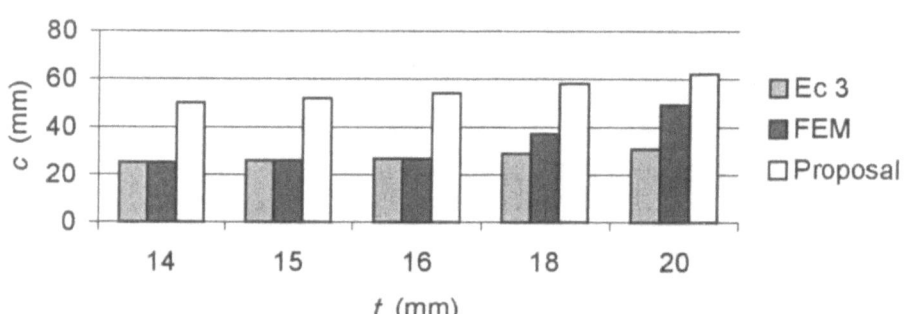

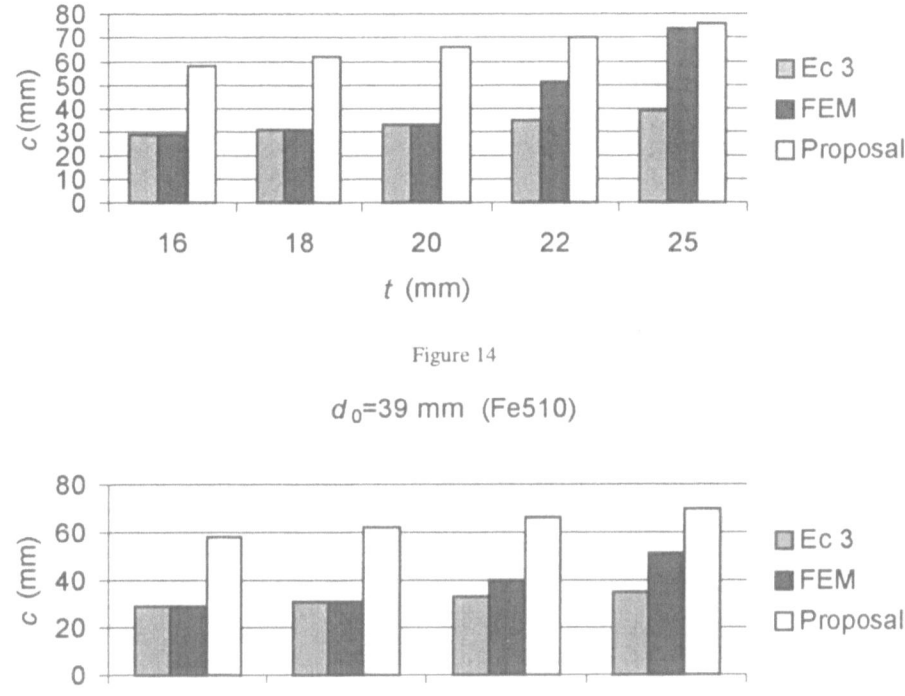

Figure 14

d_0=39 mm (Fe510)

It must be noted that the above factors are valid for both the 8.8 and 10.9 material qualities of the pin, with the only difference that some of the cases are excluded for the first one, because of the smaller strength of the pin material.

Some indicative results of the procedure developed in order to achieve the aims of this paper are presented above. Furthermore, in order to ensure the qualitative and quantitative comprehension of the results, diagrams of the implicated parameters are also provided.

5. References

1. Pasternak, H. and Komann, S. (1999) Shear behaviour of rosette joints –Tests and finite element analysis, in Studnička, J., Wald, F. and Macháček J. (eds.), *CD of the Proceedings of the 2nd European Conference on Steel Structures*, Czech Technical University in Prague, paper Nr 077.

2. ENV1993-1-1 (1992) Eurocode 3: Design of Steel Structures - Part 1.1: General Rules and Rules for Buildings, CEN, Brussels.
3. Efraimides, S.A., Sakellariadou, H.I., Bisbos, C.D. and Panagiotopoulos, P.D. (1996) On the computation of the stress state in the plate of a steel pin connection taking into account the unilateral contact between the plate and the pin, in D.A. Sotiropoulos and D.E. Beskos (eds.), *Proceedings of 2nd National Congress on Computational Mechanics*, Technical University of Crete, Vol. 1, pp. 454-461.
4. Sap 2000 NonLinear (1997) Manuals Version 6.1, Computers and Structures Inc.
5. Panagiotopoulos, P.D. (1985) *Inequality Problems in Mechanics and Applications. Convex and Nonconvex Energy Functions*, Birkhäuser Verlag: Basel, Boston.
6. Avdelas, A.V. (1986) The unilateral contact problem. A treatment by the use of the Keller direct method, in *Proceedings of 1st National Congress on Mechanics*, T.E.E., Athens, Vol. 1, pp. 207-234.
7. Spyrakos, C.C. (1994) *Finite Element Modeling in Engineering Practice*, Algor Inc., Pittsburg.
8. H.M. Ramadan, S.A. Mourad, A.A. Rashed and H. Bode (1998) Finite element modeling and experimental testing of single shear bolted joints, in B.H.V Topping (ed.), *Advances in Civil and Structural Engineering Computing for Practice*, Civil-Comp Press, Edinburgh , pp. 117-124.
9. Stark, J., Bijlaard, F. and Sedlacek, G. (1988) Bolted and welded connections in steel structures, in Bjorhovde, R., Brozzetti, J. and Colson, A. (eds.), *Connections in Steel Structures*, Elsevier, London, pp. 8-17.
10. Chung, K.F. and Ip, K.H. (1999) Finite element modeling of cold-formed steel bolted connections, in Studnička, J., Wald, F. and Macháček J. (eds.), *CD of the Proceedings of the 2nd European Conference on Steel Structures*, Czech Technical University in Prague, paper Nr 232.

RECENT ADVANCES IN THE NUMERICAL MODELLING OF THE RESPONSE OF STRUCTURAL STEELWORK TAKING INTO ACCOUNT SOFTENING IN CONNECTION BEHAVIOUR

M. J. KONTOLEON & C.C. BANIOTOPOULOS
Aristotle University, Department of Civil Engineering
GR-54006, Thessaloniki, Greece

1. Introduction

The influence of joints to the safe response of steel structures is of primary importance. The severe damages these structures sustain during seismic events, makes even more demanding the need to re-evaluate the design and construction of pinned, semi-rigid and rigid connections and to develop reliable techniques of repairing and seismically upgrading them. We may mention the using of polymer composites, high-strength adhesives, high-strength nuts and bolts, and steel stiffeners to repair welded steel frame connections that have been fractured. In this framework, new numerical methods have been recently developed taking into account the exact mechanical characteristics and material properties, fulfilling the task to calculate accurate solutions in complex, real life, engineering problems.

In the present paper, the resistance of steel structures using stress-displacement curves with softening branches corresponding to the partial damage of the connections is investigated. The ultimate strength of adhesively repaired connections, as well as, the bearing strength of bolted joints in flanges are studied, employing nonmonotone, multivalued laws with one or more vertical descending branches. This type of constitutive laws macroscopically describe the mechanical behavior of the connection and take into account phenomena such as local damage, crushing, slipping and debonding that deteriorate the strength of the material [1]. Macroscopic laws provide information on the brittle crack initiation and whether the failure is progressive or abrupt. Therefore, the use of such laws seems advantageous from an engineering point of view, as they avoid the complexity micromechnanical laws introduce. Due to the lack of monotonicity, such constitutive laws lead to a nonconvex energy problem and give rise to a new type of variational forms, called hemivariational inequalities that express the principle of virtual or complementary virtual work in inequality form (cf. e.g. [2]).

The theory of hemivariational inequalities leads to the result that local minima of the potential or the complementary energy of the structure represent equilibrium

C.C. Baniotopoulos and F. Wald (eds.), The Paramount Role of Joints into the Reliable Response of Structures, 317–326.
© 2000 *Kluwer Academic Publishers.*

positions of the structure. However, certain solutions of the problem may not be local minima, but more general types of points that render the energy 'substationary'. An analogue substationarity problem can be formulated in terms of the stresses and corresponds to the complementary energy of the structure. Therefore, the substationarity points, i.e. all the local minima, certain local maximal or saddle points constitute all the possible solutions of a hemivariational inequality [3].

In the present paper, for the numerical calculation of the steel connections under investigation, the discrete problem is first formulated leading to an optimization problem with respect to the energy function. An algorithm, based on the NSOLIB proximal bundle algorithm is then applied to the optimization of the nonconvex nonsmooth energy superpotentials [4]. The algorithm takes into account material nonlinearities in the case of an elastic-plastic analysis, finding a solution that renders the energy substationary.

The proposed method is illustrated by means of two numerical examples combing the principles of Nonsmooth Mechanics with the Finite Element Method (FEM). In the first example a steel frame is studied for the case of an elastic analysis. The second one performs an elastic-plastic analysis, calculating the ultimate yield bearing strength a bolted beam-flange connection can sustain.

2. Formulation of the Discrete Problem

We consider a steel connection that consists of solid structural elements. We make the following assumptions: a) the elements obeys to an elastic or elastoplastic law, b) certain parts of the structure may be bonded by means of an adhesive material that after reaching its maximum strength presents softening phenomena and c) interface interaction is taken into account where active, non active and sliding regions are predefined. The static behavior of the structure within a small strains and displacements framework is described by the following relations:

1. Equations of equilibrium:
$$\mathbf{G}\,\mathbf{s} = \mathbf{P} \tag{1}$$

2. Strain-displacement relations:
$$\mathbf{e} = \mathbf{G}^\mathrm{T}\mathbf{u} \tag{2}$$

3. Material law:
$$\mathbf{s} \in \partial q(e) \tag{3}$$

4. Macroscopic strain softening law:
$$\mathbf{s} \in \overline{\partial}\,j(e) \tag{4}$$

where we denote by $\mathbf{G}$ the equilibrium matrix, $\mathbf{s}$, $\mathbf{e}$, $\mathbf{u}$ and $\mathbf{P}$ the stress, strain, displacements and load vectors respectively, q is a convex function that describes the elastic or elastoplastic material behavior, j is a nonconvex nonsmooth energy function that describes the nonmonotone possibly multivalued macroscopic strain softening constitutive law, ∂ is the subdifferential of convex analysis and $\bar{\partial}$ is the generalized gradient [5].

Relation (4) for the structural member i is by definition equivalent to the inequality:

$$j(e_i, e_{i-} - e_i^*) \geq s_i\,(e_i - e_i^*), \quad i = 1, \ldots, n \tag{5}$$

where n denotes the number of elements exhibiting a macroscopical softening strain behavior law. Inequality (5) is called hemivariational inequality and generalizes the well known variational inequalities of convex analysis for the case on nonmonotonicity.

Combining the principle of virtual work for the discrete structure with relations (1)-(5) it has been proven that we may render the following substationarity problem that reads [6]:

Find $u \in V$ such as the potential energy of the structure expressed by the formula

$$\Pi(u) = q(e_i) + \sum_{j=1}^{n} j(e_j) - \mathbf{p}^T \mathbf{u} \tag{6}$$

becomes substationary.

The nonconvexity introduced by relation (4) renders the whole potential function Π nonconvex. The obtained substationarity points constitute all the possible solutions of problem (6). Although the determination of the full set of solutions of a substationarity problem remains as yet, an open problem, the complexity it implies is not always present in structural applications. This is due to the fact that loading is often imposed along a predetermined loading path. Therefore, solutions caused by external actions belonging to this path are most probably the actual solutions of the considered mechanical problem [7].

3. Application of an Iterative Elastoplastic Nonconvex Optimization Algorithm

In order to calculate the solutions of the potential energy Π for an elastic analysis, the substationarity problem (6) has to be solved whereas for the case of an elastic–plastic problem an algorithm that appropriately updates the tangential stiffness matrix must be applied.

In the present paper, we propose an iterative algorithm that converges within each iteration (i) to a solution of a field of displacements that physically corresponds to an equilibrium position of the considered structure. We provide the algorithm for the initial step, with the potential energy function, that consists of the initial tangential stiffness matrix $\mathbf{K}^{(0)}$ and the load vector $\mathbf{P}^{(0)}$, and the data that describe the elastoplastic law by curve $C^{ep}(u)$ (Fig. 1). The algorithm in each iteration calls the NSOLIB proximal bundle nonsmooth optimization procedure and checks using the information from curve $C^{ep}(u)$ the area in which the obtained solution $\mathbf{u}^T_{(i)}$ belongs and then updates the tangential stiffness matrix. The algorithm terminates as soon as the norm difference between two subsequent iterations is smaller than a given tolerance $\varepsilon < 10^{-5}$. The following steps describe the algorithm:

Algorithm

Step 1: Set the initial displacement field $\mathbf{u}^0$ equal to zero.

Step 2: Call the nonconvex, nonsmooth optimization routine NSOLIB and compute a displacement field $u^{(i)}$ that renders the potential energy substationary (6). If the optimization procedure does not converge, then end the algorithm.

Step 3: Update the new tangential stiffness matrix $K^{(i)}$ of the structure for the displacement field obtained in the previous step using the slope information of curve $C^{ep}(\mathbf{u})$.

Step 4: Calculate the norms N_{i-1}, N_i of the matrices $K^{(i-1)}$, $K^{(i)}$. If the absolute difference between the two norms $N_i - N_{i-1}$ is smaller than a given tolerance ε, then end the algorithm. If not, set $i = i+1$ and return to step 2.

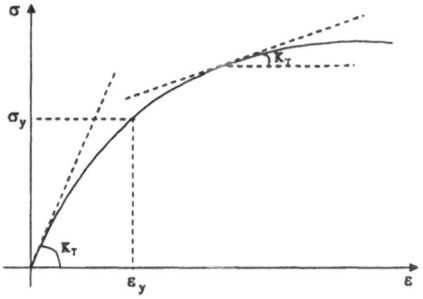

Figure 1. Elastic-plastic diagram.

4. Numerical Examples

4.1 STEEL FRAME

The frame shown in Fig. 2 is studied. The structure consists of two square tube columns 200x200x6.3, 3 meters high, two base plates 200x200x10 glued on top of each column and a IPBv 200 beam with length L=5m. The steel frame is modeled by means of three-dimensional finite elements. Specifically, the discretization of the plates and the beam was performed using 4-node plate elements, whereas the columns were discretized using 2-node Bernoulli beam elements. The discretization of the base plates with quadrilateral plane elements permits the study of delamination phenomena that appear in the area where the adhesive is applied. The material is considered linear elastic with modulus of elasticity E= $2.1 \times 10^8 N/mm^2$ and Poisson ratio v=0.3. Between the base plates and the lower flange of the beam that comes in contact with the plates, an adhesive material bonds them together. A plane projection of the three dimensional law of the adhesive material and the respective superpotential is depicted in Fig. 3.

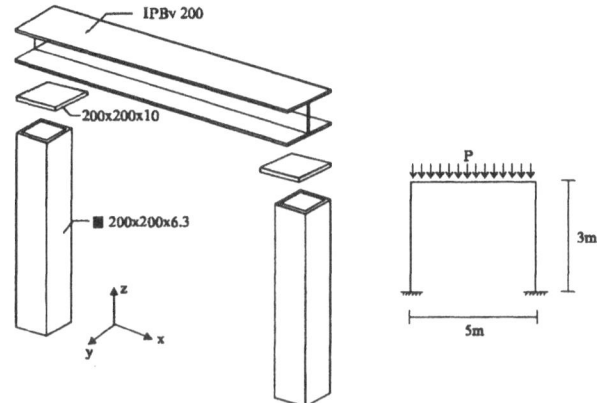

Figure 2. Steel frame subjected to loading.

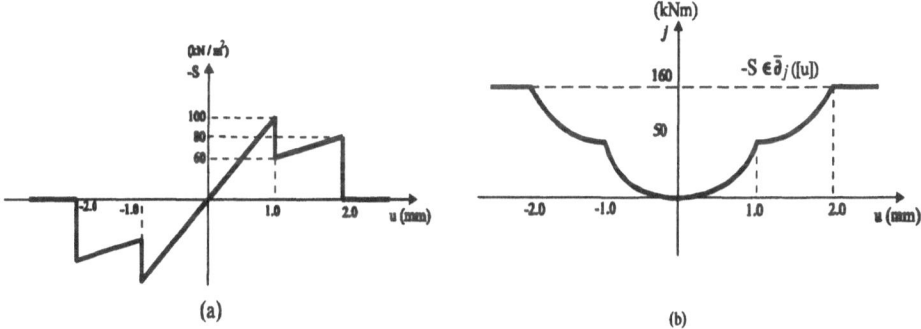

(a)

(b)

Figure 3. (a) Nonmonotone adhesive material law, (b) Plane projection of the 3D-superpotential.

The nonmonotone law S(u) (Fig. 3a) contains complete vertical jumps and has the form:

$$S(u) = \begin{cases} 100[u] & 0 <[u]<1 \\ [60,100] & [u] = 1 \\ 20[u]+40 & 1 <[u]<2 \\ [0,80] & [u]=2 \\ 0 & [u]>2 \end{cases} \tag{7}$$

The respective superpotential j(u) (Fig. 3b) has the form:

$$j(u) = \begin{cases} 1/2100[u]^2 & 0 < [u] < 1 \\ 50 & [u] = 1 \\ 1/220[u]^2 + 40[u] + 40 & 1 < [u] < 2 \\ 160 & [u] \geq 2 \end{cases} \tag{8}$$

The direction of loading on the structure is shown in Fig. 2 where the following loads: P=7000, 10000, 15000, 20000 kN/m² are applied. The relative displacements that develop in the adhesive are depicted in Fig. 4. As the vertical loading increases, the adhesive material fails after the first two loading cases when it exceeds its maximum displacement capacity of 2mm. For the remaining two loading cases, the debonding of the interface nodes can be observed. The adhesive fails for nodes 1,2,3 in the left area of the base plate and nodes 13, 14, 15 in the right area.

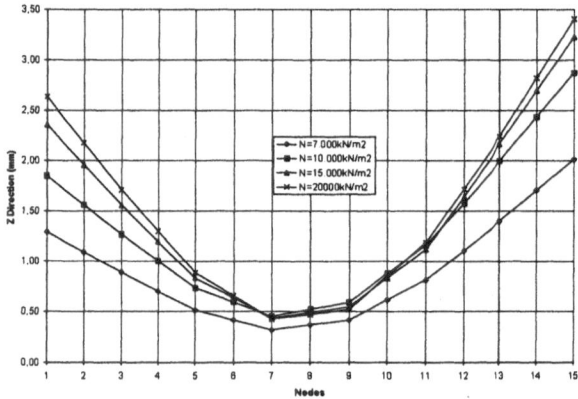

Figure 4. Relative displacements along the nodes of the interface in the normal direction (Z axis).

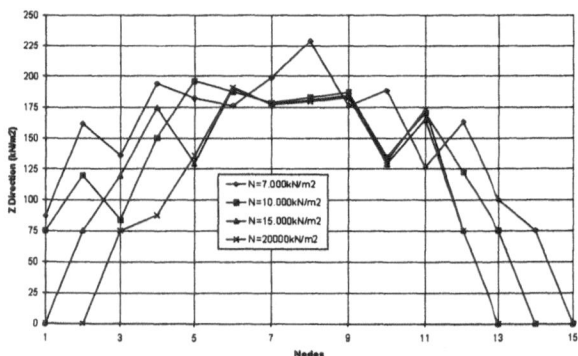

Figure 5. Adhesive stresses along the nodes of the interface in the normal direction (Z axis).

The softening behavior of the adhesive along the nodes of the interface is depicted in Fig. 5. The adhesive stresses decrease as we move from the center of the base plate towards its edges. Indeed, near the edges of the base plate the adhesive forces decreases as the adhesive enters the region where the material stiffness deteriorates due to its softening behavior. Therefore, the debonding of the nodes is easily interpreted from the stress-node diagrams due to the zero values the adhesive forces develop in this area, where the adhesive no longer holds.

4.2 BOLTED BEAM-FLANGE PARTIAL CONNECTION

The bolted beam-flange partial connection depicted in Fig. 6 is considered. The nonmononotone behavior of the resistance in bearing strength of the bolts and the respective superpotentials for the connection are shown in Fig. 7. The bearing resistance is calculated according to the Eurocode 3 (cf. e.g. [8]). The connection consists of eight M20-6.8 bolts. Due to the symmetry of the connection and of the applied external loading, we study the one fourth of the connection. The elastic-plastic analysis is performed using the algorithm described in section 3. The material properties of the steel are considered to have yield stress 480 N/mm^2 and ultimate stress 430N/mm^2. The reaction-displacement diagram (Fig. 8) is obtained by a single bolted finite element model through an elastoplastic analysis for a total compulsion of 2mm at each fixed edge of the flanges. Using this diagram we can calculate the diagonal terms the bolted connection stiffness matrix contains.

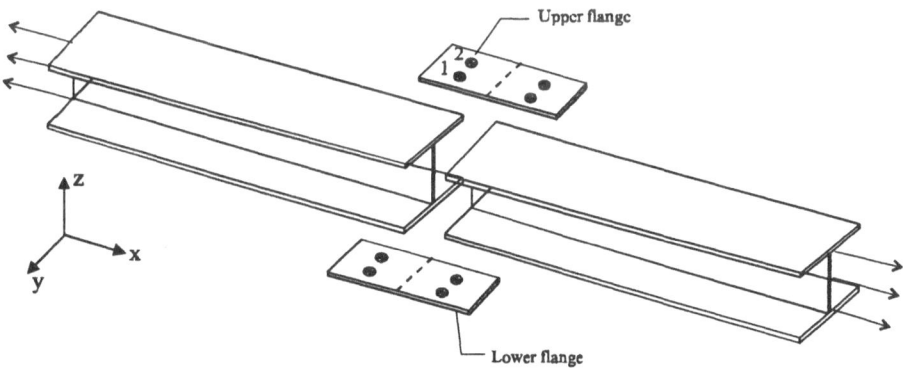

Figure 6. The bolted beam-flange partial connection.

The non-diagonal terms can be calculated from the interaction of each bolt with the others. The relation between the stiffness terms $K_{b,ij}$ and the correlation factors is given by the relation:

324

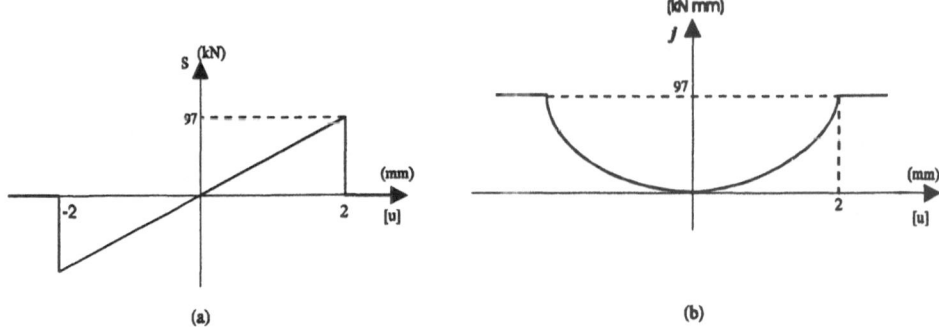

Figure 7. a) The considered nonmonotone reaction-displacement law for the resistance in bearing strength of the bolt and b) the respective energy superpotential.

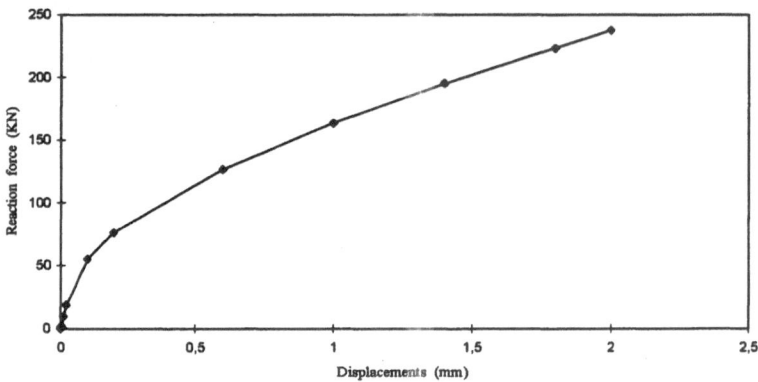

Figure 8. Reaction-displacement response for a total compulsion of 2mm for an incremental elastoplastic analysis.

$$K_{b,ij} = \lambda_{b,ij} \, K_{b,ii} \qquad (9)$$

where $\lambda_{b,ii} = 1.0$ for $i=1,2,...,n$.

The non-diagonal correlation factors $\lambda_{b,ij}$, $i \neq j$ are calculated applying unit relative displacements along the **x** direction at the top and bottom surface of one bolt and restraining the remaining relative displacements of the considered and the other bolts. The reactions $R_{b,i}$ that develop in each bolt $i=1,...,n$, determine the correlation factors from the relation:

$$\lambda_{b,ij} = R_{b,ij} / R_{ii} \qquad (10)$$

325

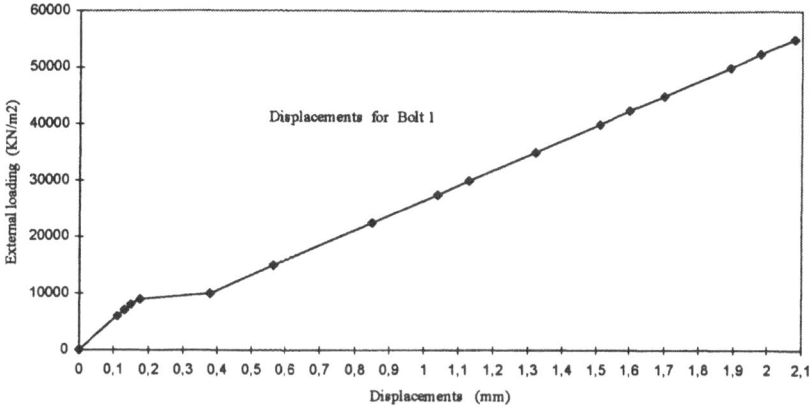

Figure 9. Bearing strength resistance-displacement diagram for bolt 1.

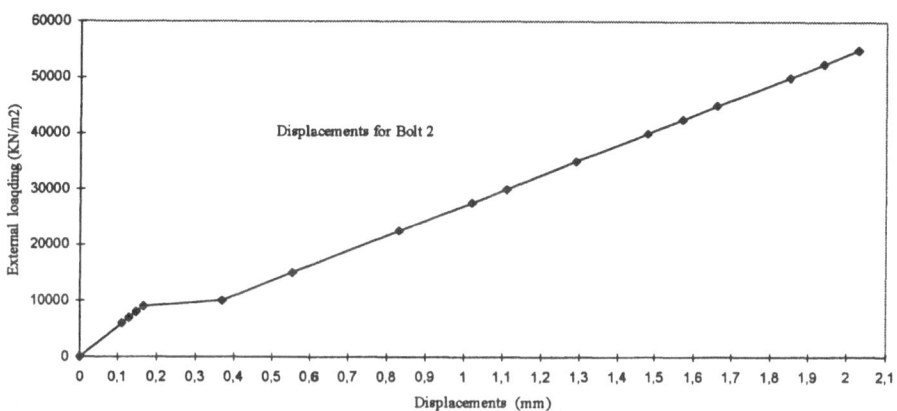

Figure 10. Bearing strength resistance-displacement diagram for bolt 2.

Applying the algorithm proposed in section 3, we obtain the diagrams of external loading and relative displacements on the opposite interfaces of the lower and upper flange with the bolt (Fig. 9,10). The results show that as soon as the external stresses reaches 55.000 kN/m², the resistance in bearing strength is exceeded. For the loads onwards the connection is in complete failure condition.

5. Conclusions

A unified method is herein presented for the numerical treatment of steel structures with both adhesive and bolted connections. The method covers both the cases of elastic and elastic-plastic material behavior leading to a substationarity optimization problem. In particular, for the more complex case of elastic-plastic analysis, an iterative algorithm that solves subsequent substationarity problems of the potential energy function is presented. The algorithm is based on the proximal bundle optimization procedure NSOLIB and calculates at least one substationarity point. The results obtained show that the adhesive connection although seriously damaged can still retain a certain amount of strength capacity due to its partial failure, whereas the bolted one fails with the initiation of the brittle crack of the bolts, immediately after their ultimate strength is exeeded.

6. References

1. Baniotopoulos, C.C. (1985) *Analysis of Structures with Complete Stress-Strain Laws*, Ph.D. Thesis, Aristotle Univ, Dept. Civil Eng.,Thessaloniki.
2. Panagiotopoulos, P.D. (1983) Non convex energy functions. Hemivariational inequalities and substationary principles. *Acta Mechanica* **42**, 160-183.
3. Panagiotopoulos, P.D. and Baniotopoulos, C.C. (1984) A hemivariational inequality and substationarity approach to the interface problem: Theory and prospects of applications, *Engineering Analysis*, **1**, 20-31.
4. Mäkelä, M.M. (1997) Issues of Implementing a Fotran Subroutine Package NSO-LIB for Nonsmooth Optimization, University of Jyväskylä.
5. Clarke, F.H. (1983) *Optimization and Nonsmooth Analysis.* John Wiley, New York.
6. Panagiotopoulos, P.D. (1993) *Hemivariational Inequalities. Applications in Mechanics and Engineering*, Springer-Verlag, Berlin.
7. Mistakidis, E.S. and Stavroulakis, G.E. (1998) *Nonconvex Optimization in Mechanics Algorithms, Heuristics and Engineering Applications by the F.E.M.*, Kluwer, Boston.
8. Ivanyi, M. and Baniotopoulos, C.C. (2000) *Semi rigid Connections in Steel Structures*, CISM Lecture Notes 215, Springer Verlag, Wien-New York.

FINITE ELEMENT ANALYSES OF STEEL-CONCRETE SUBASSEMBLAGE WITH PARTIAL SHEAR CONNECTION

B. BELEV
Dept. of Steel and Timber Structures
University of Architecture, Civil Engineering and Geodesy
1 Chr. Smirnenski Blvd., 1421 Sofia, BULGARIA

1. Introduction

In many practical design applications the steel frame girders are connected to the concrete floors by a small amount of shear connectors required for transferring the diaphragm forces and the slab contribution to the frame strength and stiffness is usually neglected. The benefits of the partial composite action in unbraced moment resisting frames (MRF) cannot be fully exploited due to lack of detailed code provisions and certain design and analysis issues.

2. Brief Review of Previous Studies

Fahmy and Robinson [1] pointed out that the steel MRFs in multi-storey buildings do not represent the complete structural system and design should take advantage of the composite action provided by the concrete floor slabs. They proposed curves for evaluation of two effective slab widths - for strength and stiffness calculations.

Tagawa et al. [2] reported results obtained from testing of steel subassemblage with floor slab under cyclic lateral loading. It was noted that the bearing surface of the concrete slab at the steel column face suffered substantial local plastic deformation and a gap opened up during load reversals.

Leon [3] concluded that the effect of the floor slab on the lateral strength and stiffness of the steel frame could be very large. The variation of the effective slab width along the girder centerlines was attributed to several factors. For the customary case of partial shear connection difficulties were found to arise in design because the effective slab width and the degree of interaction are coupled. In the negative moment regions the contribution of the slab reinforcement was reported to increase with the increase of the storey drift. For the case of full interaction, a formula was proposed that defined the equivalent beam stiffness in unbraced frames as a linear combination of the moments of inertia under positive (I^+) and negative (I^-) bending moments:

$$I_{eq} = 0.6 I^+ + 0.4 I^-$$ (1)

C.C. Baniotopoulos and F. Wald (eds.), The Paramount Role of Joints into the Reliable Response of Structures, 327–336.
© 2000 *Kluwer Academic Publishers.*

Leon et al. [4] and Hajjar et al. [5] reported results of cyclic quasi-static testing of cruciform steel and partially composite specimens indicating that the bottom flange region near the beam-to-column connection of the specimens with a slab sustained significantly larger strains and more damage compared to the top flange region. Shifting of the neutral axis location was also observed. It was more pronounced under positive bending and tended to increase as the storey drift level increased. Ignoring the slabs contribution might lead to unexpected strong beam - weak column failure mechanism. Even if the bearing at the column faces were prevented, the increased girder stiffness would attract larger bending moments that could overstress the beam-to-column connections.

Based on quasi-static cyclic testing of frame subassemblages, Bursi and Ballerini [6] concluded that the partially composite beams with ductile shear stud connectors could perform satisfactorily in terms of strength, ductility and energy dissipation if splitting and pull-out of concrete had been prevented. The shear lag effects were found to be important and should be incorporated in the two-dimensional (2-D) frame model for conventional design.

3. Motivation and Objectives of the Study

The reviewed research indicates that the gains in lateral strength and stiffness could be substantial if the concrete slab contribution is accounted for in the design of steel MRFs under gravity and wind loads. However, the practical evaluation of this contribution is not straightforward because many factors are involved. In seismic design situations, neglecting the effect of the floor slabs even for low degrees of shear connection may not be on the safe side. Doubts on the reliable cyclic performance of the partially composite girders still remain.

The traditional partial-interaction analysis has limited capabilities to handle all force transfer phenomena that are inherently non-linear. A comprehensive and consistent approach based on three-dimensional (3-D) finite element method (FEM) simulations could be more successful because the slab involvement in the lateral load resisting system, bending moment redistribution, bearing at the column faces, finite stiffness of the partial shear connection and interstorey drift level are all interrelated.

The main objectives of the now reported computational study were to demonstrate the fact that the floor slab participates structurally even at low degrees of shear connection and to quantify the slab contribution for a selected building frame subassemblage. The responses of a bare steel model and its partially composite peer to proportionally increased gravity and lateral loads were numerically evaluated and compared. The stress-and-strain state of the frame joints and the longitudinal shear transfer along the girder-slab interface were also studied. For the composite model two different spacings of the shear connectors and two options concerning the slab-column interaction (bearing effective/prevented) were simulated.

4. Model Description and Modelling Assumptions

The steel subassemblage was taken from a 6-storey MRF included by Vogel in the set of the so-called European calibration frames [7, 8]. It comprises the third floor girder and column portions extending to the mid-height of the adjacent stories (Fig. 1). Additional rigid tie-beams were employed at the column tips to distribute the story shear force proportionally to the column stiffnesses. Preliminary 2-D analyses of the entire frame for the specified set of loads were carried out to estimate the story shear F_2 and axial thrust forces F_A, F_B and F_C. The structural steel has characteristic yield strength $f_{yk} = 235$ N/mm^2. Full-strength welded beam-to-column connections were assumed and four column web stiffeners (continuity plates) were added in each joint in order to reduce the sources of joint semirigidity.

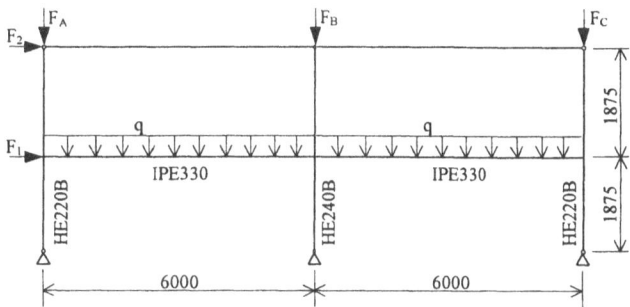

Figure 1. Steel subassemblage layout

The companion partially composite subassemblage incorporates the same steel members plus a 130-mm thick, 4-m wide concrete slab connected to the girders by 19x100 mm headed shear studs in a single row at uniform 400-mm spacing. The end studs are located on 90 mm and 80 mm from the exterior and interior column faces, respectively. An alternative case of 600-mm spacing was also analysed. Both cases should provide degree of shear connection η of 20-30 % conceived typical for MRFs.

Concrete to strength class C25/30 was adopted for the slab [9]. The reinforcement comprises top and bottom welded meshes of Ø10 bars spaced at 150 mm with characteristic yield strength $f_{sk} = 400$ N/mm^2. Concrete top and bottom cover of 20 mm was assumed. The resulting 0.4 % reinforcement ratios were conceived realistic as the amount required for crack control.

The following combination of actions was assumed in the reported study

$$1.35G + 0.9(1.5Q + 1.5W) \qquad (2)$$

where G, Q and W denote the nominal dead, imposed and wind loads, respectively. Consequently, the nominal (unfactored) load values were obtained by reducing the specified for the original 6-storey frame by 35 %. Hereafter this load level will be referred to as nominal load level.

330

Two finite element models representing the bare steel and composite subassemblages were created. The symmetry about a vertical plane allowed only half of the structure to be included in the model. Fig. 2 shows an isometric view of the composite model. To enforce the symmetry, appropriate boundary conditions were imposed to all nodes located in that vertical plane ($z = 0$) and to the slab edge at $z = 2.0\,m$. The concrete slab was line-supported against vertical displacement to simulate the hypothetical transverse beams framing into each column. All steel members except the infinitely rigid tie-beams were modelled with 4-node reduced integration shell elements S4R5 using 5 integration points across the thickness [10]. Multipoint constraints (MPC) were imposed on the column ends in order to define the support conditions and maintain the cross sections planar.

The concrete slab was modelled with 4-node full integration shell elements S4 employing 9 integration points across the thickness. The mesh reinforcement was defined using the special REBAR option of ABAQUS [10]. This is essentially a smeared rebar approach, based on the concept of reinforcement layer with equivalent thickness and uniaxial strain theory.

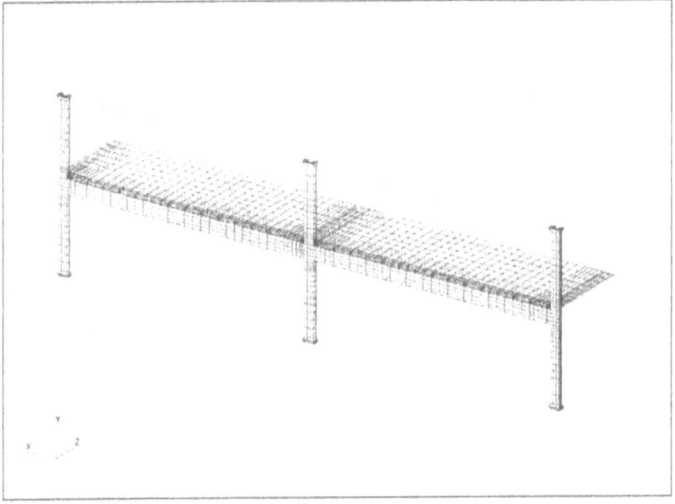

Figure 2. Finite element model

The interaction between the column flanges and slab faces was approximately modelled with unidirectional GAPUNI contact elements [10], connecting the pairs of nodes located on both sides of each contact plane. Additional nodes at the level of the top and bottom slab surfaces were defined and constrained to the basic slab nodes located at the edges. The shear connectors were modelled explicitly with a set of discrete non-linear springs resisting longitudinal shear and uplift. The SPRING2 elements [10] were placed at 400 mm spacing and "split up" to pairs in order to reduce the local overstressing. The elevation of these springs above the top steel flange was chosen based on the equivalent connector height concept [11].

An isotropic hardening model was adopted for the structural steel. The piecewise linear stress-strain relationship was taken from the original 6-storey frame model. The plain concrete model is essentially based on a smeared crack approach which assumes that cracking occurs when the principal stresses reach the predefined failure surface. In uniaxial compression the stress-strain diagram is piecewise linear, approximating the parabolic curve for short-term loading of EC2 [12]. The concrete behaviour in uniaxial tension was defined according to Fig. 3. This model is based on the exponential-decay curve proposed by Stevens et al. [13] that relates the tension stiffening effect to the reinforcement ratio and distribution. Cracking is assumed to take place at stress equal to the characteristic tensile strength $f_{ct,k}$ but the shape of the post-cracking branch was constructed using the mean tensile strength $f_{ct,m}$. Shear retention factor of 0.05 was assigned in order to avoid spurious overstiff response [14].

Exponential load-slip relation is commonly used for defining the stud shear connection behaviour. Comprehensive comparison of different approximating curves has been performed by Johnson and Molenstra [15]. For 19-mm studs in solid slabs the slip capacity is usually 6 to 7 mm. However, evidence of highly ductile response in push-out tests under displacement control is also available [16] and the increased slip capacity is attributed to the amount and detailing of the transverse reinforcement.

The now described model adopted piecewise linear load-slip relationship as shown in Fig. 4. The characteristic resistance P_{Rk} of a 19-mm headed stud in solid slab was estimated to 92 kN according to [9]. Three shear values, namely $0.5P_{Rk}$, $0.8P_{Rk}$ and P_{Rk} were considered, the first two corresponding to the average working load and design strength, respectively. A similar but tri-linear approximation was used by Dissanayake et al. [17]. Herein the slope of the initial branch was specified based on a simplification proposed by Wang [18], from which it follows that the connector stiffness in (kN/mm) equals its characteristic strength in (kN). The slip at $0.8P_{Rk}$ was estimated to 1.6 mm by averaging the predictions of Johnson's A and B and Aribert's exponential curves described in [15]. The slope of the falling branch was approximated based on the push-out test results reported in [16]. The 600-mm stud spacing was modelled by scaling the strength and stiffness of the 400-mm spaced springs to 2/3 of their actual values.

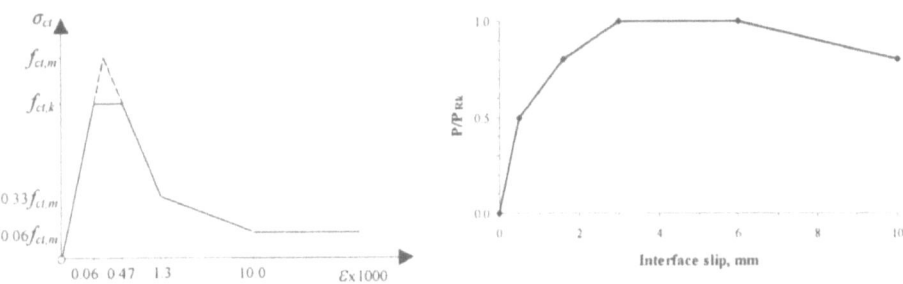

Figure 3. Uniaxial tension model of concrete   *Figure 4.* Shear connection model

5. Numerical Evaluation

Accompanying 2-D models were created both for the complete 6-storey frame and the selected steel subassemblage. The degree of shear connection was quantified based on the capacity of the steel girder section and the distance between the zero-moment points in the bare steel 2-D model. For the positive moment regions equivalent span of 370 cm and shear span of 185 cm were estimated, that gave η = 25 % and η = 31 % for the 600 mm and 400 mm stud spacing, respectively.

All 3-D analyses were carried out using ABAQUS/Standard software [10]. Parallel computations were performed using the elastic 2-D frame model.

Under nominal loads the panel zones of the interior and right-hand columns yielded severely (von Mises criterion). To avoid this premature yielding it was decided to strengthen all column web panels using two symmetric 5-mm thick doubler plates. So modified, the steel subassemblage remained elastic under nominal loads. The storey drift was approximately 1/300 thus implicating that the assumed nominal load level corresponds to serviceability limit state (SLS).

The following model notations have been used throughout the study: STEEL - bare steel subassemblage; COMPNO4 and COMPNO6 - composite subassemblages with slab isolated from column faces (no bearing) and 400-mm and 600-mm stud spacing, respectively; COMPBE6 – composite subassemblage with 600-mm stud spacing and bearing effects included. The non-linear analyses extended up to the ultimate limit state (ULS) associated with either the true limit of the load-carrying capacity (STEEL) or storey drift limit of 1.5 % (COMPNO6). For COMPNO4 the analyses terminated due to convergence problems at about 1% drift.

6. Results and Conclusions

Fig. 5 summarises the load-displacement curves computed for the different models considered. Load factor of 1.0 corresponds to the nominal load level. All models except COMPBE6 went well into the inelastic range. For the latter case convergence problems led to premature analysis termination. For load factor producing 1 % storey drift in the STEEL model, the partially composite COMPNO6 model did not show any significant yielding. As a result, the lateral displacement was reduced by almost a half.

Fig. 6 depicts the slab equivalent force for the COMPNO6 model at nominal (SLS) and ultimate (ULS) load level. It could be concluded that load increase by a factor of about 2 led to increase in the slab compression force in the positive bending regions approximately by a factor of 3. Despite of this, the stud shear forces remained below the design resistance per connector.

The column-slab interaction resulted in larger lateral stiffness and the bearing forces transmitted to the slab produced different connector shear distribution compared to the isolated slab model for the same degree of shear connection under nominal loads (Fig. 7). However, under ultimate loads this discrepancy may not be so pronounced due to internal stress redistribution.

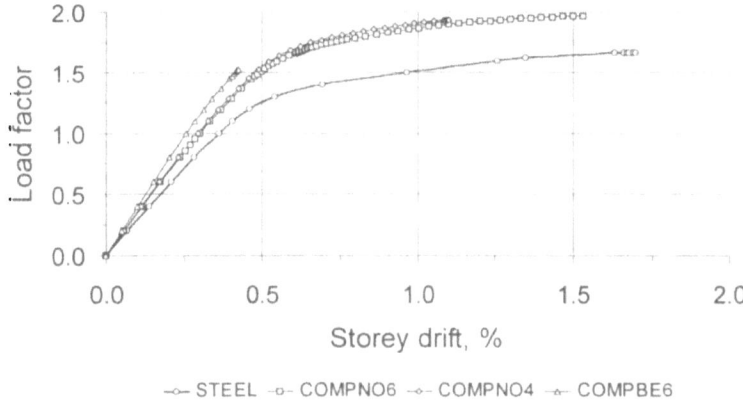

–○– STEEL –□– COMPNO6 –◇– COMPNO4 –△– COMPBE6

Figure 5. Comparison of global behaviour

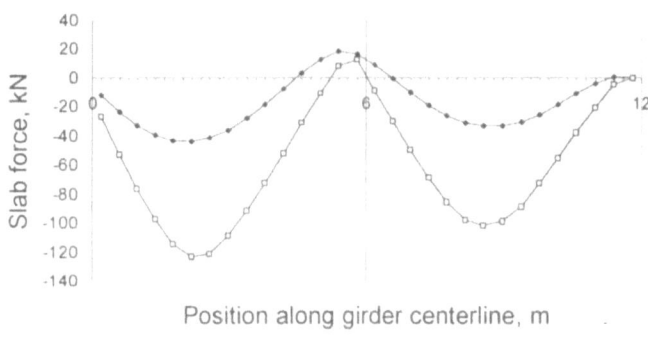

–●– SLS –○– ULS

Figure 6. Slab equivalent force in COMPNO6 model

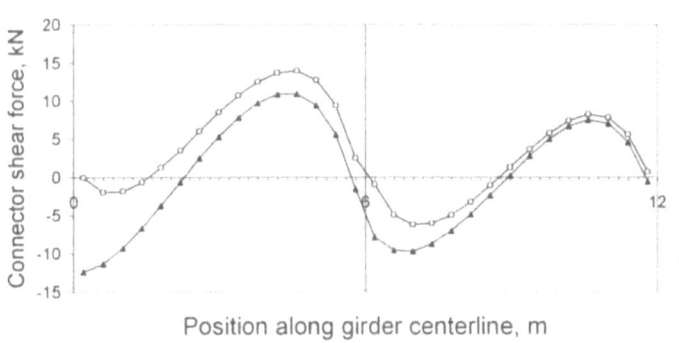

–▲– COMPNO6 –○– COMPBE6

Figure 7. Influence of bearing effects on connector shear force

Fig. 8 provides comparison of the plastic equivalent strain (PEEQ) contours (see [10] for PEEQ definition) in the interior frame joint of the STEEL and COMPNO6 models at their ULS. The partial composite action resulted in shifting of the plastic zone from the beam end to the column flange and web panel. As a general observation, the slab contribution delayed the onset of yielding in the steel members but reduced the size of beam plastic zones thus worsening the subassemblage ductility.

It is proposed herein to extend the application of (1) to partially composite MRFs by evaluating I^+ using a formula that accounts for the degree of shear connection. The following one recommended in [19] for calculation of beam deflection was adopted:

$$I_{ef} = I_s + 0.85\eta^{0.25}\left(I_t - I_s\right) \tag{3}$$

where I_s is the moment of inertia of the steel beam and I_t is the transformed moment of inertia of the composite beam assuming complete interaction. Neglecting the slab reinforcement in the negative moment regions and substituting (3) in (1) led to

$$I_{eq} = SF \times I_s \quad , \quad SF = \left[1 + 0.51\eta^{0.25}\left(\frac{I_t}{I_s} - 1\right)\right] \tag{4}$$

where SF stands for stiffness modification factor. Thus the conventional 2-D frame elastic global analysis could be used. The trial application of the proposed approach under nominal loads was performed as follows. First, the 2-D steel frame model was "tuned" to the exact lateral displacements of the STEEL model by adjusting the bending stiffness of the equivalent beam stubs representing the column web panels. Assuming effective slab width equal to a quarter of the equivalent span, the proposed approach gave $SF=2.27$ for the 600-mm stud spacing. The substitute 2-D model employing $SF=2.27$ for the girder moment of inertia underestimated the lateral displacements of the COMPNO6 model by 10.5 %. Equal displacements were reached for $SF = 1.60$.

7. Final remarks

The results of the numerical analyses showed that the floor slab participates structurally even at low degrees of shear connection. Significant overstrength and greater stiffness are available when the partial composite action is taken into account. The shear studs did not reach their deformation capacity even at large storey drifts of 1% to 1.5 %. The effect of slab bearing against the column faces resulted in different shear and slip profile along the girder centerlines. The influence of the panel zone flexibility on the lateral frame stiffness that is rarely accounted for by the designers may remain hidden due to being offset by the partial composite action.

A better interaction descriptor based on the connection stiffness is needed for incorporating the partial composite action in the design of MRFs.

Further tests and numerical simulations for investigating the adequacy of combining formulas (1) and (3) into (4) are needed.

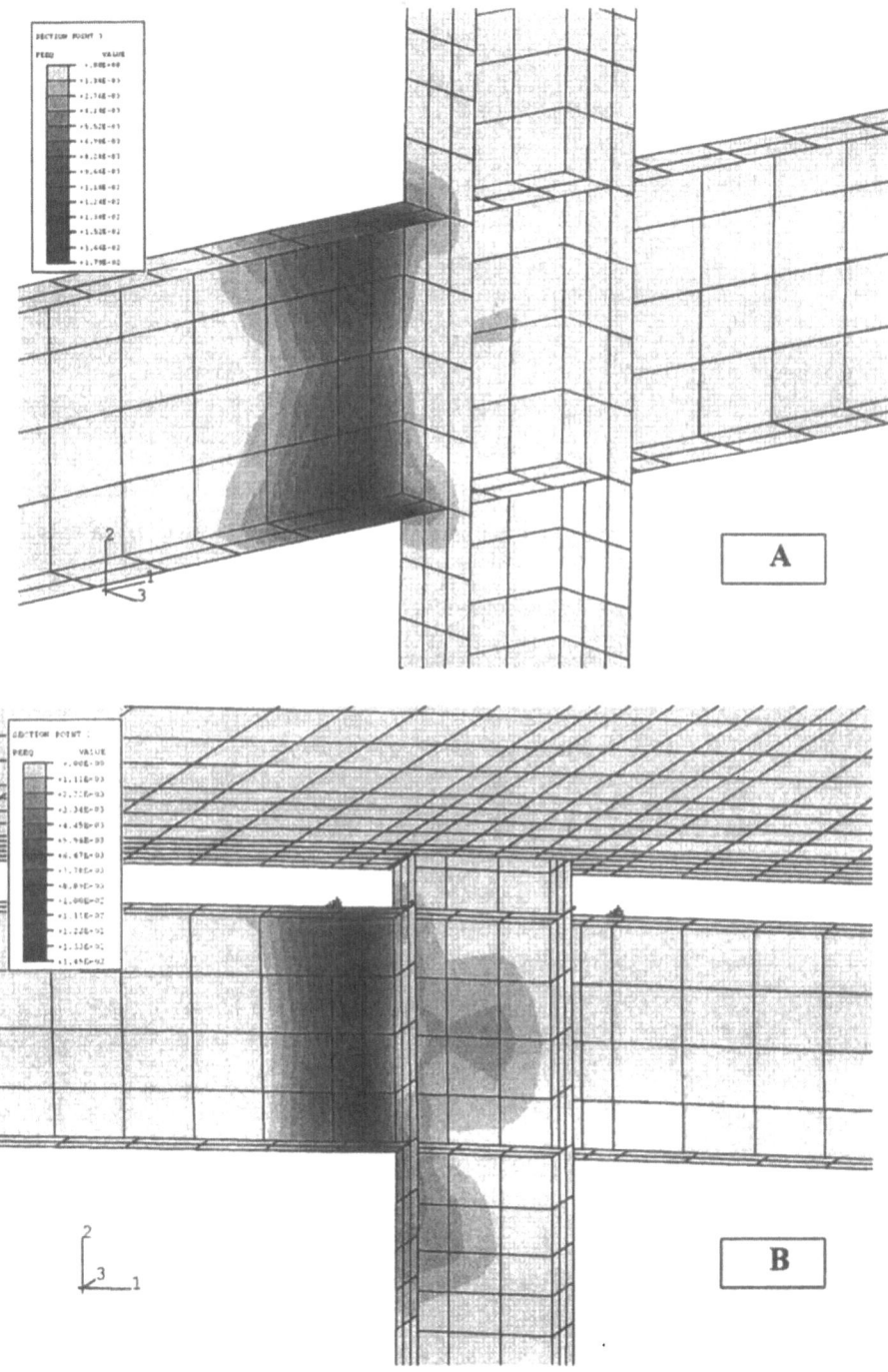

Figure 8. PEEQ Contour Plots for Interior Joint at ULS
A – STEEL Model **B** – COMPNO6 Model

336

8. Acknowledgements

The major part of the reported study was carried out at Chalmers University of Technology, Göteborg, Sweden under the supervision of Prof. Bo Edlund. The research scholarship provided by Svenska Institutet is gratefully acknowledged. The author wishes to thank Prof. Bo Edlund and his colleagues in the Department of Structural Engineering of the same university for their valuable advice and discussion.

9. References

1. Fahmy, E.H., and Robinson, H. (1986) Analyses and tests to determine the effective widths of composite beams in unbraced multi-storey frames, *Can. Journal of Civil Engineering* **13**, 66-75.
2. Tagawa, Y., Kato, B., and Aoki, H. (1989) Behavior of composite beams in steel frame under hysteretic loading, *Journal of Structural Engineering, ASCE* **115**, 2029-2045.
3. Leon, R. (1998) Analysis and design problems for PR composite frames subjected to seismic loads, *Engineering Structures* **20**, 364-371.
4. Leon, R.T., Hajjar, J.F., and Gustafson, M.A. (1998) Seismic response of composite moment-resisting connections. I: Performance, *Journal of Structural Engineering, ASCE* **124**, 868-876.
5. Hajjar, J.F., Leon, R.T., Gustafson, M.A., and Shield, C.K. (1998) Seismic response of composite moment-resisting connections. II: Behavior, *Journal of Structural Engineering, ASCE* **124**, 877-885.
6. Bursi, O.S. and Ballerini, M. (1996) Behaviour of a steel-concrete composite substructure with full and partial shear connection, Paper No. 771, *Proceedings of 11 WCEE*, Elsevier Science, London, (CD-ROM format).
7. Vogel, U. (1985) Calibrating frames, *Der Stahlbau* **54**, 295-301.
8. Toma, S., and Chen, W.F. (1992) European calibration frames for second-order inelastic analysis, *Engineering Structures* **14**, 7-14.
9. European Committee for Standardisation. (1992) *Eurocode 4 – ENV 1994-1-1: Design of Composite Steel and Concrete Structures - Part 1-1: General Rules and Rules for Buildings*, Brussels.
10. Hibbit, Karlson & Sorensen, Inc. (1997) *ABAQUS/Standard User's Manual, Version 5.7*.
11. Oehlers, D.J. and Bradford, M.A. (1995) *Composite Steel and Concrete Structural Members. Fundamental Behaviour*, Pergamon, New York.
12. European Committee for Standardisation. (1991) *Eurocode 2 – ENV 1992-1-1: Design of Concrete Structures - Part 1-1: General Rules and Rules for Buildings*, Brussels.
13. Stevens, N.J., Uzumeri, S.M., Collins, M.P., and Will, G.T. (1991) Constitutive Model for Reinforced Concrete Finite Element Analysis, *ACI Structural Journal* **88**, 49-59.
14. Rots, J. (1988) *Computational Modeling of Concrete Fracture*. Ph.D. Dissertation, Delft University of Technology, Delft, Netherlands.
15. Johnson, R.P., and Molenstra, Ir N. (1991) Partial shear connection in composite beams for buildings. *Proceedings of the Institution of Civil Engineers, Part 2* **91**, 679-704.
16. Li, A., and Cederwall, K. (1991) *Push test on stud connectors in normal and high-strength concrete*, Report 91:6, Chalmers University of Technology, Göteborg, Sweden.
17. Dissanayake, U.I., Davison, J.B., and Burgess, I.W. (1999) Composite beam behaviour in braced frames, *Journal of Constr. Steel Research* **49**, 271-289.
18. Wang, Y.C. (1998) Deflection of steel-concrete composite beams with partial shear interaction. *Journal of Structural Engineering, ASCE* **124**, 1159-1165.
19. Jayas, B.S., and Hosain, M.U. (1989) Behaviour of headed studs in composite beams: full-size tests. *Can. Journal of Civil Engineering* **16**, 712-724.

BUTT-WELDED ALUMINIUM JOINTS: A NUMERICAL STUDY OF THE HAZ EFFECT ON THE ULTIMATE TENSION STRENGTH

M. J. KONTOLEON, F. G. PREFTITSI & C.C. BANIOTOPOULOS
Aristotle University, Department of Civil Engineering
GR-54006, Thessaloniki, Greece

1. Introduction

The use of aluminium alloys structural members represents nowdays a modern trend in structural applications exhibiting a lot of advantages. The corrosion resistance and the high strength to weight ratio are certain characteristics of such an alloy which are of great significance in the design of lightweight and transportable bridges, constructions in a marine environment, curtain-wall systems, etc. Although the strength of pure aluminium is very low for structural applications, the strength and other characteristics of aluminium are improved when it is alloyed with other elements, and its mechanical properties can be further improved as soon as heat treatment is applied.

As a matter of fact the heat input in welds removes some of the beneficial effects due to the above mentioned treatments and leads to a decrease in the elastic limit $f_{0.2}$ near the welding, whereas the ultimate limit strength f_u is not so hardly affected. The limited knowledge about the structural response of aluminium structural members exhibiting a nonlinear character and the weakening of the metal around welds, known as HAZ (Heat-Affected Zone) softening has induced hesitation about the use of aluminium alloys as a structural material. This effect is taken into account through the definition of a reduced-strength zone, which extends immediately around the weld, beyond which the strength properties rapidly recover to their fully unwelded values.

The severity of this type of softening and the extent of this zone has to be in any case considered in the calculation and design of welded connections. The severity is largely a function of the parent material used, while the extent is affected by the control of temperature during fabrication; so it depends mostly upon the welding procedure (velocity, voltage, number of passes, thickness of the joint). In general, the extent of the HAZ is greater for TIG than for MIG welding procedure, both being the most commonly used welding techniques in aluminium connections. In addition, the TIG process is more dependent on the operator's technique, as compared to the semi-automatic MIG process [1].

The most common method for the estimation of the extent of the HAZ is the 'one-inch rule' that assumes that the HAZ extends within the parent metal 25mm in all the

C.C. Baniotopoulos and F. Wald (eds.), The Paramount Role of Joints into the Reliable Response of Structures, 337–346.
© 2000 *Kluwer Academic Publishers.*

338

directions around the weld and has been widely used since the decade of '60s [2]. The EC9 relates the extent of the HAZ with the thickness of the connection members. In both methods, the subject of the extent of the HAZ is not based on exact scientific results. In many cases it has been found that the effect of the HAZ on the resistance is small enough, making the 'one-inch rule' quite acceptable. However, in other cases experimental tests have shown that the HAZ softening significantly reduces the resistance and therefore, economy can be made by the use of more refined methods. Such a method that has been proposed in order to estimate the extent of the HAZ in a more precise way is the RD Method, based on the classic heat flow equations of Rosenthal [3].

In the present paper, a butt weld connection is numerically studied by means of the Finite Element Method (FEM). The HAZ is limited to a region in the vicinity of the weld where reduced mechanical properties are considered. The extent of the HAZ is a parameter that is determined using the 'one-inch rule', the EC9 and the RD Method. The obtained results are compared and discussed, showing that the extent of the HAZ does have a significant influence on the ultimate yield tension strength that the weld connection can sustain.

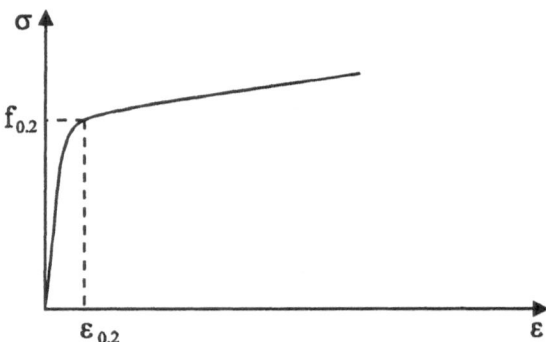

Figure 1. Stress-strain curve for aluminium (unaffected material). Ramberg-Osgood model.

2. Stress-strain Law for the Unaffected Material

The nonlinear Ramberg-Osgood law that predicts the actual material behaviour of the structural aluminium alloys with good accordance is herein used to describe material nonlinearity [4]. This law can be expressed in the following exponential form:

$$\varepsilon = \frac{\sigma}{E} + 0.002 \left(\frac{\sigma}{f_{0.2}} \right)^n \tag{1}$$

where E is the Young Modulus of Elasticity (E=70,000N/mm^2) and $f_{0.2}$ is the conventional yield limit stress used for aluminium alloys and corresponds to a stress that leaves a permanent deformation of 0.2 percent. The exponent n is different for each alloy as it is a function of $f_{0.2}$ and $f_{0.1}$ and it determines the slope of the σ-ε law as it is shown in Fig.1. For n →∞ the material behaves as an elastic-perfectly plastic one, whereas for n = 1 the above equation becomes linear. As it has been confirmed from experimental testing, non-heat-treated alloys have a value of exponent n below 20, whereas for heat-treated alloys n ranges between 20 and 40 [5].

3. Patterns of Softening in the Heat-Affected Zone

The strength of an aluminium alloy near the welds depends on the type of treatment the alloy sustains. It has been experimentally determined that in general, two patterns of softening may arise: a) softening on heat-treated materials and b) softening on work-hardened materials. A typical hardness plot at a weld in heat-treated aluminium is shown in Fig. 2. It can easily be defined that the HAZ can be divided into two regions, i.e. in region 1 where the metal attains solution-treatment temperature and thus it is able to reage to some extent on cooling, and in region 2, where this temperature can not be reached, and the metal is over-aged. The hardness approaches its minimum at the boundary between the two regions (point A), and then raises steadily as we move out to point B. Beyond B, the heat of welding has negligible effect and full parent properties can be assumed to apply. The extent of these regions depends mostly upon the composition and the type of the alloy.

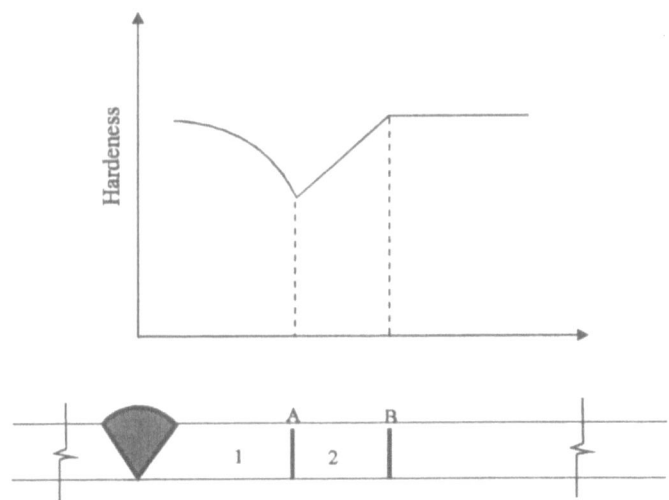

Figure 2. Pattern of softening in a heat-treated alloy.

The pattern of the HAZ softening in work-hardened materials is depicted in Fig. 3 where it can be clearly seen that also two regions are identified. The difference is that, in region 1 the hardness is now uniform and corresponds to the properties of the alloy in the annealed condition. The extent of the softening also depends on the type of the alloy, but the strength in the HAZ is still very low in comparison to its unwelded value.

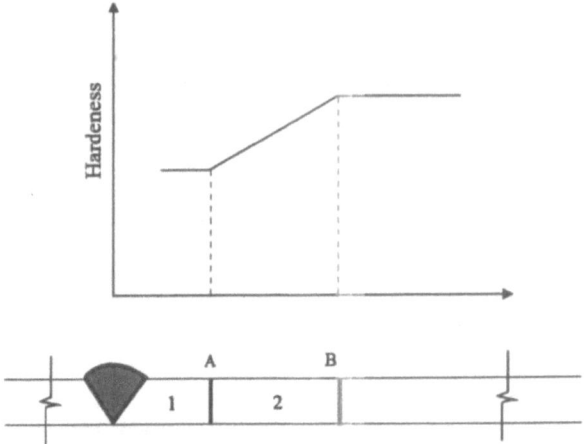

Figure 3. Pattern of softening in a work-hardened alloy.

Typical stress-strain curves that might be obtained using coupons from the HAZ and from parent material are given in Fig. 4. As is shown, the HAZ curve has a more rounded knee, with a lower proof / ultimate ratio.

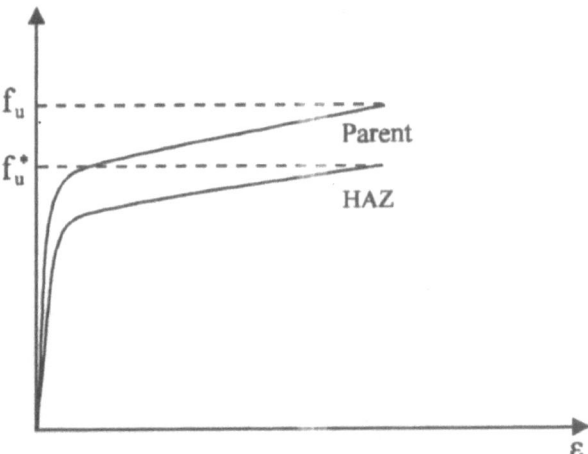

Figure. 4 Stress-strain curve in the HAZ zone.

4. Methods Estimating the Extent of the HAZ

Several methods have been recently developed in order to estimate the extent of the HAZ. The well-known 'one inch rule' is a successful simplification of a complex

problem and a good design tool for all preliminary calculations, but the final check must be done using more sophisticated methods (RD Method).
The reduced strength-zone can be considered in computation by means of the aforementioned method, which is one of the methods used in the herein presented model. The basic characteristics of the reduced strength zone are depicted in Fig. 5.

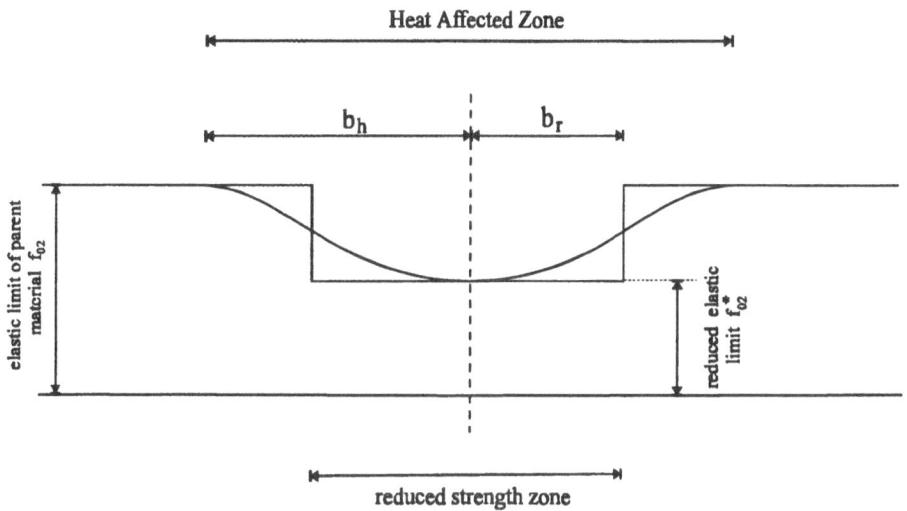

Figure 5. Reduced strength zone.

The width of the HAZ around the weld is given by the following formula:

$$b_r = \left[\int_0^{b_h} f(x)\,dx \right] \Big/ f_{0.2}^* \tag{2}$$

where

$f_{0.2}$ is the elastic limit of the unaffected base metal,
$f_{0.2}^*$ is the elastic limit in the welded region, and
b_h is the semi-width of the HAZ, from which the semi-width b_r of the reduced-strength zone is derived.

5. Numerical Applications

In the present paper a butt-welded aluminum connection under tension forces (Fig. 6), is numerically studied by means of the F.E. Method. The wrought aluminum alloy

EN-AW-6754 (H24/H34), with proof strength $f_{0.2} = 200$ N/mm^2 and ultimate tensile strength $f_u = 270$ N / mm^2 is first considered. In the area of the HAZ the strength of the alloy is considered to be reduced by a softening factor $k_z = 0.65$. The extent of the HAZ, b_{HAZ}, is a parameter that varies according to the method applied for its calculation. Applying the 'one-inch rule', b_{HAZ} is equal to 25mm. According to the EC9 design procedure [6], it takes the following values that depend on the flange thickness:

$b_{HAZ} = 20$ mm, $0 < t \leq 6$ mm,
$b_{HAZ} = 3$ 0mm, $6 < t \leq 12$ mm, and
$b_{HAZ} = 35$ mm, $12 < t < 25$ mm.

According to the RD Method the length of the HAZ is a function of the weld size, the alloy type and the applied thermal control. As mentioned in the previous section, the RD Method seems to be more realistic in comparison to the 'one-inch rule' and the EC9 design procedure, and the respective results resembles very much to the experimental ones.

The welding metal is assumed to be stronger than the material in the HAZ zone and therefore, joint failure in the weld metal is not possible to occur. The connection consists of two flanges welded to each other. The thickness of each flange is considered as a parameter that takes the following values: 5, 10, 15, 20 and 25mm covering the cases of small (t<8mm), medium (8<t<20mm) and big welds (t ≥20mm). Geometrical and mechanical imperfections are not considered.

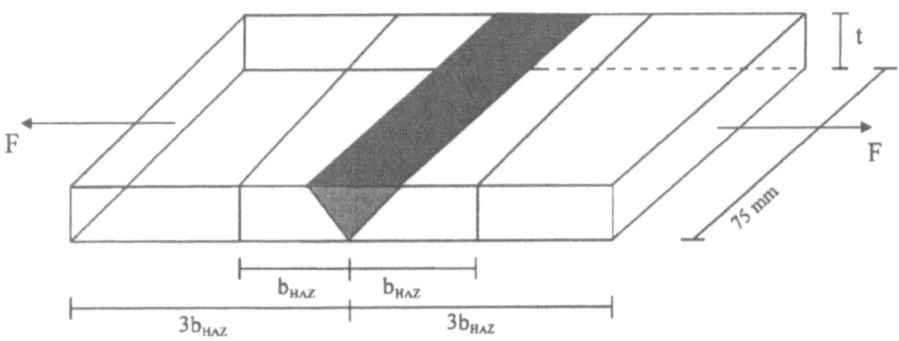

Figure 6. Geometry of the butt weld connection and direction of applied elongation forces.

The welded connection is idealized by applying the finite element code CASTEM 2000. For the discretization of the connection, three-dimensional 8-node solid elements with three translational degrees of freedom at each node are employed. The stress-strain law mentioned in section 2 (Equation 1) is used for the present elastic-plastic analysis of the connection. The obtained results (Figures 7- 10) show the strain elongation of the connection in the case that a tension force that reaches the ultimate resistance force F_u is applied, which can be calculated by the following relation:

$$F_u = k_z \, A_f \, f_u \qquad (3)$$

where A_f is the cross section of the flange.

The obtained force-strain diagrams, normalize the force axis by the ratio F / F_u making the results easily interpretable in order to determine the ultimate resistance of the connection for the three cases of extent of the HAZ. The strain elongation $\varepsilon = 0.2\%$ is considered as the strain limit from where on, yielding begins.

TABLE 1. Comparison of results according to the 'one-inch rule', EC9 and RD Method for the five cases of flange thickness.

$F_{0.2}/f_u$ 200/270 (kN/mm^2)	Method	b_{HAZ} (mm)	$F_{0.2}/F_u$ (kN)	ε_{max} $^o/_{oo}$	Reduction %
t = 5 mm F_u= 65	One-inch rule	25	1.10	2.22	-
	EC9	20	0.30	1.94	7
	RD Method	10	4.09	1.18	-
t=10mm F_u = 130	One-inch rule	25	0.83	2.52	17
	EC9	35	0.73	2.82	28
	RD Method	8,5	1.23	1.38	-
t= 15 mm F_u = 197	One-inch rule	25	0.77	2.80	23
	EC9	35	0.58	3.86	42
	RD Method	13	0.97	2.12	3
t= 20 mm F_u = 263	One-inch rule	25	0.70	3.00	30
	EC9	35	0.58	4.08	43
	RD Method	17	0.82	2.66	19
t = 25 mm F_u = 330	One-inch rule	25	0.68	3.56	32
	EC9	35	0.56	4.41	44
	RD Method	17	0.82	2.90	19

The ultimate force the connection resists under tension, the developed maximum strain and the force that coresponds to the yielding strain $\varepsilon_{0.2}$ denoted by $F_{0.2}$ are given for the five cases of flange thickness in TABLE 1.

For the case of flange thickness 5mm, the 'one-inch rule' gives the most conservative results underestimating the connection resistance assuming the largest extent of the HAZ in comparison to the other two methods (Fig. 7). For the cases of flange thickness over 5mm (Fig. 8-10), it comes out that the results from the EC9 design procedure are the most conservative. The elongation force needed in order to reach the strain $\varepsilon_y= 0.2\%$, results to the lowest strength resistance the connection can take. On the other hand, the RD Method predicts the largest resistance and the 'one-inch rule' gives results between the other two methods. Considering the effect of the flange thickness we can observe that as the flange thickness increases, the extent of the HAZ increases for the EC9 and RD Method and therefore, the extent of the area with reduced strength properties, significantly affects the stiffness of the connection. This can be verified from the increase of the maximum strain elongation as we move from the thinnest (t=5mm) to the thickest flange (t=25mm). We observe from the force-strain diagrams that the strain of the connection for the thinnest flange thickness reaches 0.22%, whereas for the thickest flange 0.41%.

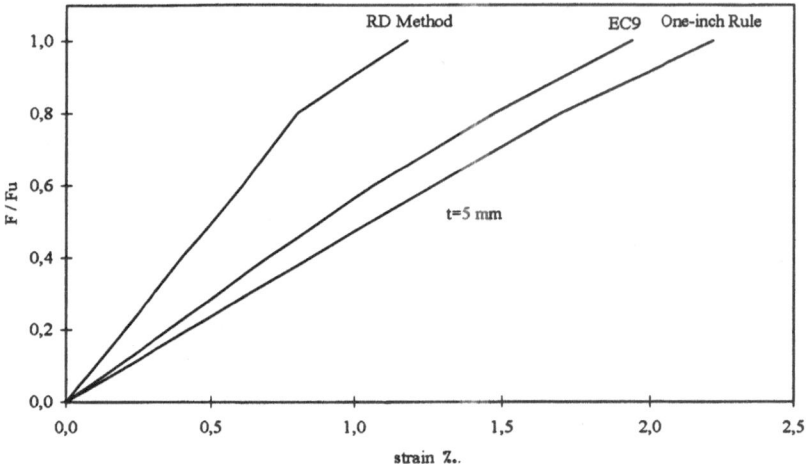

Figure 7. Comparison of methods regarding the extent of the HAZ for flange thickness 5 mm.

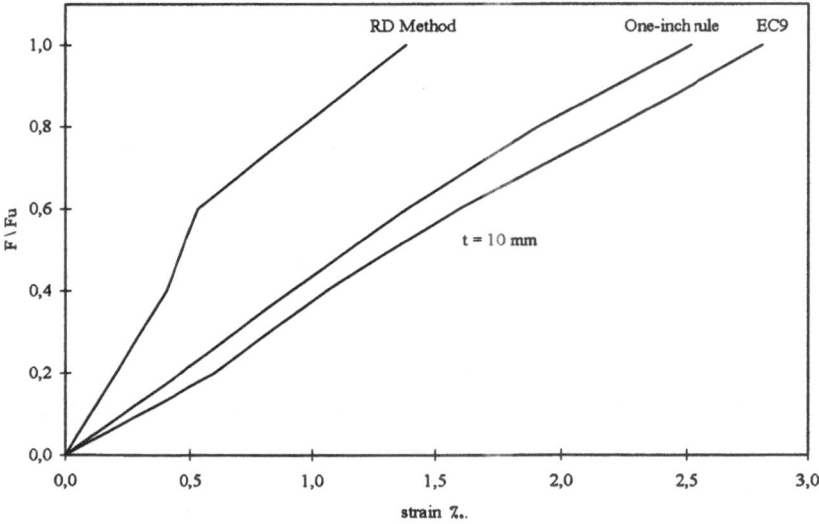

Figure 8. Comparison of methods regarding the extent of the HAZ for flange thickness 10 mm.

Comparing the results we conclude that the extent of the HAZ significantly affects the strength of the weld connection. The loss in resistance below the softened value in the HAZ is greatest when applying the EC9 and depends on the flange thickness. For the 5mm flange, the loss is 7% and reaches 44% for the 25mm flange. We must also notice that the numerical results of the FEM, when applying the RD Method for all the case of flange thickness, predict a greater yield resistance force in comparison to the calculated resistance force from relation (3). The same occurs only for flange thickness 5mm when using the 'one-inch rule' where the ratio $F_{0.2} / F_u$ is 1.1.

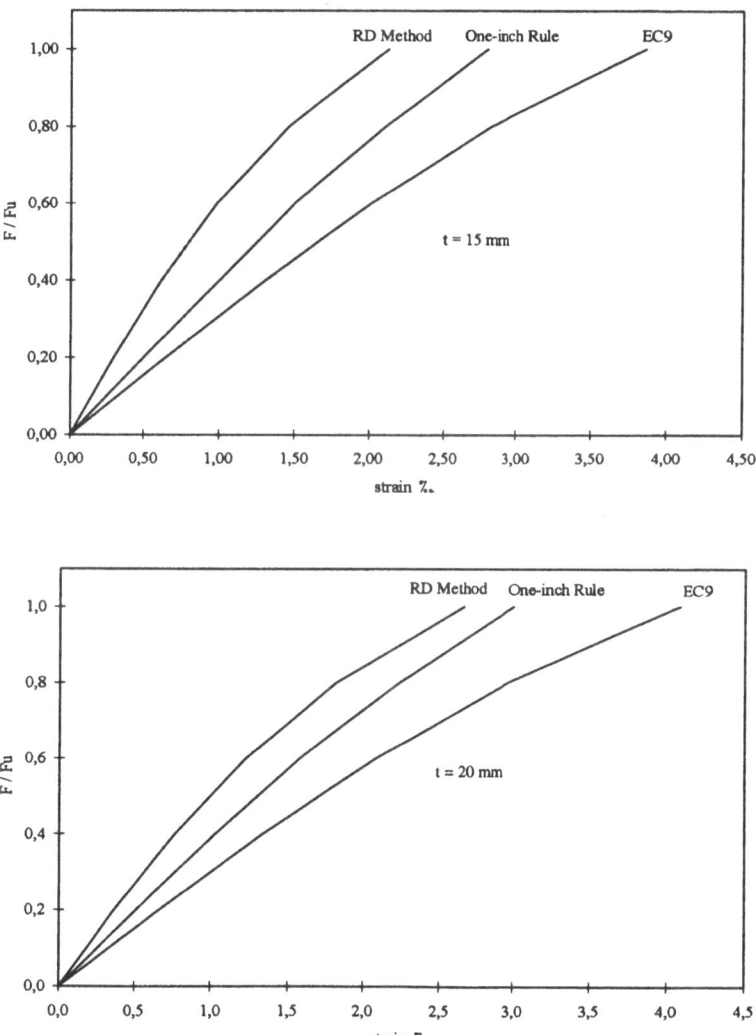

Figure 9. Comparison of methods regarding the extent of the HAZ for flange thickness 15 and 20 mm.

346

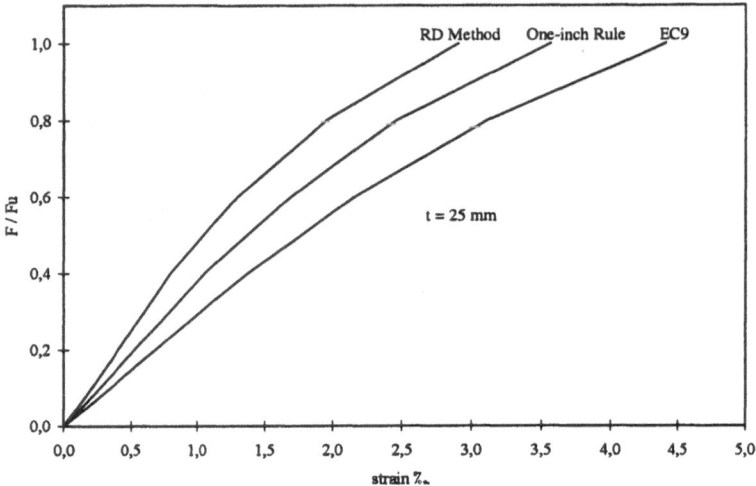

Figure 10. Comparison of methods regarding the extent of the HAZ for flange thickness 25 mm.

Acknowledgements

The support of the Greek Secretariat for Research and Technology within the Greek-Czech Research and Technology collaboration framework (Research project: "Prediction of the critical parameters affecting the structural response of welded elements in aluminium structures") is gratefully acknowledged.

6. References

1. TALAT (1995) F. Ostermann (ed.), ATP Aluminium Training Partnership, Brussels
2. Hill, H. N., Clark, J. W. and Brungraber, R. J. (1962) Design of welded aluminum structures, *Trans.* ASCE, **127**, Part II.
3. Dwight, J. (1999) Aluminium Design and Construction, E &FN Spon, London.
4. Ramberg, W. and Osgood, W. R. (1939) Description of stress-strain curves by three parameters, NACA Report 656.
5. Mazzolani F. M. (1988) Aluminum Alloy Structures, Pitman Publishing Inc, London.
6. Eurocode 9 (1997) Design of Aluminium Structures.

V. ANALYTICAL MODELS FOR JOINTS AND RELIABILITY

DESIGNING STRUCTURAL JOINTS ACCORDING TO EUROCODES

J.P. JASPART
Department MSM, University of Liège,
1, Chemin des Chevreuils, B-4000 Liège 1, Belgium.

1. Introduction

In Eurocode 3 [1] on steel buildings, the design of structural joints is covered by Chapter 6 and, for joints between H or I profiles, by Annex J. During the recent revision of Annex J [2], a new comprehensive design approach for the design of joints has been implemented and the so-called "component method" has been introduced as a basic procedure for the derivation of the stiffness and strength properties of the structural joints, whatever is the joint configuration (single-sided or double-sided beam-to-column joints, beam splices, ...) and the connection type (welded connections, bolted connections with end-plates, flange cleats, ...). More recently, the application of the component method has been extended to base plates configurations [3] and composite steel-concrete joints [4].

In the present paper, the new design approach suggested by Eurocode 3 is presented and the advantages that the component method offers in comparison with traditional design approaches are illustrated by describing or referring to practical situations where economic and safe solutions have been made possible.

2. The concept of joint representation

During many years, the research activity in the field of joints mainly concentrated on two aspects:
- the evaluation of the mechanical properties of the joints in terms of rotational stiffness, moment resistance and rotation capacity;
- the analysis and design procedures for frames including joint behaviour.

But progressively it has been understood that there were intermediate steps to consider in order to integrate in a consistent way the actual joint response into the frame analysis; this is known as the *joint representation*.

The joint representation includes four successive steps respectively named:
- the joint characterisation
  i.e. the evaluation through appropriate means of the stiffness, resistance and ductility properties of the joints (full M-φ curves or key values);
- the joint modelling
  i.e. the way on how the joint is physically represented in view of the frame analysis;

349

C.C. Baniotopoulos and F. Wald (eds.), The Paramount Role of Joints into the Reliable Response of Structures, 349–362.
© 2000 *Kluwer Academic Publishers.*

- the joint classification
  i.e. the tool providing boundary conditions for the use of conventional types of joint modelling (e.g. rigid or pinned);
- the joint idealisation
  i.e. the derivation of a simplified moment-rotation curve so as to fit with specific analysis approaches (e.g. linear idealisation for an elastic analysis).
  These four items are discussed in the next pages.

2.1. JOINT CHARACTERISATION

The procedure adopted in Revised Annex J for the characterisation of mechanical properties of the structural joints is based on the "component method".

Roughly speaking this one may be presented as the application of the well-known finite element method to the calculation of structural joints.

In the characterisation procedures, a joint is generally considered as a whole and is studied accordingly; the originality of the component method is to consider any joint as a set of "individual basic components". In the particular case of Figure 1 (single-sided beam-to-column joint with an extended end-plate connection subject to bending), the relevant components are the following :

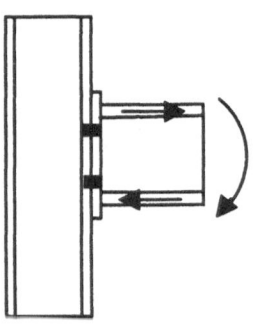

- compression zone :
  - column web in compression
  - beam flange and web in compression
- in shear zone :
  - column web panel in shear
- tension zone :
  - beam web in tension
  - column web in tension
  - column flange in bending
  - bolts in tension
  - end-plate in bending

Figure 1. Joint with end-plate in bending

Each of these basic components possesses its own level of strength and stiffness in tension, compression or shear. The coexistence of several components within the same joint element - for instance, the column web which is simultaneously subjected to compression (or tension) and shear - can obviously lead to stress interactions that are likely to decrease the strength of each individual basic component [5]; this interaction affects the shape of the deformability curve of the related components but does not call the principles of the component method in question again.

The application of the component method requires the following steps :
a) identification of the active components for the studied joint;
b) evaluation of the mechanical characteristics of each individual basic component (specific characteristics - initial stiffness, design strength, ... - or the whole deformability curve);

c) "assembly" of the components in view of the evaluation of the mechanical characteristics of the whole joint (specific characteristics - initial stiffness, design resistance, ... - or the whole deformability M-φ curve).

These three steps are schematically illustrated in Figure 2 in the particular and simple case of a beam-to-column steel joint with a welded connection.

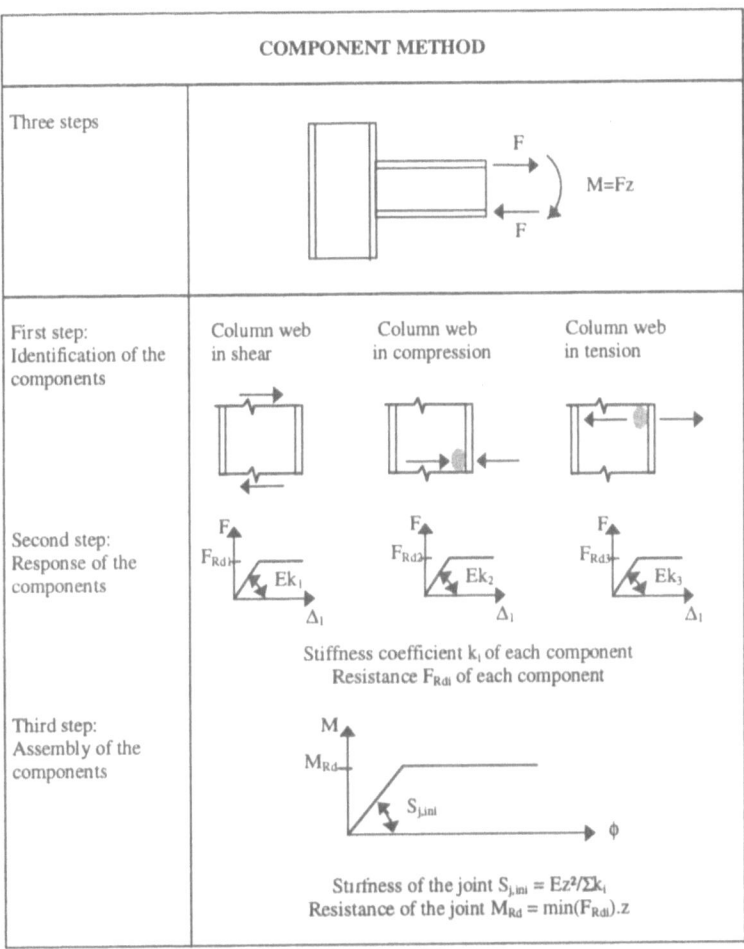

Figure 2. Application of the component method to a welded steel joint
(simplified bi-linear component and joint deformability curves)

The assembly is based on a distribution of the internal forces within the joint. As a matter of fact, the external loads applied to the joint distribute, at each loading step, between the individual components according to the instantaneous stiffness and resistance of each component. Distributions of internal forces may be obtained through different ways as discussed in [5].

The application of the component method requires a sufficient knowledge of the behaviour of the basic components. To review all the publications devoted to this subject would be quite long and therefore references are made here to recent synthesis documents.

The components active in the traditional steel joints have been deeply studied and recommendations for their characterisation are given in the Revised Annex J of Eurocode 3. The combination of these components allows to cover a wide range of joint configurations, what should largely be sufficient to satisfy the needs of practitioners as far as beam-to-column joints and beam splices in bending are concerned. The application is however limited to joints between H or I hot-rolled profiles or built-up ones with similar dimensions.

But some new fields of application have been recently investigated :

- Weak axis joints where the beam is connected to the web of a H or I column profile [5];
- Joints between I beam profiles and tubular columns [8];
- Joints subject to bending moment (and shear) and axial compression or tension forces [5];
- Column bases [3];
- Stell-concrete composite joints [4];
- Steel joints with beam haunches, end-plate stiffeners or high strength steels [5];
- Joints in pitched-roof portal frames where beams and columns form an angle higher than 90° [5];
- Joints between slender built-up welded profiles [5].

Preliminary works also indicate that the component method seems also suitable for the characterisation of joints subjected to extreme loading conditions as earthquakes or fire. Besides that, first attempts have been made in the COST C1 European Action on "Control of the Semi-Rigid Behaviour of Civil Engineering Connections" (1992-1996) to apply the component method to joints in pre-cast and timber construction. So it may reasonably be thought nowadays that an unified characterisation procedure for all structural joints is now developing and will be the common basis for the future design codes whatever is the material or the combination of materials used. This is likely to lighten the work of the designers, in particular when composite construction is of concern (the composite action between the materials is or not effective according to the erection stages).

The framework of the component method is sufficiently general to allow the use of various techniques of component characterisation and joint assembly. In particular, the stiffness and strength characteristics of the components may result from experimentations in laboratory, numerical simulations by means of finite element programs or analytical models based on theory. Often experimentation and numerical simulations are used as references when developing and validating analytical models. The latter may be developed with different levels of sophistication:

- expressions as those presented in PhD theses cover the influence of all the parameters which affect significantly the component behaviour (strain hardening, bolt head and nut dimensions, bolt prestressing, ...) from the beginning of the loading to collapse (plasticity, instability, ...);
- rules such as those which have been introduced in the Revised Annex J of Eurocode 3 are more simple and therefore more suitable for hand calculations;

- as an ultimate step in the simplification process, simplified calculation procedures and design tables for standardised joints or components have been produced [6]; they allow a quick and nevertheless accurate prediction of the main joint properties. Besides that, design software for joints are also available on the market.

Similar levels of sophistication exist also for what regards the joint assembly.

2.2. JOINT MODELLING

Joint behaviour affects the structural frame response and shall therefore be modelled, just as for beams and columns, for the frame analysis and design. Traditionally, the following types of *joint modelling* are considered :

For rotational stiffness :
- rigid
- pinned

For resistance :
- full-strength
- partial-strength
- pinned

When the joint rotational stiffness is of concern, the wording *rigid* means that no relative rotation occurs between the connected members whatever is the applied moment. The wording *pinned* postulates the existence of a perfect (i.e. frictionless) hinge between the members. In fact these definitions may be relaxed. Indeed rather flexible but not fully pinned joints and rather stiff but not fully rigid joints may be considered respectively as effectively pinned and perfectly rigid. The stiffness boundaries allowing one to classify joints as rigid or pinned are discussed in the next section.

For joint resistance, a *full-strength joint* is stronger than the weaker of the connected members, which is in contrast to a *partial-strength joint*. In the everyday practice, partial-strength joints are used whenever the joints are designed to transfer the internal forces but not to resist the full capacity of the connected members. A *pinned joint* is considered to transfer only a limited moment. Related classification criteria are also expressed in the next section.

Consideration of rotational stiffness and joint resistance properties leads traditionally to three significant joint models: rigid/full-strength, rigid/partial-strength and pinned.

However, as far as the joint rotational stiffness is considered, joints designed for economy may be neither rigid nor pinned but semi-rigid. There are thus new possibilities for joint modelling: semi-rigid/full-strength and semi-rigid/partial-strength.

With a view to simplification, Eurocode 3 accounts for these possibilities by introducing three joint models (Table 1) :

- *continuous* :   covering the rigid/full-strength case only;
- *semi-continuous* :   covering the rigid/partial-strength, semi-rigid/full-strength and semi-rigid/partial-strength cases;
- *simple* :   covering the pinned case only.

The following meanings are given to these terms :

- *continuous:*   the joint ensures a full rotational continuity between the connected members;

- *semi-continuous:* the joint ensures only a partial rotational continuity between the connected members;
- *simple:* the joint prevents from any rotational continuity between the connected members;

Table 1. Types of joint modelling

STIFFNESS	RESISTANCE		
	Full-strength	Partial-strength	Pinned
Rigid	Continuous	Semi-continuous	*
Semi-rigid	Semi-continuous	Semi-continuous	*
Pinned	*	*	Simple
* : Without meaning			

Table 2. Joint modelling and frame analysis

MODELLING	TYPE OF FRAME ANALYSIS		
	Elastic analysis	Rigid-plastic analysis	Elastic-perfectly plastic and elastoplastic analysis
Continuous	Rigid	Full-strength	Rigid/full-strength
Semi-continuous	Semi-rigid	Partial-strength	Rigid/partial-strength Semi-rigid/full-strength Semi-rigid/partial-strength
Simple	Pinned	Pinned	Pinned

Table 3. Simplified modelling of joints for frame analysis

JOINT MODELLING	BEAM-TO-COLUMN JOINTS MAJOR AXIS BENDING	BEAM SPLICES	COLUMN BASES
SIMPLE			
SEMI-CONTINUOUS			
CONTINUOUS			

The interpretation to be given to these wordings depends on the type of frame analysis to be performed. In the case of an elastic global frame analysis, only the stiffness properties of the joint are relevant for the joint modelling. In the case of a rigid-plastic analysis, the main joint feature is the resistance. In all the other cases, both the stiffness and resistance properties govern the manner in which the joints should be modelled. These possibilities as well as their physical representation for frame analysis are illustrated in Table 2 and Table 3 respectively.

2.3 JOINT CLASSIFICATION

2.3.1. *Stiffness Classification*
The stiffness classification into rigid, semi-rigid and pinned joints is performed by comparing simply the design joint stiffness to two stiffness boundaries (Figure 3). For sake of simplicity, the stiffness boundaries are derived so as to allow a direct comparison with the *initial design* joint stiffness, whatever the type of joint idealisation that is used afterwards in the analysis.

2.3.2. *Strength Classification*
The strength classification simply consists of comparing the joint *design* moment resistance to "full-strength" and "pinned" boundaries (Figure 4).

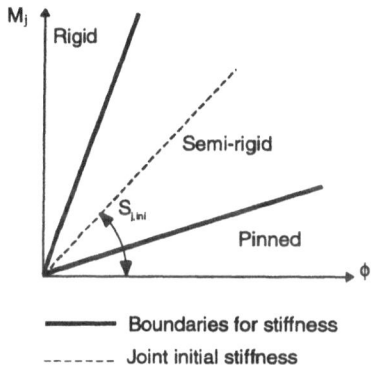

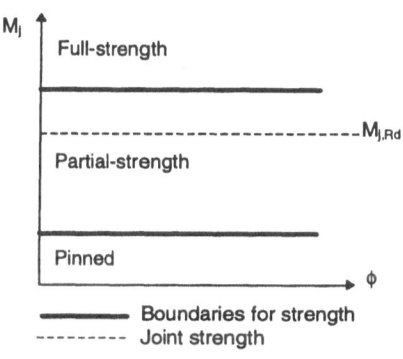

Figure 3. Stiffness classification boundaries

Figure 4. Strength classification boundaries

2.3.3. *Ductility Classes*
Experience and proper detailing result in so-called *pinned* joints which exhibit a sufficient rotation capacity to sustain the rotations imposed on them. For moment resisting joints the concept of ductility classes is introduced to deal with the question of rotation capacity.

Rather few studies have been devoted to the evaluation of the rotation capacity of joints. This is clearly illustrated in Eurocode 3 Revised Annex J where only a quite limited amount of information is given. Criteria should therefore be established to distinguish between "ductile", "semi-ductile" and "brittle" joints. Ductile joints are suitable for plastic frame analysis while brittle ones do not allow any redistribution of internal forces. The use of semi-ductile joints in a plastically designed frame can only result from a preliminary comparison between the available and required rotation capacities.

356

2.4. JOINT IDEALISATION

The non-linear behaviour of the isolated flexural spring (see Table 3) which characterises the actual joint response for frame analysis does not lend itself towards everyday design practice. However the moment-rotation characteristic curve may be idealised without significant loss of accuracy. One of the most simple possible idealisations is the elastic-perfectly plastic relationship (Figure 5.a). This modelling has the advantage of being quite similar to that used for the modelling of member cross-sections subject to bending (Figure 5.b).

The moment $M_{j,Rd}$ that corresponds to the yield plateau is termed the *design moment resistance* in Eurocode 3. It may be considered as the *pseudo-plastic moment resistance* of the joint. Strain-hardening effects and possible membrane effects are henceforth neglected, which explains the difference in Figure 4.a between the actual M-ϕ characteristic and the *yield plateau* of the idealisation.

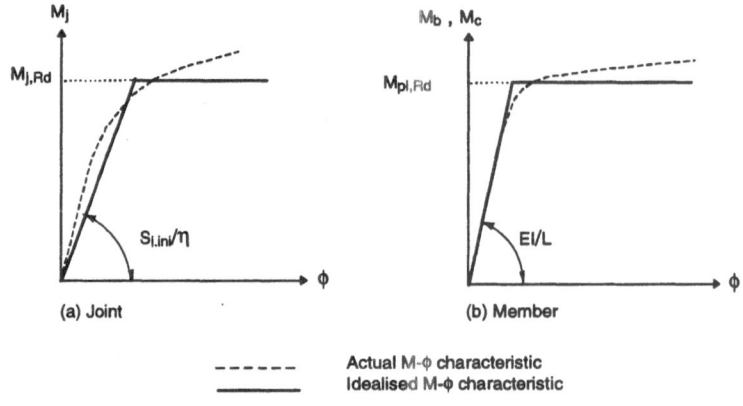

Figure 5. Bi-linearisation of moment-rotation curves

The value of the joint constant stiffness $S_{j.ini}/\eta$ is discussed in [5] and practical values are given in Eurocode 3 Revised Annex J. This coefficient results from the high non-linearity of the joint M-φ curves in comparison to those of the members.

In fact there are different possible ways to idealise a joint M-ϕ characteristic. The choice of one of them is dependent upon the type of frame analysis which is contemplated:

- Elastic idealisation for an elastic analysis;
- Rigid-plastic idealisation for a rigid-plastic analysis;
- Non-linear idealisation for an elastic-plastic analysis.

3. The merits of the consistent approach for structural joints

Both the Eurocode 3 requirements and the desire to model the behaviour of the structure in a more realistic way leads to the consideration of the semi-rigid behaviour when necessary.

Many designers would stop at that basic interpretation of the code and hence would be reluctant to confront the implied additional computational effort involved. Obviously a crude way to deal with this new burden will be for them to design joints that will actually continue to be classified as being either pinned or fully rigid. However such properties will have to be proven at the end of the design process and, in addition, such joints will certainly be found to be uneconomical in a number of situations.

It should be noted that the concept of rigid and pinned joints still exists in Eurocode 3. It is accepted that a joint which is *almost rigid*, or *almost pinned*, may still be considered as being *truly rigid* or *truly pinned* in the design process. How to judge whether a joint can be considered as rigid, semi-rigid or pinned depends on the comparison between the joint stiffness and the frame stiffness (see section 2.3).

The designer is strongly encouraged to go beyond this "all or nothing" attitude. Actually it is important to consider the benefits to be gained from the semi-rigid behaviour of joints. Those benefits can be brought in two ways :

➢ The designer decides to continue with the practice of assuming -sometimes erroneously- that joints are either pinned or fully rigid. However, proper consideration has to be given to the influence that the actual behaviour of the joints has on the global behaviour of the structure, i.e. on the precision with which the distribution of forces and moments and the displacements have been determined. This may not prove to be easy when the joints are designed at a late stage in the design process since some iterations between global analysis and design checking may be required. Nevertheless, the following situations can be foreseen:

- So that a joint can be assumed to be rigid, it is common practice to introduce web stiffeners in the column. Eurocode 3 now provides the means to check whether such stiffeners are really necessary for the joint to be both rigid and have sufficient resistance. There are practical cases where they are not needed, thus permitting the adoption of a more economical joint design.

- When joints assumed to be pinned are later found to have fairly significant stiffness (i.e. to be semi-rigid), the designer may be in a position to reduce beam sizes. This is simply because the moments carried by the joints reduce the span moments in the beams.

➢ The designer decides to give consideration, at the preliminary design stage, not only to the properties of the members but also to those of the joints. It may be shown [6] that this new approach is not at all incompatible with the sometimes customary separation of the design tasks between those who have the responsibility for conceiving the structure and carrying out the global analysis and those who have the responsibility for designing the joints. Indeed, both tasks are very often performed by different people, or indeed, by different companies, depending on national or local industrial habits. Adopting this novel early consideration of joints in the design process requires a good understanding of the balance between, on the one hand, the costs and the complexity of joints and, on the other hand, the optimisation of the structural behaviour and performance through the more accurate consideration of joint behaviour for the design as a whole. Two examples are given to illustrate this:

- It was mentioned previously that it is possible in some situations to eliminate column web stiffeners and therefore to reduce costs. Despite the reduction in its stiffness and, possibly, in its strength, the joint can still be considered to be rigid and be found to have sufficient strength. This is shown to be possible for industrial portal frames with rafter-to-column haunch joints in particular, but other cases can be envisaged.
- In a more general way, it is worthwhile to consider the effect of adjusting the joint stiffness so as to strike the best balance between the cost of the joints and the costs of the beams and the columns. For instance, for braced frames, the use of semi-rigid joints, which are not necessarily more costly than the pinned joints, leads to reducing the beam sizes. For unbraced frames, the use of less costly semi-rigid joints, instead of the rigid joints, leads to increased beam sizes and possibly column sizes.

Of course the task may seem a difficult one, and this is why a design handbook devoted to this new concept has been recently published [6]. The whole philosophy could be termed as *"Because you must do it, take advantage of it"*. The designer has therefore the choice between a *traditionalist attitude*, where however something may often be gained, and an *innovative attitude*, where the most economical result [7] may best be sought.

4. Practical applications of the Eurocode 3 design approach

In the next pages a practical application is shown where one of the design opportunities described in the previous section is illustrated.

4.1. BEAM-TO-COLUMN JOINT IN AN INDUSTRIAL PORTAL FRAME

The beam-to-column joint represented in Figure 6 is extracted from an industrial pitched-roof portal frame with an internal span of 20 m and a height of 7m. Its design has been initially achieved by a Belgian constructor on the basis of a traditional "rigid" approach.

In a second step, the evaluation of the mechanical properties of this joint has been performed by means of Eurocode 3 Revised Annex J; this gives:
- *Initial rotational stiffness*: $S_{j.ini} = 114.971$ kNm/rad
- *Moment resistance*: $M_{Rd} = 281,6$ kN

To fit with the designer's expectations, these two values have to be respectively higher than:
- The "rigid" boundary stiffness value (see 2.3.1) which, in this particular case [2], equals 85.627 kNm/rad.
- The maximum bending moment transferred by the joint; this one results from a global frame analysis and it amounts 171 kNm.

As the two design conditions are fulfilled, the joint design may be considered as fully satisfactory.

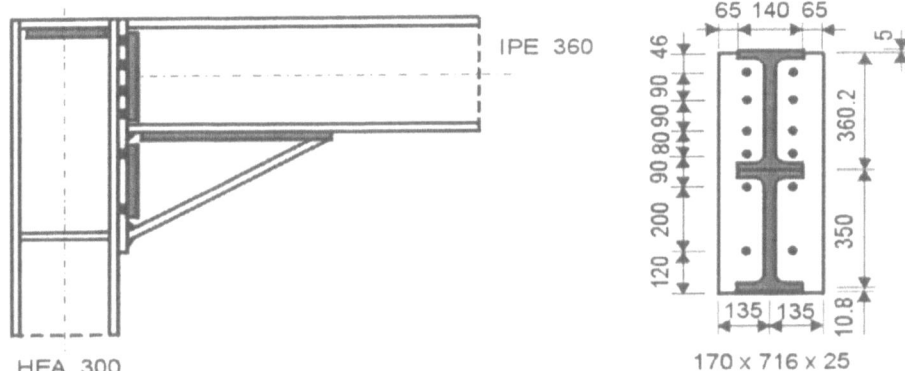

IPE 360

HEA 300

65 140 65

46
90 80 90 90
200
120

360.2

350

10.8

135 135

170 x 716 x 25

Figure 6. Geometry of the joint

Nowadays the stiffness verification requested by Revised Annex J is not achieved and the joint is simply considered as "rigid" on the basis of the designer's experience. The extra work required by Eurocode 3 in order to demonstrate that the joint is rigid could therefore be considered as unnecessary.

But before discussing it further, the following statements has first to be made:

- The lake of knowledge in the field of joint design has led designers to systematically use transverse column stiffeners (in the compression and tension zones of the joints) in combination with thick end-plates when defining the geometry of rigid joints.
- Because of this "over-stiffening", the distribution of internal forces in the joints is of a linear type; this explains why an elastic approach has been traditionally followed by the designers to derive the bolt forces.
- In such a design procedure the bolts are the weak elements in the joints and no check of the plate elements (end-plates, column flanges, web flanges, ...) is required, what is the case in many national codes.

As it is seen this traditional design approach is fully consistent but often leads to over-designed joints and uneconomical configurations as the fabrication are highly dependent on the degree of stiffening. Furthermore the stiffening usually prevents from an easy erection on site. Finally the lack of ductility associated to the bolt failure is far from being satisfactory for what regards ductility aspects.

As an alternative Eurocode 3 Revised Annex J proposes design rules where no predefined collapse mode is selected. The stiffness and resistance properties of all the constitutive components are integrated in the calculation. In other words, the actual properties of the joints are evaluated and it is up to the designer to modify the joint geometry according to his needs: select a ductile mode of collapse, make the joint more resistant, more stiff, ...

In the present case, the properties of the joint are beyond those required by the code (the resistance is larger than needed and a rotational stiffness of 85.627 kNm/rad is enough to ensure that the joint is rigid, as assumed in the frame analysis). Therefore a simplification of the joint geometry is likely to reduce the fabrication costs of the joints, and therefore of the whole structure.

In Table 4 three new joint configurations are considered:

- N°1: Initial geometry, but without the column transverse stiffener in the compression zone.
- N°2: Configuration N°1, but without the column transverse stiffener in the tension zone.
- N°3: Configuration N°3, but with three bolt-rows instead of four between the beam flanges.

For each of these configurations, the stiffness and resistance properties are listed and the result of the design checks is indicated.

In Table 5 the reduction of the fabrication costs for each of these configurations is given, the price reference being that of the initial stiffened configuration.

In conclusion it may be seen that, even for rigid joints, a quite significant benefit may be obtained by referring to the new design concept introduced in Eurocode 3. This one results from a better definition of the word "rigid" and the use of accurate calculation models for the prediction of the stiffness and resistance joint properties.

Table 4. Joint properties and design checks

Configuration	M_{Rd} (kNm)	$S_{j.ini}$ (kNm/rad)	Satisfactory ?
N°1	255,0	92.706	O.K.
N°2	250,6	89.022	O.K.
N°3	247,8	87.919	O.K.

Table 5. Relative fabrication costs

Joint configuration	Relative fabrication cost
Initial configuration	100%
Alternative N°1	87%
Alternative N°2	73%
Alternative N°3	72%

4.2. JOINT IN SWAY AND NON-SWAY BUILDING FRAMES

The interested reader will find many others involving pinned, semi-rigid and rigid joints in [6].

5. Conclusions

A new consistent design approach for structural joints is now provided by Eurocode 3. It is based on research works performed in the two last decades. For what concerns the joint characterisation, practical guidelines are now available for most of the steel joints (beam-to-column joints, beam splices, column bases, ...) in building frames made of I or I profiles and subjected to static loading. Its extension to composite joints has already been achieved and researches are in progress for a further application to other types of profiles and to other loading situations.

This new design procedure has been applied successfully in the last years to some different projects where substantial benefit resulting from a decrease of structural weight or a reduction of the fabrication and erection costs has been made possible. Such an example is described in the paper.

6. References

1. Eurocode 3 (1992), Design of Steel Structures. Part 1.1: General Rules and Rules for Buildings, European Prestandard – ENV 1993-1-1.
2. Revised Annex J of Eurocode 3 (1998), *Joints in building frames*, European Prestandard ENV 1993-1-1:1992/A2, CEN, Bruxelles, Belgium.
3. COST C1 (1999), *Column bases in steel building frames*, COST C1 report edited by K. Weynand, European Commission, Bruxelles, Luxembourg.
4. COST C1 (1999), *Composite steel-concrete joints in frames for buildings: design provisions*, COST C1 report edited by D. Anderson, European Commission, Bruxelles, Luxembourg.
5. Jaspart, J.P. (1997), Recent advances in the field of steel joints – Column bases and further configurations for beam-to-column joints and beam splices, Professorship Thesis, Department MSM, University of Liège, Belgium.

362

6. Maquoi, R & Chabrolin, B. (1998), *Frame design including joint behaviour*, ECSC Report 18563, Office for Official Publications of the European Communities, Luxembourg.
7. Weynand, K (1997), *Sicherheits- und Witschaftlichkeits-untersuchungen zur Anwendung nachgiebiger Anschlüsse im Stahlbau*, Heft 35, Shaker Verlag, Aachen, Germany.
8. Vandegans, D. (1996), Use of the threaded studs in joints between I-beams and RHS columns, Proceedings of the Istanbul Colloquium on Semi-Rigid Connections held on September 25-27, pp. 53-62, IABSE, Zürich, Switzerland.

ROTATION CAPACITY OF STEEL JOINTS: VERIFICATION PROCEDURE AND COMPONENT TESTS

U. KUHLMANN & F. KÜHNEMUND
Institut für Konstruktion und Entwurf I
Pfaffenwaldring 7, 70569 Stuttgart, Germany

1. Introduction

Modern design of steel structures has recently developed two main tendencies: The implementation of semi-rigid joints and the application of plastic analysis methods. To combine both aspects for the design of a structural system special view has to be given to the rotation capacity of semi-rigid joints. This paper suggests a verification procedure for the check of sufficient rotation capacity of semi-rigid joints. Based on experimental and theoretical investigations deemed-to-satisfy criteria can be derived which replace an explicit check of sufficient rotation capacity. For welded joint configurations available rotation capacity values are estimated with reference to component tests. For these tests special attention is given to the determination of the load deformation behaviour.

2. Definition of Rotation Capacity

Rotation capacity characterises the ability of a plastified joint to rotate while maintaining its design moment resistance. Figure 1 illustrates this definition. Following the M-Φ-curve the purely elastic rotation Φ_{el} is reached at the level of the elastic design moment resistance. To get to the plastic design moment resistance $M_{pl,Rd}$ another rotation Φ_{tr} is necessary. Due to strain hardening effects the real moment exceeds the level of $M_{pl,Rd}$. Only the plastic rotation Φ_{pl} from the point the real moment rotation curve reaches the level of $M_{pl,Rd}$ first to the point the real curve reaches this level again is decisive for the moment redistribution in a system. Consequently the available rotation capacity of a joint is defined as the difference $\Phi_{pl} = \Phi_{Cd} - \Phi_{Xd}$.

Though EC 3, revised Annex J [1] demands the check of sufficient rotation capacity of joints to allow for rigid-plastic analysis, so far no general method of verification exists. Investigations presented in this paper therefore aim at the derivation of simplified rules similar to those of members [2,3,4], which substitute an explicit check of rotation capacity values. A verification procedure for the check of sufficient rotation capacity is given in the following.

C.C. Baniotopoulos and F. Wald (eds.), The Paramount Role of Joints into the Reliable Response of Structures, 363–372.
© 2000 *Kluwer Academic Publishers.*

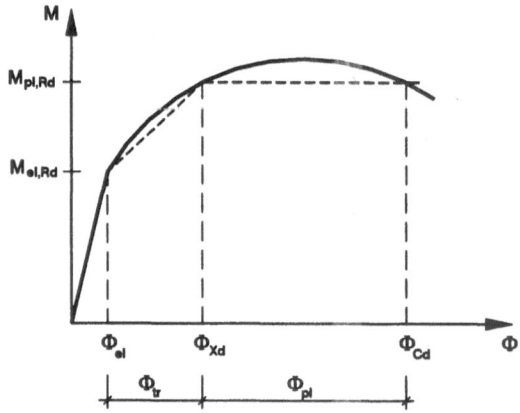

Figure 1. Definition of rotation capacity for a joint

3. Verification of Sufficient Rotation Capacity of Steel Joints

Figure 2 shows an overview over the verification procedure. The deemed-to-satisfy criteria are the results of a thorough scientific investigation. Starting with the check of individual joint components, research work aims at the determination of their load-deformation behaviour. To characterise a joint component's load-deformation behaviour different possibilities exists: Load-deformation curves gained by **tests** include all kind of realistic material and system influences as strain-hardening, material overstrength, post-limit stiffness, imperfections or residual stresses. But of course the number of variable parameters which can be looked at is limited because the important expenses for each test specimen. By help of **Finite-Element studies** the parameter field can be extended. Test results are used to calibrate the numerical models to achieve an optimum approximation of the real behaviour. With the help of **mechanical models** an analytical description of the load-deformation behaviour is possible. This is particularly important because the formulas describing the load-deformation behaviour of the single components form the basis for the model for the overall moment-rotation behaviour of the joint: The load-deformation curves of the components are identified as springs, which can be added parallel or in series to the moment-rotation curve of the whole joint [5]. These joint curves have to be checked and calibrated against tests on corresponding joint configurations.

As illustrated in chapter 2 available rotation capacity values of joints can be gained from these moment-rotation curves, whereas required rotation capacity values result from thorough investigations on structural systems, considering the semi-rigidity of the joints. As the system calculation according to plastic hinge theory only considers the rotation of a hinge after it has formed i.e. after reaching the maximum design moment resisistance only the plastic rotation value of the joint design curve Φ_{pl} is of interest, see Figure 1.

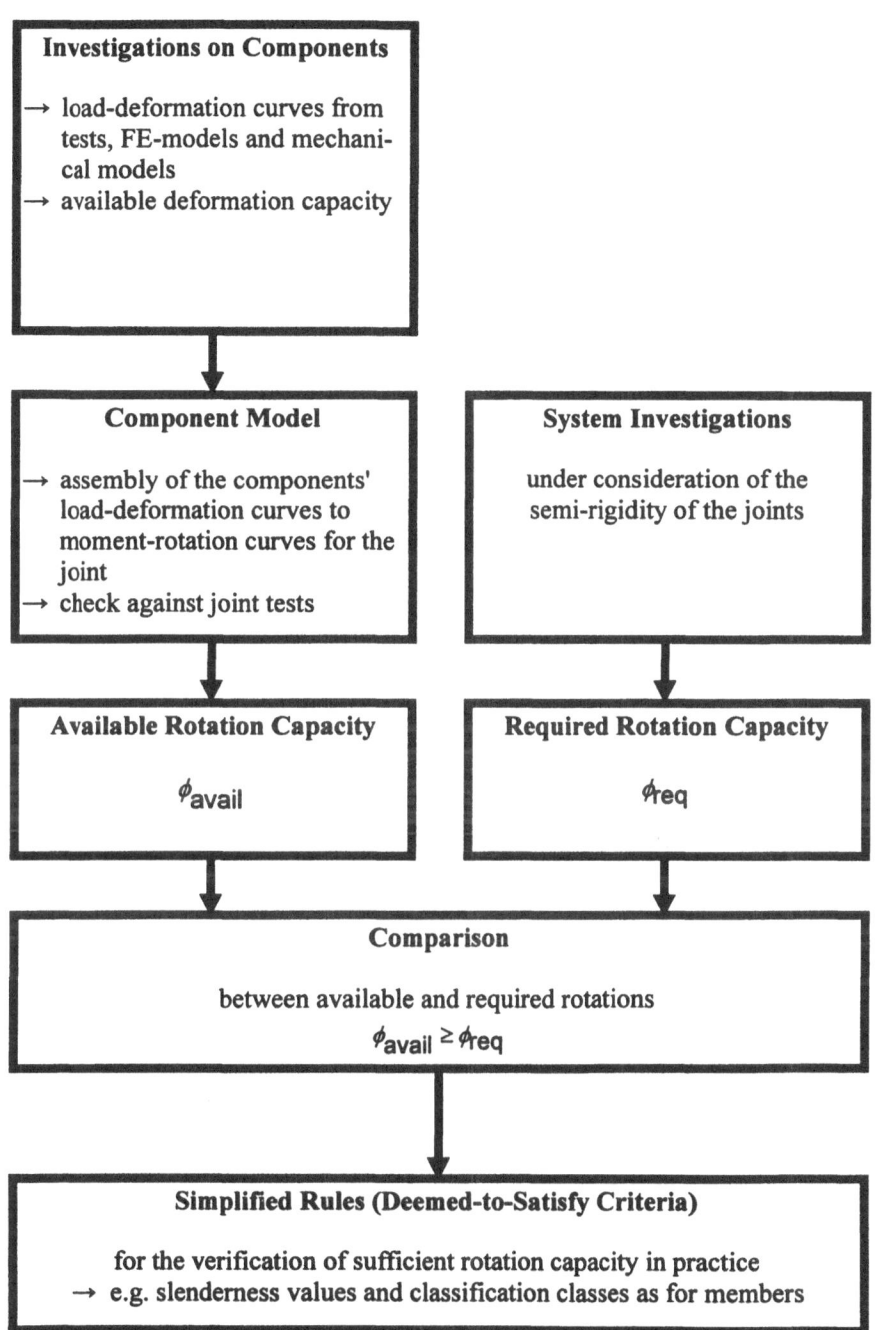

Figure 2. Procedure for the check of sufficient rotation capacity

Sufficient rotation capacity is provided if the available rotation capacity Φ_{avail} is higher than the required one Φ_{req}. But this procedure is too complicated for practical use. So systematical investigations for a range of typical systems and load situations have to be executed, which lead to boundary values for required joint rotations. These values are compared to the available joint rotations to define limits which joints have to fulfil to meet the demands of sufficient rotation capacity.

As mentioned above one possibility to gain load-deformation curves of components are tests. In the following tests on the joint component "column web in compression" and corresponding available deformation capacity values are shown. The contribution of this component to the overall rotation capacity of a joint is estimated, choosing a welded joint configuration (see chapter 5).

4. Tests on the Component "Column Web in Compression"

4.1. INTRODUCTION

In 1997 a first test series of six tests (Series I: Specimen A1 - A3 and B1 - B3) were performed at the University of Stuttgart, see [6]. The aim was
- to investigate not only strength and stiffness of the component "column web in compression", but also the deformation capacity especially in the post-critical region and
- to investigate the influence of axial forces on the load-deformation behaviour of the component "column web in compression" which is essential for the rotational behaviour of typical semi-rigid joints.

In 1999 the first test series of 1997 was supplemented by a second test series of 10 tests (Series II: Specimen A4 - A8 and B4 - B8), see [7,8]. Again the axial force in the column was the main parameter. Table 1 gives an overview on all 16 tests. To have a reference value without the influence of an axial force also in each series reference tests without axial force in the column (A3, B1 and A8, B8), one for each profile type HE 240A and HE 240B, were executed.

4.2. TEST SET UP

The test set up is shown in Figure 3. Whereas the axial force N of the column was applied by an horizontal hydraulic jack built within an horizontal frame of end-plates and 4 rods, the compression force F for the web was applied by the vertical testing machine. All tests were controlled by deformation in order to follow the decreasing part of the curves. Furthermore about ten times the tests were stopped for relaxation.

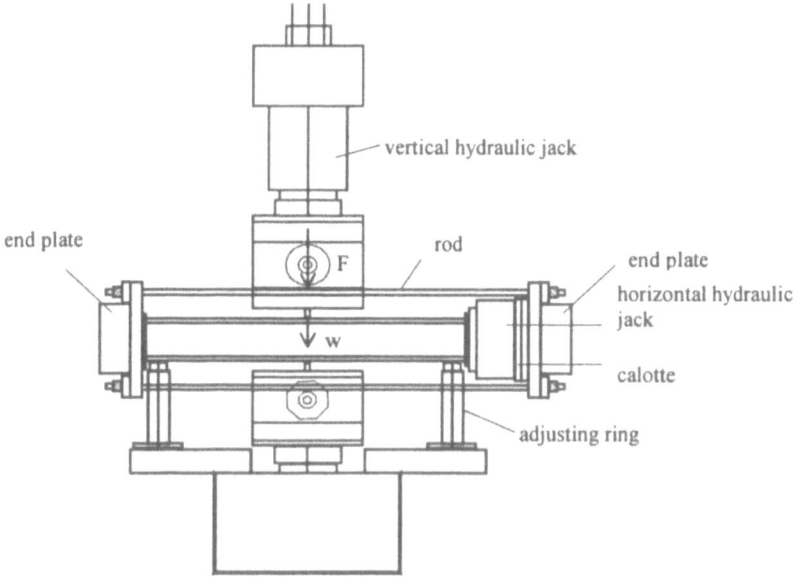

Figure 3. Test set up

4.3. LOAD-DEFORMATION BEHAVIOUR

As main results the applied compression force F is referred to the measured deformation w of the web in direction of the force F, see Figure 4.

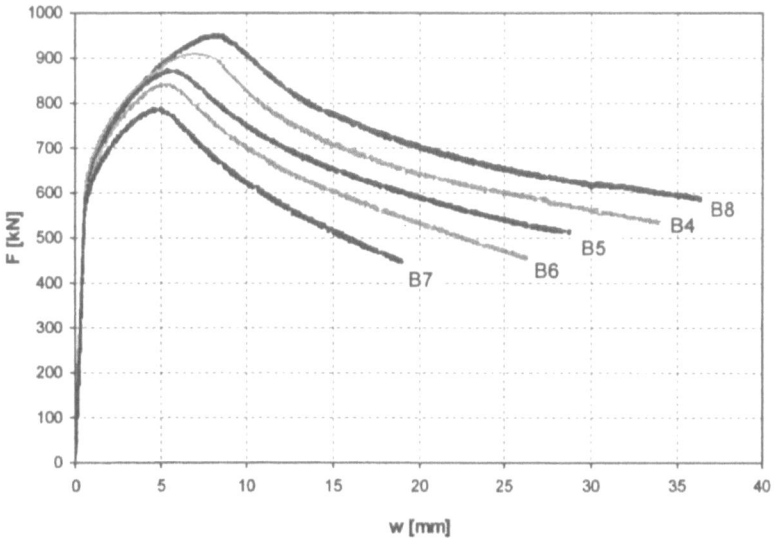

Figure 4. Load-deformation curves for test series II (1999); profiles HEB 240

The load-deformation curves clearly show the typical very ductile behaviour of the component "column web in compression". In Figure 4 the test results of five HEB 240 profiles of the second test series of 1999 are compared to each other. Whereas test specimen B8 had no axial force, B4, B5, B6 and B7 were simultaneously loaded by axial forces of 11%, 24%, 36% and 58% of N_{pl}. The diagram clearly shows the strong influence of the axial force on the load-deformation behaviour of the web.

4.4. DEFORMATION CAPACITY

The main objective of these tests was to evaluate the deformation capacity of the component "column web in compression" at the level of the plastic resistance. In contrast to earlier tests this time also the behaviour of the specimen after the vertical force F had passed its maximum was considered. To reach the characteristic plastic resistance of the component F_{pl} a deformation w_{pl} is needed. Only the additional deformation ($w_{max} - w_{pl}$) may be used for a sufficient rotation capacity for the redistribution of internal forces within the system without reducing the resistance of the component under its plastic value. So the deformation capacity is defined as distance ($w_{max} - w_{pl}$), which corresponds to the definition of the rotation capacity, see chapter 1. As can be seen by the test curve in Figure 5, the contribution of the descending part of the load-deformation curve w_{decr} is of higher importance than the increasing part w_{incr}. Table 1 contains deformation capacity values of all executed tests. Also the contribution of the increasing and decreasing parts are given. The last column of Table 1 shows the ratio w_{decr}/w_{avail}. Percentage values of more than 50 % support the importance of the decreasing part for the available deformation capacity of the component "column web in compression".

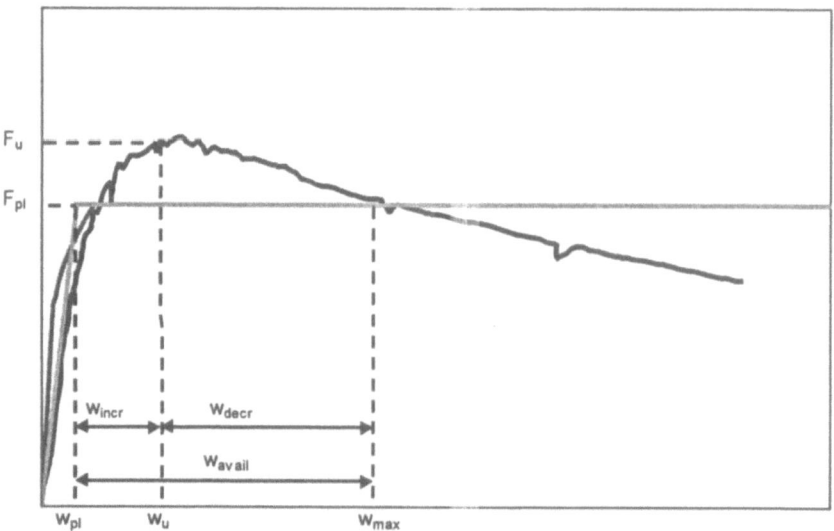

Figure 5. Definition of deformation capacity

For each of the two profiles HEA 240 and HEB 240 the values of deformation capacity w_{avail} of the tests with axial forces are divided by the deformation values $w_{avail,ref}$ of the reference test specimen without axial force. As shown in Figure 6 for the profile HEB 240 the deformation capacity decreases nearly linearly with increasing axial force.

Table 1. Deformation capacity values of the tests

Test specimen	Profile	N/N_{pl}	W_{pl} [mm]	W_u [mm]	W_{avail} [mm]	W_{incr} [mm]	W_{decr} [mm]	W_{decr}/W_{avail} [%]
A1	HEA 240	0,67	1,24	2,12	2,42	0,88	1,54	64
A2	HEA 240	0,60	1,33	2,21	2,14	0,88	1,26	59
A3	HEA 240	0,00	1,07	2,83	11,85	1,76	10,09	85
A4	HEA 240	0,11	0,51	1,94	7,99	1,43	6,56	82
A5	HEA 240	0,22	0,49	1,36	6,79	0,87	5,92	87
A6	HEA 240	0,33	0,66	1,86	5,63	1,20	4,43	79
A7	HEA 240	0,47	0,74	1,84	3,57	1,10	2,47	69
A8	HEA 240	0,00	0,62	2,31	10,05	1,69	8,36	83
B1	HEB 240	0,00	1,40	4,73	11,46	3,33	8,13	71
B2	HEB 240	0,46	1,64	3,28	4,81	1,64	3,17	66
B3	HEB 240	0,65	1,61	2,94	3,05	1,33	1,72	56
B4	HEB 240	0,11	0,67	7,01	21,03	6,34	14,69	70
B5	HEB 240	0,24	0,89	5,35	15,52	4,46	11,06	71
B6	HEB 240	0,36	0,93	5,05	12,61	4,12	8,49	67
B7	HEB 240	0,58	1,01	4,95	8,79	3,94	4,85	55
B8	HEB 240	0,00	0,88	8,30	26,34	7,42	18,92	72

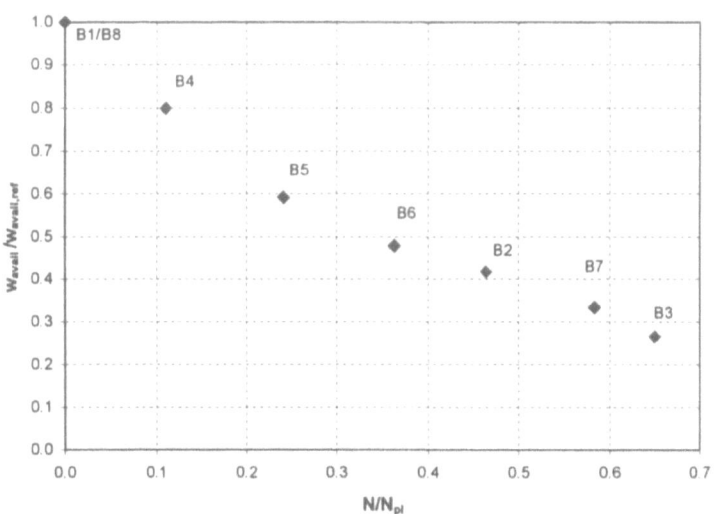

Figure 6. Ratio of deformation capacity in dependence of the axial force (HEB 240)

5. Estimation of Rotation Capacity Values for Welded Joints

To gain an estimation of the contribution of the component "column web in compression" to the rotation capacity of a joint, a welded steel joint configuration is chosen. Whereas for bolted end plate connections the point of rotation has to be found by an iterative calculation procedure [5,9] for welded configurations the point of rotation is to be assumed in the centre line of the joint, see Figure 7. As a consequence the deformation contribution of the tension component of the welded joint is about the same as that of the component "column web in compression". Consequently the rotation capacity is found by the following:

$$\Phi = \frac{W_{tension} + W_{compression}}{z} = \frac{2 \cdot W_{compression}}{z} \tag{1}$$

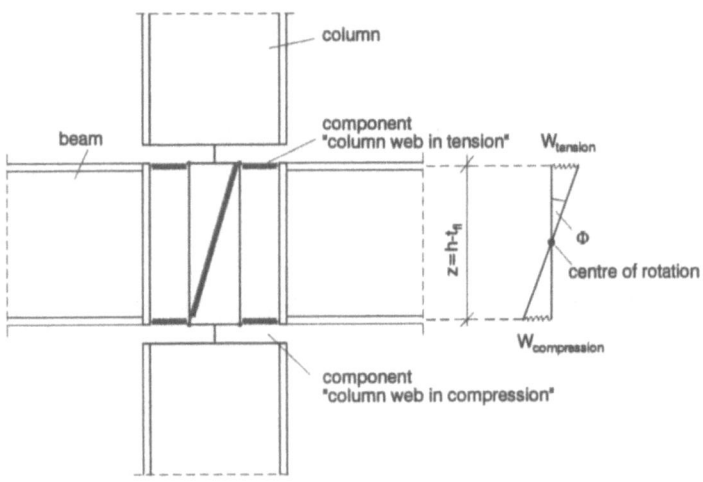

Figure 7. Component model for a welded steel joint

Table 2 shows the results of a study for welded joints. The column profiles are chosen corresponding to the executed tests (profiles HEA 240 and HEB 240). For the beams four different types of IPE profiles are taken into account. The lever arm z is derived from the following expression:

$$z = h_{beam} - t_{fl} \tag{2}$$

As can be seen from Table 2 all joint types without an axial force in the column show a high rotation capacity. But even with an axial force of about 30% of the plastic axial force the available rotation capacity is still of a sufficiently large amount, also for the lever arm reaching values up to about 600 mm. Joints with an ratio of N/N_{pl} of more than or equal to 0,6 dispose a rather small rotation capacity especially for adjacent beams with large height. As the latter configuration is of practical importance, extensive investigations for the rotation capacity of joints still have to be done.

A comparison between the two column profiles show the decisive influence of the flange and web slenderness. The compact profile HEB 240 provide about 2.5 times larger rotation capacity values than the slender HEA 240 profile.

Both influences are now thoroughly investigated also by Finite-Element studies and mechanical models are derived to describe the behaviour [10].

Table 2. Estimation of rotation capacity values based on the component test results

Column profile	ratio N/N_{pl}	w_{avail} [mm]	Beam profile	Lever arm z [mm]	Rotation capacity Φ [mrad]
HEA 240	0,00	10,05	IPE 200	191,5	105
			IPE 300	289,3	69
			IPE 450	435,4	46
			IPE 600	581,0	35
HEA 240	0,33	5,63	IPE 200	191,5	59
			IPE 300	289,3	39
			IPE 450	435,4	27
			IPE 600	581,0	19
HEA 240	0,60	2,14	IPE 200	191,5	22
			IPE 300	289,3	15
			IPE 450	435,4	10
			IPE 600	581,0	7
HEB 240	0,00	26,34	IPE 200	191,5	275
			IPE 300	289,3	182
			IPE 450	435,4	121
			IPE 600	581,0	91
HEB 240	0,36	12,61	IPE 200	191,5	132
			IPE 300	289,3	87
			IPE 450	435,4	58
			IPE 600	581,0	43
HEB 240	0,65	3,05	IPE 200	191,5	32
			IPE 300	289,3	21
			IPE 450	435,4	14
			IPE 600	581,0	10

6. Conclusions

To allow plastic analysis for systems with semi-rigid joints the rotation capacity of the joints has to be checked to be sufficient. The available rotation capacity for a certain type of joint has to be compared with the required rotation capacity values derived from system investigations. But as this procedure is too complicated for daily use a verification procedure which leads to simple design rules was presented. Special attention was given to tests for the component "column web in compression". Among other results the deformation capacity of the component was derived from these tests. By that it has been shown that the deformation capacity tends to decrease nearly linearly with an increasing axial column load. Finally based on these tests rotation capacity values for welded joints on basis of the component tests were estimated.

7. Acknowledgements

The authors like to thank the "Arbeitsgemeinschaft industrieller Forschungs-vereinigungen "Otto von Guericke" e.V. (AiF)" for the financial support of the second tests series within the frame of a research project, which the authors undertake together with Prof. Dr.-Ing. G. Sedlacek of RWTH Aachen and his team [8]. The authors thank them and the accompanying group of practitioners for the valuable discussions and collaboration.

8. References

1. Revised Annex J of Eurocode 3 (1997) Joints in Building Frames, Edited approved draft, CEN Document CEN/TC 250/SC 3 - N 671 E.
2. Eurocode 3 (1992) - Design of steel structures. Part 1-1: General rules and rules for buildings, European prestandard.
3. Kuhlmann, U. (1997) Verification procedure for rotation capacity of joints, Document COST C1/WD2/97-24, Innsbruck.
4. Kuhlmann, U. and Kühnemund, F. (2000) Procedures to Verify Rotation Capacity, in M. Ivanyi and C. C. Baniotopoulos (eds.), *Semi-rigid Connections in Structural Steelwork*, CISM Lecture Notes 215, Springer Verlag (in print).
5. Huber, G. (1999) Nonlinear calculations of composite sections and semi-continuous joints, Doctoral thesis, University of Innsbruck.
6. Kuhlmann, U. and Fürch, A. (1997) Deformation capacity of the component "column web in compression", Internal test-report, University of Stuttgart.
7. Kuhlmann, U. and Kühnemund, F. (1999) Deformation capacity of the component „column web in compression", Internal test report, University of Stuttgart.
8. Kuhlmann, U.; Sedlacek, G.; Kühnemund, F.; Stangenberg, H. (2000). Verformungsverhalten der Komponenten von wirtschaftlichen steifenlosen Anschlußkonstruktionen für die Anwendung plastischer Bemessungskonzepte im Stahlbau. Research Project, Arbeitsgemeinschaft industrieller Forschungsvereinigungen "Otto von Guericke" e.V., not yet finished.
9. Tschemmernegg, F., Huber, G., Huter, M., and Rubin, D. (1997) Komponentenmethode und Komponentenversuche zur Entwicklung von Baukonstruktionen in Mischbauweise, *Stahlbau* 66, 624 - 638.
10. Kühnemund, F. (2001). *Zum Rotationsnachweis nachgiebiger Anschlüsse im Stahlbau*, Doctoral thesis in preparation, Universität Stuttgart.

REVIEW OF DEFORMATION CAPACITY OF JOINTS RELATED TO STRUCTURAL RELIABILITY

C.M. STEENHUIS, H.H. SNIJDER & F. VAN HERWIJNEN
Eindhoven University of Technology
Faculty of Architecture, Building and Planning
Department of Structural Design, Postvak 7
P.O. Box 513
5600 MB Eindhoven
The Netherlands

1. Introduction

Design rules in modern design codes on steel and composite construction, like the Eurocodes, are nowadays calibrated with reliability studies. The reliability studies focus very much on the resistance of the structures. The structural reliability however also depends on the rotation capacity of the (semi-rigid) joints in the structure.

In this paper, a review is given of the state of knowledge regarding rotational capacity of joints in steel and composite structures. Secondly, a literature review is presented regarding the effect of the rotational capacity of joints on the reliability of the structure. The paper concludes with recommendations for further research.

The structural response of buildings is -in general- dependent on three different mechanical characteristics: resistance (in the sense of bearing capacity), stiffness and deformation capacity. The resistance represents the capability of a structure to resist the loads working on the structure (self weight, live loads, wind loads, etc.). The stiffness is of importance in view of the serviceability of a building structure (for instance, deformations or vibrations should be within certain limits so people feel comfortable in a building). Deformation capacity is the capacity of a structure or a part of a structure to deform without loss of resistance. Deformation capacity is required to allow redistribution of the loads in a structure. Deformation capacity yields ductile behaviour and will avoid sudden failure of the structure.

In design codes much attention is paid to resistance, some attention is paid to stiffness and little attention is paid to deformation capacity. Structural reliability depends on both resistance and deformation capacity and therefore, deformation capacity deserves more attention.

C.C. Baniotopoulos and F. Wald (eds.), The Paramount Role of Joints into the Reliable Response of Structures, 373–386.
© 2000 *Kluwer Academic Publishers.*

This paper gives an overview of the state-of-the art on rotation capacity of (semi rigid) joints in steel and composite structures. Rotation capacity of joints is a special case of deformation capacity. It refers to the capability of a joint to rotate without loss of resistance. First, attention is paid to rotation capacity as a design issue. Then an overview is given of the kind of code provisions, which are currently present in the European Standards on rotation capacity. Definitions of rotation capacity are reviewed. The current state of knowledge is reported on models for the prediction of rotation capacity. Special attention is paid towards the effect of the rotation capacity on the reliability of the structure. Finally, conclusions are drawn.

This state of the art overview is part of a PhD research to the relation between the reliability of structures and deformation capacity. The research takes place at Eindhoven University of Technology.

2. Rotation Capacity as a Design Issue

A structure should be designed such that failure of a structure due to lack of rotation capacity does not occur. The reason is that the failure mode should be ductile and the users of the building are "warned" with the occurrence of large deformations before collapse.

There are various possibilities how to deal with this design issue:
- by performing a check whether the available rotation capacity is larger than the required rotation capacity;
- by "deemed to satisfy" rules;
- by a classification system.

The first possibility, checking if the available rotation capacity is larger than the required rotation capacity, is for instance already included in some design standards. No guidance is given on how to calculate the required and available deformation capacity. Therefore, for a designer in practice, this rule is difficult to apply.

Deemed to satisfy rules are much applied in design standards because of their practical applicability. For instance, Eurocode 3 Annex J [1] states that provided that in a beam-to-column joint shear of the column web panel is the governing failure mode, there is sufficient deformation capacity. In that case, no further check of rotation capacity is needed. Deemed to satisfy rules allow structural engineers to design for rotation capacity, in other words, selecting the layout of a joint such that ductility is achieved.

The use of a classification system is a third option for design for rotation capacity. Kuhlmann [2, 3] gives a description of a possible classification system for ductility of joints based on the classification system of deformation capacity for steel sections applied in beams, columns or beam-columns. The basic idea of the classification system for steel members is that based on certain geometrical and material properties, a structural element is able to undergo deformations allowing:

- plastic frame design and design of the section based on its full plastic resistance (class 1);
- elastic frame design and design of the section based on its full plastic resistance (class 2);
- elastic frame design and design of the section based on its full elastic resistance (class 3);
- elastic frame design and design of the section based on its partial elastic resistance (class 4).

However, the criteria to classify joints similarly to sections do not exist at the moment.

3. Code Provisions in European Standards for Steel and Composite Joint Design

Steel joints transferring bending moments are treated in Eurocode 3 Annex J. This is an extensive annex giving all guidance required for the determination of the resistance, stiffness and rotation capacity of steel joints. Figure 1 shows some typical joint configurations covered by Annex J. Only one page of the total size (approx. 50 pages) is devoted to rotation capacity. This is due to the limited knowledge available on this topic and its complexity.

Figure 1. Typical joint configurations covered by Annex J.

In Eurocode 3 Annex J, first a general statement is given concerning rotation capacity: joints should have sufficient rotation capacity when plastic global analysis is used, except when plastic rotations form in the connected members.

Then, clauses are given which can be categorized as follows:

- clauses stating that if in a joint a certain failure mechanism occurs, there is insufficient rotation capacity for plastic global analysis;

- clauses stating that if some other failure mechanisms govern the behaviour of the joint, there may be assumed to be sufficient deformation capacity.
- clauses giving for some types of joints and failure mechanisms, a specific value of the minimum available rotation.

In Eurocode 4, dealing with composite structures, no specific part is included for the design of joints. However, in ECCS publication 109 [4], model clauses have been proposed for Eurocode 4. Figure 2 shows joint configurations covered in this publication. In these model clauses the same rules apply as those given in Eurocode 3. Additionally, it is explicitly stated that the rotation capacity in a joint may be determined based on tests or models based on tests, provided that variations in material and geometrical properties have been taken into account.

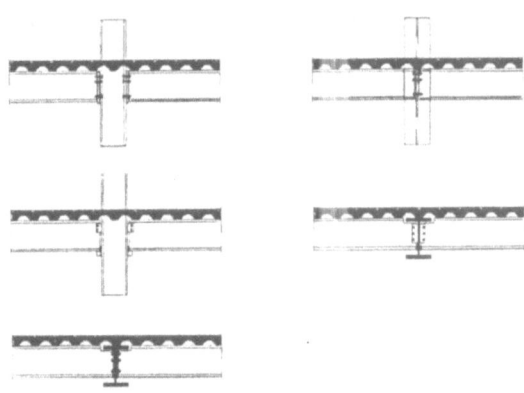

Figure 2. Joint configurations covered in ECCS publication 109.

4. Definitions

In literature, different definitions for rotation capacity are given. It is important to consider two aspects relating to the definitions:
- the way the rotation capacity is defined in a design moment rotation diagram;
- the relation between the design moment rotation diagram and a test result.

For the design curve given in Eurocode 3 [1] and in the model clauses for Eurocode 4 [4], the definitions of the moment resistance $M_{j.Rd}$ and the rotation capacity of a joint ϕ_{Cd} are given in Figure 3.

Kuhlmann et al. [2] give other definitions for the design rotation capacity, see Figure 4.

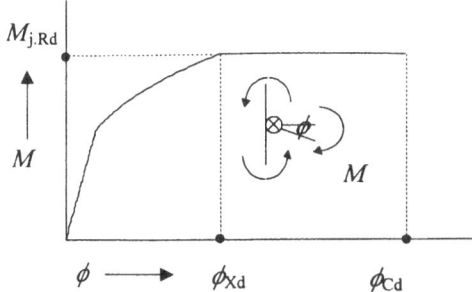

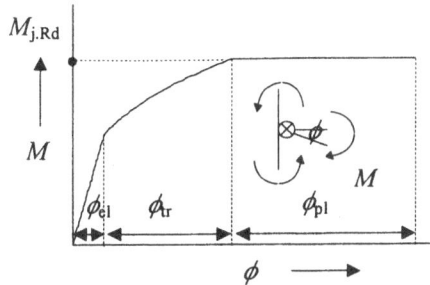

Figure 3. Rotations ϕ_{Xd} and ϕ_{Cd} according to Eurocode 3 and 4

Figure 4. Rotations ϕ_{el}, ϕ_{tr}, ϕ_{pl} according to [2]

There is a direct relationship between the definitions of rotations given in Eurocode 3 and 4 and the ones given by Kuhlmann et al. [2].

$$\phi_{Xd} = \phi_{el} + \phi_{tr} \tag{1}$$

$$\phi_{Cd} = \phi_{el} + \phi_{tr} + \phi_{pl} \tag{2}$$

In other words, the rotation ϕ_{Xd}, occurring when the rising part of the M-ϕ curve reaches $M_{j.Rd}$, consists of an elastic rotation ϕ_{el} and a transition rotation ϕ_{tr}. The total rotation capacity ϕ_{Cd} additionally also includes a plastic rotation ϕ_{pl}.

The definitions of Eurocode 3 and 4 and those of Kuhlmann et al. [2] are therefore interchangeable. Which definitions should be used seems therefore a matter of taste and practicality.

Huber and Tchemmernegg [5] relate rotation capacity to a bi-linear modelling of the moment rotation characteristics of a joint, see Figure 5. According to Huber and Tschemmernegg, the rotation capacity can be expressed dimensionless as ϕ_{pl} / ϕ_{el}.

Focussing on the moment-rotation curve according to Eurocode 3 and the model clauses of Eurocode 4, the question arises how to determine the moment resistance $M_{j.Rd}$ and the rotations ϕ_{Xd} and ϕ_{Cd} from a specific test result. Jaspart [6] gives an overview of approaches how to determine the moment resistance $M_{j.Rd}$ from a test. Figure 6 shows how the design moment resistance $M_{j.Rd}$ can be read from a test curve by drawing a line through the part of the test curve with post-limit stiffness K_{st}. Where this line crosses the vertical axis of the M-ϕ curve, the level of $M_{j.Rd}$ is defined.

Alternatively, the definition of $M_{j.Rd}$ as given in Figure 7 can be adopted [6, 7] . In this case, $M_{j.Rd}$, corresponding to the intersection of the initial stiffness $S_{j.ini}$ and the strain hardening stiffness K_{St} will yield a higher level of the moment resistance than in case of Figure 6. Differences are about 5%.

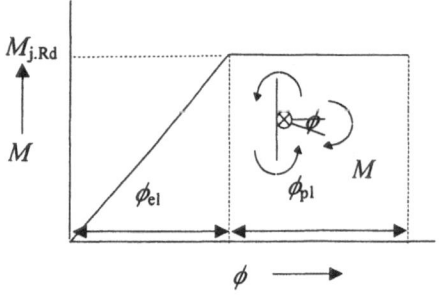

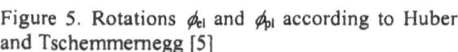

Figure 5. Rotations ϕ_{el} and ϕ_{pl} according to Huber and Tschemmernegg [5]

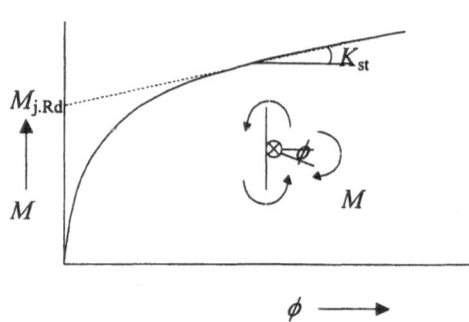

Figure 6. $M_{j.Rd}$ according to Jaspart [6]

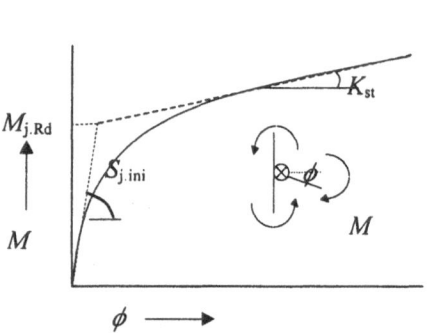

Figure 7. Alternative definition for $M_{j.Rd}$

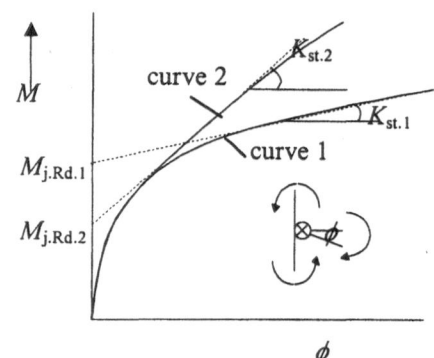

Figure 8. Comparison of two experimental curves

There are some points that need attention in the definitions of $M_{j.Rd}$ from Figure 6 and Figure 7. The first one is that the level of the moment resistance $M_{j.Rd}$ is dependent on the angle of the post-limit stiffness K_{st}. Compare for instance the two curves in Figure 8. The design moment resistance $M_{j.Rd}$ of curve 2 is smaller than the moment resistance of curve 1. The peak moment of curve 2 with stiffness $K_{st.2}$ is however greater than the peak moment of curve 1 with stiffness $K_{st.1}$ but the deformation capacity of curve 2 is smaller than that of curve 1. Therefore, the relation between the "design" safety of the structure based on a calculation with $M_{j.Rd}$ and the "real" safety based on the non-linear curve is not obvious.

A second point is that in some cases, it may be difficult to assess the post-limit stiffness K_{st}. This is especially the case in joints with relative low ductility, where a post-limit branch does not form. It is not clear from the definition what to do in such cases, see Figure 9.

Another possible definition is suggested by Weynand [8]. In this case, the initial stiffness of the joint is assessed in the M-ϕ curve, as being the elastic stiffness.

Then, a secant stiffness is determined as the initial stiffness divided by a fixed factor. Where this secant stiffness crosses the experimental curve, the level of the moment resistance $M_{j.Rd}$ is defined. In agreement with the design model in Eurocode 3, Weynand proposes that the secant stiffness at the level of $M_{j.Rd}$ is taken equal to one third of the initial stiffness $S_{j.ini}$, see Figure 10.

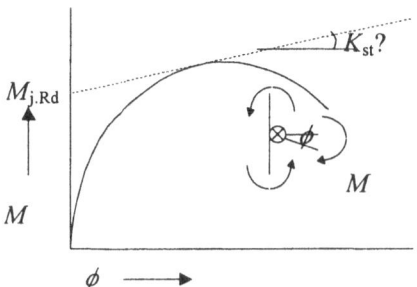
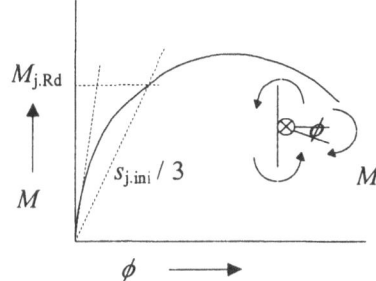

Figure 9. Determination of post-limit stiffness K_{st}

Figure 10. Alternative assesment of $M_{j.Rd}$ according to Weynand

Weynand used this definition to validate the design models in Eurocode 3 Annex J for beam-to-column joints with tests.

Once the moment resistance $M_{j.Rd}$ is defined in a consistent way, the rotation capacity can be determined for a certain test. The definition adopted by Kuhlman et al. [3] is consistent with the definition used for deformation capacity of members as adopted in Eurocode 3. Assume now that an experimental curve is available as given in Figure 11. The rotation capacity $\phi_{Cd} = \phi_{el} + \phi_{tr} + \phi_{pl}$ is now defined as the point where the descending branch of the experimental curve and the plastic branch at the level of $M_{j.Rd}$ intersect.

In case the curve has a rising post-limit branch, the rotation capacity ϕ_{Cd} is taken equal to the length of the branch.

Crisinel and Kattner [9] use a similar definition to derive the rotation capacity from a test result.

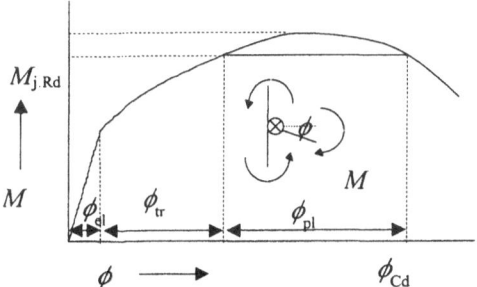

Figure 11. Experimental curve and definition of resistance and rotation capacity

5. Models for the Prediction of Rotation Capacity

5.1 CLASSES OF MODELS

Models for the prediction of rotation capacity of joints can be distinguished into two classes. The first class consists of models to determine the required rotation capacity of a joint in a frame; the second class consists of models for the prediction of the available rotation capacity in a joint.

5.2 MODELS FOR REQUIRED ROTATION CAPACITY

The required rotation capacity of joints can be determined with help of frame analysis programs for the determination of the frame response. In case non-linear calculations are adopted the required rotations can be calculated.

Alternatively, design models can be adopted to determine the required deformation capacity. The advantage of design models is that it is not required to perform a sophisticated frame analysis to determine the required rotation capacity.

Boender et al. [10] proposed a model based on a beam-line approach. The model of Boender is suitable for joints in steel structures and has an elastic perfectly plastic mechanical frame model as background.

Li et al. [11] proposed a model for the determination of the required rotation capacity of joints in a composite frame. Equations are derived for different load cases and different frame arrangements. Since the behaviour of the attached composite beams is taken into consideration, the equations have a rather sophisticated nature. In Nethercot et al. [12], these equations have been simplified.

5.3 MODELS FOR AVAILABLE ROTATION CAPACITY

Different researchers have developed models for the prediction of the available rotation capacity in steel joints and composite joints.

Voorn [13] performed tests on welded steel beam-to-column joints. From the test observations it was concluded that in case shear of the column web panel is the predominant failure mode, rotation capacity of the joints is sufficient for "normal" plastic design. In case column web buckling is the governing failure mode in an unstiffened joint, the minimum rotation obtained in the joint is equal to 0,015 radians. In case the web is stiffened in compression and unstiffened in tension and shear is not the dominant failure mode, the rotation may be assumed to be at least 0,025 h_c/h_b radians, where h_c is the column depth and h_b is the beam depth. These rules have been implemented in Eurocode 3.

Weynand [8] developed a model for the prediction of the available rotation capacity of steel joints by extending the design model of Eurocode 3 Annex J. Two types of components are distinguished in his thesis. The first category consists of components with a low ductility. These components are not assumed to contribute to the deformation capacity of the joint. An example of such a component is a bolt in tension. Secondly, components are distinguished that possess a certain ductility. By combining load deformation curves of all components in a joint, the moment-rotation curve of the joint is determined. The model assumes a fixed ratio between the ultimate resistance and the design resistance of a component. Furthermore, a fixed ratio is taken by between the initial stiffness and the post limit stiffness. The plastic deformation of the component is then calculated as the post limit stiffness multiplied by the difference between the ultimate and design resistance. The rotation capacity determined with this model therefore corresponds to the peak level of the moment rotation curve. In Figure 12, this process is illustrated. All deformations and resistances of the components are expressed in terms of contributions to rotations and moment resistances.

Deformation capacity in composite joints is more a design issue than in steel joints, due to the nature of the materials. Therefore, research on composite joints focuses on the deformation capacity of the reinforcement bars in the joints and the deformations due to slip between the steel beam and the concrete slab.

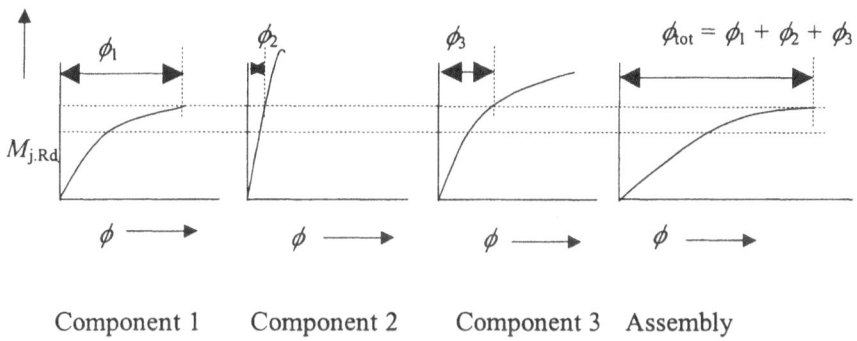

Component 1   Component 2   Component 3   Assembly

Figure 12. Determination of rotation capacity by Weynand [8]

Anderson et al. describe in [14] the state of the art concerning the prediction of the rotation capacity of composite beam-to-column joints. A model is presented for the prediction of the rotation capacity of such joints on the following basis:

$$\phi_u = (\Delta_{u,s})/(D + D_t) + s/D \tag{3}$$

where: ϕ_u is the resulting rotation capacity;
$\Delta_{u,s}$ is the deformation in the slab reinforcement;
s is the slip in the shear connectors connecting slab and steel beam;
D is the depth of the beam;
D_t is the distance between the top of the steel beam to the centroid of the reinforcement.

The calculation of $\Delta_{u,s}$, the deformation in the slab reinforcement, is based on a the CEB-FIB Model code [16].

Concerning the determination of the slip s between steel beam and concrete slab, which is dependent on the shear connectors between both, reference is made to research carried out by Aribert [17].

From the references it is not clear whether the rotation capacity calculated in this way includes the elastic deformations. Furthermore, it is not indicated if the rotation capacity found corresponds to the peak level of the moment rotation curve or another point.

Ren and Crisinel [18] also developed a model for the prediction of rotation capacity in composite joints. In this model, the stiffness of a joint can be determined based on stiffness coefficients of joint components, in a quite similar fashion, as is the case for the model clauses for Eurocode 4. The rotation capacity is calculated by adopting the secant stiffness of the components in the plastic cracked state of the slab. The resulting rotation is then equal to:

$$\phi_u = M_{j.Rd}/S_j \tag{4}$$

where S_j is the stiffness of the joint in the plastic cracked state of the slab.

The rotation found in this way corresponds to the rotation capacity occurring at the peak of the moment rotation curve and includes elastic deformations.

6. Effect on the Reliability of the Structure

The reliability of a structure is dependent on the rotation capacity of its joints. When there is rotation capacity, the structure failure modes are for instance insufficient section strength or instability. Insufficient deformation capacity in the joints, may be a potential failure mode, which may lead to sudden collapse. The design rules in modern design standards are nowadays calibrated based on reliability studies. Vrouwenvelder and Siemens reported on the calibration process of the Dutch Codes in [19]. These studies, however, focus on the reliability of structural elements like members and joints. For the effect of the rotation capacity on the reliability it is required to assess the reliability of the structure as a whole. For an introduction on reliability theory it is referred to [20].

Weynand [8] investigated the reliability of steel frames taking into consideration rotation capacity of the joints. Based on M-ϕ curves generated with the model described in paragraph 5.3, Monte Carlo simulations were carried out to determine the reliability index of some frames. The following procedure was followed:

1. Design of a structure including joints based on Eurocode 3. A certain ratio between vertical and horizontal loading on the frame is assumed;

2. Determination of a load factor λ^* (factor to multiply the loads with) based on the design values of all parameters including partial safety factors according to Eurocode 3. This factor is determined with a non-linear frame analysis program;
3. Determination of a load factor λ based on simulated values of all stochastic parameters;
4. Determination of $Z = \lambda - \lambda^*$. If $Z < 0$, failure occurs;
5. Repetition of steps 3 and 4 for a large number of simulations
6. Determination of the average value μ_Z, the standard deviation σ_Z and the reliability index $\beta_Z = \mu_Z / \sigma_Z$ of Z.
7. The so calculated reliability index is compared to a target reliability index of $0,8 \cdot 3,8$. The factor 0,8 reflects that loading is taken as a deterministic parameter and 3,8 is the target reliability index used for members and joints in Eurocode 1 [21].

According to the simulations, the model of Eurocode 3 Annex J to predict an M-ϕ curve of joints in conjunction with the proposed safety factors leads to a reliability greater than the target reliability. It should be noted that a limited set of parameters in the design model are taken as stochastic variables; others were taken as deterministic. The effect of this choice on the calculated reliability is not investigated. Weynand concludes that the problem of rotation capacity needs further investigation, especially when joint components possess limited deformation capacity.

There are some issues concerning the relation between rotation capacity of joints and the structural reliability that have not been addressed in literature. Of importance is the impact of the definition of the design curve on the calculated reliability in comparison with the "real" reliability of a structure. This needs to be investigated.

A second question is how to adopt a safety concept in practical design rules. For instance, Kemp [22, 23] proposes for composite and steel beams to adopt partial safety factors on the available deformation capacity. The proposed level of the safety factors γ is taken as dependent on the failure mode of the beam: γ is equal to 1,3 for ductile failure modes and γ is taken equal to 3 for non-ductile failure modes. The partial safety factor only applies to the inelastic deformation in the beam. The values of these partial safety factors have been established by direct comparison of a prediction model with test results. A reliability study to determine these values has not been carried out.

The research carried out at Eindhoven University of Technology will address these issues.

7. Conclusions

Design rules in modern codes for structural steel and composite structures are calibrated on a probabilistic basis. Calibration studies normally focus on the resistance of the structure. Concerning joints in steel and composite structures, the rotation capacity is a characteristic, which has an effect on the reliability of the structure.

The knowledge on rotation capacity of joints is increasing but still limited compared to the knowledge on resistance. Studies have been performed to gain knowledge on the required rotation capacity in steel and composite frames and the available rotation capacity of joints. These studies have a deterministic nature.

In literature, a variety of definitions for rotation capacity of joints are used. These definitions seem to be rather arbitrary, although it is already a step forward if the definitions are used in a consistent way. In a research program carried out at the University of Eindhoven the effect of these definitions on the reliability of the frame will be studied. In that case, a more justified choice can be made regarding the definitions to be used for resistance and rotation capacity of joints in frames.

The final purpose of the research carried out in Eindhoven is to derive practical, simple and reliable guidance for structural engineers to design joints with sufficient rotation capacity in steel and composite structures.

8. References

1. Eurocode 3 (1998) Design of Steel Structures Part 1.1. General Rules for Buildings. Revised Annex J, *ENV 1993-1-1/A2*, CEN, Brussels.

2. Kuhlmann, U., Davison, J.B. and Kattner, M., (1998) Structural Systems and Rotation Capacity, *proceedings COST C1 international conference*, Liège, pp 167-176.

3. Kuhlmann, U., in J.P. Jaspart (1999) *Recent Advances in the Field of Structural Steel Joints and their Representation in the Building Frame Analyses and Design Process*, COST C1 Brussels.

4. ECCS (1999) Composite Steel-concrete Joints in Frames for Buildings: Design Provisions, *ECCS Publication 109*, Brussels.

5. Huber G. and Tschemmernegg, F. (1998) Classification and Assessment of Joints, COST C1/WD2/98-02, Thessaloniki.

6. Jaspart, J.P. (1991) Etude de la Semi-rigidité des Nœuds Poutre-Colonne et son Influence sur la Résistance et la Stabilité des Ossatures en Acier, *Ph-D Thesis*, University of Liège, Liège.

7. Zanon, P. and Zandonini, R. (1988) Experimental Analysis of End Plate Connections, *Proceedings of the state of the art workshop on connections and the behaviour of strength and design of steel structures*, Cachan, pp. 41-51.

8. Weynand, K. (1997) Sicherheits- und Wirtschaftlichkeitsuntersuchungen zur Anwendung Nachgiebiger Anschlüsse im Stahlbau, *Heft 35*, Shaker Verlag, Aachen.

9. Crisinel, M. and Kattner, M. (1999) Verfügbare Rotation von Verbundsknoten, in G. Huber and Th. Michl, *Festschrift Prof. Dr. Ferdinand Tchemmernegg*, University of Innsbruck, Innsbruck. pp. 419-432.

10. Boender, E., Steenhuis, C.M. and Stark, J. (1996) The Required Rotation Capacity of Joints in Braced Steel Frames, *proceedings of the IABSE Semi-Rigid Structural Connections Colloquium*, Istanbul, pp. 259-268.

11. Li, T.Q., Choo, B.S. and Nethercot, D. (1995) Determination of Rotation Capacity Requirements for Steel and Composite beams, *Journal of Constructional Steel Research*, No 32, Malta, pp. 303-332.

12. Nethercot, D., Li, T.Q. and Choo, B.S. (1995) Required Rotations and Moment Distributions for Composite Frames and Continuous Beams, *Journal of Constructional Steel Research, No 35*, Malta, pp. 121-163.

13. Voorn, W.J.M. (1971) Welded Beam-to-Column Joints in Braced Frames (in Dutch), *Rapport IBBC TNO BI-71-24*, Delft.

14. Anderson, D., Aribert, J-M. and Kronenberger, H.-J. (1999) Rotation Capacity of Composite Joints, in R. Maquoi, *Control of the Semi-Rigid Behaviour of Civil Engineering Structural Connections*, COST C1, Brussels, pp. 177-186.

15. Xiao, Y. and Anderson, D. (1994) Review of the Research in the United Kingdom on Composite Semi-rigid Joints, in F. Wald (ed.) *Proceedings of the second state of the art workshop* COST C1, Brussels.

16. CEB-FIB (1990) *Model Code 1990*, Lausanne.

17. Aribert, J.-M. (1996) Influence of Slip of the Shear Connection on Composite Joint Behaviour, in R. Bjorhovde et al. (ed), *Connections in Steel Structures III: Behaviour, Strength and Design*, Pergamon, pp. 11-22.

18. Ren, P. and Crisinel, M. (1996) Prediction Method for the Moment Rotation Behaviour of Composite Beam to Steel Column Connection, in R. Bjorhovde et al. (ed), *Connections in Steel Structures III: Behaviour, Strength and Design*, Pergamon, pp. 11-22.

19. Vrouwenvelder, A.C.W.M. and Siemes, A.J. (1987) Probabilistic Calibration Procedure for the Derivation of Partial Safety Factors for The Netherlands Building Codes. *Heron No. 32(4)*, Delft,.pp. 9-29.

20. Schneider, J. (1997) Introduction to Safety and Reliability of Structures, *IABSE Structural Engineering Document 5*, Zurich

21. Eurocode 1 (1992) *ENV 1991-1-1, Basis of design*, CEN, Brussels.

22. Kemp, A.R. (1999) A limit State Criterion For Ductility of Class 1 and Class 2 Composite and Steel Beams, in D. Dubina (ed.) *Stability and Safety Of Steel Structures,* Elsevier Science Ltd, pp. 291-298.

23. Kemp, A.R., Dekker, N.W. (1991) Available Rotation Capacity in Steel and Composite Beams, *The Structural Engineer,* 65:5, London, pp. 96-101.

MODELLING OF JOINTS OF SANDWICHES PANELS

J. MAREŠ , F. WALD, Z. SOKOL
Department of Steel Structures, Czech Technical University in Prague
Thákurova 7, 166 29 Praha, Czech Republic

1. Introduction

The lightweight sandwiches panels are produced by a modern and environmental controlled technology. A typical sandwich panel consist of two covering steel sheets, generally made of the trapezoidal or light profiled steel sheets, and of a weak lightweight core. The increasing use of lightweight sandwiches panels in industrial and civil buildings requires the development of suitable methods for analysis. The response of single panel as well as behaviour of whole structure is influenced by joints between sandwiches panels and framework. The stiffening effect of the cladding panels on the structural behaviour of the structure, the stress skin procedure, with the respect to the connections was evaluated in [3] to observe the behaviour under the seismic loading. The published analysis [1] neglects stress distribution between the both metal faces. Only inner sheet is assumed to carry the applied loads and the bearing capacity of the connection, see [2].

One-sided joint of the self-tapping screws of diameter 6,3 mm and length according to the sandwiches panel thickness was observed experimentally and simulated by FE to develop an engineering analytical prediction model. The procedure is based on decomposition into the shear and bending components. The experiments show significant influence of the flexibility of the screw on the stress distribution between two covering metal steel sheets. The presented study is taking into account the bending stiffness of the screw and the stress distribution over both covering metal sheets of the three-layered sandwich.

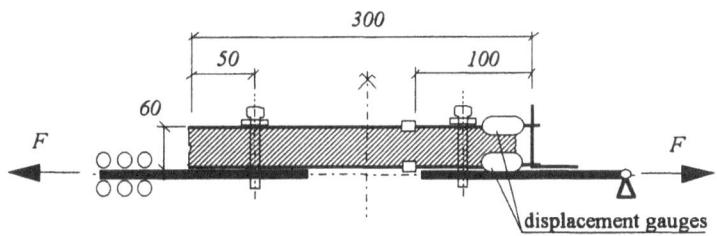

Figure 1   The arrangement of the test

C.C. Baniotopoulos and F. Wald (eds.), The Paramount Role of Joints into the Reliable Response of Structures, 387–394.
© 2000 *Kluwer Academic Publishers.*

388

2. Experimental Program

Simple tests with the one-sided sandwich panel's joints have been performed see Fig. 1. The force deformation diagrams were measured for both covering steel sheet on the tests with the monotonically increasing force and with the repeated loading. The gauges records displacement changes of both openings. Test set-up with measurement device is shown on Figure 2. The gauges are positioned on axis in the planes of both covering plates to avoid rotation of the sandwich cross-section. Total 18 specimens ware tested in the presented series.

Figure 2   The test specimen with the displacement gauges

The failure depends on the material and the diameter of fasteners and the material and the thickness of inner metal sheet. The fasteners of the austenitic stainless steel tend to be tougher. The failure is caused than by ovalization of the inner hole [5] only. The previous sets of test with fasteners of the carbon steel failed by fracture of the bolts near the steel plate with folding and tearing of the inner opening. However the failure by tearing of inner hole, see Figure 2, is most common for all performed tests. The fracture of the fastener has to be taken into consideration. The material characteristics of the covering steel plates were determined by the standard coupon tests.

TABLE 1   Measured material characteristics

Metal faces	Core
Yield strength f_y = 339,1 MPa	Polyurethane foam with density = 30 kg/m^3
Ultimate strength f_u = 366,4 MPa	Fasteners - Self-tapping screw
Elongation A_{50} = 24,2 %	Type TDB-T-T16-6,3 x 76
Elongation A_{80} = 21,9 %	Diameter 6,3 mm with washer 16 mm

A linear part till the displacement of *0,6 mm* about was observed on the load - deflection curves, see Figure 7. The corresponding loading is 1 kN about. Behind this limit a non-linear part is visible until the plateau of yielding is reached. The stiffening effect occurs by folding of the metal sheet around the fastener.

The serviceability is limited by tearing of inner sheet, which arise after small deformation. The permanent deformations are observed also in the linear part of the load-deflection curve. The progress of the damages of the inner and outer plate is visualised on graph of Figure 4, where hysteretic loops of the test with the cyclic load history are presented.

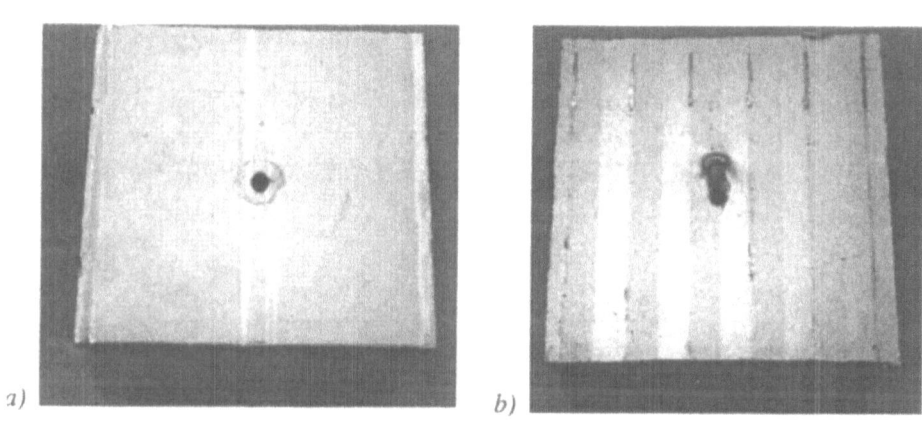

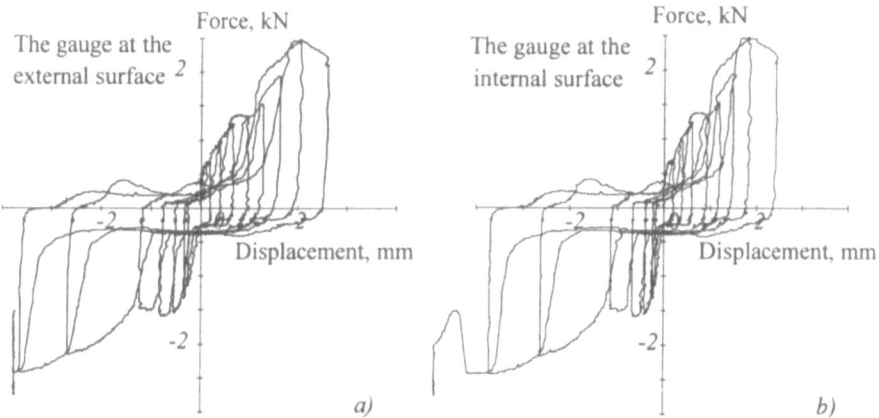

Figure 4 The behaviour of the outer a) and inner b) part of connection under cyclic loading

3. Mechanical Model of the Connection

3.1. Failure Modes

The acting force is either caused by the perpendicular loading of the panels by framework, see Figure 10 a), c), by the temperature difference of the outer and inner sheet or Figure 10 b), d). The failure modes typical for the sandwich panel connections are developed in addition to the standard failure modes, see [2]. The failure by tearing of the inner hole only see Figure 10 a, b) is taken into account as well as the fracture of the screw shank together with the ovalization of the inner hole, see on Figure10 c, d). The shank of the fasteners is relatively flexible so the inner sheet transmits the predominant part of a reaction, but not whole. The distribution depends on the relative bending stiffness based on the panel thickness and on the diameter of the fastener.

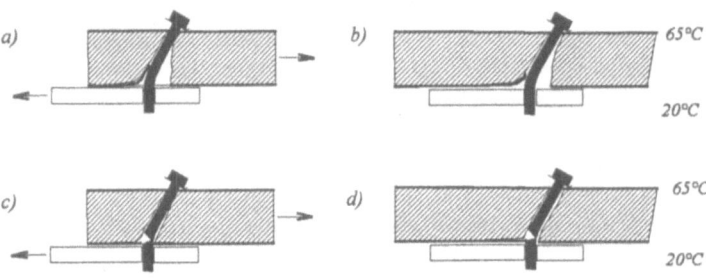

3.2. Bending Stiffness of the Screw

The behaviour of the screw can be modelled simple as a cantilever beam. The forces are transferred to the screw by the covering sheets of the sandwich panel. The steel sheet of the panel is considered not stiff enough out of its plane to support the head of the screw, which can be observed at Figure 3b. The spring model of the joint is shown in Figure 6.

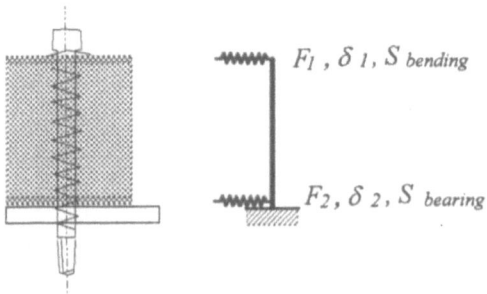

Figure 6   Model of connection

$$\delta = \frac{F \, l^3}{6 \, E \, I} \, ,$$

(1)

The screw stiffness can be defined as

$$S_{bending} = \frac{F}{2 \, \delta} = \frac{3 \, E \, I}{l^3} \, ,$$

(2)

where for the screw is $I = \pi \, d^4 / 64$ with d the nominal diameter of the screw. The screw stiffness becomes

$$S_{bending} = \frac{3 \, \pi \, E \, d^4}{64 \, l^3} \, .$$

(3)

3.3. Bearing Stiffness of Steel Plate

In order to define the bearing stiffness, a formula determined for plates from revised Annex J of Eurocode 3 [6] is used. It fits for prediction of thin plates well even, it was developed based on tests of steel thick plates. The Eurocode stiffness is in form of

$$k_8 = \frac{24 \, n_b \, k_b \, k_t \, f_u \, d}{E} \, ,$$

(4)

where n_b is the number of bolt rows connecting the plate,

$$k_b = min\left(0{,}25 \frac{e_b}{d} + 0{,}5; \ 0{,}25 \frac{p_b}{d} + 0{,}375; \ 1{,}25\right)$$

$$k_t = min\left(1{,}5 \frac{t_j}{d_{M16}}; \ 2{,}5\right)$$

d is the nominal bolt diameter of the applied bolts,
d_{M16} is the nominal bolt diameter of an M16 bolt,
e_b is the end distance from the bolt-row to the free edge plate,
p_b is the pitch between two bolt-rows in plate,
f_u is the ultimate tensile strength of the plate,
t_j is the thickness of that component.

For the bearing stiffness of the screw can be the formula (4) rewritten as

$$S_{bearing} = k_8 \, E = 24 * 1{,}25 \frac{1{,}5 \, t_j}{d_{M16}} d \, f_u \, .$$

(5)

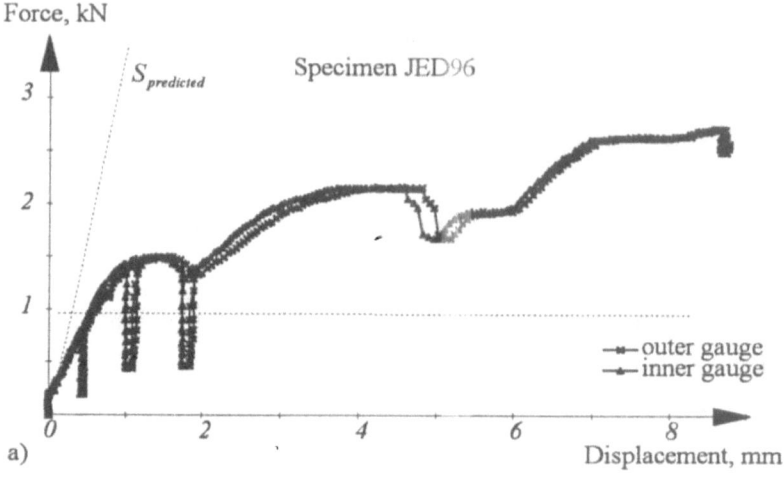

a)

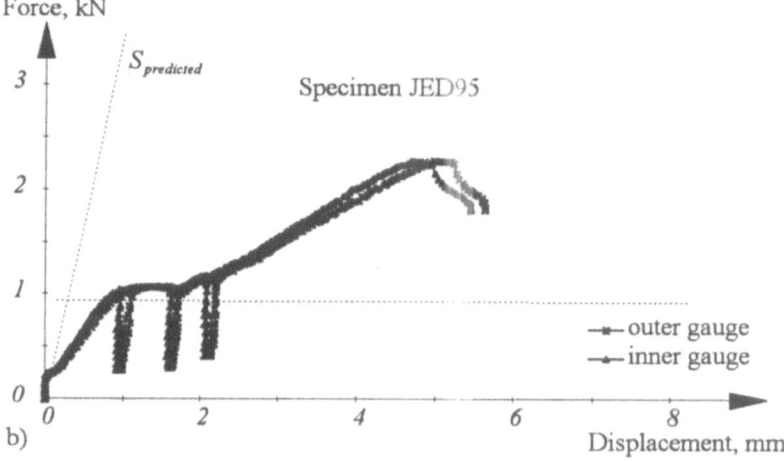

b)

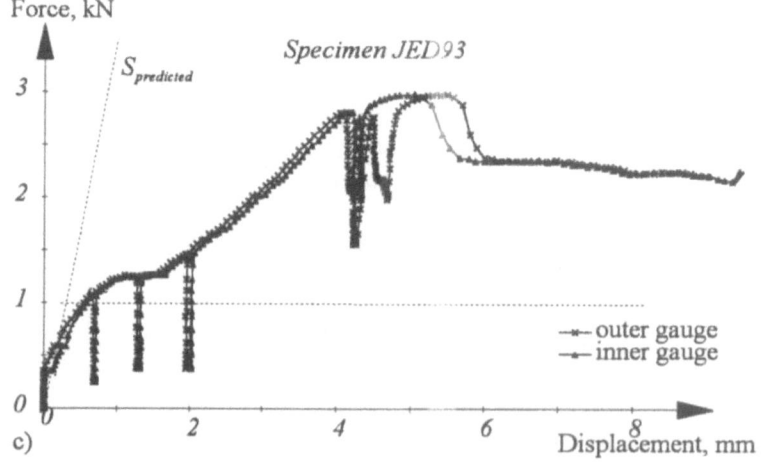

c)

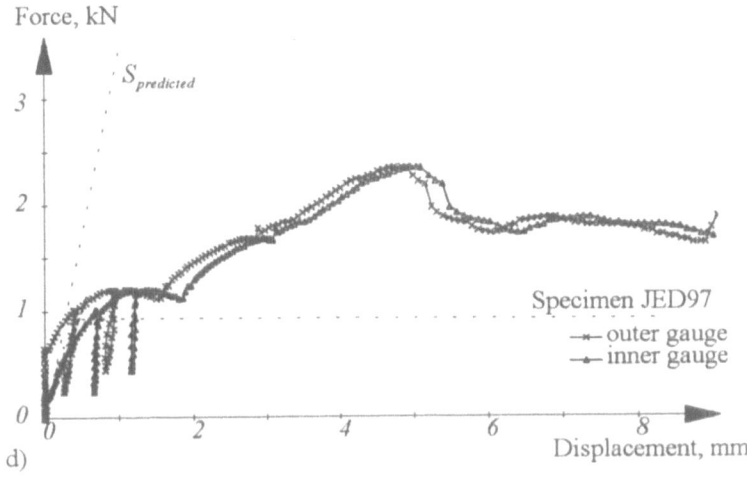

Figure 7   An Example of load-deflection curves of both surfaces of panel in connections

4. Validation to Experiments

Validation of the prediction to the experiments is shown on the bearing component of inner hole on Figure 7. Good agreement was achieved. The outer component, which is represented by the bending stiffness of the screw shank, is about 15 times less stiff compare to the bearing component of inner hole. The model is neglecting the shear stiffness of the polyurethane foam.

The sensitivity study was performed using finite-element 3D model by Ansys 5.3 code. The contact elements were simulated in the area where bearing between the screw and the metal sheeting's occurs. The stress distribution along both covering sheets was investigated. The linear part only of the model shows satisfactory agreement to the calibrating experiments.

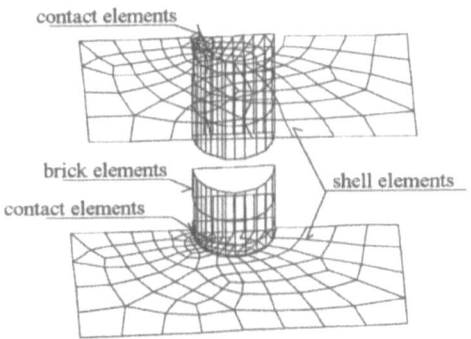

Figure 8   Meshing for a FEM simulation of the joint for the sensitivity study

5. Conclusion

The presented prediction model simulates the behaviour with good accuracy. The relative stiffness of this joint has valuable influence on the collapse mode.

The prediction under monotonic loading is the first step. The prediction of behaviour under cyclic loading by the component method is under finalising [8].

The application of reached knowledge of the joint stiffness shows by stress skin method the reduction of vertical drift of the structure of the simple industrial hales under serviceability limit state of 10% according to the number of fasteners per panel and the geometry of the structure [7].

6. Acknowledgement

This work has been supported by the grant of Czech Ministry of Education No. MŠM 21 000 000 1.

7. References

1. ECCS (1991) *Preliminary European Recommendations for Sandwich Panels*, ECCS - TC7, N°66, Brussels.
2. ENV 1993-1-1 / A1 (1998) Design of steel structures *Cold formed members and sheeting*, Eurocode 3 / A1, Brussels.
3. Mazzolani, F. M., Matteis, G. D. and Landolfo, R. (1994) *Shear test on sandwich panel connections*, in Semi-rigid behaviour of civil engineering structural connections, ed. Wald F., Workshop COST C1, Prague, pp. 481 - 491.
4. Baehre R. and Ladwein T. (1994) Diaphragm Action of Sandwich Panels, *Journal of Construct. Steel Research*, Vol. 31, London, pp. 305 – 316.
5. Mareš, J. (2000) *Joints of sandwiches panels, the state of the art review* (in Czech), Prague, p. 46.
6. ENV 1993-1-1 (1992) *Design of Steel Structures*, Eurocode 3, European Prenorm, CEN, Brussels.
7. Dini, L. (1999) *Stressed skin design of steel structures*, Diploma thesis, Clermont Ferrand, p. 119.
8. Mareš, J. (2000) *Modelling of joints of sandwiches panels*, ČVUT in Praha, Ph.D. thesis (in Czech), in printing, Prague, p. 129.

NEW DESIGN CONCEPTS FOR STRUCTURAL TIMBER CONNECTIONS

P. HALLER
Dresden University of Technology
Institute of Structural Design and Timber Engineering
Mommsenstr. 13, 01062 Dresden, Germany

1. Introduction

Wood is an anisotropic and brittle material which makes it difficult to connect. Whereas traditional joints in carpentry suffer from reduced sections that weaken the joint significantly, the structural performance of engineered connections – mostly done with dowel type fasteners (DTF) - is limited by weak strength in shear and perpendicular to grain. In order to improve the load bearing behaviour, a thorough understanding of the local (single DTF) and global (connecting area) failure mechanisms are required as a prerequisite.

A good joint is characterised by high load bearing and with respect to deformation by the absence of clearance, high initial and creep stiffness as well as by ductility and energy dissipation in case of cyclic and seismic loading.

The precise knowledge of the phenomena involved in the load displacement curve are crucial for further technological improvements of timber joints and an important step towards a more realistic structural analysis, which already models construction elements with good accuracy but still considers the connection as pinned or rigid although the real behaviour is in between or semi-rigid.

2. Structural behaviour of timber joints under static loading

To study the structural performance of joints one has to study different physical phenomenon of the load displacement curve (figure 0) . The slip of phase I result from the clearance due to the manufacturing tolerances leading to a progressive loading of all dowels. In phase II the joint behaves linear elastic followed by the formation of plastic hinges in the DTF and damage of the wood within the next two phases. To model the load displacement curve, it is necessary to distinguish the transition from one phase to the other by means of an appropriate criterion and to dispose of a law for the evolution of the variable in consideration.

C.C. Baniotopoulos and F. Wald (eds.), The Paramount Role of Joints into the Reliable Response of Structures, 395–406.
© 2000 *Kluwer Academic Publishers.*

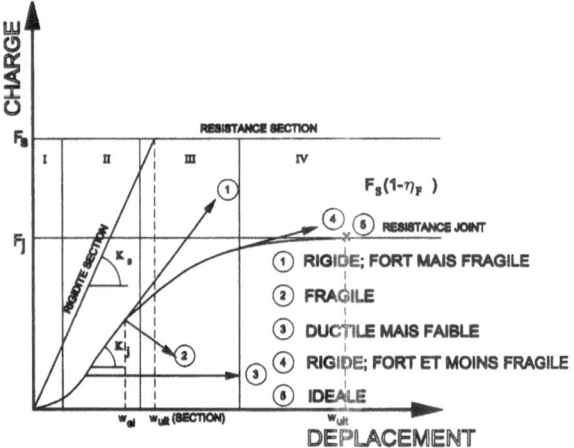

Figure 0. General load displacement curve of timber joint with

3. Design principles to improve the structural performance of DTF connections

The different sections of the load displacement curve are linked to the overall design goals – high strength, stiffness and ductility. Whereas phases I and II determine stiffness, phases III and IV concern strength and ductility. The principles to improve structural behaviour are quite limited, however, the means to achieve these objectives are various and will be presented in the following.

1. Phase I - Slip

Generally speaking, phase I has to be avoided. This can be done by means of a tight fit of the fastener. Nails behave stiffer than bolts due to the manufacturing tolerances for the drilling hole. The use of computer aided manufacturing is an appropriate approach to get rid of the slip problem.

A further possibility consist in the injection of an epoxy resin to fill up the clearance between dowel and drilling hole. The resin provide better embedding characteristics with respect to stiffness and strength. Rodd (Haller) has investigated this method successfully within the framework of the European Cost C1 action "Control of the Semi-Rigid Civil Engineering Connection" (Haller).

Finally, the DTF itself can fill the gap in the drilling hole. Leijten (Haller) used gas tubes and expanded them by means of a special device operated by a hydraulic jack. The load displacement curve from this fastener present no slip and behave very ductile since the tube can deform in its transverse direction.

2. Phase II – Stiffness

The stiffness of the joint depends on the embedding stiffness of the single DTF and its elastic deformation depending especially on the slenderness. The modulus of the elastic foundation or the embedding stiffness can be improved in different way. In the previous chapter it was reported that resin injection lead to better embedding characteristics. Lateral reinforcement of the connecting area can also be considered as an effective method to cope with high local stresses around the DTF. Different solutions are proposed in the literature (Haller):

- plywood reinforcement (Rodd)
- punched metal plate (Kevarinmäki et al.)
- densified veneer wood (Leijten)
- glass fibre fabrics (Haller, Chen)
- Densified solid wood (Haller, Wehsener).

The latter both were also investigated in combination. Glass-fibre reinforcement with commercial fabrics does not much improve embedding strength and rigidity but essentially improves ductility (Haller et. al. 1998). The embedding behaviour of wood decisively depends on its density (Ehlbeck, Werner 1992) so that the use of hard wood or densification of soft wood (Haller et. al. 1996) results in a further improvement of load carrying behaviour. Textile structures and densified wood present two techniques that produce very stiff, strong and ductile joints.

Technical textiles prove to be very promising and universally applicable. Especially transverse and shear strength of wood can be improved very effectively by the application of textile structures with adhesives. Common technical textiles are reasonable in price, can be draped and varied in almost every way concerning their weight, kind of fibre and structure (Offermann, Franzke 1996) (Hufenbach, Müller 1995). The following therefore focuses on the use of densified solid wood and technical textiles as a new concept for the design of structural timber connections.

3. Phase III and IV – Strength and ductility

The load bearing capacity depends on the embedding strength which is affected by the use of a stronger material or the lateral application of various reinforcements. As wood is a brittle material, ductility may either result from the formation of plastic hinges in the DTF or reinforcements that makes wood more ductile as in the case of textiles.

4. State of the art

According to the objectives timber elements can be reinforced by textiles in two different ways.

On the one hand the timber element is reinforced by means of prefabricated unidirectional laminates glued to it parallel to the grain. Thus e. g. flexural rigidity and strength of beams are increased (Tingley 1998), (Dagher 1998), (Davalos, Qiao 1998) what is applied with great success in wooden buildings rehabilitation (Kempe 1997). In this case mainly stiff high-strength carbon and aramide fibres are used. The especially good mechanical properties of the laminated wood have to be considered together with its small cross section so that its ultimate load and tensile strength are only approximately equivalent to those of an ordinary board.

On the other hand textile structures can be glued on the wet or dry surface (Gougeon Brothers 1998) producing the composite directly on the wood. This method is applied in boat, propeller and timber construction (Colling et. al.1987), (Larsen 1994), (Sucz 1991), (Haller 1994), (Chen 1999). This does not primarily aim at improving the lateral strength of the wood but its transverse and shear strength as well as the protection against environment and weather (Haller, Wehsener 1998).

Nowadays wood is reinforced by means of commercial fabrics of different weight whereby the applied fabric is not always being oriented in dependence of the occurring stresses.

At present the authors intent a further use of textile engineering methods that allow an orientation of the fibre flow according to the local stress state and the shape of the construction element [see also (Kriechbaum 1994)].

4.1. OBJECTIVES

The objective of the present study is to increase the load carrying capacity of timber joints for a better exploitation of the timber cross section as well as to increase their rigidity and ductility.

Starting with a longitudinal joint with inserted steel plates and dowels its manufacture is being varied making use of glass-fibre fabrics and densified wood (Birk 1999). The cross section geometry for bending and shear amounts to 120 x 200 mm. For the tensile test there was chosen a smaller geometry (110 x 100 mm) for technical reasons due to the testing device.

The influence of the type of reinforcement on the load carrying behaviour is determined separately as well as in combination for different loading - tension, shear and bending. Therefore one sample of common laminated wood; (glass-fibre) reinforced laminated wood; reinforced and densified laminated wood; reinforced by oriented textiles and densified laminated wood are tested respectively. The tests also serve as basis to compare with future tests on textiles which are optimally adapted to occurring stresses.

5. Material and Methods

5.1. MATERIAL

The samples were made of spruce boards (2000 x 135 x 26 mm) of grade GK II according to [DIN 1052]. Prior to the tests these boards were industrially dried and a part of them was densified in a heating press to 50 % of the initial cross section. These boards - normal and densified - were connected by means of finger joints and glue laminated. For the bending specimens the densified part of the laminates was shifted by 100 mm respectively (fig. 1c) what resulted in an improved conical connecting area. As an adhesive there was used formaldehyde-free polyurethane glue (manufacturer Collano: Purbond HB 110). Table 1 contains the average values of wood moisture content and density before and after densification.

Moreover, samples were made of commercial laminated wood of grade GK II which serve as a reference with and without reinforcement.

TABLE 1. Density and wood moisture of spruce laminates used

Materials	Density [kg/m^3]	Moisture [%]
laminated wood, GK II	370 ... 430	10 ... 14
spruce laminate before densification and	450 ... 480	14 ... 16
after densification (to 50%)	880 ... 980	6 ... 8

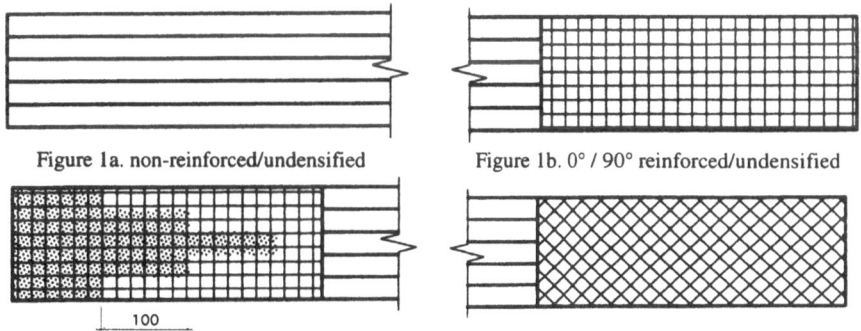

Figure 1a. non-reinforced/undensified   Figure 1b. 0° / 90° reinforced/undensified

100

Figure 1c. 0° / 90° reinforced/densified   Figure 1d. +45° / - 45° reinforced/undensified

5.2. MAKING OF SAMPLES

5.2.1. Densification

The laminates were densified in three steps: heating up, densification and recooling. During the first phase the laminates were heated up by the panels of the press using a low contact pressure of 0.2 to 0.3 MPa. The solid wood heating was calculated with an assumed value of 1 to 2 mm/min. After the temperature in the centre of the sample had reached 140 °C the densification process started. With a further heating to 150 °C pressing power was continually increased to 2.5 MPa. In order to prevent excessive drying there was chosen a relatively high closing speed of the press of 1 mm/min. After the densification process of about 40 minutes the recooling of the wood samples to 60 °C began. This lasted approximately as long as the heating (fig. 2).

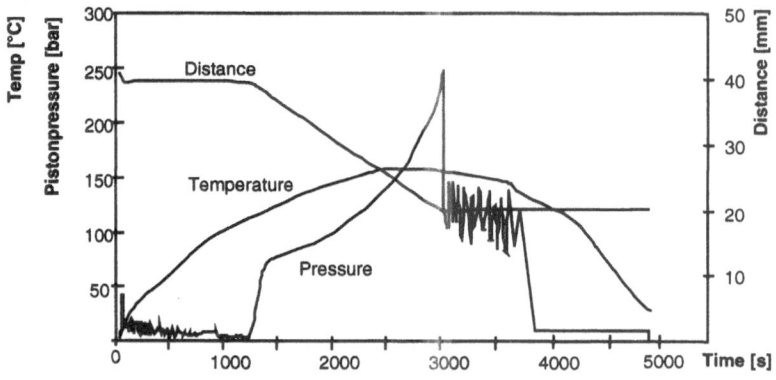

Figure 2. Densification of spruce board [section: 120 x 40mm]

5.2.2. Lamination

The timber components were reinforced by glass-fibre fabrics of type E and a weight of 200 g/m². Previously an adhesive activator (manufacturer: Vosschemie; G4 - primer) was applied to enhance bonding strength. Two layers were applied manually using a two-component epoxy resin (manufacturer: Vosschemie; LN-1 Epoxy A + B). Here The „wet method" was used, where the adhesive is spread on the timber surface and the fabric is pressed on it without prior impregnation. Using epoxy resin avoids an additional surface pressure. For complete hardening the components had been stored under ambient climate for 10 days. Rigidity and characteristic values of the fabric and the adhesive are presented in table 2.

TABLE 2. Material values according to (Gay 1991) *) and (Chen 1999) **)

Characteristic values	Glass fibre Type E*)	Epoxy resin	Composite **)
Young-modulus [MPa]	73000	3700	25 ... 30000**)
Density [kg/m³]	2600	1040 (hardener) 1150 (resin)	approx. 2000
Bending strength [MPa]		123	
Compression strength [MPa]		130	
Tensile strength [MPa]	2400		150... 200**)

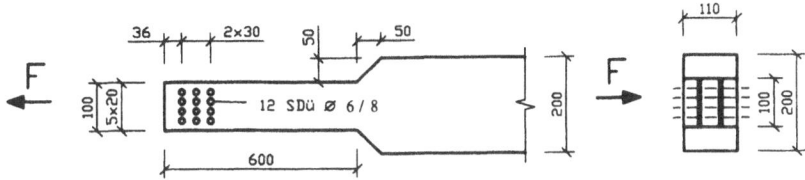

Figure 3 a. Tension specimen

5.3. TEST SET-UP

For an assessment of load-carrying behaviour the joints were tested for three different forces according to EN 26981. Per reinforcement and load one sample of non-reinforced laminated timber, (glass-fibre) reinforced laminated timber, reinforced and densified laminated timber were tested respectively. The geometry of samples and test set-up are shown in (fig. 3a - c).

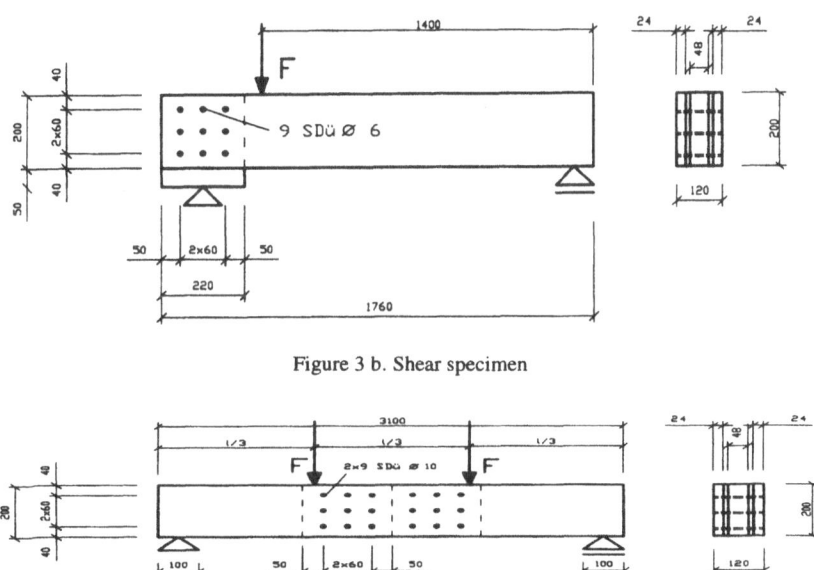

Figure 3 b. Shear specimen

Figure 3 c. Bending specimen

6. Tests and Test Results

6.1. BENDING

Load bearing behaviour was tested in a four point bending test. The timber cross section was of 120 x 200 x 1550 mm and had 9 dowels of 10 mm diameter arranged in rectangular shape. Textile reinforcement lead to an increase in failure to 149 % compared to the non-reinforced joint that serves as reference (100 %). Using densified laminate tim-

402

ber with oriented (-45°/+45°) reinforcement 211 % could be reached. Non-reinforced timber joints showed brittle failure in the lower row of dowels. The fabric of the glass fibre reinforced sample failed at the end of the metal plate. The fasteners showed large deformations (ductility) when the ultimate load was reached. The densified and reinforced joints showed the same deformations at higher load. Contrary to undensified joints the fasteners showed large plastic deformation.

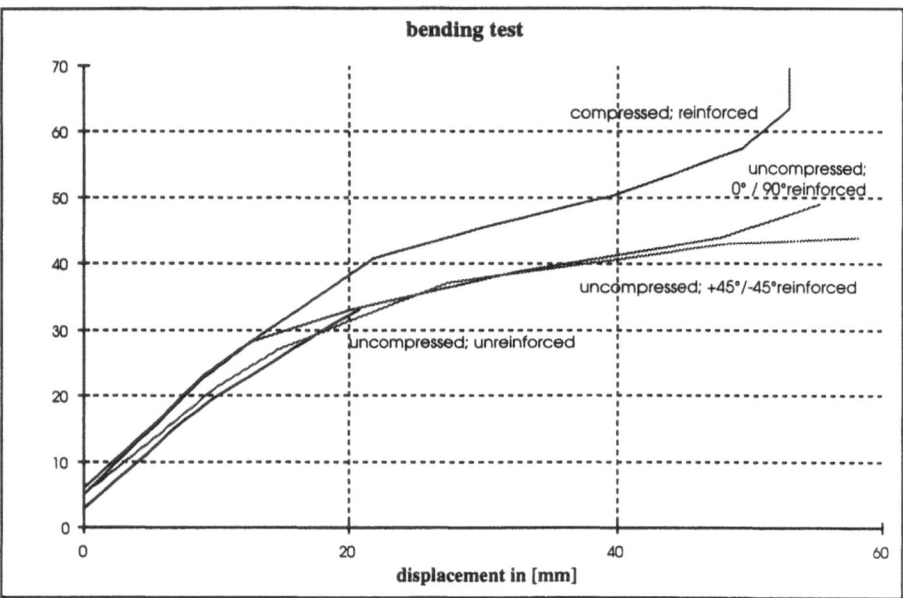

Figure 4. Load - slip - relationship of bending stress

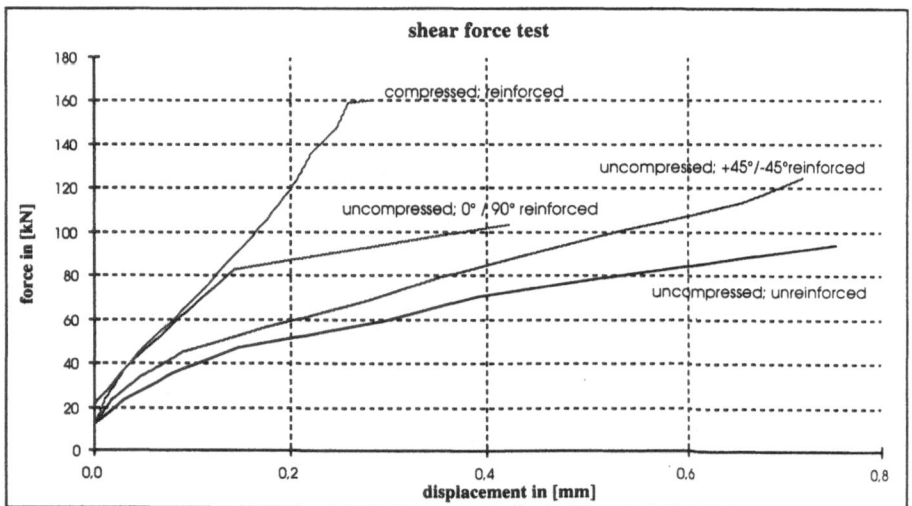

Figure 5. Load - slip - relationship of shear stress

6.2. SHEAR FORCE

The shear force samples were of the same dimensions as the above mentioned samples. Differences were in the arrangement of dowels and the load transfer. Load was applied at the bearing via the projecting metal plates. The fasteners have the minimum distances required in [DIN 1052]. The failure of the non-reinforced joint occurred in the lower row of fasteners. The reinforced samples broke in the upper and lower row of dowels at the same time.

6.3. TENSILE FORCE

The tension specimen measured 110 x 100 x 1100 mm and consisted of gluelam made from densified boards. In order to guarantee the load transfer at the grips of the testing device the section was widened to 200 x 100 mm. The joint itself presented two steel plates of 6 mm thickness fixed with dowels of Ø 8 mm arranged in a rectangular shape. End distance and spacing of the dowels amounted to 20 and 30 mm respectively. The unreinforced tension specimen which was built with two steel plates (St 37) of 12 mm thickness and whose dowels had a diameter of 6 mm made an exeption. Both facts significantly influence the timber joint ductility. End distance and spacing of the dowels hereby stayed constant. The glassfibre reinforcement led to an increase of the failure load to 127% compared with the reference specimen. A further increase to 149% was observed with +45°/ - 45° orientation of the fabric. Unfortunately, the densified specimens failed outside the joint area in the grip at 133% so that only the stiffness of the joint can be utilized from this test.

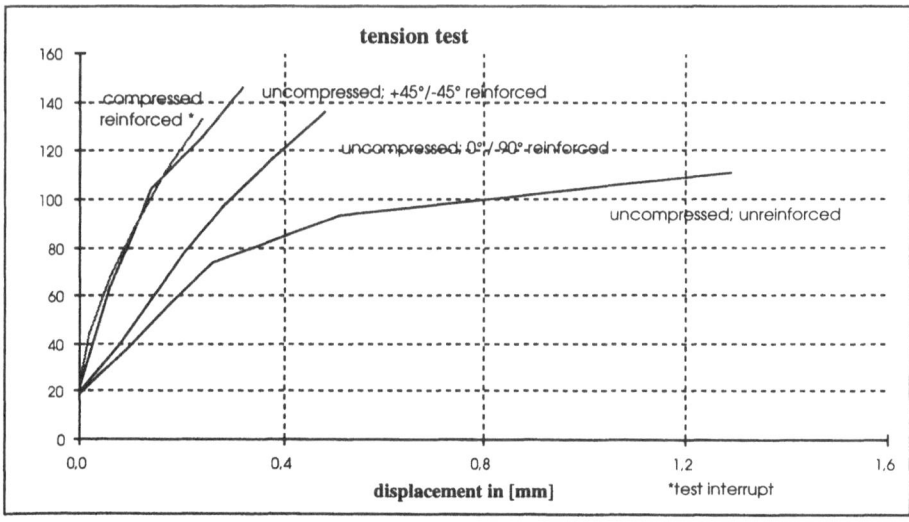

Figure 6. Load - slip - relationship of tension stress

TABLE 3. Summery of test results

Test	Geometry	Fastener	Failure load [kN]	Failure load [%]	Deforma- tion [mm]
bending test		St 37			
unreinforced 0% compressed	120 x 200 x 3100	9 x ∅10mm	30,0	100	20,7
0° / 90° reinforced 0% compressed	120 x 200 x 3100	9 x ∅10mm	44,2	149	47,7
+45°/- 45° reinforced 0% compressed	120 x 200 x 3100	9 x ∅10mm	39,0	134	58,1
0° / 90° reinforced 50% compressed	120 x 200 x 3100	9 x ∅10mm	63,6	211	52,9

Test	Geometry	Fastener	Failure load [kN]	Failure load [%]	Deforma- tion [mm]
shear force test		St 37			
unreinforced 0% compressed	120 x 200 x 1760	9 x ∅6mm	92,2	100	0,7
0° / 90° reinforced 0% compressed	120 x 200 x 1760	9 x ∅6mm	103,4	112	0,4
+45°/- 45° reinforced 0% compressed	120 x 200 x 1760	9 x ∅6mm	124,0	135	0,7
0° / 90° reinforced 50% compressed	120 x 200 x 1760	9 x ∅6mm	160,5	173	0,3
tension test		St 37 / C 45			
unreinforced 0% compressed	110 x 100x 1100	12 x ∅6mm	139,0	100[1]	1,5
0° / 90° reinforced 0% compressed	110 x 100x 1100	12 x ∅8mm	142,9	127	0,5
+45°/- 45° reinforced 0% compressed	110 x 100x 1100	12 x ∅8mm	167,3	149	0,5
0° / 90° reinforced 50% compressed	110 x 100 x 1100	12 x ∅8mm	149,5	133*	0,3

[1]) 6mm St 37 dowel diameter;   *) failure out of joint

7. Conclusion

The experimental study on glassfibre reinforced and densified timber joints has shown that the load carrying capacity could be increased up to two times. Apart from the ultimate load, stiffness and ductility are improved considerably so that construction elements and parts could be prevented from sudden failure. The structural use of densified wood arise questions about the bonding to wood and glassfibre fabrics. Particularly, the connection of densified and undensified boards which was done in this study by means of finger joints present a weak point and must be improved.

In order to fully benefit from new technical textiles, more especially from those having an optimal orientation regarding the stress state of the construction element, fundamentals in mechanics of the wood-textile-composite are required as a prerequisite.

8. References

1. Birk, T. (1999) *Textilbewehrte Holzverbindungen mit Pressholz*, TU Dresden; Institut für Baukonstruktionen und Holzbau, Diplomarbeit
2. Chen, C.J.(1999) *Study of mechanical behaviors and optimization of fiber glass reinforced timber joints*, Dissertation, EPF-Lausanne, Schweiz
3. Colling, F., Siebert, W., Belchior-Gaspard, P. (1987) Tragfähigkeit glasfaserverstärkter Brettschichtholzträger, Karlsruher Forschung im Ingenieurholzbau, *Bauen mit Holz* 8/87
4. Dagher, H. J. (1998) *Creep behavior of FRP-reinforced glulam beams*, Proceedings, Vol. 1, S. 161-165, 5th World Conference of Timber Engineering Montreux, Schweiz
5. Davalos, J. F., Qiao, P. (1998) *Development of a prototype composite-reinforced timber railroad crosstie*, Proceedings, Vol. 1, S. 494-498, 5th World Conference of Timber Engineering Montreux, Schweiz
6. *DIN 1052* (1988) Holzbauwerke; Beuth Verlag
7. Ehlbeck, J., Werner, H. (1992) *Tragfähigkeit von Laubholzverbindungen mit stabförmigen Verbindungsmitteln*, Forschungsbericht, Universität Karlsruhe
8. Gay, D. (1991) *Materiaux composites*, Editions Hermes, Paris
9. Gougeon Brothers (1998*) Moderner Holzbootsbau*, Deutsche Ausgabe; Verlag M. und H. von der Linden, Wesel, Germany
10. Haller, P. (1994) *Experimentelle Untersuchungen an glasfaserverstärkten Verbindungen*, Ehrenkolloquium Prof. Dr.-Ing. habil. M. Gruber, Schriftenreihe des Instituts für Baukonstruktionen und Holzbau; Heft 1, TU Dresden, S. 96-102
11. Haller, P., August (1998) *Progress in the development and modelling of timber joints*, Proceedings, 5th World Conference of Timber Engineering, Montreux, Switzerland, Bd. 1, S. 337-344
12. Haller, P., Wehsener, J. (1998) *Faserverstärkte Hochleistungsverbindungen aus Pressholz*, Aif- Zwischenbericht; AiF - Nr: 11164 /B1
13. Haller, P., Chen, C. J., Natterer, J. (1996) *Experimental Study of Glass-Fibre Reinforced and Densified Timber Joints*, Proceedings, International Wood Engineering Conference, New Orleans, Louisiana, USA
14. Haller, P., Wehsener, J., Chen, C. J., (1998) *Development of joints by compressed wood and glassfibre reinforcement*, Proceedings, Final Conference Report, COST C1, Control of the semi rigid behaviour of connections in civil engineering, Liege, Belgium
15. Hufenbach, W., Müller, C. (1995) *Hochgeschwindigkeits-Leichtbaurotoren*, Chemische Industrie 118, S.27-32
16. Kempe, O. (1997) Ertüchtigung von Deckenbalken; Mitteilungen des Landesamtes für Denkmalpflege Sachsen, S. 46 – 52

17. Kriechbaum, R., (1994) *Ein Verfahren zur Optimierung der Faserverläufe in Verbundwerkstoffen durch Minimierung der Schubspannungen nach Vorbildern der Natur*, Forschungsbericht, Kernforschungszentrum Karlsruhe; KfK 5406

18. Larsen, H. (1994) *Experimental Study on Fiber Reinforced Gluelam Beams*, Proceedings; Pacific Timber Engineering Conference; Gold Coast; Australia

19. Offermann, P., Franzke, G. (1996) *Innovative Textilstrukturen für den Leichtbau*, In: Vortragsband zur DLR-Tagung, TU Dresden

20. Sucz, C. A. (1991) *Etude du renforcement du bois par fibre de verre*, Dissertation; Universität Metz, Frankreich

21. Tingley, D. A. (1998) *FRP reinforcement glulam performance: A case study of the Lighthouse Bridge*, Proceedings, Vol. 2, S. 177-181, 5th World Conference of Timber Engineering Montreux, Schweiz

22. Wald, F. [ed.] (1994) *Control of semi-rigid behaviour of civil engineering structural connections*, COST C1 Proceedings of the second state of the art workshop, Prague, published by the European Commission

TIMBER JOINTS LOADED PERPENDICULAR TO THE GRAIN:

LONG-TERM STRENGTH, THEORY AND EXPERIMENTS

H. J. LARSEN
Danish Building Research Institute and Lund University
Postboks 119, DK 2970 Horsholm, Denmark

P. J. GUSTAFSSON
Structural Mechanics, Lund University
Box 118, S 221 00, Lund, Sweden

1. Introduction

Joints play a very great role in timber structures and it is important to have effective methods to calculate their load-carrying capacity and the deformations.

With a few exceptions, joints are overlap joints where dowel-type fasteners (nails, screws, bolts and tight fitting dowels) transfer the forces. The necessary contact areas are large and often decisive for the member sizes.

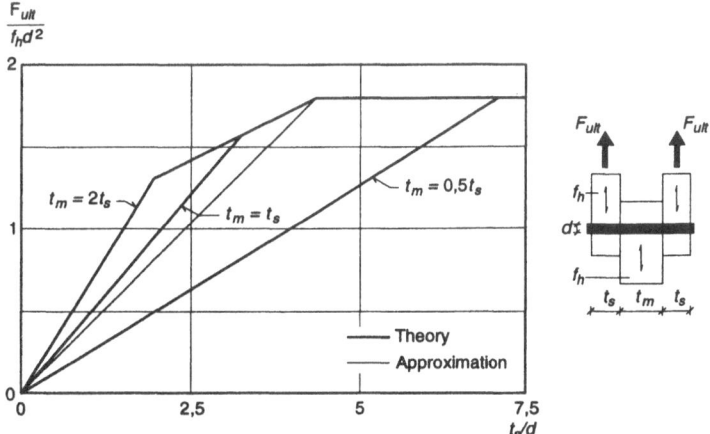

Figure 1. The load-carrying capacity F_{ult} as a function of the slenderness ratio l_s/d, where l_s is the thickness of the side member and d is the diameter of the bolt. The thickness of the middle member is l_m. A ratio of $f_y/f_h=10$ between the yield strength of the steel, f_y, and the embedding strength of the wood, f_h, has been assumed. The thin line shows the approximation proposed in the 1999 draft for Eurocode 5 [7].

C.C. Baniotopoulos and F. Wald (eds.), The Paramount Role of Joints into the Reliable Response of Structures, 407–416.
© 2000 *Kluwer Academic Publishers.*

Timber joints are generally rather flexible. The contribution to deformations from slip in the joints is often more important than the deformations of the timber members themselves and may have great influence on the force distribution, see e.g. [1] and [2].

Joint failures can be ductile. This is for instance often the case for slender fasteners with sufficient spacing and distances to end and edges. It has been demonstrated in many papers that in this case the so-called European Yield Model (first proposed by Johansen [3] in 1949) gives an excellent prediction of the load-carrying capacity, see e.g. [4], and [5]. A typical example of the load-carrying capacity as a function of the bolt slenderness is shown in figure 1.

In some cases, however, failure is brittle, caused by shear, splitting or tension perpendicular to the grain, see figure 2. This is of course most common when the load acts perpendicular to the grain, but can also be the case for load parallel to the grain, even with spacing etc. in accordance with normal design practice. This may for instance be the case with several fasteners in line in the force direction, especially for non-slender fasteners.

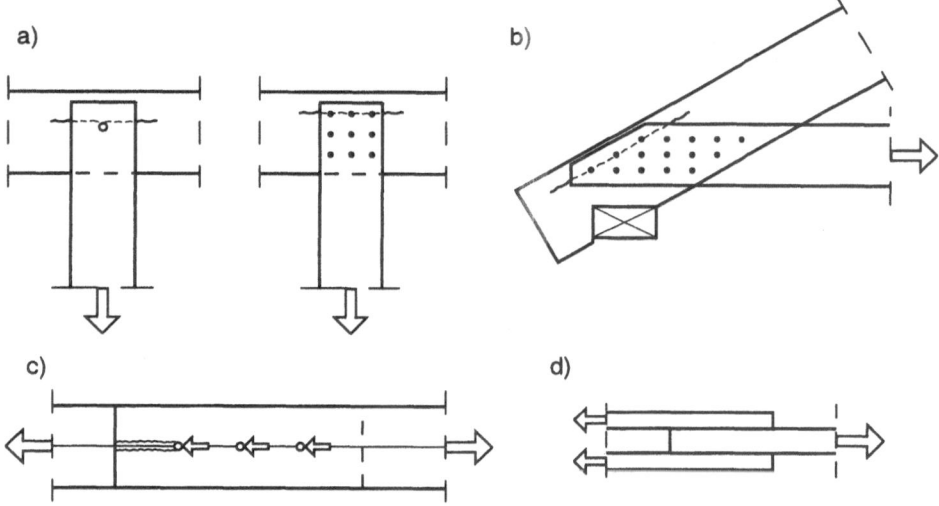

Figure 2. Examples of tension and shear failure of joints. a) and b) Load applied perpendicular (at an angle)to the grain. c) Splitting due to wedge action. d) Wood failure of glued joint.

For brittle failure, only rather crude theoretical expressions exist, and only for very special cases. Design is, therefore, based either on simple empirical or semi-empirical rules, e.g. those of Eurocode 5 [6], Ehlbeck et al. [7] or van der Put [8], or on reducing the theoretical load-carrying capacity according to the European Yield Model rather arbitrarily. The 1999 draft for Eurocode 5 [6] is an example of the latter approach. In this it is proposed that the yield values for slender bolts ($l_m/d >$ about 4) be used, but with reduced values as shown in figure 1 for rigid bolts. There are two problems with this approach. The first is that it is very much on the safe side for many normal joint configurations – bolts are very often used with l_m/d-ratios around 1.5–3, thus reducing the competitiveness of timber structures. The second is that it is unsafe for some configurations.

The empirical rules mentioned have been evaluated by Ballerini [9] on the basis of experiments. His conclusion is that the rules proposed by Ehlbeck et al. [7] give the best prediction but that they often overestimate the strength.

2. Theory

2.1 INTRODUCTION

To design timber structures effectively it is necessary to have design methods that can correctly predict the load-carrying capacity for both ductile and brittle failure taking into account the influence of the loading time.

Jorissen [10] has derived an expression for the shear failure load for a single rigid dowel loaded parallel to the grain based on fracture mechanics. He concludes that the fracture mechanics model results in a more accurate prediction of the load-carrying capacity of a single fastener connection with rigid bolts than the European Yield Model.

Yasumura and Daudeville have analysed bolted timber joints loaded perpendicular to the grain, see [11], which contains further references. A finite element model has been used to analyse the stresses around the bolt holes and to follow the crack propagation. They conclude that

- a linear elastic finite element method is an appropriate tool,
- the load-carrying capacities obtained from the simulations agree well with experimental results, and
- in general, bolted joints loaded perpendicular to the grain fail due to brittle failure at loads smaller than corresponding to the yield strength.

2.2 METHODS FOR STRENGTH ANALYSIS

For a theoretical strength analysis, the properties of the wood must be defined in terms of a constitutive model and a failure criterion (if fracture modelling is not a part of the constitutive model). Moreover, to enable stress analysis, kinematic constraints of some kind are commonly adopted, such as those of beam or plane strain analysis. In this respect, the finite element method is a very powerful tool since all such constraints can in principle be avoided if a fine mesh of 3D elements is used in the analysis. On the other hand, constraints can make is possible to derive simple explicit strength equations not requiring numerical solution of a system of equations.

For wood, the basic constitutive model is that of an orthotropic linear elastic continuum. Depending on loading and environmental conditions, there may be a need to complete this basic model by considering effects of moisture variations and duration of load.

Failure criteria for shear or tension perpendicular to grain are commonly defined in terms of the shear and tensile stresses in the most stressed point, e.g. according to the criterion of Norris [12]. There are however difficulties in applying such a criterion: Fracture starting from a crack cannot be analysed because of the stress singularity and it is difficult to get good results if the high stress region is small and with strong stress gradients. The opposite kind of rational fracture criterion, that of conventional linear elastic fracture mechanics, requires on the other hand a pre-existing crack in order to make strength analysis possible.

Other kinds of criteria are, therefore, needed for general strength analysis based on linear elastic stress analysis. Two such criteria are "the average stress method" and "the initial crack method", see [13]. In the average stress method, some criterion like the one of Norris is applied, using the average stresses acting over a certain volume or area, determined from the basic material properties of the wood. In the initial crack method, some conventional fracture mechanics crack propagation criterion is applied for an assumed (fictitious) crack located in the most stressed region. The length of the crack, being very important for the strength prediction, is determined from the basic strength, stiffness and fracture energy properties of the wood.

Another method for strength analysis is non-linear fracture mechanics [14, 15], where the development and growth of a fracture zone and crack are explicitly simulated by a constitutive relation that defines the gradual damage and fracture softening of the material as deformations increase beyond the limit strain.

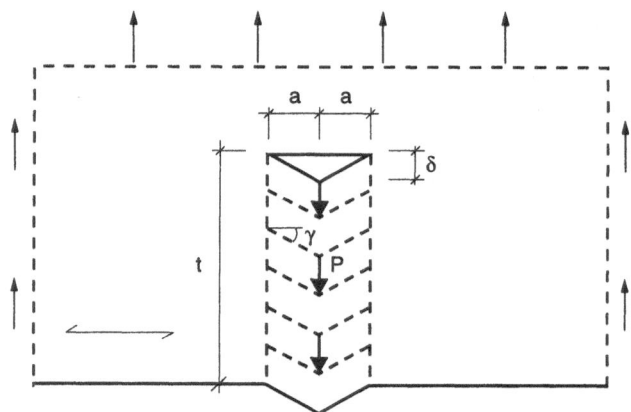

Figure 3. Specimen for which a simple, explicit strength equation is derived.

2.3 DEVELOPMENT OF A SIMPLE THEORETICAL STRENGTH EQUATION

Before making a detailed finite element analysis, a simple explicit strength equation is developed for the specimen shown in figure 3. It is based on a fracture mechanics energy balance equation and a chosen displacement field – fulfilling the conditions of geometrical compatibility – for the deformations of the wood during crack propagation.

The specimen considered in figure 3 corresponds to those used in the ongoing tests, see figure 4. The total load, P, is applied through one or more dowels placed in line. The distance to the innermost dowel hole is l. The holes are formally regarded as pre-existing cracks located along the grain and having a length $2a$ equal to the diameter of the dowel hole. The strains in the load-carrying region are assumed to be dominated by shear strains, γ, and the assumed deformation pattern is as shown in figure 3. The displacement of the load P is

$$\delta = a\gamma = a\tau/G = 0.5aP/(lbG) \qquad (1)$$

where b is the thickness of the specimen and G is the shear modulus of the wood, assumed to be linear elastic. The potential energy of the system, W, is the sum of the strain energy in the wood and the potential energy of the load. Since the material is regarded as linear elastic:

$$W = -P\delta/2 = -P^2a/(4lbG) \qquad (2)$$

At the limit load where the crack starts to propagate, there is balance between the decrease in the potential energy, $-dW$, and the fracture energy required for propagation of the crack. The energy required for increasing the crack length by $2da$, i.e, increasing a by da, is $G_fb(2da)$, where G_f is the perpendicular to grain fracture energy of the wood. This energy balance:

$$-dW = 2G_fbda \qquad (3)$$

together with equation (2) gives the failure load:

$$P_{ult} = 2b\sqrt{2lGG_f} \qquad (4)$$

Typical material parameter values for softwoods are $G = 500$ N/mm^2 and $G_f = 0.3$ N/mm (= 300 J/m^2), see [16]. It is noteworthy that the predicted P_{ult} is not affected by the diameter of the hole or by the tensile strength of the wood. Instead, the decisive material parameters are shear stiffness and fracture energy, and the decisive geometrical parameter is the edge distance, l, for the innermost dowel. It must, however, be recalled that the derivation of equation (4) is based on a simplified assumption in terms of a constrained displacement field. Since the chosen displacement field fulfils the compatibility conditions, the stiffness is overestimated and the strain energy for a given load P accordingly underestimated. This means that equation (4) can generally be expected to overestimate the absolute value of the failure load. This can be expressed by means of an efficiency factor, η, for the displacement field, giving

$$P_{ult} = 2b\eta\sqrt{2lGG_f} \qquad (5)$$

where, in general, $\eta \leq 1.0$.

3. Test specimens and testing

3.1 SPECIMENS

Six types of specimens were tested, see figure 3. The specimens were double symmetrical. The dowel specimens are named after the number of dowel lines and the number of dowels in the line. The distances $4d$ are the minimum values according to Eurocode 5. DCB stands for Double Cantilevered Beam specimen. T denotes tension specimens.

The specimens were made from 40 mm thick Laminated Veneer Lumber (LVL) from Malarply, Sweden. LVL was chosen to get a uniform material and because of the

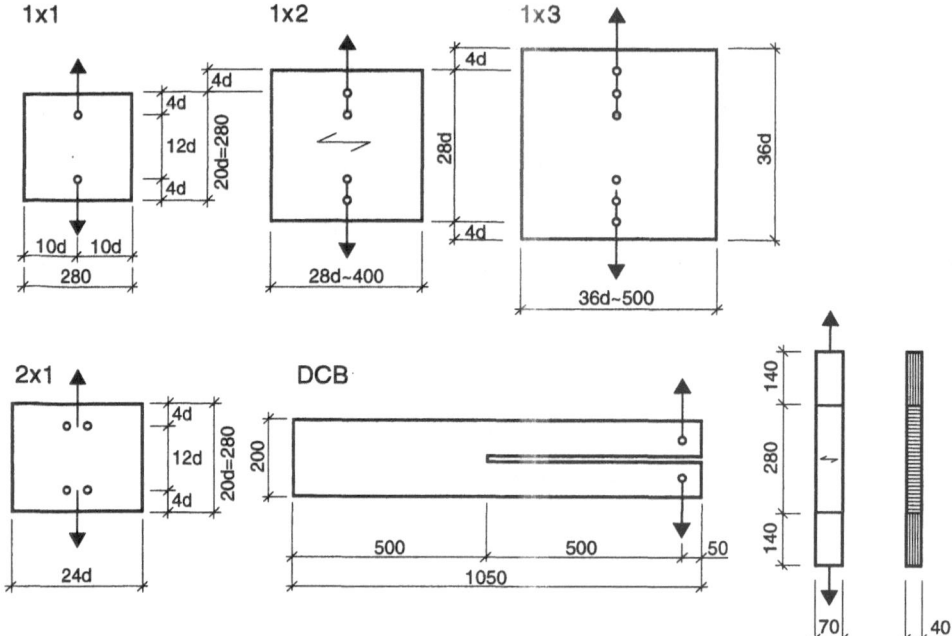

Figure 4. Test specimens. The T-tests were made by gluing LVL with the grain parallel to the axis to the ends of the test section that is loaded perpendicular to the grain.

size of the specimens. The dry density was about 450 kg/m^3 with a coefficient of variation of 2.4 per cent.

The dowel diameter was 14 mm and the steel quality ensured that the stresses in the dowels were in the elastic range.

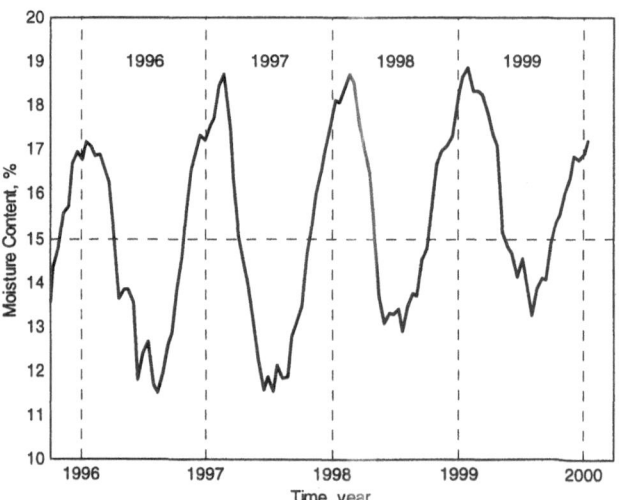

Figure 5.Moisture variations in wood in the test barn.

The specimens are stored in an open barn in southern Sweden. To have a realistic moisture variation in the specimens the end grain is sealed with an asphalt tape. The moisture content of the specimens is controlled by weighing 3 pieces of LVL every two weeks. The air temperature and the relative humidity are recorded. The typical variation of the moisture content in wood is shown in figure 5.

3.2   TEST SET-UP

Figure 6. Testing of DCB specimens.   Figure 7. Testing of 1x3 specimens.

The load was transferred by two 4 mm thick steel gusset plates to a 16 mm dowel. The 17 mm hole for the dowel was predrilled in both plates as were the 15 mm holes for the 14 mm dowels in one of the steel plates. This plate was placed on top of the joint and used as lead when drilling 15 mm holes through the LVL and the bottom steel in one operation, thus ensuring a perfect fit. The specimens were loaded with dead load through chains and/or shackles.

The DCB (Double Cantilever Beam) specimens were loaded directly as shown in figure 5. The other five types were tested in rigs as shown in figure 6. A more detailed description of the test rigs, which have been used for several EU-projects, can be found in [16] and [17].

TABLE 1. Test results: minimum – **mean** – maximum – (coefficient of variation in per cent)

	1×1	1×2	1×3	2×1	T	DCB
Standard, SBI 23 °C/65 % RF						
short-term strength (kN)	7.24-**8.12**-8.73 (6.7%)					1.33-**1.44**-1.54 (5.5%)
time to failure (s), 80 %						
time to failure (s), 65 %[1]						
Standard, ASA 23 °C/65 % RF						
short-term strength (kN)	7.23-**7.55**-8.34 (5.1%)	9.49-**11.05**-11.90 (6.1 %)	12.71-**14.53**-16.35 (7.9 %)	6.56-**7.50**-8.34 (7.1 %)	1.85-**2.49**-3.02 (14.2 %)	0.80-**1.05**-1.26 (12.7%)
time to failure (s), 80 %						
time to failure (s), 65 %[2]						
Winter 2000 moisture: 14.4 %						
short-term strength (kN)	7.29-**7.64**-7.91 (3.7 %)	9.72-**10.42**-11.17 (4.7 %)	10.67-**12.81**-14.22 (9.7 %)	6.54-**7.20**-8.45 (10.1 %)	0.62-**1.12**-1.49[3] (35 %)	1.10-**1.22**-1.25 (5.1 %)
time to failure (s), 80 %						
time to failure (s), 65 %						
Spring 2000 moisture						
short-term strength (kN)	6.84-**7.41**-7.94 (5.2 %)	11.52-**12.35**-13.53 (7.1 %)	11.62-**14.35**-15.59 (10.2 %)	7.52-**8.06**-8.50 (4.2 %)	1.02-**1.43**-2.62 (43 %)	0.97-**1.12**-1.26 (8.8 %)
time to failure (s), 80 %						
time to failure (s), 65 %						
$F_{ult, theory}$ kN	10.36	14.66	17.95	14.66		
γ (effectivity factor)	0.75	0.75	0.81	0.72	—	

[1] For DCB: 75 %
[2] For DCB: 75 %
[3] One value (3.19) is disregarded. If this value is included the mean becomes 1.25

3.3   TEST PROGRAMME

The main element is the test packages described below tested/to be tested February 2000 (winter), May 2000 (spring), August 2000 (summer), November 2000 (autumn) and February 2001 (winter).
A test package consists of the following:
- short-term ramp-loading (time to failure about 3 minutes) of 6 specimens of each type, 36 specimens in all
- long-term loading of 5 specimens of each type, 30 specimens in all, under constant load with a load level of about 80 per cent
- long-term load of 5 specimens of each type, 30 specimens in all, under constant loading with a load level of about 65 per cent
- Determination of the moisture profile for two specimens.

The load level is estimated from the short-term results from the same package, which may vary as a result of the varying moisture content (maximum in winter, minimum in summer, and with maximum variations in spring and autumn).
Further, the test programme consists of determination of the following properties after conditioning in standard climate (65 per cent relative moisture content and 23 $^\circ$C):

- short-term ramp-loading of 10 specimens of each type, 60 specimens in all
- long-term loading of 5 specimens of types 1×1, T and DCB , 15 specimens in all, under constant load with a load level of about 80 per cent
- long-term loading of 5 specimens of types 1×1, T and DCB, 15 specimens in all, under constant load with a load level of about 65 per cent.
- determination of the compression strength and the embedding strength perpendicular and parallel to the grain for 20 specimens, 80 specimens in all.

The short-term strength was also determined for specimens 1×1, 1×2 and 1×3 after conditioning in 35, 65 and 85 per cent relative humidity and 23°C. Each series consisted of 18 specimens, 54 in all.

4.   **Preliminary test results**

The first test results (from February and end of April 2000) are shown in table 1.
The different conditioning conditions have little effect on the DCB and the 1×1 specimens. The T-specimens were much more straight when conditioned in standard climate than in natural climate and the strength and especially the variation were much smaller.

416

5. References

1. Larsen, H.J. and Jensen, J.L. (2000) Influence of semi-rigidity of joints on the behaviour of timber structures, *Progress in Structural Engineering and Materials* **vol 2, no. 3**. In print

2. Maquoi, R. (editor) (1999). *Control of the semi-rigid behaviour of civil engineering structural connections*, European Communities EUR 18854 EN.

3. Johansen, K.W. (1949). Theory of timber connectors, *Publication 9, International Association of Bridge and Structural Engineering*, Basel.

4. Whale, L:R:J., Smith, I and Larsen, H.J. (1987) Design of nailed and bolted joints., Paper CIB-W18A/20-7-1 in *Proceedings of Meeting Twenty in CIB Working Commission W18A – Timber Structures*.

5. Hilson, B:O. (1995) Joints with dowel-type fasteners – Theory in Blass, H.J. et al (editors), *Timber Engineering STEP*, Centrum Hout, Almere, The Netherlands.

6. Draft prEN 1995-1-1 (1999) Design of Timber Structures – Part 1-2: General Rules and Rules for Buildings. European Committee for Standardisation.

7. Ehlbeck, J., Görlacher, R. and Werner, H. (1999) Determination of perpendicular-to-grain tensile stresses in joints with dowel-type fasteners. Paper CIB-W18/22-7-2 in *Proceedings of Meeting Twenty-two in CIB Working Commission W18 – Timber Stuctures*.

8. van der Put, T.A.C.M. (1990) Tension perpendicular to grain at notches and joints Paper CIB-W18/23-10-1 in *Proceedings of Meeting Twenty-three in CIB Working Commission W18 – Timber Stuctures*.

9. Ballerini, M. (1999) A new set of experimental tests on beams loaded perpendicular-to-grain by dowel-type joints. Paper CIB-W18/32-7-2 in *Proceedings of Meeting Thirty-two in CIB Working Commission W18 – Timber Structures*.

10. Jorissen, A. (1998) Double shear timber connections with dowel type fasteners. *Diss. Civil Engineering Department of Delft University of Technology*, The Netherlands.

11. Yasumura, M. and Daudeville, L. (1999) Design and analysis of bolted timber joints under lateral force perpendicular to grain. Paper CIB-W18/32-7-3 in *Proceedings of Meeting Thirty-two in CIB Working Commission W18 – Timber Structures*.

12. Norris, C.B. (1962) Strength of orthotropic materials subjected to combined stresses. *Report No. 1816*, Forest Products Laboratory, Madison, USA.

13. Gustafsson, P.J., Petersson, H. and Stefansson, F. (1996) Fracture Analysis of Wooden Beams with Holes and Notches. *Proceedings of the International Wood Engineering Conference*, New Orleans 1996, pp 4281-4287.

14. Gustafsson, P.J. (1985) Fracture Mechanics Studies of Non-Yielding Materials Like Concrete, *Diss. TUBM-1007, Division of Building Materials, Lund University*, Sweden.

15. Boström, L. (1992) Method for Determination of the Softening Behaviour of Wood and the Applicability of a Non-linear Fracture Mechanics Model, *Diss. TUBM-1012, Division of Building Materials, Lund University*, Sweden.

16. Gustafsson, P.J. (1988) A Study of Strength of Notched Beams. Paper CIB-W18A/21-10-1 in *Proceedings of Meeting Twenty-one in CIB Working Commission W18 – Timber Structures*. Parksville, Canada

17. Gustafsson, P. J. (1997) Report of Lab5, in Consolidated Progress *Report of Project AIR2-CT94-1057*

COMPONENT METHOD FOR HISTORICAL TIMBER JOINTS

F. WALD, J. MAREŠ, Z. SOKOL
Czech Technical University, Department of Steel Structures
166 29 Praha, Czech Republic

M. DRDÁCKÝ
Institute of Theoretical and Applied Mechanics of the Academy
of Sciences of the Czech Republic, Prosecká 76, 190 00 Praha 9,
Czech Republic

1. Introduction

Timber structures represent one of the most important ancient engineering works spanning over considerable distances. They involve not only an evidence of structural knowledge and creativity of their makers but also a good deal of structural beauty, Figure 1.

When studying historical timber roofing frames we have to deal with several problems. Firstly, the roofing frameworks have passed a long way of development of their structural schemes and improvements of their layout. Secondly, the development of roofing frameworks and also building frames was accompanied by changes in joints and their structural behaviour [1]. Let us start our short study with remarks concerning typical structural carpentry joints and their load – deformation behaviour as it was acquired by experiments.

Figure 1   The joint configuration of a frame of the Castle Horní Branná [1]

C.C. Baniotopoulos and F. Wald (eds.), The Paramount Role of Joints into the Reliable Response of Structures, 417–424.
© 2000 *Kluwer Academic Publishers.*

2. Experiments

Three sets of tests were performed to simulate the behaviour of historical timber joint – experiments of a component of the joint (test of timber in compression), tests of replica of historical carpenter joint and test of authentic historical joint taken from the reconstructed structure, see Figure 2. The experiments with replicas were carried out on specimens made from several hundred years old timber.

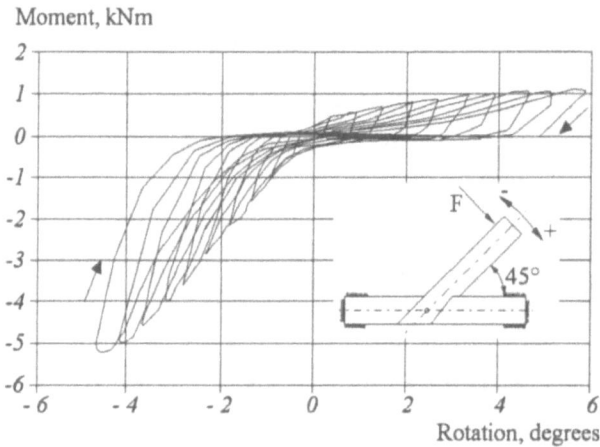

Figure 2   Example of moment rotational diagram of test H-4-45

After the experiments with the joints were finished, material characteristics were determined by tests on coupons taken from the joint assemblies. The evaluated material properties were modulus of elasticity in bending, in compression parallel to fibres and an indentation response to the concentrated load introduced by narrow wooden or steel indenters, see Figure 3.

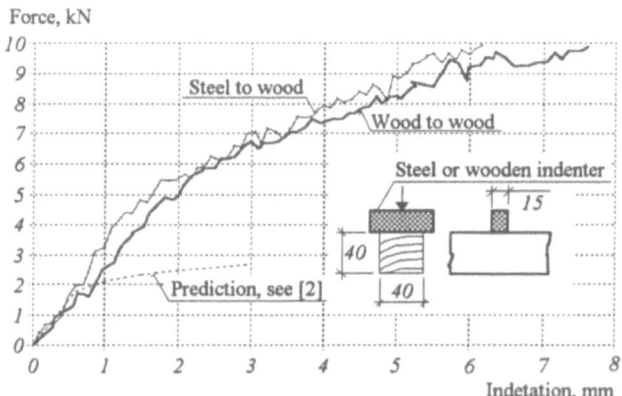

Figure 3   Comparison of indent tests, steel to wood and wood to wood

Modulus of elasticity in four-point bending varied from 11,53 GPa to 25,91 GPa. Modulus of elasticity in compression varied from 12,47 GPa to 17,35 GPa and the compression strength between 35 and 57,1 MPa. Typical indentation test results are presented on Figure 3. The dashed line shows a literature reference [2]. As a conclusion to the test we can state the response depends on the direction of annual rings and it can be approximated by a bilinear function. Tension tests proved tension strength parallel to fibres between 74 MPa and 170 MPa and modulus of elasticity from 11,2 GPa to 16,2 GPa, (at moisture content of about 9%).All tests were carried out under moderate temperature of about 15°C and moisture content of about 11%.

The test results show rather good quality of ancient wood, which has been experienced several times by other researchers. Nevertheless, the degradation can reduce material characteristics substantially and influence computational modelling.

3. Mechanical model of joints

The component method was applied to the joint to derive simple model of joint behaviour subjected to static and cyclic loading, see [2]. The joint is decomposed into components, which are represented by force - deformation diagram, see Figure 3. It is supposed the dowel resists to shear force and clearly fixes the position of the centre of rotation of the connection.

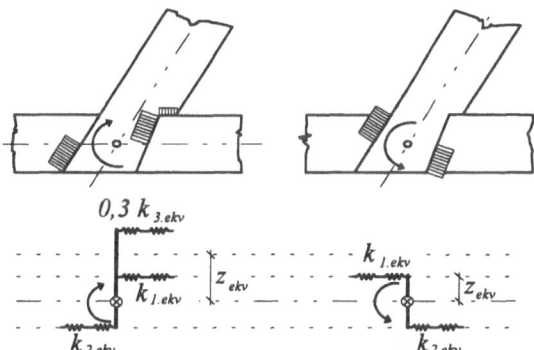

*Figure 4   The contact surfaces and the component model
for loading in clockwise and counter-clockwise directions*

Finite element simulation was performed to verify assumptions about contact areas of the tested connections, see Figure 5. The model was created from two bodies (8-noded SOLID 45 elements, 5307 nodes) connected by point to surface CONTACT 49 elements. Initial gap 2 mm was used to simulate initial slip in the connection.

Principal stresses for clockwise loading are plotted on the surface as contour lines to show compression stresses of the studied connection.

420

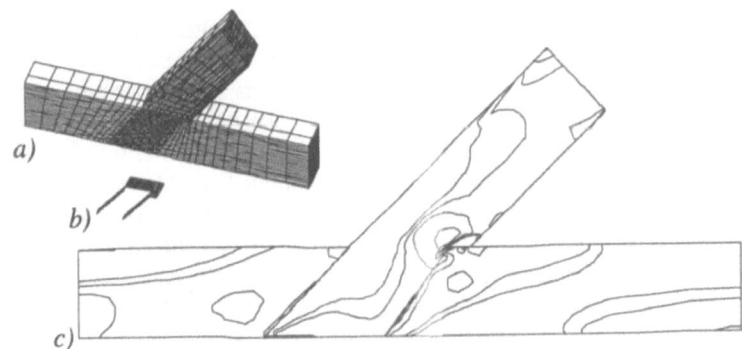

Figure 5   The finite element model, a) the mesh, b) the contact elements ,
c) isolines of compression stress

3.1. Component Timber in Compression

The presented connection is composed from the one type of component only –timber in compression. The resistance of the material in compression with respect to the direction β of fibres of the wood is

$$f_{c,\beta,d} = \frac{f_{c,0,d}}{\dfrac{f_{c,0,d}}{f_{c,90,d}} \sin^2 \beta + \cos^2 \beta} .$$ (1)

The resistance of the component $F_{u,i}$ can be derived from area in compression A_i.

$$F_{u,i} = A_i \, f_{c,\beta,d} .$$ (2)

Stiffness of this component is calculated on basis of theory of deformation of elastic half space.

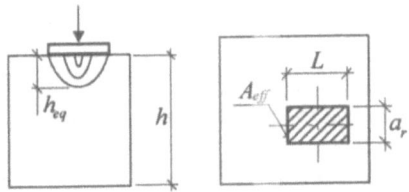

Figure 6   *Deformation of a timber block on surface zone*

The deformed zone in the joint can be predicted using concept of effective height. The deformation of the block under a rectangular rigid plate was solved by several authors. Solution by Lambert and Whitman [3] gives deformation

$$\delta_c = \frac{F_c \, \alpha \, a_r}{E_\alpha \, A_r} ,$$ (3)

where F_c is the applied compressed force, a_r is width of the rigid plate, E_α is the Young's modulus of wood, A_r is area of the plate, L is length of the plate, see Figure 6.

The factor α depends on ratio between L and a_r and on Poison's ratio v of the material. The value of the factor and α was calculated with $v = 0,09$, see Table 1. The table also gives an approximation of the factor α.

Table1   Factor α

$\dfrac{L}{a_r}$	Factor α according to Lambert and Whitman [3]	Approximation $\alpha \approx 0,80\sqrt{L/a_r}$
1	0,88	0,85
1,5	1,09	1,04
2	1,23	1,20
3	1,45	1,47
5	1,73	1,90
10	2,14	2,69

The formula (3) describing the displacement under the plate can be simplified by substituting approximate formula for α from Table 1

$$\delta_c = \frac{0,80\,F_c}{E_\alpha\sqrt{L\,a_r}} \tag{4}$$

The stiffness of the component expressed in form suitable for component method gives

$$k_c = \frac{F_c}{E_{90}\,\delta_c} = \frac{E_\alpha\sqrt{L\,a_r}}{0,80\,E_{90}} \tag{5}$$

3.2. Assembly for monotonous loading

The assembly procedure of the components into behaviour of the joint is based on geometry of the connection and properties of the components see [4]. The bending moment resistance is given by

$$M_u = \sum_i F_{u,i}\,z_i \tag{6}$$

The initial bending stiffness can be calculated from stiffness of the components

$$S_j = \frac{M}{\varphi} = \frac{E\,z_{ekv}^2}{\sum\limits_i \dfrac{1}{k_{i.ekv}}}, \tag{7}$$

where the components in series should be replaced by a single spring with stiffness k_i

$$\frac{1}{k_i} = \frac{1}{k_{i.1}} + \frac{1}{k_{i.2}}. \tag{8}$$

422

The stiffness $k_{i.1}$ and $k_{i.2}$ belong to components representing two contact surfaces with different fibre orientation and therefore with different stiffness.

The effective stiffness of the components $k_{i.eff}$ takes into account elastic force distribution of components in parallel configuration

$$k_{i.ekv} = \frac{\sum_i \frac{1}{k_i} z_i}{z_{ekv}} \qquad (9)$$

where

$$z_{ekv} = \frac{\sum_i k_i z_i^2}{\sum_i k_i z_i} . \qquad (10)$$

The predicted behaviour of the joint is compared to experiment H-4-45, see Figure 7.

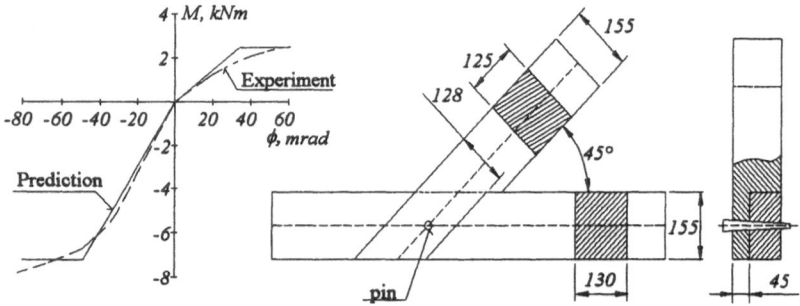

Figure 7   Comparison of the model to experimental results of the second loading step and geometry of the connection (test H-4-45)

3.3.   Assembly for cyclic loading

Two major types of components can be distinguished, see Figure 8. One directional component represents the contact between two surfaces. The component resists in compression, but offers no resistance in tension. The other type is two directional component. It represents behaviour of components resisting in tension and compression, i.e. plate or bolt.

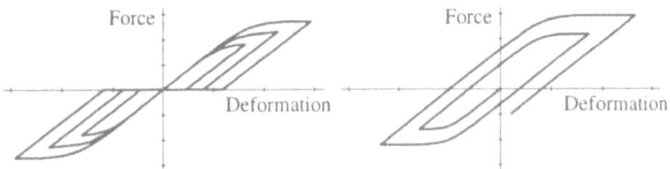

Figure 8   Basic types of the component behaviour;
a) one directional (contact), b) two directional

The component method is extended to simple model for cyclic loading. The model consists from one directional components only. When the character of the bending moment changes, the other components are taken into account, see Figure 4. The resistance $F_{u,i}$ and the initial stiffness k_i of the components are based on the above prediction. Response of the components is described by a non-linear curve. The loading branch of force-displacement curve is replaced by power function, see [5],

$$\delta_i = \frac{F_i}{E_\alpha\, k_i} \frac{1}{1 - \left(\dfrac{F_i}{F_{u,i}}\right)^n}\,, \tag{11}$$

where shape factor $n = 4,5$ is based on experimental results. The unloading branch is linear with unloading ratio defined by Penserini [5]. The value of $\kappa = 0, 82$ is based on observations of the test results of this component.

The step-by-step procedure was applied to calculate the moment-rotation curve of the joint. Because of the non-linear nature an iterative procedure was necessary. Only last two steps were used for convergence check at every load step.

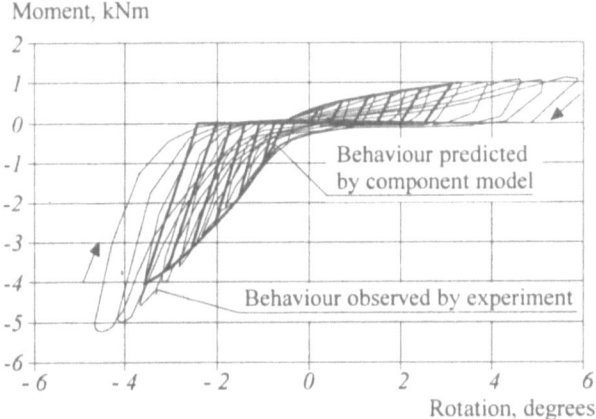

Figure 9   Comparison of the predicted model for cyclic loading to the moment-
rotation diagram of the joint H-4-45, starting point at rotation -0,5°

The model exhibits good agreement to the joint behaviour, see Figure 9. However, the model is less accurate than sophisticated curve fitting procedures [6] and [7]. The method is limited to connections with small number of precisely defined components [8].

4. Conclusion

The presented results of study oriented to modelling of response of historic timber joints approved the necessity to investigate the influence of deformability of classical carpentry connections [9]. Their rigidity in general sense plays an important role in computation of the global deformation and force distribution of historic roofing frames. Those computations are necessary for prediction of behaviour of historic structures during reconstruction works or in cases of assessment of their safety and remaining life.

5. Acknowledgement

This work has been supported by the grant No. 107/97/S051 by Grant Agency of Czech Republic.

6. References

1 Drdácký M. F., Wald F. and Mareš J. (1999) *Modelling of real historic timber joints*, v Proceedings of the STREMAH 99 Sixth International Conference Structural Studies of Historical Buildings, Dresden, pp. 169 - 178, ISBN 1 853 12 690 X.

2 Vergne, A. (1998) *Testing and modelling of load carrying behaviour of timber joints,* in Proceedings of International Conference Control of Semi-Rigid Behaviour of Civil Engineering Structural Connections, Liege, p. 4-4.

3 Lambert, T. W. & Whitman, R. V., *Soil Mechanics,* MIT, John Wiley & Sons, Inc., New York, 1969.

4 Leiten A., Ragupathy P. and Virdi, K. S. (1995) *Rotation capacity of semi-rigid joints in timber frames,* in Proceedings of Second State-of-the-Art Workshop COST C1 Semi-Rigid Behaviour of Civil Engineering Structural Connections, ed. F. Wald, Brussels, pp. 335 - 376.

5 Penserini, P. (1991) *Caracterisation et modelisation du comportement des liasons structure metallique-fondation,* These de doctorat de l'Universite Paris 6, E.N.S. de Cachan - C.N.R.S. - Université Pierre et Marie Currie, p. 254.

6 Mazzolani, F.M. (1988) *Mathematical Model for Semi-Rigid Joints Under Cyclic Loads,* in Connections in Steel Structures: Behaviour, Strength and Design, ed. R. Bjorhovde et al., Elsevier Applied Science Publishers, London, pp. 112 - 120.

7 Ermopoulos J., Stamatopoulos J., Wald, F. and Sokol Z. (1997) *Mathematical Modelling of Semi-Rigid Connections in Steel Column-Bases under Cyclic Loading,* Research report, CZ-GR Gr. No. 5/95, Athens, p. 33.

8 De Martino, A., Faella, C. and Mazzolani, F. M. (1984) Simulation of Beam-to-Column Joint Behaviour Under Cyclic Loads. *Construzioni Metalliche,* 6, pp. 346 - 356.

9 Drdácký, M., Wald, F. and Sokol, Z. (1999) *Sensitivity of historic timber structures to their joint response,* Proceedings of the 40th Anniversary Congress of the IASS, Madrid.

APPLICATION OF THE COMPONENT METHOD TO STEEL JOINTS UNDER FIRE LOADING

LUÍS SIMÕES DA SILVA
Department of Civil Engineering, University of Coimbra
Polo II, Pinhal de Marrocos, 3030-290 Coimbra, Portugal

ALDINA SANTIAGO
Department of Civil Engineering, University of Beira Interior
Rua Marquês d'Avila e Bolama, 6200 Covilhã, Portugal

PAULO VILA REAL
Department of Civil Engineering, University of Aveiro
Campo de Santiago, 3600 Aveiro, Portugal

1. Introduction

Recent research [1] on the behaviour of steel structures under fire loading highlighted the influence of joint behaviour on the overall response of the structure. The lack of experimental results on the response of steel joints under fire conditions and the use of numerical models relying on empirical relations established from tests either at room temperature, or for a limited range of (low) temperatures, has led to a simplistic specification from the current codes of practice. In fact, according to Eurocode 3, Part 1.2 [2] and Annex J [3], the concentration of mass within the joint area, when compared to the connecting members, delays its temperature increase, therefore suggesting that joints could be disregarded under fire conditions. However, in contrast to the EC3 specification, recent experimental results [4,5] have highlighted the need to evaluate the behaviour of steel joints at elevated temperatures, since they exhibit a pronounced reduction of strength and stiffness that clearly affect the global response of the structure.

Naturally, any attempt at predicting the behaviour of a steel joint under fire loading, already complex at room temperature [6], is further complicated by several phenomena:

(i) variation of material properties of steel with temperature;
(ii) accurate prediction of time-temperature variation within the various joint components;
(iii) differential elongation of the various joint components because of increasing temperature
(iv) proper definition of fire development models within the building envelope and subsequent time-temperature profiles reaching the joint;

Clearly, item (iv) involves the architectural layout of the building and the particular fire event and lies outside the scope of this paper, a thorough treatment being found elsewhere [7]. Items (i) to (iii) are strictly required to predict the moment-rotation response of steel joints and are discussed in the following sections, in the context of an

C.C. Baniotopoulos and F. Wald (eds.), The Paramount Role of Joints into the Reliable Response of Structures, 425–434.

approach based on the so-called "component method" [8], proposed in this paper to predict the behaviour of steel joints under fire loading, described next.

2. Component model

2.1   COMPONENT METHOD

To overcome the need to implement complex non-linear finite element analysis in the prediction of the moment-rotation response of steel joints, a simpler approach was developed in the form of the so-called component method [9-11]. Briefly described, the method consists of modelling a joints as an assembly of extensional springs and rigid links, whereby the springs (components) represent a specific part of a joint that, dependent on the type of loading, make an identified contribution to one or more of its structural properties [8]. The application of the method to a typical flush end-plate beam-to-column joint is illustrated in Figure 1, the various components contributing to the overall response of the joint being: (1) column web in shear, (2) column web in

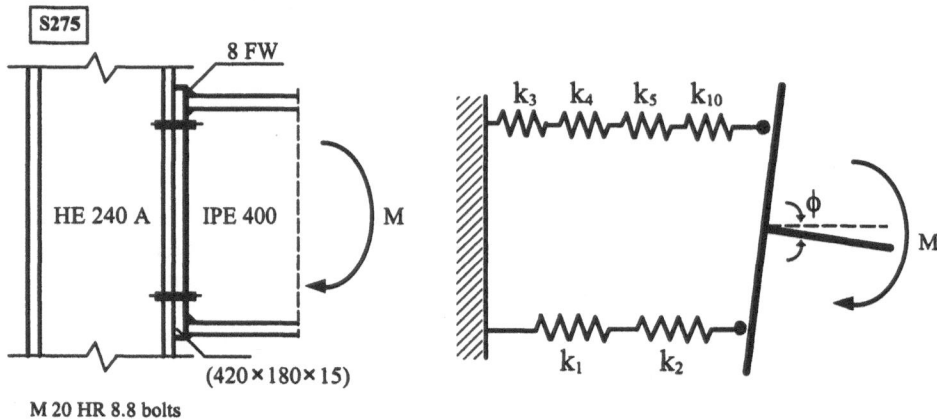

Figure 1. Typical flush end-plate beam-to-column joint (Bradran 123.001).

compression, (3) column web in tension, (4) column flange in bending, (5) end-plate in bending, (6) flange cleat in bending, (7) beam flange in compression, (8) beam web in tension or compression, (9) plate in tension or compression, (10) bolts in tension, (11) bolts in shear, (12) bolts in bearing and (13) welds.

2.2   COMPONENT CHARACTERISATION

A key aspect to the component method relates to the characterisation of the force-deformation curves for each individual extensional spring. Following the review presented in [12], the various components relevant for steel joints are classified in three main groups: (a) components with high ductility, (b) components with limited ductility and (c) components with brittle failure. Common to all is the identification of four properties, namely elastic stiffness (k_e), post-limit stiffness (k_p), limit load (F^y) and limit displacement (Δ^f), as seen in Figure 2.

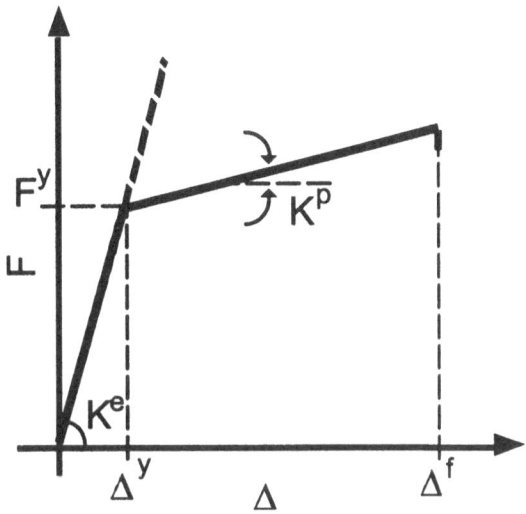

Figure 2. Bi-linear characterisation of component behaviour.

3. Behaviour at elevated temperatures

3.1 THERMO-MECHANICAL PROPERTIES OF STEEL AT ELEVATED TEMPERATURES

Steel is characterised by a reduction of yield stress, ultimate stress and Young's modulus with increasing temperature. Coupled with the thermal dilation of steel, this results in steel joints quickly reaching yield under fire conditions, even at constant mechanical loading. Analytical expressions for these properties can be found in Part 1.2 of EC3, a review of thermo-mechanical properties of steel being found in [7].

3.2 COUPLED THERMO-MECHANICAL LOAD CHARACTERISATION

Being two independent processes, mechanical and thermal loading may present arbitrary time histories. For generality, the following situations may arise:
(i) mechanical loading takes place before the fire event starts;
(ii) mechanical loading and fire event take place simultaneously (totally or partially);
(iii) mechanical loading takes place after fire event has reached its maximum temperature, at sustained temperature conditions.
Naturally, situations (i) and (iii) are much simpler and correspond, respectively, to an incremental temperature analysis at constant load level (anisothermal analysis) or an incremental mechanical analysis at constant temperature (isothermal analysis). Here, the

more realistic situation of simultaneous mechanical and thermal loading (transient analysis) will be considered, corresponding, for example, to the load redistribution that inevitably takes place during a fire event.

4. Analytical prediction of the fire behaviour of steel joints

4.1   INTRODUCTION

As explained above, the evaluation of the fire response of steel joints requires the continuous change of mechanical properties of steel as temperature increases. In the context of the component method, this is implemented at the component level, as shown in Fig. 3. Noting that the elastic stiffness, K^e, is directly proportional to Young's modulus of steel and the resistance of each component depends on the yield stress of steel, equations (1) to (3) illustrate the change in component force-deformation response with increasing temperature, for a given temperature variation.

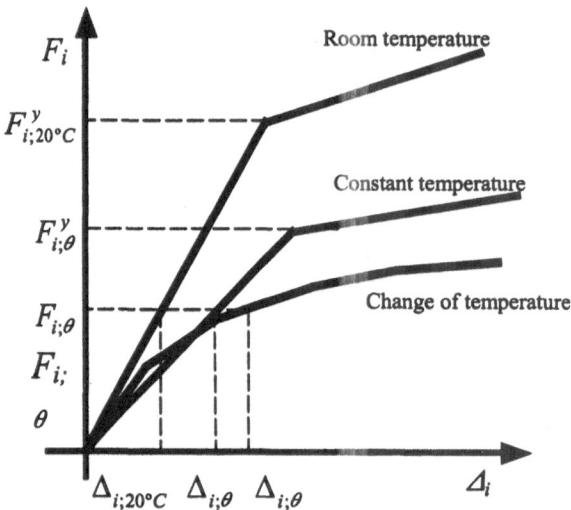

Figure 3. Force-deformation response for typical component: room temperature versus constant elevated temperature and variable temperature

$$F_{i;\theta}^{y} = k_{y,\theta} \times F_{i;20°C}^{y} \qquad (1)$$

$$K_{i;\theta}^{e} = k_{E,\theta} \times K_{i;20°C}^{e} \qquad (2)$$

$$K_{i;\theta}^{p} = k_{E,\theta} \times K_{i;20°C}^{p} \qquad (3)$$

Introducing equations (1) to (3) for the corresponding (constant) values of K^e, K^p and F in any evaluation of moment-rotation response of steel joints at room temperature yields the required fire response. Implementation of this procedure in a practical way requires an incremental procedure with sufficiently small temperature increments so that the mechanical properties of steel can be kept constant within each temperature interval, the detailed procedure being explained in the next two paragraphs.

4.2   ISOTHERMAL RESPONSE

Assuming that the moment rotation response of a steel joint at room temperature is known, either using a non-linear numerical procedure or the closed-form analytical procedures developed by Simões da Silva and Coelho [11], the isothermal response of a steel joint loaded in bending can be obtained as follows, for a constant temperature level, θ.

For a given level of applied force $F_i < F^y_{i;\theta}$ the component deformation $\Delta_{i;\theta}$ is given by:

$$\Delta_{i;\theta}(F_i) = \frac{1}{k_{E;\theta}} \times \Delta_{i;20°C}(F_i) \tag{4}$$

so that the yield deformation becomes

$$\Delta^y_{i;\theta} = \frac{F^y_{i;\theta}}{K^e_{i;\theta}} = \frac{k_{y;\theta}}{k_{E;\theta}} \Delta^y_{i;20°C} \tag{5}$$

From equilibrium considerations, it can be shown [13] that:

$$M_{i;\theta} = k_{y;\theta} \times M_{i;20°C} \tag{6}$$

so that

$$M^y_{i;\theta} = k_{y,\theta} \times M^y_{i;20°C} \tag{7}$$

$$M_{j,máx;\theta} = k_{y,\theta} \times M_{j,máx;20°C} \tag{8}$$

and

$$S_{j,ini;\theta} = \frac{M^y_{1;\theta}}{\phi^y_{1;\theta}} = k_{E,\theta} \frac{M^y_{1;20°C}}{\phi^y_{1;,20°C}} \tag{9}$$

$$\phi^y_{i;\theta} = \frac{M^y_{i;\theta}}{S^y_{i;\theta}} = \frac{k_{y,\theta}}{k_{E,\theta}} \times \phi^y_{i;20°C} \tag{10}$$

430

giving the generic moment-rotation curve at constant temperature θ, where the yielding sequence of the various components is identified, as shown schematically in Fig. 4.

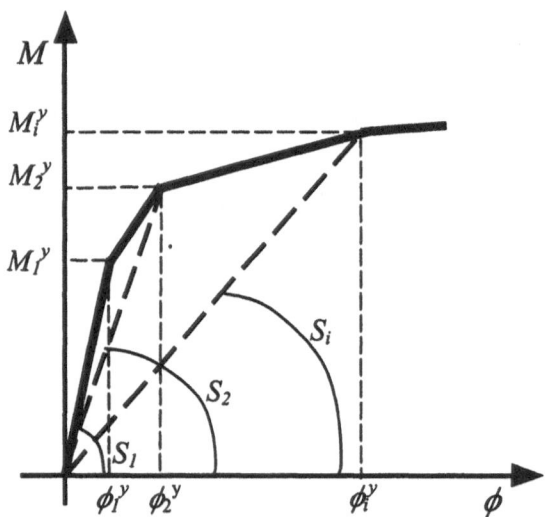

Figure 4. Isothermal moment-rotation curve at temperature θ.

4.3 TRANSIENT RESPONSE

As explained above, whenever the variation of loading and temperature takes place simultaneously, a transient analysis is required. To reflect the continuous variation of Young's modulus and yield stress of steel with increasing temperature, an incremental temperature procedure is proposed [13]. In this methodology, the mechanical properties of steel are evaluated at the average temperature of each temperature interval. Consequently, for each temperature interval, the joint behaves as in the previous case of isothermal loading, so that the total joint response can be obtained by superimposing the results from each temperature interval, taking as initial conditions the stress and deformation state at the end of the previous step.

It can be shown [13] that a component reaches its yield force whenever the total applied force on the component reaches its yield force, at temperature θ_r, as described in eq. (11):

$$\frac{F_{i;20°C}^{y}}{M_{i;20°C}^{y}} \times \sum_{\theta=\theta_1}^{\theta_r} \Delta M_{\theta} = k_{y,\theta_r} \times F_{i;20°C}^{y} \tag{11}$$

Analogously, the yield deformation of each component is given by

$$\Delta_{i;\theta}^y = \frac{\Delta_{i;20^\circ C}^y}{M_{i;20^\circ C}^y} \times \sum_{\theta=\theta_1}^{\theta_r} \frac{\Delta M_\theta}{k_{E;\theta}} \qquad (12)$$

while eq. (13) describes the post-yield deformation of component i, for temperature θ_n:

$$\Delta_{i;\theta} = \Delta_{i;\theta}^y + \frac{\Delta_{i;20^\circ C}^f - \Delta_{i;20^\circ C}^y}{M_{i;20^\circ C}^f - M_{i;20^\circ C}^y} \times \sum_{\theta=\theta_1}^{\theta_n} \left(\frac{\Delta M_\theta}{k_{E;\theta}} \right) \qquad (13)$$

The joint rotation at yielding of the first component, $\phi_{1;\theta}^y$, is given by:

$$\phi_{1;\theta}^y = \frac{\phi_{1;20^\circ C}^y}{M_{1;20^\circ C}^y} \times \sum_{\theta=\theta_1}^{\theta_m} \left(\frac{\Delta M_\theta}{k_{E;\theta}} \right) \qquad (14)$$

while the rotation at yield of component i, $\phi_{i;\theta}^y$, is given by the sum of the contribution of the previously yielded components:

$$\phi_{i;\theta}^y = \sum_{i=1}^{q} \left(\frac{\phi_{i;20^\circ C}^y - \phi_{(i-1);20^\circ C}^y}{M_{i;20^\circ C}^y - M_{(i-1);20^\circ C}^y} \times \sum_{\theta=\theta_1}^{\theta_{m+p+s}} \frac{\Delta M_\theta}{k_{E;\theta}} \right) \qquad (15)$$

θ_{m+p+s}, being the corresponding temperature. Finally, the corresponding bending moment $M_{j;\theta}$ is given by:

$$M_{j,\theta} = \Delta M_{\theta_1} + \Delta M_{\theta_2} + ... + \Delta M_{\theta_n} = \sum_{\theta=\theta_1}^{\theta_n} \Delta M_\theta \qquad (16)$$

4.4 APPLICATION TO BOLTED END-PLATE BEAM-TO-COLUMN STEEL JOINTS

In order to illustrate the evaluation of the fire response of steel joints, two alternative configurations for bolted end-plate beam-to-column joints were selected: (1) flush and, (2) extended end-plate with backing plates. The geometry for these examples was taken from the database of steel joints SERICON II [14] and is summarised in Table 1.

Starting with the flush end-plate joint (123.001) under isothermal loading, the moment-rotation curve of Fig. 5 is obtained, which shows yielding of the first three components (tension zone of the connection), column web in tension (k_4) and end-plate in bending (k_5), and yielding and simultaneous failure of the third component, bolts in tension (k_{10}), As the temperature increases, the moment resistance of the joint progressively decreases, the strength of the joint becoming negligible for temperatures

TABLE 1. Joint configurations

Identification	Test 123.001	Test 109.003
Column	HEA 240	HEB 180
Beam	IPE 400	IPE 300
Mat. Properties	Column, beam and plate: S275 Bolts: 8.8	Column, beam and plate: S275 Bolts: 8.8
Geometry	Flush end-plate connection (420x180x15) and one bolt row in tension	Extended end-plate connection (370x165x17) with backing plates and two bolt rows in tension

above 1000°C. Similar results are obtained for the extended end-plate connection (109.003), the compression zone of the connection now being critical, yielding occurring for the two components, column web in compression (k_1) and column web in shear (k_2).

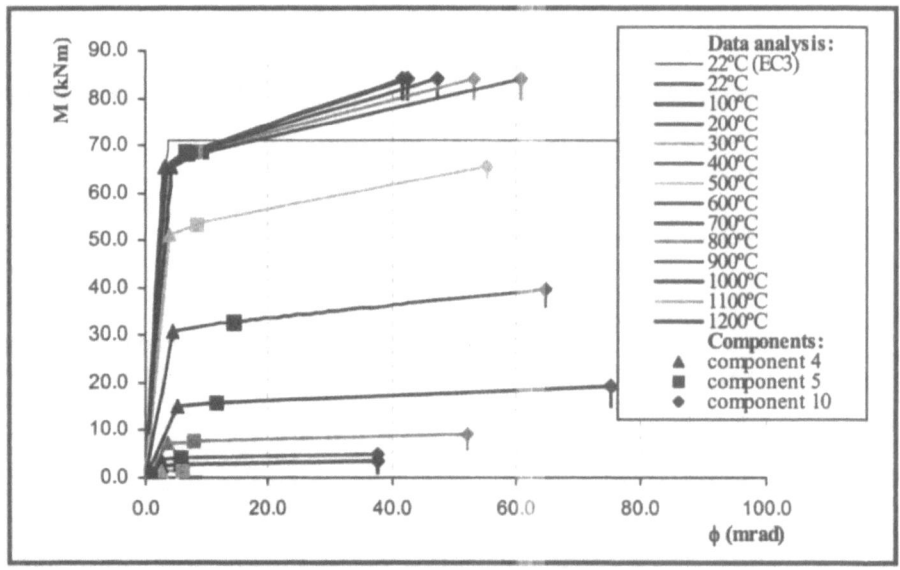

Figure 5. Isothermal moment-rotation response for Bradran joint

The same joints analysed under transient loading exhibit the results illustrated in Figs. 6 and 7, both compared with the reference analysis at room temperature for three distinct scenarios: (i) the mechanical loading increases at twice the speed of the temperature, *High velocity*, (ii) the mechanical loading increases at the same speed as the temperature, *Medium velocity*, and (iii) the mechanical loading increases at half the speed of the temperature, *Slow velocity*. For both joints, the larger the loading speed:

(i)   the bigger the moment resistance;
(ii)   the greater the joint rotation
(iii) the lesser the collapse temperature of the joint
(iv)   the bigger the rotational stiffness

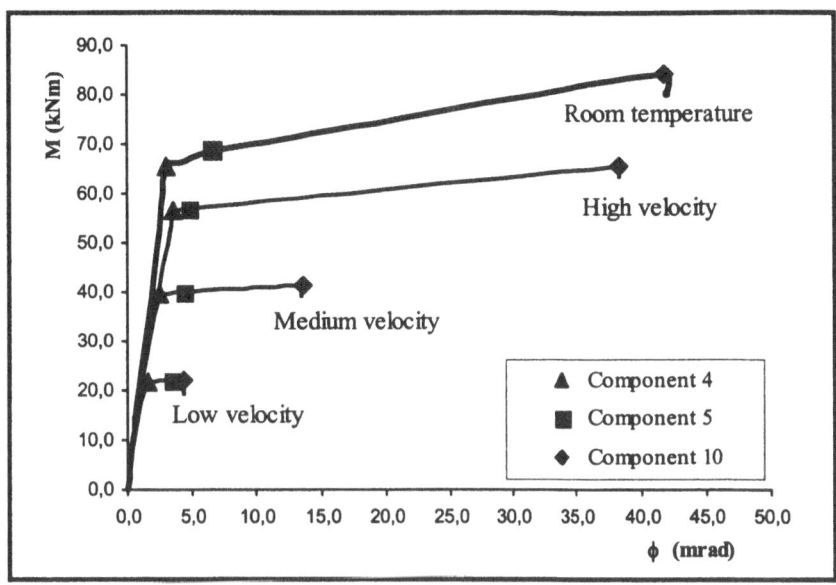

Figure 6. Transient moment-rotation response for Bradran joint

It is noted that, as for the isothermal analysis, the sequence of yielding of the various components remains unchanged from the analysis at room temperature.

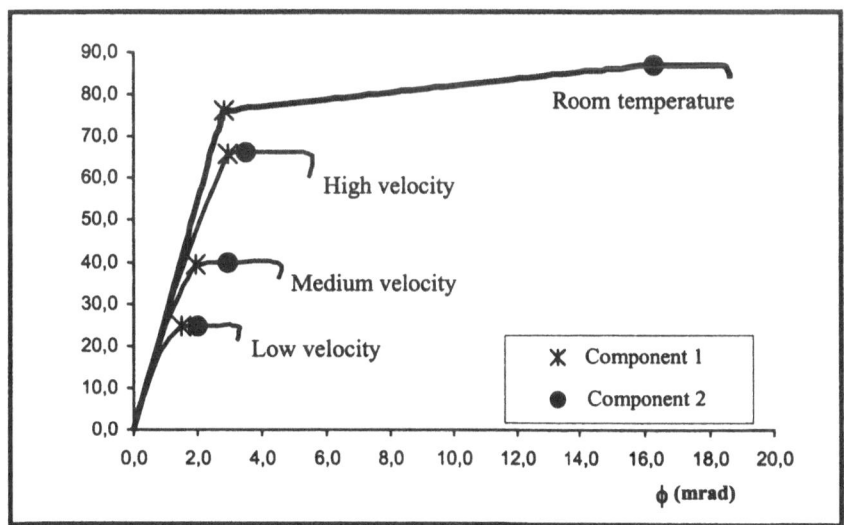

Figure 7. Transient moment-rotation response for Humer joint

5. Concluding remarks

An analytical procedure for the evaluation of the behaviour of steel joints under fire loading was proposed in this paper. Coupled with recently developed closed-form analytical procedures [6,11] for the evaluation of the non-linear moment-rotation response of steel joints in the context of the component method, this should provide a useful solution to an otherwise untractable problem.

However, a few shortcomings still remain to be solved: (i) thermal elongation of steel was not considered in the analysis, (ii) proper definition of the time-temperature variation within the joint, (iii) component characterisation parameters, Ke, Kp and F were assumed as directly proportional to, respectively, Young's modulus of steel and yield stress, and, most important, (iv) thorough calibration with experimental test results at elevated temperature.

6. Acknowledgments

Finantial support from "Ministério da Ciência e Tecnologia" - PRAXIS XXI research project PRAXIS/P/ECM/13153/1998 is acknowledged.

7. References

1. Al-Jabri, K.S., Burgess, I.W., Lennon, T., Plank, R.J. (1999) The Performance of Frame Connections in Fire, in Proceedings of the Conference Eurosteel'99, CVUT, Praha, 1999.
2. Eurocode 3, ENV - 1993-1-2 (1995) Design of Steel Structures - Part 1.2: General Rules - Structural Fire Design. CEN, European Committee for Standardisation, Document CEN/TC 250/SC 3, Brussels.
3. Eurocode 3, ENV - 1993-1-1:1992/A2, (1998) Annex J, Design of Steel Structures – Joints in Building Frames. *CEN, European Committee for Standardization, Document CEN/TC 250/SC 3*, Brussels.
4. Al-Jabri, K.S., Lennon, T., Burgess, I.W., Plank, R.J. (1998) Behaviour of Steel and Composite Beam-column Connections in Fire. *Journal of Constructional Steel Research 46*, 1-3.
5. Leston-Jones, L.C., Burgess, .W., Lennon, T. and Plank, R.J. (1997) Elevated Temperature Moment-rotation tests on steelwork connections, in Proceedings of Institution of Civil Engineers. Structures & Buildings, 122 (4), pp 410-419.
6. Simões da Silva, L., Girão Coelho, A. and Neto, E. (2000) Equivalent post-buckling models for the flexural behaviour of steel connections, *Computers & Structures*, in print.
7. Drysdale, D.D. (1999) *An Introduction to Fire Dynamics*. 2nd ed, West Sussex, England, John Wiley & Sons.
8. Weynand, K., Jaspart, J-P., and Steenhuis, M. (1995) The stiffness model of revised Annex J of Eurocode 3, in R. Bjorhovde, A. Colson and R. Zandonini (eds), *Connections in Steel Structures III*, Proceedings of the 3rd International Workshop on Joints, Trento, Italy, May 8-31, pp. 441-452.
9. Zoetemejer, P. (1974) A design method for the tension side of statically-loaded, bolted beam-to-column joints, *Heron*, 20(1): 1-59.
10. Jaspart, J.P. (1991) Etude de la semi-rigidité des noeuds poutre-colonne et son influence sur la resistance des ossatures en acier (in french), Ph.D. Thesis, Department MSM, University of Liége, Liège, Belgium.
11. Simões da Silva, L. and Girão Coelho, A. (2000) A ductility model for steel connections, *Journal of Constructional Steel Research*, in print.
12. Simoes da Silva, L., Santiago, A. and Vila Real, P. (2000) Ductility of steel connections, *Canadian Journal of Civil Engineering* (submitted for publication).
13. Santiago, A. (2000) Behaviour of steel joints under fire loading (in portuguese), MSc. Thesis, Department of Civil Engineering, University of Coimbra, Coimbra, Portugal.
14. Cruz, P.J.S., Simões da Silva, L.A.P., Rodrigues, D.S. and Simões, R.A.D., Database for the Semi-Rigid Behaviour of Beam-to-Column Connections in Seismic Regions. *Journal of Constructional Steel Research*, 46, pp 233-234, 1998.

STIFFNESS OF LAP JOINTS WITH PRELOADED BOLTS

A.M. GRESNIGT
Delft University of Technology
Faculty of Civil Engineering and Geosciences
Delft, The Netherlands

C.M. STEENHUIS
Eindhoven University of Technology
Department of Structural Design
Eindhoven, The Netherlands

1. Introduction

In 1997, fatigue cracking was found in the deck of the movable part of an important highway bridge near Rotterdam (one of the two Van Brienenoord Bridges, figure 1). Figure 2 gives an impression of the cross-section and the layout of the bridge where the cracks were found.

Several methods of repair and strengthening of the bridge deck were considered. A very important factor in the decision process was the demand to limit the time of closure as much as possible. The work was to be started and completed within one week in a "low traffic" period of the year (beginning of August 1998). Repair and strengthening of the deck in situ would take too much time.

It was decided to replace the damaged part just before the main hinges (pivots). This meant the cutting of the two main girders, the removal of the old deck, the positioning of the new deck and the connection of its main girders to the remaining part (the joint in figure 2). Figure 3 gives an impression of the putting into place of the new bridge deck.

At the joint, the main girders are about 8100 mm high, the web has a thickness of 40 mm and the bottom flange has a cross-section of 1000 x 70 mm. Welding would take far too much time. The other possibility was bolting. But bolting at the bridge deck level, the top flange, was not considered a real option because of demands on flatness of the bridge deck.

It was decided to investigate the possibility of a partly welded, partly bolted joint. For the fatigue resistance it is important to carefully consider the point where the weld in the web stops and the bolted joint begins. In order to avoid stress concentrations, the stiffness of the welded part and the bolted part should be about equal, with a preference for a slightly stiffer bolted joint, to be discussed later.

Therefore, a study was carried out into the stiffness of bolted lap joints, to be compared with the stiffness of welded joints. Figure 4 shows the design of the joint

C.C. Baniotopoulos and F. Wald (eds.), The Paramount Role of Joints into the Reliable Response of Structures, 435–448.
© 2000 *Kluwer Academic Publishers.*

where the top flange and a part of the web below (500 mm) is welded and the remainder of the web and the bottom flange are bolted with high strength friction grip (HSFG) bolts M36-10.9.

Figure 1. The two Van Brienenoord Bridges in the main highway North-South near Rotterdam. The wording **wegens vermoeiing** means because of fatigue. The deck of one of the two bridges was replaced because of fatigue damage in the deck structure.

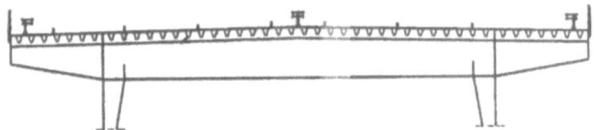

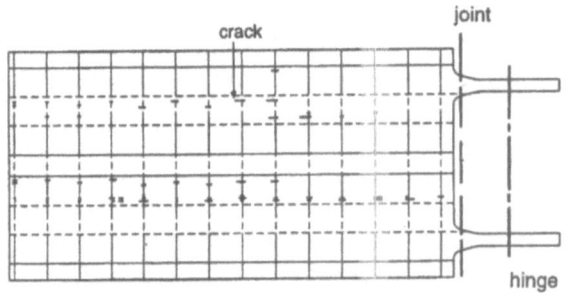

Figure 2. Cross-section and layout of the movable bridge. The length of the deck is 54 meter. The bridge has 2 * 3 lanes.

Figure 3. Putting the new bridge deck into place.

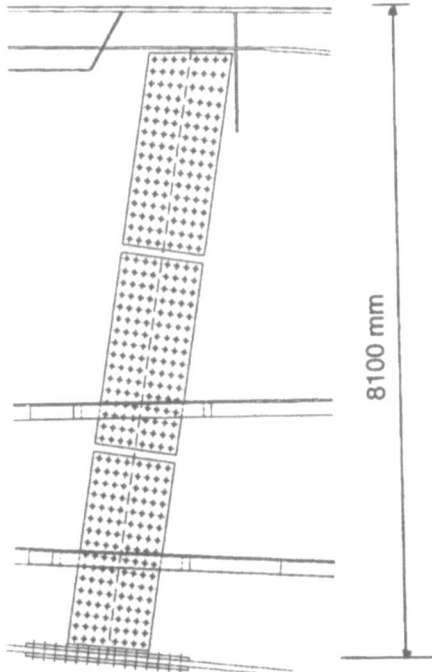

8100 mm

Figure 4. The partly welded and partly bolted joint in the two main girders.

For the bolted part of the web, lap joints were proposed with 1040 x 25 mm plates and for the bottom flange lap joints with 1200 x 40 mm plates.

In this paper, a summary is presented of the study into the parameters governing the stiffness of HSFG bolted lap joints. A model for the determination of the load distribution over the bolts and the stiffness of the lap joint is given.

2. Design model

Figure 5 gives the geometry of a bolt row in the web of the Van Brienenoord Bridge. Also the geometry of a welded joint with the same "measuring length" is given.

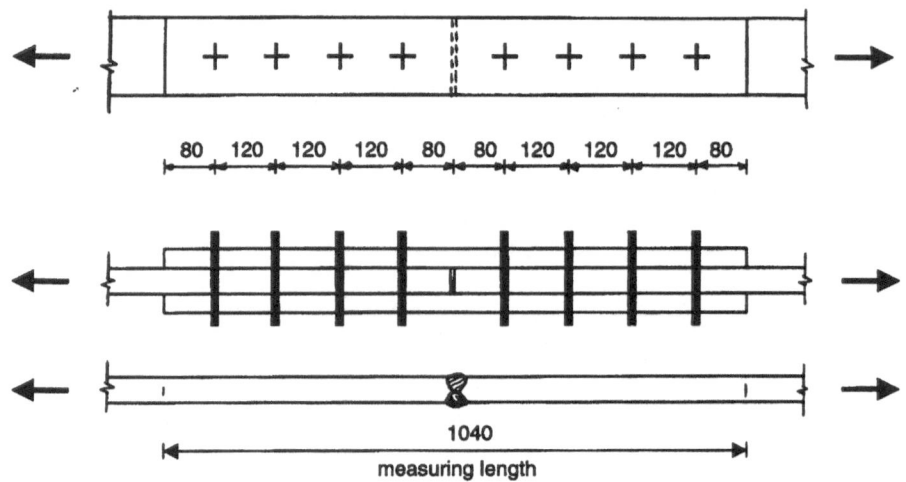

Figure 5. Bolted lap joint and welded joint in the Van Brienenoord Bridge.

The web thickness is 40 mm. The thickness of the cover plates is 25 mm. The standard distance between the bolt rows is 150 mm. At the top end of the bolted part, the bolt rows are not parallel (distance between about 150 and 190 mm). For the calculations a distance of 175 mm is adopted. HSFG bolts M36 – 10.9 are applied.

The joint can be conceived as systems of springs, each spring having its own characteristic stiffness, see figure 6. For the bolts the most important factor is the deformation caused by the load transfer by friction. Because of symmetry, half of the bolted lap joint is taken into account. In the model, the following notations are used.

A_h	=	Cross-sectional area of the main plate: $A_h = t_h * b$
A_s	=	Cross-sectional area of the cover plate: $A_s = t_s * b$
b	=	Width of the plate (distance between bolt rows)
p	=	Pitch (distance between bolts)
ℓ_{begin}	=	Edge distance cover plates
ℓ_{end}	=	Edge distance main plate
$B_1\ B_2\ B_3\ B_4$	=	Forces transmitted by HSFG bolts
$H_0\ H_1\ H_2\ H_3$	=	Parts in the main plate
$S_1\ S_2\ S_3\ S_4$	=	Parts in the cover plates
$\sigma_{h0}\ \sigma_{h1}\ \sigma_{h2}\ \sigma_{h3}$	=	Stresses in the main plate
$\sigma_{s1}\ \sigma_{s2}\ \sigma_{s3}\ \sigma_{s4}$	=	Stresses in the cover plates
$d_{b1}\ d_{b2}\ d_{b3}\ d_{b4}$	=	Displacements in bolts B_1, B_2, B_3, B_4
$d_{h0}\ d_{h1}\ d_{h2}\ d_{h3}$	=	Displacements in parts H_0, H_1, H_2, H_3 of the main plate.

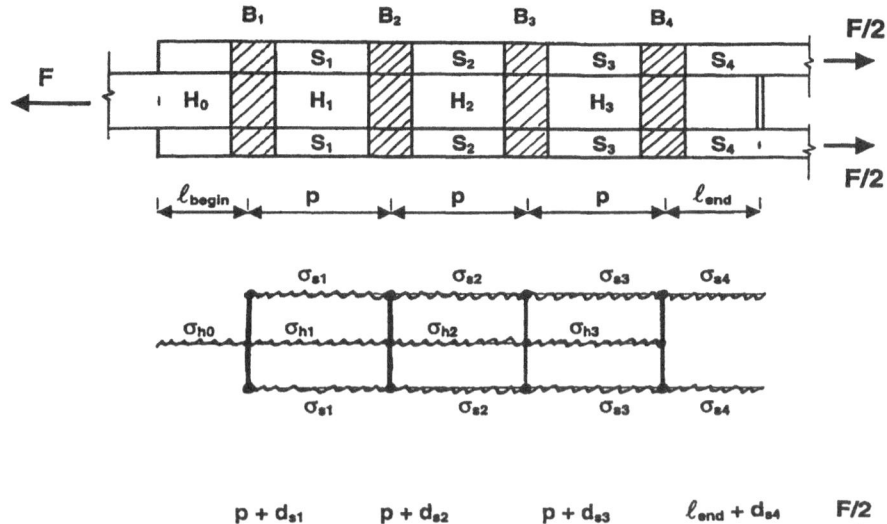

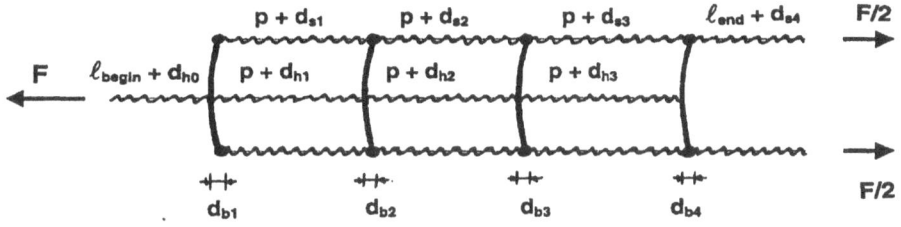

Figure 6. Spring model for a lap joint with four high strength friction grip bolts (HSFG bolts); stresses and deformations are indicated.

The displacements in the bolts are a function of the load to be transmitted by the bolt and the properties (condition) of the plate surfaces at the friction plane. Also the geometry of the bolt-plate assembly will have some influence. The relation between the bolt forces B_1, B_2, B_3, B_4 and the displacements d_{b1}, d_{b2}, d_{b3}, d_{b4} will be discussed in section 3. The stresses in the main plates and cover plates are:

$$\sigma_{h1} = \frac{F - B_1}{A_h} \qquad (1a)$$

$$\sigma_{s1} = \frac{B_1}{2 \cdot A_s} \qquad (2a)$$

$$\sigma_{h2} = \frac{F - B_1 - B_2}{A_h} \qquad (1b)$$

$$\sigma_{s2} = \frac{B_1 + B_2}{2 \cdot A_s} \qquad (2b)$$

$$\sigma_{h3} = \frac{F - B_1 - B_2 - B_3}{A_h} \qquad (1c)$$

$$\sigma_{s3} = \frac{B_1 + B_2 + B_3}{2 \cdot A_s} \qquad (2c)$$

$$\sigma_{h0} = \frac{F}{A_h} \qquad (1d)$$

$$\sigma_{s4} = \frac{B_1 + B_2 + B_3 + B_4}{2 \cdot A_s} \qquad (2d)$$

The elongation's in the main plates and cover plates can be calculated as follows:

$$d_{h1} = \frac{\sigma_{h1}}{E} \cdot p \qquad (3a) \qquad\qquad d_{s1} = \frac{\sigma_{s1}}{E} \cdot p \qquad (4a)$$

$$d_{h2} = \frac{\sigma_{h2}}{E} \cdot p \qquad (3b) \qquad\qquad d_{s2} = \frac{\sigma_{s2}}{E} \cdot p \qquad (4b)$$

$$d_{h3} = \frac{\sigma_{h3}}{E} \cdot p \qquad (3c) \qquad\qquad d_{s3} = \frac{\sigma_{s3}}{E} \cdot p \qquad (4c)$$

$$d_{h0} = \frac{\sigma_{h0}}{E} \cdot \ell_{begin} \qquad (3d) \qquad\qquad d_{s4} = \frac{\sigma_{s4}}{E} \cdot \ell_{end} \qquad (4d)$$

The compatibility conditions are:

$$d_{b1} + p + d_{s1} = p + d_{h1} + d_{b2}$$
$$d_{b2} + p + d_{s2} = p + d_{h2} + d_{b3} \qquad (5)$$
$$d_{b3} + p + d_{s3} = p + d_{h3} + d_{b4}$$

These can also be written as:

$$d_{h1} - d_{s1} = d_{b1} - d_{b2}$$
$$d_{h2} - d_{s2} = d_{b2} - d_{b3} \qquad (6)$$
$$d_{h3} - d_{s3} = d_{b3} - d_{b4}$$

The total elongation is:

$$\delta_{tot} = d_{h0} + d_{h1} + d_{h2} + d_{h3} + d_{b4} + d_{s4} \qquad (7a)$$

also:

$$\delta_{tot} = d_{h0} + d_{b1} + d_{s1} + d_{s2} + d_{s3} + d_{s4} \qquad (7b)$$

The total load F is:

$$F = B_1 + B_2 + B_3 + B_4 \qquad (8)$$

With these equations, the unknowns can be solved and the elongation can be calculated for various values of F, if also the load-deformation behaviour of single bolts is known, see next section. For solving the equations, an iterative procedure is necessary. The Microsoft computer program Excel was used for this purpose.

3. The load-deformation behaviour of a single bolt

As indicated before, it is necessary to know the load-deformation behaviour of a single bolt as an input parameter in the spring model. A limited literature study has been carried out to collect relevant test data. Only a rather small number of test series were found, where the elongation of lap joints was measured over the measuring length as defined in figure 5.

In 1966-1967, the Otto-Graf-Intitute carried out many tests on HSFG bolted lap joints for the "Office for Research and Experiments of the International Union of Railways" (ORE, [2]). Several tests were lap joints with two bolts at each side of the lap joint. Figures 7 and 8 give the test specimen and the test set-up. It is noted that in this test set-up, there was a direct measurement of the deformation at the bolts. Figure 9 gives, as an example, the measured load- deformation diagrams of one of the tests.

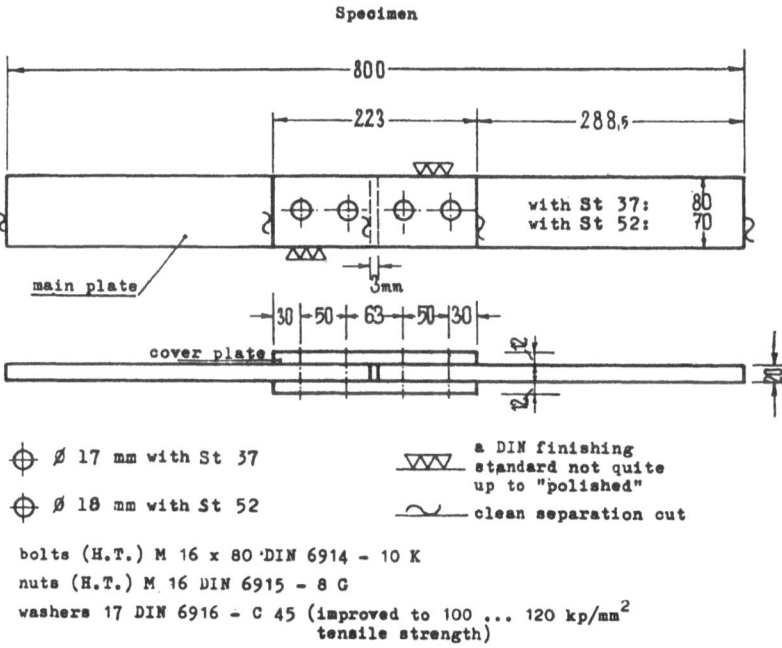

*Figure 7. Test specimen ORE-tests with 2 * 2 bolts [2].*

Further to figure 7, the ORE-report gives the following details.
- The surfaces were blasted with chilled iron grit grain size 34, sharp edged, hardness HVI = 700-800 kg/mm^2, grain size mixture: 75% of grain size 0,3 to 0,5 mm, 25% of oversized grain and/or undersized grain.
- Steel grade St 52-3 (DIN 17100) and St 37-2. The steel grade had a minor influence on the results (less than the scatter in the tests with the same steel grade).

Because of the fact that the thickness of the main plate was 20 mm, while the total thickness of the cover plates was 24 mm, the load transfer of the inner bolts was slightly lower than of the outer bolts. With the model, the bolt forces ($B_1 + B_2 = F$) were calculated. The following values serve as an example: $60,0 + 56,8 = 116,8$ kN and $120,0 + 118,2 = 238,2$ kN.

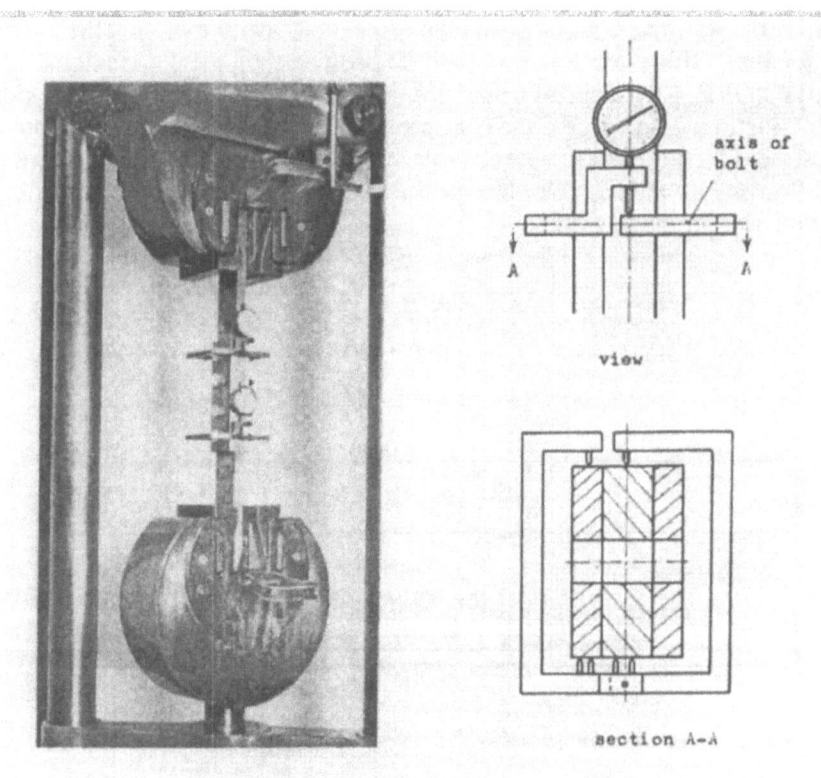

*Figure 8. Test set-up of the ORE-tests with 2 * 2 bolts [2].*

From figure 8 it can be seen that the displacements between both plates were measured at the bolt level. In some other test series the deformations were measured over the measuring length as defined in figure 5, but in most test series found, the displacements were measured between the end of the cover plate and the adjacent main plate. These measurements are not very useful for the present model, because they only contain the deformation of the outer bolt (B_1) and the elongation of a part of the main plate (d_{ho}). Figure 9 gives the measured load-deformation diagrams of one of the ORE tests.

The pre-load in the M16 bolts was 100,0 and 100,0 kN (upper bolts) and 99,4 and 110,5 kN (lower bolts). The friction coefficient for the upper bolts was 0,667 and for the lower bolts 0,683.

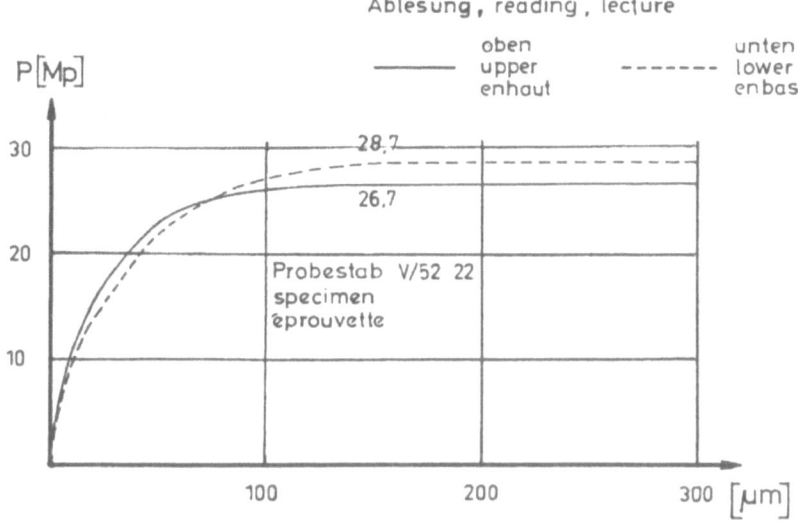

Figure 9. Measured load-deformation diagrams of test V/52-22 [2].

In figure 10 the results of four tests are placed in one figure where the vertical axis is made non-dimensional by dividing the load by the slip-load. It can be seen that some scatter is evident, despite the careful way of preparing the test specimens and execution of the tests.

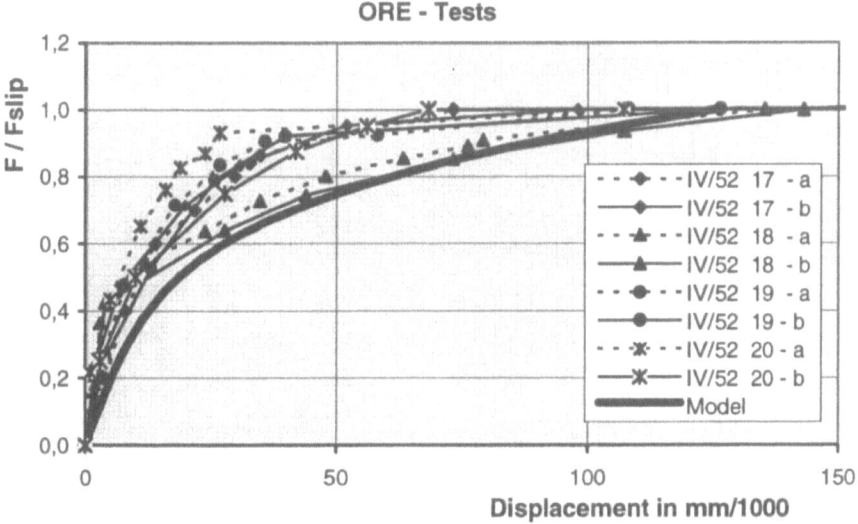

Figure 10. Dimensionless load-deformation diagrams of bolts in tests IV/52-17, IV/52-18, IV/52-19 and IV/52-20 [2]. Also the load-deformation behaviour according to the model (formula 9) is given.

444

From the ORE tests and also from other tests it appears that the load-deformation characteristics are non-linear. For the spring model we developed the following formula:

$$\delta = \left(\frac{B}{F_{slip}}\right)^{\alpha} \cdot 100 + \frac{B}{F_{slip}} \cdot 25 \qquad (9)$$

with:

δ = displacement (mm/1000)
α = coefficient (α = 4)
B = force transferred by the HSFG bolt (kN)
F_{slip} = slip force (kN)

This formula gives a reasonable fit with the available test results, see figure 10.

4. Validation with other tests

In the literature, a number of test series is found where the load deformation behaviour for test specimens with more bolts was measured. The tests by Foreman-Rumpf [3] and Klöppel-Seeger [4] are most suited for this comparison. Figure 11 gives one of the test specimens and test results according to figure 7 in Foreman-Rumpf [3]. Figure 12 gives the comparison of these test results with the model.

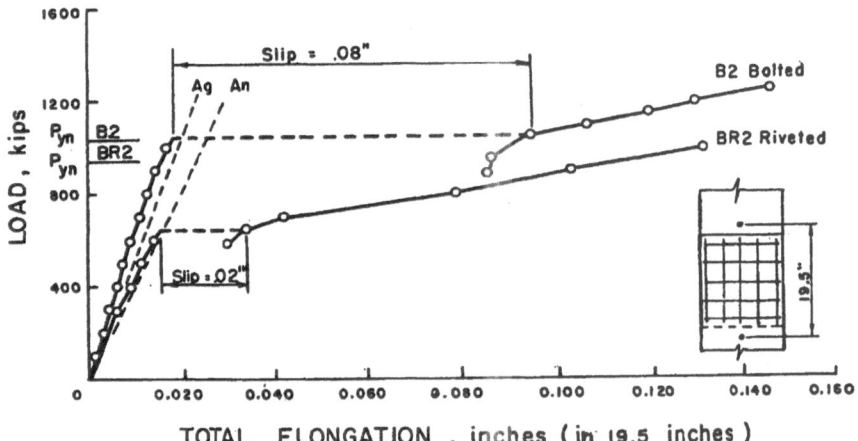

Figure 11. One of the test specimens and test results according to figure 7 in Foreman-Rumpf [3]. Riveted and HSFG bolted joints were compared. Ag is the load-deformation behaviour of a plate, while An is the load-deformation behaviour of the main plate, taking into account the influence of the bolt holes.

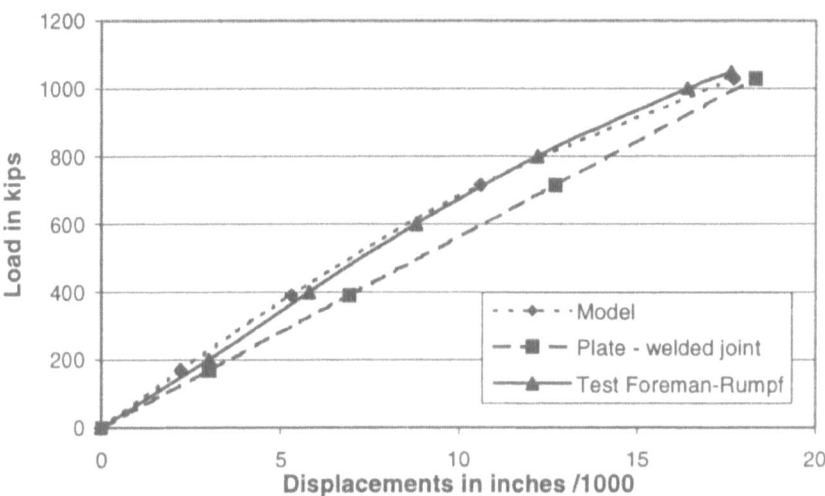

Figure 12. Load-deformation behaviour of the HSFG bolted joint and welded joint (= plate only), according to figure 7 in Foreman-Rumpf [3], compared with calculated behaviour with the model in this paper.

Figure 13 gives one of the test specimens and test results according to Klöppel-Seeger [4]. Figure 14 gives the comparison of these test results with the model.

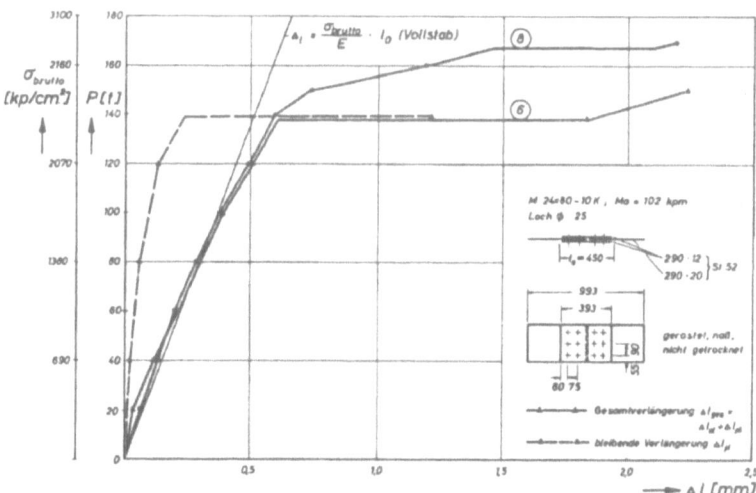

Figure 13. Test specimens and two test results according to "Bild23" on page 65 of Klöppel-Seeger [4]. In these two tests the surfaces were sandblasted, then during 6 weeks, the plates were subjected to corrosion in a wet condition and finally brushed before assembling the test specimens.

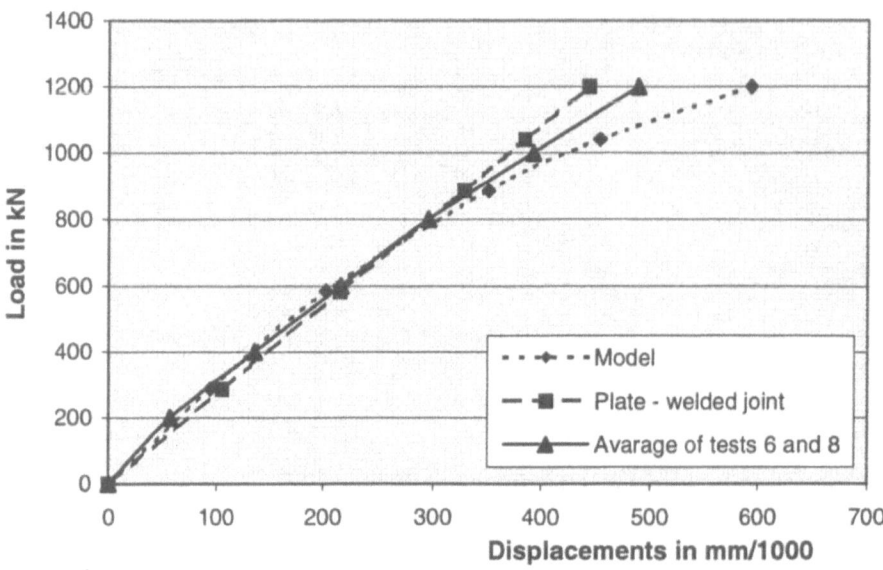

Figure 14. Load-deformation behaviour of the HSFG bolted joint and a welded joint according to figure 13, compared with calculated behaviour with the model in this paper.

5. Application to the Van Brienenoord Bridge

Figure 15 gives the load-deformation behaviour of the proposed bolted joint in the web of the main girders compared to the load-deformation behaviour of a welded joint with the same length. The figure contains two lines for the calculated stiffness of the bolted joint. The model with constants 25 and 100 gives the results for the load-deformation behaviour of the bolts according to formula (9). The model with constants 50 and 150 gives the results for a less stiff load-deformation behaviour of the bolt where in formula (9) the constants 25 and 100 are replaced by 50 and 150. In both cases the bolted joint is stiffer than the welded one.

The fact that the bolted joint is stiffer than the welded joint is on the safe side, since the fatigue strength of HSFG bolted joints is better than of butt welded joints, especially at the ends of such welded joints. The results also show that in case the rather heavy bolts (in comparison with those in the tests mentioned before), give somewhat larger displacements, there is not much danger that the bolted joint is less stiff than the welded one.

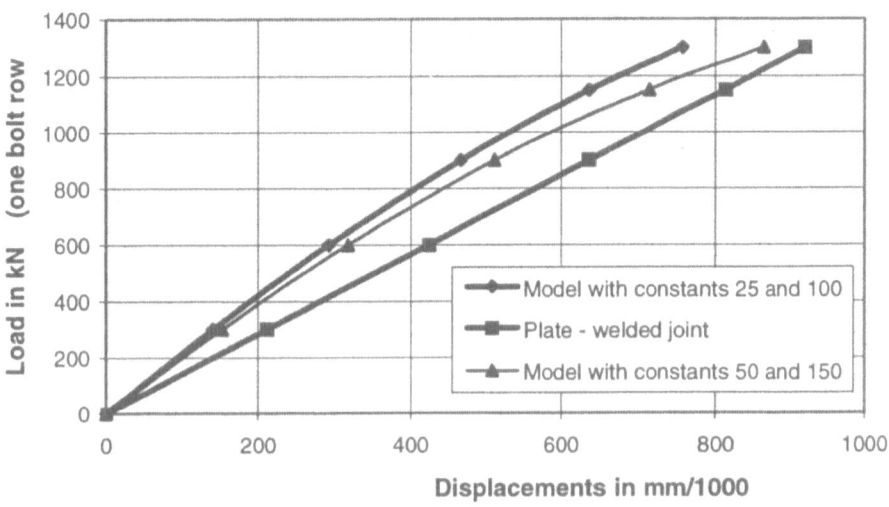

Figure 15. Comparison of the calculated load-deformation behaviour of the HSFG bolted web joint and the welded joint in the Van Brienenoord Bridge (top bolt rows near to the weld, where the distance between the rows is taken as 175 mm).

Table 1 gives the calculated bolt forces stresses and displacements for the various parts of the joint for a total load of 900 kN per bolt row. With a width over the bolt rows of 175 mm, the cross-section is 7000 mm^2 for the main plate per bolt row and 8750 mm^2 for the two cover plates per bolt row. The measuring length was 1040 mm. It is noted that at low loads, the bolt forces are more uneven than at high loads. The reason is the relatively less stiff behaviour of the bolts at higher loads.

TABLE 1: Calculated bolt forces stresses and displacements for F = 900 kN for the model with constants 25 and 100 and a distance between the bolt rows of 175 mm.

Number	F_{bolt} (B_i) (kN)	δ_{bolt} (d_{bi}) (mm/1000)	σ_{main} (σ_{hi}) (N/mm^2)	σ_{cover} (σ_{si}) (N/mm^2)	δ_{main} (d_{hi}) (mm/1000)	δ_{cover} (d_{si}) (mm/1000)
0	-	-	129	-	49,0	-
1	280	53,4	89	32	50,6	18,3
2	194	21,1	61	54	34,8	30,9
3	175	17,3	36	74	20,5	42,4
4	251	39,2	-	103	-	39,2

448

6. Acknowledgement

The authors wish to thank Ir H. van der Weijde and ir J.J. Taal of "Rijkswaterstaat" (The Netherlands Government Civil Engineering Works Department) for their support and permission to publish this paper.

7. Conclusions

a. In joints where both welds and bolts are applied, it is important to take carefully into consideration not only the strength of the fasteners, but also the differences in stiffness and deformation capacity. The stiffness properties are especially important in fatigue loaded structures.
b. A spring model has been developed for the determination of the load-deformation behaviour of HSFG bolted lap joints. The model can be used as a design tool for lap joints where the stiffness is important, e.g. because of fatigue loading in joints that are partly welded and partly bolted.
c. The spring model enables a good insight in the uneven distribution of bolt forces as a function of the joint geometry (length, width, thickness of main plate and cover plates and number, placing, size and preload of the HSFG bolts).
d. Bolt forces may be very uneven. Knowledge of bolt force distribution is important for analysis of the fatigue resistance and the analysis of possible partial slip in long joints.
e. The above insight may be very useful when estimating the fatigue resistance of long bolted lap joints. The model has recently been used for the evaluation of fractured bolted joints in some other movable bridges.
f. The extra flexibility due to load transfer by friction can be compensated by the lower stresses and strains in the cross-section because of the greater total thickness of main plate plus cover plate.

8. References

1. Van der Weijde, H. (1999) Val vervangen wegens vermoeiing (Van Brienenoord Bridge deck replaced because of fatigue), *Bouwen met Staal* **146**, January/Februari 1999, 40-44.
2. ORE (1966) Problems of high strength bolted connections in steel construction, *Report D 90/RP 1/E*, Office for Research and Experiments of the International Union of Railways, Utrecht.
3. Klöppel, K. and Seeger, T. (1965) Sicherheit und Bemessung von HV-Verbindungen aus St 37 und St 52 nach Versuchen unter Dauerbelastung und ruhender Belastung, *Veröffentlichungen des Instituts für Statistik und Stahlbau der Technischen Hochschule Darmstadt*, Heft 1 (Selbstverlag).
4. Foreman, R.T. and Rumpf, I.L. (1961) Static Tension Tests of Compact Bolted Joints, *Transactions ASCE*, Vol. **126**, Part II, 228-254.
5. Klöppel, K. and Seeger, T. (1965) Sicherheit und Bemessung von HV-Verbindungen aus St 37 und St 52 nach Versuchen unter Dauerbelastung und ruhender Belastung, *Veröffentlichungen des Instituts für Statistik und Stahlbau der Technischen Hochschule Darmstadt*, Heft 1 (Selbstverlag).
6. Steinhardt, O., Möhler, K. and Valtinat, G. (1969) Versuche zur Anwendung vorgespannter Schrauben im Stahlbau, *Berichte des Deutschen Ausschusses für Stahlbau*, Heft **25**.

AUTHORS INDEX

		paper	pp.		
Abdalla K.	Jordan	**2.7**	169		
Adany S.	Hungary	**2.5**	147	**4.2**	293
Aksogan O.	Turkey	**3.8**	267		
Aroch R.	Slovakia	**1.5**	59		
Arslan H. M.	Turkey	**3.8**	267		
Avdelas A.	Greece	**4.3**	307		
Baniotopoulos C. C.	Greece	**4.4**	317	**4.6**	337
Belev B.	Bulgaria	**4.5**	327		
Blagov D. T.	Bulgaria	**1.4**	47		
Borri C.	Italy	**1.8**	87		
Calado L.	Portugal	**2.5**	147		
Celik O. C.	Turkey	**3.2**	197		
Ciutina A.	Romania	**2.3**	129		
Cruz P.	Portugal	**3.4**	217		
Dakov D.	Bulgaria	**2.4**	139		
Dinu F.	Romania	**3.7**	257		
Dubina D.	Romania	**2.3**	129		
Dunai L.	Hungary	**2.5**	147		
Ganchev, O.	Bulgaria	**2.4**	139		
Gebbeken N.	Germany	**4.1**	279		
Geradin M.	JRC-EU-Ispra	**3.6**	237		
Grecea D.	Romania	**3.3**	207		
Gresnigt A.M.	The Netherlands	**5.9**	435		
Gustafson P. J.	Sweden	**5.6**	407		
Haller P.	Germany	**5.5**	395		
Herwijnen F. Van	The Netherlands	**5.3**	373		
Ivanyi M.	Hungary	**1.3**	31		
Jaspart J.-P.	Belgium	**5.1**	349		
Kalenov V.	Russia	**2.1**	105		
Kontoleon M. J.	Greece	**4.4**	317	**4.6**	337
Kuehnemund F.	Germany	**5.2**	363		
Kuhlmann U.	Germany	**5.2**	363		
Kvedaras A. K.	Lithuania	**1.6**	69		

450

Continuing of Authors index

SUBJECT INDEX*

---

* The selected subject can be found in the paper on the mentioned page